AF865279

DISCARDED

UNIVERSITY OF WINNIPEG
LIBRARY
515 PORTAGE AVENUE
WINNIPEG, MAN. R3B 2E9

THE

CHEMICAL INDUSTRY

1900–1930

HD
9650
.5
.H29

THE CHEMICAL INDUSTRY 1900–1930

International Growth and Technological Change

BY

L. F. HABER

CLARENDON PRESS · OXFORD
1971

Oxford University Press, Ely House, London W.1
GLASGOW NEW YORK TORONTO MELBOURNE WELLINGTON
CAPE TOWN SALISBURY IBADAN NAIROBI DAR ES SALAAM LUSAKA ADDIS ABABA
BOMBAY CALCUTTA MADRAS KARACHI LAHORE DACCA
KUALA LUMPUR SINGAPORE HONG KONG TOKYO

© OXFORD UNIVERSITY PRESS 1971

PRINTED IN GREAT BRITAIN
BY BUTLER AND TANNER LTD
FROME AND LONDON

PREFACE

THE writing of industrial history is assembly-type work: the information is available somewhere—in back issues of trade papers or learned journals, in government reports, dissertations, biographies, technical studies, company archives, and so on. Unity and continuity eventually emerge from this mass of material, but before these objectives are attained, several problems must be solved: the chief sources have to be located, they have to be accessible, and the relevant has to be culled from the general documentation. I have been very fortunate in getting so much valuable help at all stages of the work, and in particular with the first three. This book could not have been written without the patient and good-tempered assistance of many librarians and archivists in Britain and abroad, who drew my attention to significant material and obtained it for me. In the first place my thanks are due to all of them and in particular to Miss Newton, Miss Fenn, and Miss Houseman, who ran the head office library of I.C.I. for many years.

Williams Haynes's lifelong concern with the development of the American chemical industry was infectious, and it was lucky for me that he took a steadfast interest in my work and with characteristic generosity allowed me to consult his notes, typescripts, and library. His intervention opened many doors to me so that I was able to talk to people whose recollection of the industry's growth was informed and valuable. I am grateful to Dr. W. E. Bennett of Dow Chemical, Mr. B. Thomas of Monsanto, Mr. R. F. Hansen of Allied Chemical, Mr. P. A. Fabry of Du Pont, and Dr. N. F. Wilkinson of the Hagley Museum at Wilmington for much useful information on their firms' history.

The staff of the Department of Justice offices in Chicago and New York went to great trouble to pinpoint the relevant sections in anti-trust proceedings involving chemical manufacturers. Without their help I would have been totally lost in the oceans of paper which characterize major prosecutions.

Professor Edward Ames, now at the State University of New York, Stony Brook, and Mr. A. Kaufmann of the National Bureau of Economic Research drew my attention to Czarist and Soviet sources on the Russian chemical industry. They were kind enough to translate and comment upon key passages. The transliterations of Russian names follow as far as possible the usual rules.

Talks and correspondence with Dr. J. Jaenicke, formerly of Metallgesellschaft, cleared up a number of problems concerning Germany. His extensive researches into my father's life put him in touch with many people in the large chemical companies and through him and Dr. K. Holdermann I got to know Dr. G. Wolf and Miss Maria Frosch of Badische. They have put up with my questions over a period of many years and their help has been most valuable. I also wish to acknowledge the assistance received from Dr. E. Fischer of Farbwerke Hoechst and Mr. Faerber of Farbenfabriken Bayer. Nearer home, former I.C.I. colleagues were often able to contribute useful information from their experience and I am particularly grateful to Dr. D. W. F. Hardie, Mr. W. H. Cliffe, Mr. W. N. Hackett, and Mr. P. L. Morgan. Mr. W. L. Reader kindly read two chapters, and from his extensive knowledge of I.C.I.'s history was able to correct my typescript. The late Mr. E. M. Fraser, Home Sales Controller of I.C.I., intervened at an important stage in the work, allowed me short leave of absence and so enabled me to do some of the field studies for the book.

Miss Kay Andrews improved certain aspects of my work by specially reworking some of her research data on British scientists.

Editing the typescript and seeing it through the press cannot have been easy, if only on account of the frequent mention of foreign words and literature references. But the Clarendon Press and the printers rose to the occasion and I wish to thank them for all the trouble they have taken over the work in the past year. Such mistakes as remain are necessarily mine!

Finally, though by no means least, I pay tribute to my wife's forbearance. People who research and write in their spare time are a bore: she not only put up with it, but also created the conditions under which I could do the work. Thus she made the book possible and I dedicate it to her.

London, July 1970 L.F.H.

CONTENTS

ABBREVIATIONS

Annu. Lambert	*Annuaire Statistique des Engrais et Produits Chimiques déstinés à l'Agriculture* par E. & M. Lambert, 1912 edn.
BASF or Badische	Badische Anilin- & Soda-Fabrik AG
Bayer	Farbenfabriken vorm. Friedr. Bayer & Co., now Farbenfabriken Bayer AG
Ber. dt. chem. Ges.	*Berichte der deutschen chemischen Gesellschaft*
CAB followed by number	Cabinet Papers in Public Records Office
Chem. Ind., Dusseld.	*Chemische Industrie, Düsseldorf*
Chem. Trade J.	*Chemical Trade Journal*
Chemy Ind.	*Chemistry and Industry*
Chemie Verein	Verein zur Wahrung der Interessen der chemischen Industrie Deutschlands
Chim. Ind.	*Chimie et industrie*
DP followed by number	Du Pont Trial Exhibit in U.S. *v.* Du Pont (1949–59) Civil Action No. 49 (C 1071)
Enquete	Ausschuss zur Untersuchung der Erzeugungs-und Absatzbedingungen der deutschen Wirtschaft, Die deutsche Chemische Industrie, 1930
GTX followed by number	Government Trial Exhibit in U.S. *v.* Du Pont (see above DP)
Hoechst	Farbwerke Hoechst AG
HMSO	His (Her) Majesty's Stationery Office
I.C.I.	Imperial Chemical Industries Ltd.
IG	*Interessengemeinschaft* (community of interests), generally refers to association of German dyestuffs companies in 1916
IG Farben	Interessengemeinschaft Farbenindustrie AG
Ind. Chim.	*L'Industrie Chimique*
J. chem. Educ.	*Journal of Chemical Education*
J. ind. Engng Chem.	*Journal of Industrial and Engineering Chemistry*
J. Soc. Dyers Colour.	*Journal of the Society of Dyers and Colourists*
J. Soc. chem. Ind., Lond.	*Journal of the Society of Chemical Industry, London*
Norsk Hydro history	*La Société Norvégienne de l'Azote*; French. edn. of book commemorating the 50th anniversary of the foundation of Norsk Hydro Kvaelstof A/S, 1955
PP	Parliamentary Papers, followed by year and volume number
P.R.O.	Public Records Office (London). The letters and numbers that follow give the reference.
Paris Exhibition	*Les Industries Chimiques et Pharmaceutiques, Rapport du Jury International*, prepared by A. Haller, secretary to the Jury for Class 87 (Chemicals) at the Paris exhibition of 1900
TNEC	Temporary National Economic Committee (76th Congress 3rd Session: Investigation of Concentration of Economic Power, 1941)
tr.	Translated by author

USTC	U.S. Tariff Commission censuses, followed by year, Tariff Information Series number and year of publication
Z. angew. Chem.	*Zeitschrift für angewandte Chemie*

RATES OF EXCHANGE

(The following Annual averages, rounded off, have been used throughout the text.)

	U.S. $ to £	French fr. to £	Mark/RM to £
1900–14	4·9	25	20
1919	4·4	32	227
1920	3·7	52	405
1921	3·8	52	8156
1922	4·4	55	—
1923	4·6	76	—
1924	4·4	85	18
1925	4·8	102	20
1926	4·9	152	20
1927–30	4·9	124	20
1931*	3·5	94	16

* After devaluation of sterling

1

INTRODUCTION

THE chemical industry grew fast in the first thirty years of the century. In some countries its development can be compared with that of industry generally, and the growth rates from 1900 to 1930 illustrate the performance. The United States headed the list, with industrial production rising by 4·6 per cent a year and chemical production by 6·3 per cent a year. In Britain the corresponding rates were 1·5 and 2·1 per cent. Germany and France, the only other large countries for which such series are available, were not so sharply differentiated: their industries both grew by 2·3 per cent, but the German chemical industry increased by 5·2 per cent annually, the French by 4·8 per cent annually.[1] In all countries periods of vigorous industrial activity were followed by recessions. But chemical manufacture, though affected by these cycles, weathered them better, and as a rule progressed at a rate half as high again as that of industry generally, and frequently twice as high. Growth rates reflect the pace of economic development. The high rates of the chemical industry have continued, with few interruptions, to the present day. Such long-sustained expansion is a remarkable phenomenon, and its causes are a rewarding study to all who are interested in industrial history.

Chemists and chemical technologists have long paid attention to the development of applied chemistry. Their descriptions, though valuable, often lack a quantitative approach. Now, however, the behavioural sciences have been called in, and economists have applied their hypotheses to chemicals. The new approach, so far, has been only partly successful, because the available data inadequately support the framework of econometric models. More needs to be done, and one reason for writing the present book is to combine the two approaches in order to give a clearer picture.

It would be even more useful if one could explain with precise detail why the industry's growth was so rapid, and in particular isolate the role of technological factors on the one hand, and of demand on the other, at various times and in different countries. But there are formidable obstacles, first, the absence of information and second, the complex nature of the chemical

[1] The indicators and definitions of the chemical industry used to calculate the growth rates vary appreciably from country to country, so that comparisons between countries are probably not very accurate. For the United States I have calculated the rates from the indices given in *Historical statistics of the U.S., colonial times to 1957* (Bureau of the Census, 1960), p. 414. For Britain the source is the London and Cambridge Economic Service. *The British economy, key statistics 1900–1966*, n.d., pp. 5–6. The French rates have been derived from T. J. Markovitch. L'Industrie française de 1789 à 1964, sources et méthodes. *Cahiers de l'ISEA*, Série AF 4, **163** (1965), p. 216, and for Germany from W. G. Hoffmann *et al. Das Wachstum der deutschen Wirtschaft seit der Mitte des 19. Jahrhunderts* (1965), p. 361 ff.

industry's growth, which is not shaped exclusively by technology and economics, but also, and less tangibly, by advances in chemistry, by changes in the attitude to scientific education, and by the facilities for and the use of research. Another determinant of growth was the role of governments during and after the First World War, and in particular the effects of *dirigisme* on chemical manufacture. Last, though certainly not least, there was that variable factor of production, the entrepreneur.

All these aspects matter, and are sufficient justification for writing on the subject, but do not explain why I chose the first thirty years of the century. There are both practical and personal reasons. I have written on the nineteenth century,[1] and want to move on, but feel that from the early 1930s onwards the proliferation of products complicates the story too much. In any event the slump marks an important boundary: the depression that overwhelmed agriculture and destroyed the consumer-goods boom of the late 1920s also struck at chemical manufacture. Its recovery was paced by the shadow of rearmament, and the prospect of war. The Second World War caused the principal changes in recent chemical technology: the adoption of petroleum feedstock for chemical operations at one end of the scale, and the astounding growth of plastics, man-made fibres, and synthetic rubber at the other.

There is another, personal element. My father's work falls into the first thirty years of the century, and at the centre of his scientific achievement is the ammonia synthesis, which has its own scientific literature. It seems relevant to me to emphasize the role of the Haber process, the first instance of high-pressure synthesis, as one of the key elements in the modern chemical industry. Its effects were widespread indeed: it cemented the relationship between chemicals and agriculture and also between chemical companies and suppliers of process plant; it imposed novel financial requirements and so affected the management of the enterprise; and by drawing on the nitrogen of the atmosphere it ensured unlimited supplies of ammonia, which, in war as in peace, is essential to the modern state. It is fortunate that a great deal of material is available, so that the progress of the invention, from the Karlsruhe laboratory in 1908/9 to the formation of the Convention Internationale de l'Azote in 1930, can be followed. So far as I know the techno-commercial consequences of the Haber process have not yet been fully evaluated, and, in part at least, my book represents an attempt to do so.

I hope enough has been said to justify the choice of the subject, the time-span, and one instance of special emphasis. But the story of the chemical industry is a large subject consisting of many strands, which cannot all be dealt with equally, and so some selectivity is unavoidable.

In the nineteenth century the British soda industry and the German dyestuffs industry were the most conspicuous manifestations of chemical manufacture. But the picture changed importantly from about 1900 onwards.

[1] Haber, L. F. *The chemical industry during the nineteenth century* (1958, reprinted 1969).

Developments in the United States become increasingly significant, and the small businesses in the other countries emerge as important traders or as promoters of technological innovation. Examples abound: dyestuffs in Switzerland and elsewhere, electrochemicals in Norway, the fertilizer and agricultural chemical companies of France and Italy, the rapid growth of the Russian and Japanese industries. As the number of countries involved grows and the trade between them expands, so the story becomes more complex. Thus the exports of the newcomers affected at different times and in different sectors the traditional business of British and German manufacturers, thereby causing further changes in the two countries. It is now a commonplace that the chemical industry is international, but, as will be seen, this was already true at the turn of the century.

The scientific environment in which industrial change occurred was profoundly different from that which witnessed, but scarcely influenced, the pioneering days of Leblanc soda and sulphuric acid a century earlier. The novel element was that chemistry and chemical manufacture interacted. Chemistry itself was changing. About 1880 physical chemistry was not yet a formally established academic discipline; a generation later it was transforming the industry, the chief instances being nitrogen fixation and the electrolytic manufacture of chlorine and caustic soda. It is relevant to note in this context that electrochemistry was first given a separate place at an international exhibition in 1900.[1] The speed of the advance and its consequences were far-reaching, and formed part of a much wider process of change, which included electricity, the internal combustion engine, and assembly-line production, in short 'a cluster of innovations that have earned the name of Second Industrial Revolution'.[2]

The Leblanc factories were slow to mechanize their operations, and chemicals were usually handled by men rather than by machines. The introduction of a rational layout for chemical works, with purpose-designed buildings, the provision of materials-handling devices, and standardized vessels, filters, dryers, and autoclaves, came late. It spread from Germany, where the systematic approach to chemical manufacture stemmed from the fruitful cooperation of chemists and engineers, to the United States, and thence during and after the war to other countries.[3] Two other characteristics sharply differentiated the modern chemical enterprise from its nineteenth-century forerunner. One was the abandonment, whenever possible, of batch operation, which was the usual way of making Leblanc soda, fertilizers, and dyestuffs. Solvay's soda process introduced continuous-flow operation, and the contact

[1] It was Class 24. Haber, F. Die Elektrochimie auf der Pariser Weltausstellung. *Z. angew. Chem.* **14** (1901), no. 8, 184.

[2] Landes, D. S. Technological change and development in western Europe 1750–1914, in Habbakuk, H. J. and Postan, M. (Eds.) *The Cambridge economic history of Europe* (1966), vol. 6, p. 462.

[3] In Britain the Ordnance factories of the 1914–18 war made a significant contribution to the introduction of modern chemical works design.

process for sulphuric acid provided another example a few years later. The new approach to chemical engineering included gravity flow, heat exchange, and invariably the fullest possible utilization of energy. The other characteristic was the industrial application of catalysis, which was pioneered by Deacon and Winkler for chlorine and sulphuric acid towards the end of the century, and taken further by Haber and Bosch between 1908 and 1913. Their process operated at high pressure and temperature and required nitrogen and hydrogen of the highest purity. Previously operating conditions had not been a major problem. Even in the dyestuffs business the highest temperatures were 250°C, pressures rarely exceeded 50 atm, and purity, though essential, was not critical, for the batch volume rarely exceeded a few tons.[1] The requirements of the ammonia synthesis changed all that: the operating conditions became severe, the quantities scores of tons a day, and the engineering standards had to be redesigned. Continuous flow and pressure catalysis in combination gave chemical manufacture of the 1910s and 1920s a totally different aspect from that of the 1880s or even 1900s. The novelty lay not only in the products, but also in the peripheral subjects such as gas preparation and liquefaction, the systematic search for robust catalysts, and the provision of precise flow instruments, corrosion-resistant pressure vessels, and compressors.

The industry made products to replace traditional materials—for example coal-tar dyes superseded vegetable colours—and also hitherto unknown products that met a long-felt need and so created their own market. There was not always a clear line of demarcation between the two, but the end result was to raise demand. The process can be followed in the soda and fertilizer trades, where the quantities were large; other instances of growing demand were chemicals with specific end-uses, such as copper sulphate for viticulture or solvents for cellulose paints. The list lengthened in the 1920s and the effects of substitution as well as of novelty on growth have remained powerful forces to the present day. It should be observed here that the effect of rising demand on the output of the industry can only be measured for a fairly small number of products, because the necessary information is usually lacking. For alkalis, dyestuffs, and fertilizers there are some indications, but in the case of the first two they point to a slowing down in the rate of growth of demand. It is useful therefore to note how the volume of other finished chemical products was changing, and in particular how man-made fibres and plastics, which have contributed so importantly to chemical growth in the past twenty years, were evolving. Accordingly this is an opportune point at which to digress and sketch their early history, if only to show that the period of gestation was surprisingly long, and spanned almost the first thirty years of the twentieth century.

The discovery of man-made fibres was the by-product of the search for a strong carbon filament for electric lamps. Sir Joseph Swan in Britain

[1] Bernthsen, A. Die synthetische Gewinnung des Ammoniaks. *Z. angew. Chem.* **26** (1913), pt. 1, 14.

(1883–4) and Chardonnet in France (1884) thought they had the answer in nitrocellulose filament; Fremery and Urban in Germany, some years later, in cuprammonium filament. A third type of fibre, viscose, patented by Cross and Bevan in 1892, was to serve the same purpose, but metallic filaments for lamps introduced about that time proved to be far superior. All three fibres were mentioned in the report of the Paris Exhibition of 1900.[1] They were pilot-plant curiosities, manufactured under great difficulties connected with the spinning stage by Chardonnet and others in France, Germany, Switzerland, and intermittently in Britain. Commercially the fibres were of minor interest, and were used in place of silk or cotton for ribbon weaving or to make tassels. These minuscule markets at least displayed the new fibres at their best, but their significance can be judged by noting that production, then entirely confined to western Europe, was estimated in 1908 at 4000 tons, worth £1 million. The output was divided among a dozen firms and nitrocellulose accounted for nearly one-half of the total. Three years later the estimated volume was 6000–6500 tons, and viscose and cuprammonium were in the ascendant.[2] Factories to convert cellulose and spin the fibres were being opened in Russia and the United States. The manufacture of the fibres called for various chemicals such as acids or alkalis and solvents, but the scale of operations was still far too small to exert a significant demand on the products of the chemical industry. The change came after the war, and in every respect was astonishing. In 1919 world production of viscose and other fibres was about 11 000 tons; five years later it was 61 000 tons; and by 1929 it had trebled again to 197 000 tons. Artificial silk—rayon—suddenly appeared as a potential rival to cotton, though the quantities were still small compared with the 600 000 tons a year of cotton yarn spun in the United Kingdom alone in the mid-twenties.[3] The growth of viscose and later of acetate was boosted by their substitution for lisle and silk, and by the creation of a mass market for rayon hosiery. The effect on chemical manufacture now became important, and in particular the makers of viscose were soon to replace those of soap as the largest group of customers for caustic soda, a position they already held for carbon disulphide.

The variety and large volume of plastics are recent developments. Production in the first quarter of the century was small, but just how small is not precisely known, and the available estimates often conflict. According to one authority, the total tonnage grew from 20 000 in 1900 to 50 000 in 1913 and 85 000 in 1929.[4] The pre-war quantities appear very high. The only plastic

[1] *Paris Exhibition* (2), 421 ff.

[2] *Chem. Trade J.* **46** (1910), no. 1181, 32, article entitled 'The artificial silk industry'; Coleman, D. C. *Courtaulds* (1969), vol. 2, pp. 93–4; sales were about 1000 tons less.

[3] Urquhart, A. R. Cellulose derivative rayons, past, present and future. *The Textile Institute: The development of some of man-made fibres* (1952), p. 13; Committee on Industry and Trade, *Survey of textile industries*, HMSO (1928), pp. 51, 280 ff.; *Text. Organon* **29** (1958), no. 1, 24. The figures are estimates.

[4] Hopff, H. Zur Entwicklung der Kunstoffchemie. *Chem. Ind., Düsseld.* N.S. **4** (1952), no. 10, 719. The author does not give his sources, so that his figures cannot be checked.

then made on a large scale was celluloid: it had first been produced in 1869, and at the turn of the century was made in several factories in the United States, Germany, France, and Britain.[1] On the eve of the war production in these four countries was estimated at around 20 000 tons[2] and the plastic was a familiar enough material, since it was fashioned into combs, collars, cigar cases, and many novelties. A contemporary British report noted that 'the use of celluloid has steadily increased and it is now made to serve a great variety of purposes'.[3] Cinematograph film, then made from celluloid, was another important end-product, but its inflammability made it very dangerous and was a spur in the search for safe materials. Cellulose acetate film had the desired properties and was introduced before the war, but its large-scale development had to wait for the availability of suitable solvents at economic prices some years later.[4]

Another plastic, introduced about 1910, was phenol-formaldehyde. The key patents were taken out in England by L. H. Baekeland (an American of Belgian origin) on 18 February 1907, and on the following day by an electrical engineer, James (later Sir James) Swinburne. Chemische Werke Albert first prepared the resins in Germany a short time later. The coincident activities of firms and individuals in the United States, Britain, and Germany resulted in patent litigation, and the high prices charged by the few makers delayed the growth of these promising new compounds. They were used at first for insulating purposes in place of varnishes and lacquers of vegetable origin and also in the manufacture of brake linings. In the 1920s phenolic mouldings and extrusions appeared in components for radio sets and electrical accessories for cars.[5] This was the start of a period of extremely rapid growth, which took the volume of production in the United States from 2600 tons in 1922 to 14 700 tons seven years later, by which time these new plastics were surpassing the old-established celluloid, without, however, replacing it.[6] Thus, at the very end of the period, plastics, at least in the United States, and prob-

[1] Some were quite large. The British Xylonite Co. had works at Brantham in Suffolk and Hale End near London that employed about 1100 people in 1902 in making celluloid as well as articles from it. In France the Société Générale pour la Fabrication des Matières Plastiques (the successor to the Compagnie Française du Celluloid, founded in 1877) was making 1000 tons a year in 1900. Kaufman, M. *The first century of plastics*, n.d. (1964?), pp. 43–4; *Paris Exhibition* (2), 415–18.

[2] Schueller, E. Les matières plastiques, in Gérard, J. (Ed.) *Dix ans d'efforts scientifiques et industriels* (1926), vol. 1, p. 1403. It has not been possible to reconcile the figures given by Hopff and by Schueller. The latter was president of the Syndicat des Fabricants de Caséine Durcie and so presumably familiar with French celluloid production, which he puts at 4000 tons in 1913. His figure for Germany was 8000 tons, fairly close to the 9000 tons quoted by H. Ost in his *Lehrbuch der chemischen Technologie* (14th ed., 1925, p. 555). Schueller's estimate for the U.S. was 5000 tons compared with 4000 tons in 1915 mentioned in GTX 110 (Du Pont Development Department memorandum to Executive Committee dated 12 September 1915). On the whole his total of 20 000 tons would appear to be closer to the mark than the 50 000 tons mentioned by Hopff.

[3] *Report of the Departmental Committee on Celluloid*, HMSO (1913), Cd. 7158, p. 4.

[4] The development of solvents would justify a separate chapter, but cannot be pursued here for reasons of space.

[5] Kaufman, op. cit., pp. 61 f.

[6] USTC *Censuses* for the years mentioned; Hirt, F. L. Growth characteristics of the economy illustrated by the chemical industry. *Survey of current business* (September 1954), pp. 12, 14.

ably also in Germany, were becoming important articles of chemical commerce. But they had not yet been taken over by the large chemical companies, which, as is well known, have made the manufacture of plastics the cornerstone of their expansion. Because plastics were outside the mainstream of the industry and did not affect its growth in the first thirty years of the century, they have been omitted from the story.

The range and diversity of organic compounds, of which fibres, plastics, and dyestuffs are only three examples, is difficult to comprehend. The number of compounds identified and described soared from 15 000 in 1880 to around 150 000 in 1910, and more than doubled in the next thirty years. Only a fraction of the total were produced on an industrial scale, yet even here the variety was remarkable. It can be measured very crudely by counting the pages devoted to three familiar chemicals: ethanol, acetone, and formaldehyde, in Beilstein's *Handbuch*. In the third edition (1892–4) ethanol occupied 7 pages, acetone and formaldehyde 2 and 1½ respectively. In the fourth edition (1918–38) there are correspondingly 22, 16, and 16. The numbers, as well as the ratios, illustrate the significance of the compounds.[1]

Obviously a selective approach is essential. A certain degree of arbitrariness is unavoidable since retrospective market research to identify the important products at different periods is impracticable. The following chapters will therefore be concerned mainly with heavy inorganic chemicals, especially sulphuric acid, alkalis, and phosphatic fertilizers, and with dyestuffs. They were already the main products of the industry at the end of the nineteenth century, and, with the addition of nitrogenous fertilizers, calcium carbide, and solvents, remained the chief products for another generation. Where relevant the development of other industrial chemicals will be included.[2] Similarly it is necessary to restrict the history to the principal countries. Germany, Britain, and the United States occupy first place, because their chemical industries were the largest and technologically the most progressive. But their lead-time diminished, despite the ownership of patents and of know-how, owing to the increasingly rapid diffusion of applied chemistry in other countries. The process of diffusion was only one aspect of the growing industrialization of those countries, but makes it essential to look at and compare the developing chemical industries of France, Italy, Switzerland, Russia, and Japan.

[1] Richter, F. (Ed.) *75 Jahre Beilstein's Handbuch der organischen Chemie* (1957), pp. 51, 92.

[2] There is not enough material on the growth of proprietary medicines and, during the early part of the present century, of the pharmaceutical industry. Neither was connected with the heavy chemical trades, but there were many structural as well as scientific links with the manufacture of dyestuffs. R. Schmitt ('Die pharmazeutische Industrie und ihre Stellung in der Weltwirtschaft, Ph.D. thesis, Frankfurt (1932), pp. 218–19) has estimated that world trade rose from £11·5 million in 1913 to £23 million in 1928. The figures must be treated with some caution owing to imprecise definitions and incomplete returns. Nevertheless the foreign trade in those goods represents impressive growth. The significance of the business is the more striking when compared with much more widely publicized dyestuffs trade, which the League of Nations experts (*The chemical industry* (1927), p. 29) estimated at £35 million in the mid-1920s. Although it would encounter great obstacles, a study of the early years of the pharmaceutical industry in the principal countries would make a rewarding contribution to modern economic history.

The short list of products and countries suffices to illustrate the pattern of growth. Up to the Great War product specialization was usual; thereafter diversification became the chief characteristic. One of the principal reasons for the difference between the pre-1914 and post-1918 industry was the changing attitude to research and development outside Germany. By this I mean not only the chemical companies' own research efforts, which became noticeably more purposive and successful after 1914, but also the relationship of academic research to commercial development. Cause and effect cannot be measured, but can be illustrated by pointing to instances: the role of research institutes in cross-fertilization, the improvement in higher chemical education, the tendency of management in large chemical enterprises to become science-orientated. In short, research paid, and not only in Germany. Government policy also veered from *laissez-faire* and indifference to intervention and the vigorous promotion of specific activities: official research, for example, increased, though the efforts made rarely benefited chemical manufacturers. Of much greater use was the protection and occasionally the financial support of particular branches of the industry. The role of the government in the affairs of the industry forms yet another strand in the story and warrants description and evaluation.

2

THE CHEMICAL INDUSTRY IN 1900

Definition and size

PERHAPS the chief difficulty for anyone wishing to compare the chemical industry with other industries is how to define it. In 1900 there was general agreement that it included the old-established manufacture of inorganic acids, alkalis, and bleaching materials. These three groups of products made up what the English called 'the heavy chemical industry' and the French until quite recently described as 'la grande industrie chimique'. This loose term also covered the newer products such as manufactured fertilizers, calcium carbide, explosives, and cyanides. Dyestuffs, textile auxiliaries, photo-chemicals, fine chemicals, and metal salts were considered separately, and as a group were called 'petite industrie chimique'.[1] Finally there were a number of other branches, which, in the aggregate, were of considerable and growing importance. These comprised the drug, pharmaceutical, and perfumery trades, oil and tallow, paint, colour and varnish, sheep-dips, disinfectants, and the match trade. In France, the flourishing essential oil and perfumery industry was also considered part of chemical manufacture, but petroleum, then already of significance in the United States and Russia, was excluded. Some countries, for example the United Kingdom, included seed crushing and the manufacture of soap, while the Germans extended the scope of the industry to chemists' shops.[2]

The size of the chemical industry and its importance to other industries were not reliably measured; detailed foreign trade and production statistics were virtually unknown and only the United States had taken a census of production. Nevertheless sufficient data are available to present a general picture.

For Britain the 1901 Census of Population probably gives the best idea of the industry's size: 30 000 people were working in establishments classified as 'Alkali Manufacture and other Chemical Works', (then called 'Manufacturing Chemists'), and a further 1800 were employed in fertilizer works, so that there were almost 32 000 in the heavy chemical industry. Explosives and matches employed another 7300, and the group of dye, paint, and ink trades 7500. These figures are insignificant when set alongside cotton, coal, and iron and steel, and it is more meaningful to compare chemicals with that other new

[1] *Paris Exhibition* (1), 1, 297.

[2] All, except the last, formed part of Class 87 (Chemicals and Pharmaceuticals) at the Paris Exhibition of 1900. Class 87 also included polishes, glue and gelatine, plastics, and artificial silk.

industry, electrical machinery, where employment soared from a mere 13 000 in 1891 to 50 000 ten years later.

The German employment census of 1895 showed 19 300 people working in the dyestuffs industry, 10 500 in fertilizers, and 36 400 in other chemical and pharmaceutical factories, altogether about 66 000.[1] The total was certainly higher by 1900, probably around 80 000.[2] The annual return of the social security association of the chemical industry gave 153 000 workers for 1900. Staff were excluded, but the number was swollen by workers in trades that are not part of the chemical industry as defined in this book.[3]

The United States chemical industry employed about 55 000 people in 1899 and 70 000 five years later. In both years it was slightly larger than electrical machinery, but very much smaller than iron and steel.[4] In France the 'grande industrie' numbered about 21 000 employees in 1901, the 'petite industrie' 9500, and the important pharmaceutical and perfumery trades nearly 8000, so that, in round numbers, there were altogether 38 000 workers. This was nearly two and a half times the employment in electrical engineering, but less than a twentieth of the number in the textile industry.[5] In other countries employment in the chemical industry was, in terms of absolute numbers, very small; for example, in Switzerland about 2300 were working in the chemical industry and a further 1700 in allied trades in 1895.[6]

At the opening of the century Great Britain dominated the traditional heavy chemical industry, Germany the new organic industry. The position of the British manufacturers was being challenged: Leblanc soda manufacture, the mainstay of the nineteenth century, had been convulsed by the appearance of the Solvay process. Their exports fell because the Leblanc Works were seriously affected by the establishment of soda industries in former overseas markets, notably the United States and Germany. In 1890 exports of alkali were £2·6 million, in 1900 they were £1·4 million, and this disappointing performance cancelled out the growth in fertilizer and coal-tar distillates, so that total overseas sales fell from £9·0 million in 1890 to £8·6 million in 1900.

No other country had chemical exports approaching those of the United Kingdom. Germany, however, was catching up. Her exports rose from about £4½ million to £6½ million between 1890 and 1900 and four-fifths were represented by dyestuffs and intermediates. French exports, valued at £4½

[1] *Chem. Ind., Düsseld.* **19** (1897), 37.

[2] The estimate is based on the employment figures in *Paris Exhibition* (2), 32 ff. For example the dyestuffs manufacturers represented at the Exhibition employed about 21 000 people.

[3] The list includes soap, candles, mineral-oil refining, lubricating oils and greases, perfumery, pencils, resins, mastics, roofing felt, rubber goods, wood impregnation, mineral waters, lamp and other blacks, chemists' shops.

Ebert, W. *Die chemische Industrie Deutschlands* (1926), pp. 9–10; Witt, O. N. *Die chemische Industrie des deutschen Reiches* (1906), p. 18.

[4] U.S. Dept. of Commerce, 13th Census (1909), 10 (Manufactures), (1913).

[5] Ministère du Travail (France), *Resultats statistiques du Recensement Général* (1908), vol. 1, pp. 72–4.

[6] Lunge, G. *Zur Geschichte der Entstehung und Entwicklung der chemischen Industrien in der Schweiz* (1901), p. 69.

million in 1900, included even at that late date £600 000 worth of vegetable dyes. Finally, the Swiss dyestuffs manufacturers exported goods valued at £615 000 in 1900.

Apart from employment and foreign trade, sulphuric acid production was often used in those days to compare the importance of the chemical industry in one country with that in another. But this is also a somewhat unsatisfactory yard-stick, because, aside from the reliability of the figures, the consumption pattern is of greater significance than the bare output statistics. Still, the figures in Table 2.1, despite these shortcomings, indicate at least an order of importance, and the inclusion of Russia and Japan points to the spreading development of the industry.

TABLE 2.1

Estimated production of sulphuric acid, 1900[1]

	(000 metric tons 100% H_2SO_4)
Great Britain	1100
United States	940
Germany	850
France	625
Austrian Empire	200
Belgium	160
Russia	125
Japan	50
Total	4050

Western Europe

The lack of statistics and the secretiveness of the leading firms make it difficult to assess the performance of the British chemical industry, which in 1900 was probably still the world's largest, measured by the volume of production and the capital invested. While the foreign trade statistics were better presented than those of other countries, a major reclassification about 1900 prevents reliable comparison. During the 1890s the trend of exports was downwards and in 1900 they amounted to £8·6 million. The alkali and bleaching powder business had been hard hit: exports fell from £2·6 million in 1890 to £1·4 million in 1900, thus showing the effects of the higher American and Russian tariffs of the 1890s and of German competition. But fertilizers and coal-tar distillation products (including kerosene) showed a significant improvement. The former, of which two-thirds were sulphate of ammonia, rose

[1] Lunge, G. Uber den gegenwärtigen Stand der Schwefelsäurefabrikation. *Z. angew. Chem.* **16** (1903), no. 29, 694–5; Witt, op. cit., p. 66; Fauque, M. *L'Evolution économique de la grande industrie chimique en France* (1932), p. 21.

from £2·1 million in 1890 to £2·4 million in 1900, the latter from £1·4 to £1·8 million. Imports of chemicals were traditionally small, a large identifiable item being coal-tar dyes, which more than doubled from £0·3 million in 1890 to £0·7 million in 1900; this was accompanied by a sharp fall in natural indigo imports.

The trade returns reflect weaknesses and structural changes in the British chemical industry. The most important from the economic and social point of view was the accelerating decline in the Leblanc soda business. We shall examine its problems in more detail later. Suffice it to note here that the United Alkali Co. (U.A.C.) had been formed in 1890 not merely to shore up the Leblanc trade, but indeed to reanimate it. But U.A.C. was unable to achieve this objective. Between 1891 and 1901 its output of soda ash, soda crystals, and caustic soda fell from 214 000 tons to 54 000 tons. Meanwhile the manufacture of ammonia soda was prospering, and production rose from 180 000 tons in 1890 to an estimated 250 000–300 000 tons (including soda ash converted to caustic soda) ten years later.[1]

The jury of the Paris Exhibition were critical of the British soda industry, and deplored the useless struggle between alkali processes, which had caused business to be lost to continental rivals. But the jury were simplifying to excess. The technological changes in the British industry were slow in making themselves felt because the Leblanc process had been so well established and so much capital and skill had gone into it. While the Leblanc process was becoming obsolete, the Solvay process, worked by Brunner, Mond & Co., was profitable, and the company's links with other Solvay licensees throughout Europe were extremely advantageous. Solvays had reinforced their position by securing the European rights to the Castner–Kellner electrolytic process. That process was not the only one for the electrolytic decomposition of brine to chlorine and caustic soda, but it had influential backing, and was the first to be worked on a substantial scale in the United Kingdom. The Castner–Kellner Alkali Co., founded in 1895, had its works at Runcorn, opposite Widnes, then the remaining stronghold of the Leblanc process. Electrolytic caustic soda and chlorine, the latter turned into bleaching powder, competed successfully with the same products made by the Leblanc trade and so caused it further damage. The final irony was that the directors of U.A.C. had been offered a licence for the Castner–Kellner process but had rejected it!

Alongside these large firms such as U.A.C. and Brunner, Mond there were

[1] Hardie, D. W. F. *A History of the chemical industry in Widnes* (1950), pp. 146, 151; Alkali Inspectors Report for 1890, HMSO, (1891), C 6357. There are no comprehensive alkali statistics from about 1890 onwards, and my estimate for the ammonia–soda production in 1900 is very rough. Haller's estimate (*Paris Exhibition* (1), 98) of 432 000 tons of ammonia soda and 361 000 tons of Leblanc soda refers to 1896. It is grossly inflated by double-counting and includes caustic soda and sodium sulphate. Alkali statistics are important indicators of industrial growth, and their publication would make a useful contribution to economic history and also illustrate the speed with which the old process was replaced by the new one.

several relatively small businesses, which were remarkably prosperous, perhaps because their owners were alert to technical change. Among these the business of Albright & Wilson, then as now the only makers of phosphorus and its derivatives in the United Kingdom, were switching over to electric furnaces. Nearby the Oldbury Alkali Co., the chemical offshoot of the Chance glass factory, made alkali on a small scale. The firm obtained royalties from the Chance sulphur recovery process, and began to manufacture ammonia salts by working up the ammoniacal liquors of gas works. It merged in 1898 with the somewhat smaller soda and sulphuric acid business of William Hunt of Wednesbury to become Chance & Hunt, and so the largest chemical enterprise of the Midlands, with interests in the rapidly growing cyanide trade.

Entirely distinct from the alkali industry was that of fertilizers, of which Britain was among the principal European producers in 1900. Much of this consisted simply of compounding imported guano or the far more important Chilean nitrates. Britain was a leading producer of superphosphate, output being about 620,000 tons in 1900.[1] Basic slag production was estimated at 250 000 tons in 1899, of which 140 000 tons were exported.[2] The country's gas industry was the main source of sulphate of ammonia, of which Britain was then by far the biggest producer in the world, with an output of 210 000 tons.[3] Superphosphate and sulphate of ammonia required large quantities of sulphuric acid, and the fertilizer industry as a whole was growing.

In striking contrast to Germany, the manufacture of British dyestuffs, if not moribund, was in a very poor condition. No British firm was mentioned in the report of the 1900 Paris Exhibition, and the jury had this to say:

> . . . the English in general are steeped in a narrow utilitarianism which has prevented them from engaging in pure research. This is the only fruitful research if, at the same time, there are also practical men who can put to use the results obtained. There are plenty of technical schools, but they offer no higher education of any value, with the result that the English in the matter of dyestuffs are not only behind the Germans, but even the French.[4]

English observers agreed with this damning comment: Professor A. G. Green devoted his address to the British Association meeting in 1901 to a detailed account of British dyestuffs. He attributed the weaknesses partly to the reluctance of dyers to change their habits, partly to the buying practices of the recently formed combines in the textile finishing trades. The Calico Printers' Association was accused of forcing down prices to such an extent that only the Germans were able to sell dyes at a profit.[5] There were then only two

[1] *Annu. Lambert*, p. 298. Haller in *Paris Exhibition* (1), 19–20 shows 800 000 tons, which seems too high.

[2] Packard, W. G. T. *The history of the fertiliser industry in Britain* (n.d.), p. 13; *Chem. Trade J.* **20** (1897), no. 518, 285; Sir Corbett Woodall quoted in *Chem. Ind., Düsseld.* **37** (1914), no. 2, 40.

[3] *Annu. Lambert*, pp. 134, 142.

[4] *Paris Exhibition* (2), 64. tr.

[5] Levinstein, H. From within the dyestuffs industry. *J. Soc. Chem. Ind., Lond.* **50** (1931), 253 T.

makers of any significance, both of long standing: the British Alizarine Co. and Levinstein Ltd. Read Holliday and the Clayton Aniline Co. specialized in organic intermediates, and the Boer War brought them unexpectedly good business on account of the great demand for trinitrophenol, commonly referred to as picric acid or lyddite, which was then the principal explosive for shells. The military placed very large orders, and measures had to be taken to organize supplies of carbolic acid, of which Britain was then the chief producer, and provide adequate capacity for the extraction of phenol. Exports were stopped and Read Holliday successfully synthesized phenol from benzene, which was less scarce.[1]

The German chemical industry appeared to be the most dynamic in the world. Its considerable growth and manifest successes naturally attracted much attention, and some people found a ready explanation in national characteristics. The practical nature and discipline of the Germans, their self-confidence, orderliness, and attention to detail were believed to have contributed importantly to the industry's achievements. Without underrating the role that such national characteristics can play, it seems to me that the causes are simpler to find and their importance easier to assess. The high standard of scientific education at university as well as at technical college level was certainly a major influence. This will be discussed in the next chapter, but it is legitimate to anticipate slightly by pointing out that there was a great deal of cross-fertilization between universities and industry. We may generalize by noting that the overall business strategy as well as the day-to-day operation of chemical firms was directed by chemists and scientifically orientated managers. In 1897 F. Fischer estimated that 4000 chemists were employed in Germany, almost 3000 of them in industry.[2] Three years later the total had risen by another 500.[3] Large dyestuffs manufacturers were even then employing scores of chemists; BASF and Bayer had about 150 each.[4] Numbers and the quality of the chemists were by no means the sole reason for success. Managerial and financial ability also counted, and the financial basis of the larger companies was sound because of the generous depreciation provisions and the careful husbanding of resources. Profits were ploughed back and the impression of financial strength was carefully fostered; it was assisted by the practice of the banks, which often held substantial stakes in the companies that they had helped to form.

The country's geography and natural resources were also important factors: the Rhine provided a convenient means of transport and railway communications were good. Fuel was cheap, though it may be doubted whether hard coal was as cheap on the upper Rhine as it was in Lancashire. More important

[1] Whittaker, C. M. 'Some early stages in the renaissance of the British dyemaking industry 1899–1920. *J. Soc. Dyers Colour.* **72** (1956), no. 12, 557–8.
[2] Fischer, F. *Das Studium der technischen Chemie* . . . (1897), p. 45.
[3] *Chem. Trade J.* **29** (1901), no. 741, 87.
[4] *Paris Exhibition* (2), 32 ff.

perhaps was the utilization of coal for chemical purposes. The Ruhr coke ovens supplied coal tar, though not in sufficient quantities. There were abundant salt deposits in the country between the middle Elbe and Weser, and also in the Neckar valley, though they were not so conveniently situated as in Cheshire. As for sulphur, Germany had a small indigenous source of pyrites in the Siegerland just south of the Ruhr, but the deposits could not be worked economically. The zinc smelters of the Aachen region and in Silesia were generally combined with acid works to utilize the sulphur dioxide, but in the main the country was dependent on imports of pyrites and brimstone.

By 1900 potash and lignite were being produced on a large scale, and during the following years their growth was of great advantage to the German chemical industry. Potash was first mined in the 1860s and by 1900 output had grown to 3·0 million tons, while world consumption expressed in terms of active content amounted to 233 000 tons as K_2O for fertilizer and 70 000 tons for chemical use.[1] Lignite was initially worked only in the Rhineland where production soared from a mere 127 000 tons in 1880 to 5 million tons twenty years later. The central German deposits were still unimportant in 1900, but there, as in the Rhineland, the development of lignite mining became associated at an early stage with chemical manufacture. In the former the energy for electrolytic plants, which were built from the mid-1890s onwards, was obtained from lignite. In the Rhineland the manufacture of calcium cyanamide depended from the 1900s onwards on cheap thermal electricity using lignite as fuel.

In 1900 the chief chemical enterprises were located at Ludwigshafen opposite Mannheim, in the Frankfurt region, and along the lower Rhine. New growth points were emerging in central Germany, in the Berlin area, in Silesia, and on the upper Rhine above Basle, where hydro-power was available. These works employed about 80 000 people and were said, somewhat optimistically, to have a value of production of £47 million, including exports worth roughly £20 million.[2] The value of dyestuffs production was of the order of £6½–7 million, with exports about £5 million.[3]

The structure of the German chemical industry had some interesting characteristics. In the first place concentration of manufacture in the medium and large works had gone far by 1895, when 37 per cent of all workers were employed in plants with more than 200 people (in the machinery and engineering industry the proportion was 39 per cent, in textiles 35 per cent).[4] There was also a good deal of specialization within the various sectors of the industry.

[1] Krische, P. *Das Kali* (1923), pt. 1, pp. 134, 184 ff.

[2] The value figures appear in *Paris Exhibition* (1), xvi f. They seem very high, possibly because of double-counting, and certainly by reason of adding to chemicals such allied products as soap, rubber goods, and others listed on p. 9.

[3] *Paris Exhibition* (2), 31. Production has been derived from the known export statistics, it being assumed that at least three-quarters of production was exported.

[4] Schultze, H. *Die Entwicklung der chemischen Industrie Deutschlands* (1908), p. 18.

The manufacture of heavy chemicals was not beset, as in contemporary Britain, by the problem of the obsolescent Leblanc process and uneconomic works. Virtually the entire output of 300 000 tons of sodium carbonate and crystal soda was produced by ammonia–soda plants, principally those of Deutsche Solvay-Werke. During the 1890s Germany had become a large exporter, the tonnage of carbonate and crystal soda aggregating about 46 000 tons in 1900.[1] Other companies, occasionally associated with ammonia–soda makers, produced caustic soda and chlorine by the then novel electrolytic processes. The growth of these works in central Germany was extremely rapid and one expert claimed that they accounted for one-quarter of world production of chlorine, expressed as bleaching powder.[2] The claim cannot be checked but it is a fact that German exports of bleaching powder reached 26 000 tons in 1900. Germany's production of sulphuric acid, at 850 000 tons (as 100 per cent acid), placed it third in the world league. Productivity in this branch had increased sharply, principally because of the new acid plants installed by the dyestuffs companies.[3]

These companies, of which only BASF, Farbwerke Hoechst, Farbenfabriken vorm. Fr. Bayer, AG für Anilinfabrikation and Oehler at Offenbach were of major significance, stood in marked contrast to the manufacturers of heavy inorganic chemicals. They were big producers of heavy chemicals but made only for their own needs, which were large as far as sulphuric acid, chlorine, and caustic soda were concerned. By contrast the companies relied on the coal-tar distilleries, which were linked with and dependent upon the Ruhr coke industries for benzole, toluole, naphthalene, phenol, and anthracene. Since German production was insufficient, the distilleries were large importers of crude tar as well as of refined products, principally from England, France, and Belgium.[4] Difficulties arose when Britain prohibited the export of phenol in January 1900 and the Germans had to synthesize it from benzole.[5]

As far as dyestuffs were concerned some important technical advances in the 1890s deserve mention. Unquestionably the most significant was the indigo synthesis, for which the starting-point was initially naphthalene and later, by a different route, benzole. A derivative of anthracene, anthraquinone, attracted considerable attention as a starting-point for a series of acid dyes. These were first marketed between 1894 and 1897, and were later supplemented by anthraquinone wool dyes, which required a chrome mordant. Yet another great discovery was Vidal's sulphur black of 1893, which was followed within a few years by scores of patents in the same field; the dye re-

[1] *Paris Exhibition* (1), 61.

[2] Witt, O. N. *Die chemische Industrie auf der internationalen Weltausstellung zu Paris 1900* (1902), p. 128.

[3] Witt, *Chem. Ind. des deutschen Reiches* (1906), p. 66. Output per man increased as follows (000 tons 100 per cent H_2SO_4): 1891—1·4, 1893—1·5; 1895—1·8; 1897—2·0; 1899—2·3.

[4] ibid., pp. 198–201.

[5] F. Raschig of Ludwigshafen was the first company to do so. Ihle, E. A. *75 Jahre Raschig* (1966), pp. 20, 22.

quired for its manufacture aniline or phenol.[1] Though of little commercial significance, the first dye specially designed for a man-made fibre was introduced by Hoechst in 1897 under the name of 'Janus' for Chardonnet's artificial silk. Far more important, though just outside our date-line, was the first anthraquinone fast vat dye discovered by R. Bohn of BASF in 1901; his indanthren blue was soon followed by others and eleven years later there were no less than 182 dyes of this class.[2]

The preliminary investigations and the pilot-plant studies for these dyes were extremely costly and led to vertical integration of the large works. Another consequence was the formation of groups among the leading firms. BASF and Hoechst, both of whom had spent large sums on the indigo synthesis, evolved workable processes almost at the same time. The necessity of safeguarding resources, from which stemmed the attractions of price agreements, led directly to cartels and conventions among industrialists who were temperamentally disposed to market sharing rather than to unbridled competition. This was a favourable environment in which specialization could be practised to some extent, not only by the large firms but also by the smaller businesses that prospered on the patronage of the big four or five. These small firms were still numerous in 1900 and tended to concentrate on making aniline, naphthols, and other intermediates.[3]

In 1900 the pharmaceutical industry and the related branches specializing in fine chemicals and synthetic essences were already large and clearly separate from the dyestuffs industry. Among the latter only two firms, Bayer and Hoechst, were interested in the manufacture of antipyretics and febrifuges. The closing years of the nineteenth century witnessed remarkable activity in German pharmacology as a result of the pioneering work of Knorr, Behring, E. Fischer, and others. During these years the patent medicine and toilet preparation trades were being raised to the industrial scale, were lucrative, and were often associated with the manufacture of what are now called ethical preparations. Between 1870 and 1900 almost fifty companies first entered this business and a large export trade emerged.[4]

Chemical manufacture in France, though large, was extremely diffuse. In contrast to Germany only 7 per cent of all workers were employed in factories with over fifty people in 1901,[5] and the bulk of the industry consisted of chemical workshops. Nevertheless exports were rising: they had been valued at about £3·7 million in 1893 and seven years later approached £4½ million.[6] While exports of vegetable dyes were declining fast, the business in alkalis was growing, and exports of caustic soda, at 39 000 tons in 1900,

[1] I.C.I. Dyestuffs Division. *An outline of the . . . dyestuffs industry* (1954), pp. 16 ff.

[2] *Neue Zürcher Zeitung* (Tech. Suppl.), 8 May 1952, Entstehung und Entwicklung der Indanthren.

[3] Erban, F. *Kartelle und Konventionen in der chemischen Industrie* (1910).

[4] Vershofen, W. *Wirtschaftsgeschichte der chemisch-pharmazeutischen Industrie* (1958), vol. 3, pp. 50 f., 87.

[5] Fauque, op. cit., p. 188.

[6] George, F. *La rénovation de l'industrie chimique française* (n.d. –1919?), pp. 19 ff.

were beginning to approach those of the United Kingdom. Even more important was the production of and trade in superphosphate. The recently discovered North African deposits were being worked with French capital and the crushed rock shipped to France, where large superphosphate plants had been built. In 1900 the country's production of this fertilizer was 940 000 tons and France was the largest European producer: in the five years to 1900 exports doubled to 69 000 tons and imports fell to 61 000 tons.[1]

Another aspect in the development of the industry was the rapid elimination of the Leblanc process. Though a French discovery, soda made by it had not attained the economic importance it occupied in Britain. Nevertheless the quantity made had been significant; in 1870, for example, it amounted to 57 000 tons. But the opening of Solvay's factory near Nancy a few years later was followed by a swift decline, so that by 1891 only 27 000 tons of Leblanc soda were made. Nine years later output was negligible, while that from the ammonia–soda plants continued to grow and reached 205 000 tons in 1900.[2] The leading manufacturers retained some of the Leblanc equipment, but did not make alkali. Some specialized in making sodium sulphate for use in glass works, others produced hydrochloric acid, which was sold to makers of glue and gelatine, then an active trade in France. Some bleaching powder was also made. The erstwhile Leblanc soda manufacturers generally set up equipment to make superphosphate in the 1890s.[3] The role of this fertilizer at that particular juncture of the industry was most important, because the increasing demand kept the firms busy, if not prosperous, at a time when the business in Leblanc soda ash and crystals had ceased to pay. In 1899 there were at least fifty-three superphosphate works; the largest were situated in northern France, along the lower Seine, and in the Paris region. Superphosphate manufacture in contrast to the rest of the industry was less fragmented, and a few big producers held a large proportion of the business: three companies owned between them seventeen of the works, among them the St. Gobain concern with no less than ten.[4] The consequence of these developments was a remarkable growth in sulphuric acid production from 200 000 tons in 1878 to 625 000 tons in 1900.[5]

At the time of the Paris Exhibition the French electrochemical industry had made astonishing progress and was then probably the most important in the world. Henri Moissan, the discoverer of calcium carbide and the leading French electro-chemist of his day, was playing an active role in the application of his work. Much venture capital was attracted to this new branch of chemistry, which flourished in Savoy where water power could be tapped

[1] *Annu. Lambert*, p. 298; *Paris Exhibition* (1), 19–20.

[2] Fauque, op. cit., p. 21; *Paris Exhibition* (1), 16–17.

[3] Kuhlmann, the largest purely chemical business, began the manufacture of superphosphate in 1895.

Anon. *Les etablissements Kuhlmann 1825–1925* (n.d.), pp. 16 f.

[4] Guillet, L. *Etat actuel de l'industrie des produits chimiques inorganiques en France* (1902), vol. 1, pp. 326–7.

[5] Fauque, op. cit., p. 21. The quantities are in terms of 100 per cent H_2SO_4.

easily and cheaply. There were twenty-three electrochemical works in France in 1900, of which thirteen were in the neighbourhood of Grenoble. In 1889 the installed horsepower had been a mere 4000, in 1898 it was 51 000, and within two years it doubled to 109 000.[1] Most of these works were making calcium carbide and acetylene; some were in the alloy and aluminium business. The manufacture of electrolytic alkalis was still in its early stages: two firms in Savoy were making chlorates, a third near Compiègne (north-east of Paris) was producing caustic soda and bleaching powder; it made about 4500 tons of the latter, equivalent to roughly a sixth of French production. Another caustic soda works was then under construction in the Alps.[2]

The general impression of the French heavy chemical industry at the opening of the century is one of adaptability to change and acceptance of new chemical manufacturing methods. But the jury of the Paris Exhibition also drew attention to some serious weaknesses, notably the lack of scientifically minded management, high costs, and poor transport.[3] Nowhere were these defects more apparent than in the production of synthetic dyes. Within a generation dyestuffs production, once large and indeed pioneering, had shrunk into insignificance. Only three French firms exhibited in 1900. One was so small that it can be ignored, the second, Usines du Rhône, made dyestuffs merely as a sideline to its pharmaceutical business. The third, however, S.A. des Matières Colorantes & Produits Chimiques de St. Denis, could be called a dyestuffs business. It was becoming prominent, because Vidal, the discoverer of sulphur black, was its chief chemist, and the firm was successfully making this class of dyes.[4] While France was largely dependent on Germany, or the French subsidiaries of German companies, she still had a big vegetable-dye industry and was a large producer of tanning extracts, as well as of essential oils and perfumes.

In the remaining European countries chemical manufacture was small in absolute terms, and of little significance relative to other industries. From the turn of the century onwards there were signs of change, but in 1900 production consisted primarily of sulphuric acid and superphosphate, and, additionally in Belgium, the Austrian Empire, and Russia, of ammonia soda. Switzerland lacked all this, but possessed vigorous dyestuffs manufactures. In general it is more accurate to speak of groups of firms, instead of the existence of a chemical industry.

While heavy chemicals production had atrophied in Switzerland, and the country's water-power resources had not yet been harnessed to electrochemicals, there were in the city of Basle five firms making dyestuffs, which, in 1901, employed altogether 1326 people.[5] The number is considerably

[1] Haber, F. Die Elektrochimie auf der Pariser Weltausstellung. *Z. angew. Chem.* **14** (1901), no. 8, 184.

[2] Guillet, op. cit., p. 326; *Paris Exhibition* (1), 15.

[3] Among the lesser problems were the restrictions on the transport of chlorine by rail; *Paris Exhibition* (1), lxxi, lxxiii.

[4] ibid. (2), 18–24.

[5] The firms were CIBA, Sandoz, J. R. Geigy, L. Durand, Huguenin & Cie, and Basler Chemische

smaller than that of a single German dyestuffs maker. Owing to their small size the Swiss works were not technically integrated and bought all their process chemicals as well as most intermediates. Nevertheless they were not working up imported materials: at this time the Swiss companies were gradually giving up the manufacture of the cheaper, large tonnage dyes and were employing chemists and skilled manpower to prepare the more complex colours. Alizarin was abandoned, and indigo manufacture did not start until about 1910, but fast wool dyes were being made before that date.[1] The bulk was for export, which grew slowly in the late 1890s to reach £0·6 million in 1900, and then accelerated to £0·8 million five years later. Dyestuffs were relatively unimportant in Switzerland's foreign trade, and may be compared with exports of watches and embroidery, each valued at £5½ million in 1905.[2]

Swiss dyestuffs manufacture was in a difficult period of transition. Though the absence of patent protection enabled manufacturers to copy foreign discoveries with impunity, they were nevertheless in an obviously weak position vis-à-vis the powerful German producers from whom they bought intermediates. The solution was found at a fairly early date in understandings with the Germans that proved to be of mutual advantage: the Germans were able to keep copying by the Swiss to a minimum, and in return the Swiss were not squeezed out by the Germans.[3] It is perhaps not without significance that the companies that did not have links with the Germans went out of business during the next few years.

Dyestuffs were not made in Belgium, and the heavy chemical industry in 1905 consisted of fifty-seven factories with just under 4900 workers; total employment in all chemical trades (excluding soap and alcohol distilling) was of the order of 10 000. These numbers give an indication of the structure of chemical manufacture as one of many small factories, generally employing between 50 and 100 workers. Indeed only two works were employing from 400 to 600 men and were therefore comparable with British or German factories.[4] The main products were sulphuric acid and superphosphate. About 210 000 tons of the latter were made in 1900, half for export. The largest group in this branch was the Union Commerciale des Fabricants Belges de Produits Chimiques, comprising six firms with 1650 employees. Alkalis ranked next in importance to fertilizers, and a large proportion of the make was exported.

Fabrik AG, which was taken over by CIBA in 1908. F. Petersen & Cie was situated at Schweizerhalle in the Canton of Baselland. F. Hoffmann–La Roche & Cie made pharmaceuticals in Basle. Jaquet, N. *Die Entwicklung . . . der schweizerischen Teerfarbenindustrie* (1922), p. 114.

[1] Wizinger-Aust, R. Schweizerische Forschung über künstliche Farbstoffe . . . . *Neue Zürcher Zeitung* (21 July 1955), no. 198 (Fernausgabe); Erban, op. cit., pp. 26 ff.

[2] Jaquet, op. cit., appendix.

[3] There were links between CIBA and BASF, particularly for basic dyes; agreements were also reported between Hoechst and Sandoz, Bayer and Geigy, Cassella and L. Durand, Huguenin & Cie. Erban, op. cit., 32–3.

[4] Ministère de l'Industrie et du Travail. *Fabrication des produits chimiques* (1905), p. 12.

Solvay & Co. were by far the biggest producers and their two works employed about 600 people.[1]

Chemical manufacture in Sweden and Norway was on a minuscule scale, and the utilization of the water-power resources was only just beginning. In 1900 about 20 000 hp of electricity were used in making calcium carbide and potassium chlorate. Employment was around 800 and the value of production about £300 000.[2] Big projects were maturing in Norway and will be described later, but about 1900 political difficulties over the separation from Sweden and inadequate local finance delayed development.

In Italy about 250 000 tons of superphosphate were made in 1900, and together with refined sulphur and copper sulphate represented the heavy chemical industry. There were several producers of vegetable tanning extracts from chestnut and oak, and also firms making essential oils; there was no production of coal-tar dyes.

Eastern Europe

In Eastern Europe the end of the nineteenth century was a most active period in the history of chemical manufacture. Change occurred because communications improved and foreign capital was attracted to local enterprises. The investments were protected by high tariffs behind which inorganic chemical production developed vigorously.

The largest business in the Austrian empire was the Oesterreichischer Verein für Chemische und Metallurgische Production at Aussig in the Sudetenland, which employed over 2500 men at the main works and had a turnover of around £800 000 in 1900/1. The company still worked the Leblanc and associated processes and made 25 000 tons of alkali (including sodium sulphate) and 55 000 tons of hydrochloric acid, which was used for making zinc chloride, then widely employed as a wood preservative in eastern Europe, and bleaching powder.[3] Jointly with Solvay the Verein built ammonia-soda plants at Ebensee in the Salzkammergut and at Maros-Ujvár in Hungarian Transsylvania, opened in 1885 and 1896 respectively, which made about 35 000 tons of soda ash.

Among the smaller enterprises were a Leblanc factory at Hruschau, the old firm of Wagenmann, Seybel & Co, which prepared acids and sulphate of ammonia in Vienna, where it employed 500 men, and a few fertilizer works in Hungary, which used indigenous pyrites to make sulphuric acid, which was employed for making superphosphate from imported phosphate rock.[4] There was no manufacture of dyestuffs in the empire and among other organic chemicals wood distillation products, notably methanol and acetic acid, were especially important.

[1] *Paris Exhibition* (1), 88 ff.

[2] *Paris Exhibition* (1), 116; Witt, O. N. *Die chemische Industrie auf der Internationalen Weltausstellung zu Paris 1900* (1902), p. 93.

[3] *Paris Exhibition* (1), 87–8; Anon. *Der Oesterreichische Verein für Chemische und Metallurgische Production* (1906).

[4] Witt, op. cit., pp. 72 ff.

Chemical manufacture in Russia had been on a small scale until the early 1890s. The rapid development that then set in, telescoped within a matter of 10–15 years an evolution that in England and France had been spread over 50–60 years. The period coincides with Witte's tenure at the Ministry of Finance (1892–1903), where he gave the Rouble a firm rate of exchange in 1897, encouraged investment by foreign capitalists, extended the railway network, and promoted the steel industry on the Donets, and the textile industry in Moscow, both of which used Russian raw materials. Tariff protection was an essential feature and chemical manufacture in particular benefited from the higher duties of 1891, which were designed to keep out foreign products, though the prohibitively high rates positively encouraged evasion. For example, on soda ash the duty was almost £6 a ton, on bleaching powder £7. 11*s.* 8*d.* a ton and on caustic soda £9. 15*s.* 0*d.* a ton.[1]

Russian statistics of the time, though plentiful, are often contradictory and their reliability is uncertain. The figures below, which were obtained by the jury of the Paris Exhibition from the Government, nevertheless show remarkable growth during the 1890s.

TABLE 2.2

Size of the Russian chemical industry, 1890, 1897[2]

	1890	1897
Number of factories	748	769
Value of production (£ million)	3·3	6·3
Number of workers (thousands)	28	35

Note: Rate of exchange to the £, 1890 = 9 R, 1897 = 8·70 R

The above employment figures refer to chemical as well as allied trades. The former, which in Russia included paint and pharmaceuticals besides heavy chemicals, was quite small, numbering 17 000 in 1897.[3] This may be compared with over half a million in mining and metals and nearly 150 000 in pottery and glass.[4]

The country's great natural resources were not being exploited in a rational manner; for example though pyrites deposits were plentiful, production was small and the big sulphuric acid plants relied on imported pyrites and brimstone. Perhaps the chief obstacles were the distances and the climate. The extremes of heat and cold dictated a seasonal cycle of operations in the sulphuric acid and Leblanc soda factories, while the first ammonia–soda works at Beresniki had to be housed in steam-heated buildings. Further, the annual freeze-up of rivers created problems so that coal and oil (the latter an important fuel even by 1900) could not be moved except by rail and at much

[1] *Chem. Trade J.* **38** (1906), 82 ff. The famous chemist Mendeleev was president of the commission that drafted the tariff.

[2] *Paris Exhibition* (1), lxiv–v.

[3] Strumilin, S. Khimicheskaya Promyshlennost SSR. *Problem ekonomiki* (1935), no. 2, 118 ff.

[4] *Pans Exhibition* (1), lxiv.

expense. The environment was thus a big element in the geographical distribution of the industry and its sources of supply. One important area was in the west and comprised St. Petersburg, Riga, Warsaw, and Lodz: in this part of the empire foreign materials, shipborne or sent by rail from Germany, were able to compete with Russian products despite the high tariffs. Further inland, in the Moscow area, possibly the main market for soda ash and soda crystals, Russian chemicals had some advantage, while further inland still they predominated. Thus the Baku and Grozny oil refineries relied on their own supplies of sulphuric acid and caustic soda.

Plant and machinery were increasingly made in Russia and were considered of good quality. But installation and freight were expensive and one German observer thought that a chemical plant cost about four times as much as in Germany.[1] One peculiarity of the Russian chemical industry was that while its dependence on foreign techniques diminished, it was unable to compete abroad and exports were negligible. Consequently products were made exclusively for the home market and the business, at any rate in heavy chemicals, seems to have been fairly good, despite short spells of depression. In 1901, twenty-nine chemical companies were said to have averaged a net profit of 8·8 per cent on their paid up capital; but their dividends averaged only 6 per cent compared with 15·6 per cent for the booming oil industry.[2]

The manufacture of alkalis developed rapidly in the 1890s. Production of soda ash rose from 48 000 tons in 1895 to 86 000 tons in 1900, that of caustic soda from 6000 tons to 24 000 tons in the same period.[3] Consumption of soda ash was estimated at 64 000 tons in 1895 and 100 000 tons five years later.[4] Imports were thus necessary in both years, but their relative importance was significantly lower in the latter. As far as caustic soda was concerned we have no consumption figures; although output increased greatly a substantial tonnage had still to be imported, according to one Russian authority at the rate of about 6000 tons a year in the mid- and late 1890s.[5] Finally we may note that bleaching powder consumption doubled in a few years to about 7000 tons, the demand stemming not only from the textile industry, but also increasingly from public health authorities. Until the 1890s Russia was largely dependent on imports, predominantly from the United Kingdom, but by about 1900 the quantity was down to 2000 tons.[6]

[1] Hauffe, M. Die chemische Grossindustrie in Russland geschildert auf Grund eigener Erfahrungen und Beobachtungen. *Z. angew. Chem.* **13** (1900), no. 1, 27–29.

[2] *Z. angew. Chem.* **15** (1902), no. 9, 211–12.

[3] Lukianov, P. M. *Istoria Khimicheskikh Promyslov . . . Rossii do Kontsa 19 Veka* (1948–52), vol. 2, 676; Nutter, G. W. *et al.*, *Growth of industrial production in the Soviet Union* (1962), 423 f. The two sources agree on soda ash, but differ on caustic soda for which Nutter gives 18 000 tons in 1895 and 33 000 tons in 1900.

[4] Dyes, W. A. Die Production, Ein- und Ausfuhr chemischer Producte in Russland. *Chem. Ind., Düsseld.* **22** (1899), no. 17/18, 389.

[5] Lukianov, op. cit., vol. 2, 689–90.

[6] During cholera and other epidemics bleaching powder was admitted duty free: Dyes, op. cit., p. 391; and his Produktion und Rentabilität der russischen Chemischen Industrie. *Chem. Ind., Düsseld.* **28** (1905), no. 12, 338; Broun, J. Die chemische Industrie Russlands. *Chem. Ind., Düsseld.* **23** (1900), no. 24, 544.

The bulk of the Russian alkali production was then made by foreign owned or controlled firms. The reason for this was as follows: the Liubimov brothers, large entrepreneurs in the salt and timber trade on the Volga and Kama, were granted a special privilege to manufacture soda. At that time the Russian market was dependent on imports, and there seems to have been only one Leblanc works. At any rate, the Liubimovs combined with Solvay and opened their first plant at Beresniki on the middle Kama in 1883. Eight years later they opened a second plant at Lissichansk on the river Don in the Ukraine, and a third, small plant was opened at Barnaul in western Siberia in 1900. In 1895 these works made altogether 43 000 tons of soda ash and caustic soda, and extensions at Lissichansk carried the figure to 71 000 tons five years later[1] [in addition the company produced salt and limestone]. Liubimov, Solvay & Co. employed about 1500 people at their soda factories. It was closely connected with the sulphate of ammonia business and although its annual requirements of this material were only a thousand tons the quantity represented roughly half the entire Russian production. Indeed only the St. Petersburg gasworks were equipped at the time to make sulphate of ammonia. The Solvay group therefore decided to pioneer the installation of recovery ovens at some coking plants in the Ukraine.[2] (Sulphate of ammonia was not used then as a fertilizer in Russia.)

Several other alkali works were founded in the latter part of the 1890s, mainly with German capital. A company whose title in translation reads South Russian Company for the Manufacture and Sale of Soda and Other Products was established in 1897 at Slaviansk to make ammonia soda by the Honigmann process; two years later Russki–Elektron, using the Griesheim cell, built an electrolytic plant alongside. The other electrolytic soda works were built with German capital in Russian Poland near the Silesian frontier. The South Russian Co. was the largest, having a capacity of 33 tons a day of alkali; it is unlikely to have operated at capacity throughout the year. The works in Poland produced about 4000 tons of bleaching powder. The figures are not particularly reliable, but all these companies together with Liubimov, Solvay & Co. were certainly far more important than the Russian alkali makers.[3] Of these, P. S. Ushkov (or Uschkow) & Co., founded in the 1850s, employed early in the twentieth century about 1700 people in factories at Kazan and Jelabuga on the Kama. The company made most products of the Leblanc industry, notably sodium sulphate, caustic soda, and about 6600 tons of bleaching powder. It also sold a substantial tonnage of hydrochloric acid to glue and gelatine works and for making zinc-chloride wood preservative. (Creosote, then widely used in the West, was not available in Russia to any significant extent.) The Tentelev Company of St. Petersburg dated back

[1] Lukianov, op. cit., vol. 2, 673.

[2] Dyes, articles in *Chem. Ind., Düsseld.* **28** (1905), no. 13, 370, and no 14, 406; Broun, op. cit., p. 548; Fedotijev, P. P. Die chemische Grossindustrie Russlands. *Z. angew. Chem.* **15** (1902), no. 34, 865, 867.

[3] Dyes, article in *Chem. Ind., Düsseld.* **22** (1899), no. 17/18, 390.

to 1876. It employed nearly 700 people and was the largest producer of sulphuric acid in the empire with an output of about 25 000 tons a year in 1900.[1] The firm used imported pyrites and among its other activities was the manufacture of alums, nitric acid, and rather oddly, platinum refining.[2] Tentelev claimed to have discovered the contact acid process in 1886 and worked it since that date. BASF disputed the claim and although the matter was apparently never cleared up Tentelev nevertheless granted licences to Nobel Brothers at Baku and to the General Chemical Co. in the United States.[3] The small sulphuric acid works of the Moscow area and along the Kama used mainly Russian pyrites, but at Baku they relied on Sicilian brimstone. Sulphuric acid was usually sold as 'chamber acid' (less than 70 per cent H_2SO_4) and when required was concentrated in platinum retorts and vessels. The old-fashioned method was cheaper than the modern Glover–Gay-Lussac system, which had become the main source of acid (78 per cent H_2SO_4), in western Europe, but needed a good deal of expensive chemical plant.

Sulphuric acid production in Russia ranged between 110 000 and 130 000 tons (as 100 per cent H_2SO_4) in the early 1900s. There seems to have been a considerable increase in output in the preceding years, but it is not possible to be more definite. The pattern of consumption differed noticeably from that of western Europe. The largest single group of users consisted of the oil refineries in Baku which were said to require 20 000 tons (as 100 per cent H_2SO_4) annually. Next in importance came the production of sodium sulphate[4] and hydrochloric acid, occasionally associated with the preparation of soda crystals.[5] The trade in sulphuric acid was well regulated, and the makers had agreed to divide the market into two main zones, the Baltic region and Poland, and the Moscow area, within which they maintained uniform delivered prices.[6]

Fertilizers, which by the turn of the century had become the main outlet for sulphuric acid in Western countries, were of little commercial importance in Russia. We have already noted that the output of sulphate of ammonia was tiny, and superphosphate was only made on any scale in St. Petersburg, at Riga, and in Poland. The available statistics are in conflict, but even if we take the highest figures quoted, 80 000 tons made and 32 000 tons imported

[1] Fedotijev, P. P. The Russian Chemical Industry. *Chem. Trade J.* **20** (1897), no. 523, 364; Witt, op. cit., p. 99.

[2] Witt, op. cit., p. 100.

[3] BASF alleged that Tentelev bought the know-how from an agent who had obtained the particulars from a former employee. The dispute is ventilated in some acrimonious correspondence in the *Z. angew. Chem.* **18** (1905), nos. 3, 10, and 16, pages 97, 379–80, and 625–6 respectively.

[4] There was also natural sodium sulphate, which was being worked at Karabugas on the eastern shore of the Caspian. The material was shipped across the inland sea to Baku where it was treated by the traditional Leblanc process to yield soda, whence caustic soda was prepared.

[5] Dyes, articles in *Chem. Ind., Düsseld.* **22** (1899), no. 17/18, 408 and **28** (1905), no. 12, 339, 341; Nutter, op. cit., estimates that the production of sulphuric acid rose from 52 000 tons to 178 000 tons between 1895 and 1900. As there is no indication of the strength of the acid, the figures are not very meaningful.

[6] Dyes, *Chem. Ind., Düsseld.* **28** (1905), no. 12, 329

in or about 1900, we are bound to conclude that consumption was very small by comparison with western Europe.[1]

The tariff policy was designed to force foreign firms to establish dyestuffs works in Russia. At first imported colours were merely diluted and repacked, later intermediates were worked up, and still later, when some tar-distillation products became available, the works began to make simple dyestuffs. German products predominated and all the big companies owned factories in Russia. In 1899 the Swiss CIBA Company bought a half share in a dyestuffs works at Pabianice near Lodz, the cotton centre of Poland. Of Russian firms there is mention of very small factories at Moscow, St. Petersburg, and Lodz.[2] The dyestuffs business was considered the most lucrative section of the Russian chemical industry because of the absence of all competition. The only other organic chemical activity worth mentioning is wood distillation, but in 1900 it was still in its early stage.

The United States and Canada

It is not so far-fetched to compare the chemical industries of the United States and Russia at the opening of the twentieth century; they had indeed much in common. The two countries possessed abundant supplies of raw materials, and, while their exploitation was further advanced in the United States and the transport system was much better, nevertheless distances tended to create local markets supplied by distinct groups of chemical enterprises in America as well as in Russia. The works were specially numerous in the east along the Atlantic coast, particularly around Philadelphia, in northern New Jersey and close to the New England textile industry. Other centres emerged at this time in the neighbourhood of Niagara Falls, where cheap hydro-power was being developed, around Detroit, where salt had been discovered at Wyandotte, in central Michigan, for the same reason, in west Virginia, because of its coal and salt, and finally in the oil region of western Pennsylvania and the steel mills of Pittsburgh. Even in the southern states there were some interesting developments: the phosphate rock deposits, the largest of their kind in the world, led to the establishment of superphosphate factories in the Carolinas, Georgia, and Florida. Chemical manufacture was also emerging in the Mid-West, where conditions, at first sight, looked less promising. The Gulf Coast, now one of the main centres of the chemical industry, was of no significance as yet, though the Frasch process was about to transform the situation; towards the end of 1894 H. Frasch carried out his successful experiment of pumping molten sulphur. Technical difficulties and lack of cheap fuel caused delays, but the discovery of the Spindletop oilfield

[1] Dyes, art. cit., 344; Fedotijev's article in *Z. angew. Chem.* **15** (1902), no. 34, 867, mentions that production and imports were 50 000 and 25 000 tons respectively, but according to *Annu. Lambert*, (p. 298) production was 30 600 tons.

[2] Broun, art. cit., pp. 541 ff.; Dyes, article in *Chem. Ind., Düsseld.* **28** (1905), no. 13, 374–5.

in east Texas in 1901 solved the fuel problem, and the United States became rapidly self-sufficient in brimstone.[1]

The United States did not have a Leblanc soda works, and until the mid-1880s the country was entirely dependent on imports from the United Kingdom. The alkali industry which developed in the next fifteen to twenty years grew up behind the shelter of tariffs, which were designed, as in Russia, to safeguard it from competition. Between the McKinley (1890) and Dingley (1897) tariffs, the duty on soda ash went up from £1. 3*s*. 0*d*. to £1. 14*s*. 6*d*. a ton ($\frac{1}{4}$–$\frac{3}{8}$ cents/lb), that on caustic soda and bicarbonate fell slightly to £3. 9*s*. 0*d*. ($\frac{3}{4}$ cents/lb), while bleaching powder, once duty-free, became chargeable at about 18*s*. a ton.[2] Thus protected, four companies were making ammonia soda in 1900,[3] and their output jumped from 150 000 tons in 1895 to about 350 000 tons five years later.[4] The plants of these companies were comparable to those of large European works. For example Solvay's at Syracuse and Detroit had capacities of 450 tons and 180 tons a day respectively.[5] A large proportion of the make was converted to caustic soda by the lime–soda process, about 150 000 tons being made in this way in 1900. The manufacture of electrolytic alkali, though established at a very early date in the United States, seems at first to have made very slow progress. Le Sueur produced caustic soda and chlorine at Rumford (Me.) as early as 1892, the year in which the Mathieson Alkali Co. began experimental work with the Castner cell at Saltville. Mathieson only began large-scale production at Niagara Falls in 1897, and the following year the Pennyslvania Salt Manufacturing Company established electrolytic works at Wyandotte, where they eventually employed Gibbs cells.[6] At that time the chief object of the fairly small electrolytic plants does not seem to have been the manufacture of bleaching powder, which continued to be imported in fairly large quantities, but caustic soda, chlorine, and sodium hypochlorite, the latter two being used for treating wood pulp from the mid-1890s onwards.

As in France, there was much experimenting with various electro-thermal and electrolytic processes; both generally relied on hydro-electricity. The first large calcium carbide works was built by Union Carbide at Niagara Falls in 1898. In the same area and year unsuccessful trials were made by Bradley and Lovejoy to fix nitrogen by the arc process,[7] and the Oldbury Electrochemical

[1] A full account is given in Haynes, W. *Brimstone, the stone that burns* (1959), pp. 40–62.

[2] *Chem. Ind., Düsseld.* **20** (1897), 351; *Chem. Trade J.* **7** (1890), 240.

[3] Solvay Process Co., founded in 1884, with works at Syracuse (N.Y.) and Detroit; Michigan Alkali at Wyandotte (Mich.); Mathieson Alkali at Saltville (W. Va.); Columbia Chemical Co., which started in 1900 near Akron (Ohio).

[4] *Paris Exhibition* (1), 96. Haller confused short tons with metric tons, and his figures have been adjusted accordingly.

[5] Haynes, *American chemical industry* (1945–54), vol. 1, 273–4.

[6] Haynes, op. cit., vol. 1, 277–8. Pennsylvania Salt had previously made caustic lye by a process which used cryolite from Greenland as raw material. So far as I know it is the only company ever to have done so.

[7] Hempel, E. H. *The economics of chemical industries* (1939).

Co. a subsidiary of Albright & Wilson, was producing white phosphorus. In 1897 chlorates became dutiable at £11. 10*s*. 0*d*. a ton (2½ cents/lb), and so the company took up the production of potassium chlorate and later sodium chlorate.[1] Another British-owned company, the North American Chemical Co., a subsidiary of U.A.C., was also set up in 1898 to make chlorates, but its works at Bay City (Mich.) relied on thermal electricity.

So far the development of the American Chemical industry had been greatly influenced by foreign capital and to a rather lesser extent by foreign technology. But there was no lack of indigenous chemical enterprise, indeed American chemists were pioneers of applied electrochemistry, and this can be illustrated by briefly looking at two mid-Western entrepreneurs: H. H. Dow, with backing from Cleveland business men, made his reputation by a process for extracting bromine from the brines at Midland (Mich.), and thus broke what had been a German monopoly. His electrolytic process also yielded chlorine, and by 1902 his company was said to be making about 80 tons a day of bleaching powder; if true it might well have been the largest manufacturer in the world.[2] Dow's contemporary J. F. Queeny of St. Louis (Mo.) incorporated the Monsanto Chemical Co. in 1901 to make saccharine. It remained a small struggling company for some time and was less important than the old established Mallinckrodt Company of the same town, which specialized in fine- and photochemicals and was also supplying ammonia to refrigerating plants.[3]

At the beginning of the century the range of products made by the United States chemical industry was broadening quickly. Even more striking was the increase in plant size. We have noted this in connexion with the ammonia-soda works, and the manufacture of sulphuric acid showed the same characteristics. The country was then the second largest producer of sulphuric acid in the world, and within a few years was to overtake the United Kingdom. Production of chamber acid increased spectacularly owing to its use in petroleum refining and even more so for superphosphate manufacture.[4] This fertilizer was then being made on a huge scale along the Atlantic coast for sale to cotton and tobacco growers. By contrast other fertilizers were only used to a small extent, a mere 55 000 tons of Chilean nitrates in 1900,[5] and the consumption of sulphate of ammonia was insignificant. As in contemporary Russia the recovery and concentration of ammoniacal liquors from coke ovens and gas works was still in its early stages, and likewise the Solvay

[1] Threlfall, R. E. *The story of 100 years of phosphorus making 1851–1951* (1951), pp. 260 f.

[2] Campbell, M. and Hatton, H. *H. H. Dow* (1951), p. 55. The Dow diaphragm cell liberated bromine at low voltages. The debrominated brine was then electrolysed again at a different voltage and chlorine was obtained as the principal product. The residual slurry from this second operation contained calcium, magnesium, and weak caustic soda, and was run to waste.

[3] Haynes, W. *Chemical pioneers* (1939), pp. 235 and 158–9.

[4] Haynes, *American chemical industry*, vol. 1, 254 states that 1 lb of acid was used for refining 8 U.S. gal. (say 50 lb) of kerosene and one ton of 50° Bé acid (62½ per cent H_2SO_4) for every ton of phosphate rock to make superphosphate.

[5] Federal Trade Commission, *The fertilizer industry* (1950), p. 20.

Co. had to build recovery ovens in order to safeguard its essential ammonia supplies.

The sulphuric acid and superphosphate trades provided the first instances of amalgamations in the United States chemical industry. In 1895 the Virginia–Carolina Chemical Co. combined many of the southern fertilizer works, and four years later the American Agricultural Chemical Co. united twenty plants in the northern and eastern states. The General Chemical Co. was set up, also in 1899, by twelve heavy chemical producers; it was the most important of the three and technically very competent, having developed its own contact acid process.[1]

Wood distillation and the working up of the products derived therefrom was then the most important organic chemical business. The United States was then probably the world's largest producer of methanol, output of which increased to about 15 000 tons in 1900, almost a fifth being exported. A co-product from these activities was calcium acetate of which 27 000 tons were exported in 1901, the material being the chief source for making acetone, a solvent then becoming important in Europe. The manufacture of dyes had flourished for a short time until a change in the tariff in 1883 had knocked the bottom out of this business. The duties had been further reduced in 1890 but were raised to 30 per cent *ad val.* by the Dingley Act of 1897. In this case protection came too late: only one small firm, Schoellkopf Aniline and Chemical, survived German and Swiss competition. For all practical purposes the United States was wholly dependent on imports.[2]

Some significant developments may be observed in Canada where hydro-electricity was being used to make calcium carbide in the late 1890s under the direction of T. L. Willson, whose role in North American electro-thermal technology was as important as Moissan's in France. At about the same time potassium chlorate began to be made, and in 1896 the first electric phosphorus furnace was built at Buckingham (P.Q.); a few years later this business, the Electric Reduction Co., became a subsidiary of Albright & Wilson.[3]

Forerunners of chemical change

There is no clear dividing line in industrial chemistry marking the end of one phase in its development and the beginning of another. Still, anyone looking back from Paris in 1900 to the Chicago Exhibition of 1893 must have noted some profound changes: the emergence of electrochemistry, the contact acid process, the manufacture of industrial gases. And looking around there were signs that chemical synthesis was spreading beyond dyestuffs: the first attempts at fixing atmospheric nitrogen were taking place, man-made fibres

[1] The General Chemical Co., as has already been noted, also obtained a licence from Tentelev, but I do not know what role the Russian (or German?) know-how played.

[2] Haynes, op. cit., vol 1, pp. 308–9.

[3] Warrington, C. J. S. and Nicholls, R. V. V. *A history of chemistry in Canada* (1949), pp. 160 ff.

were appearing, and essences and drugs were being created out of coal-tar derivatives.

Some see all these changes as the outcome of systematic research translated from the laboratory to the factory. Others view it in more dramatic terms: 'the chemical phase of the industrial revolution', it has been said, began in the 1890s and its characteristic aspect was 'the reshaping of materials'.[1] It is certainly not possible to ascribe the changes to any single cause, but it is proper in this context to digress slightly and look briefly at what was happening outside applied chemistry.

At the beginning of the century, the most significant aspect was the interpenetration of chemistry, physics, and biology and the emergence of the new disciplines of physical chemistry and biochemistry. We shall be primarily concerned with the former. Some scientists had been able to encompass them all —Faraday and Liebig are examples—but E. von Meyer, himself one of the younger men of science, was speaking for the majority when he wrote:

> . . . it has come to be more recognized especially within the last 40 and more especially within the last 20 years that chemical investigation runs the danger of becoming one-sided without a free use of physical aids. Chemists have perceived the necessity for their science of physico-chemical methods.[2]

What were these physico-chemical methods? Put at its simplest, they applied the laws of physics to chemistry with the object of explaining the course of chemical reactions. We may take as an example the application of the laws of heat to gases, that is thermodynamic reactions. Progress in this area was as follows: by about 1880 the behaviour of molecules in a gaseous state was fairly well understood, and the next step was to consider the effect of heat and pressure on gases. Here J. W. Gibbs, J. C. Maxwell, and particularly Van't Hoff laid the basis on which the theoretical studies of gas reactions were built by Le Chatelier, Nernst, and Arrhenius. Chemical thermodynamics is primarily concerned with the application of the three laws of thermodynamics to reactions, notably the time and temperature conditions under which such reactions proceed and the circumstances in which they terminate. One outcome of this work was the determination of the critical temperatures of gases. The next step led to the liquefaction of gases, and so to their separation: by the late 1890s Dewar and Linde had developed appropriate processes. Most gas reactions require a catalyst, that is a substance that accelerates the reaction without being changed by it. Catalysts had been known and used empirically for many years, but the theory and indeed the principles of selection came much later, and are associated with the names of Ostwald at the end of the nineteenth century and of Langmuir in the 1910s. It was found that catalysis plays a key role in biochemistry through the action of enzymes on starch and sugar, and in organic chemistry by the action of hydrogen on hydrocarbons,

[1] T. J. Kreps in an essay contributed to *Economics, sociology and the modern world* (N. E. Himes, ed.) (1935), pp. 147 f.

[2] *A history of chemistry* (3rd Eng. Edn. 1906), p. 392.

oils, and fats. Catalytic gas reactions quickly became of the greatest interest to academic chemists and were shortly to have a profound effect on industrial chemistry; among the most important examples were the work of Sabatier and Senderens on catalytic hydrogenation in the late 1890s, and of Nernst and Haber on ammonia in the following decade.

We must now turn back to another branch of physical chemistry, which is concerned not with gases but with solutions. Here three physical chemists, Arrhenius, Van't Hoff, and Ostwald, evolved during the 1880s the ionic theory, or theory of electrolytic dissociation. It remains of key importance to analytical chemistry and to the understanding of electrolytic processes. The latter illustrate how physical chemistry interacted with chemical technology: electrochemical processes had been used for many years: electroplating since the 1840s, copper refining since the late 1860s, and the reduction of alumina to aluminium since the mid-1880s. By 1900 two distinct branches had emerged: electrothermal methods for making phosphorus, carbide, and graphite, and electrolytic methods for making aluminium, caustic soda, chlorine, caustic potash, chlorates, hydrogen, and hydrogen peroxide.[1]

The utilization of gas reactions and electrochemistry was dependent on progress in mechanical and electrical engineering. Until the 1880s the rotating or 'revolver' furnace of the Leblanc works was the most elaborate piece of chemical plant. In terms of chemical engineering it had become a historical curiosity twenty years later. The contributions made by Rankine, Claude, and Linde cannot be sufficiently emphasized; the last named resigned the chair in machine design in Munich to devote himself to the building of an air-liquefaction machine. He achieved the object in 1895, and a few years later devised a continuous process for obtaining oxygen and nitrogen, which is still used. This pioneering work led to other gas separation processes that are of fundamental importance in chemical technology. Electrochemistry made no headway until the dynamo had been perfected, and the rotary converter introduced so that direct current for the cell systems became available.

At the turn of the century nitrogenous fertilizers, and pharmaceuticals and veterinary products were on the eve of major industrial developments. In both cases the driving forces were the new chemical disciplines, respectively physical chemistry and biochemistry. The stepwise progression from theory via the laboratory to the pilot plant and finally to factory-scale production was common to both and without the new machines neither would have progressed. As these developments gained momentum they had far-reaching effects on the structure of the chemical industry. Until about 1900 this can be summarized by dividing the manufacturers into three clearly distinct groups: makers of heavy inorganic chemicals, dyestuffs producers, and compounders who bought, mixed, and resold; the last were characteristic of the fertilizer, pharmaceutical, and toilet preparation trades. But during the next twenty-five years the lines blurred, the groups overlapped, and production became

[1] Sherwood Taylor, F. *A history of industrial chemistry* (1957), pp. 384 ff.

integrated. More will be said later on this subject; here it is only necessary to observe that integration is seen at its best in fertilizer manufacture, specifically nitrogenous fertilizers made by the new synthetic processes. Nitrogen is the third of the essential plant-nutrient components. The other two were even by 1900 available on a large scale. Phosphorus was used in the form of superphosphate and Thomas meal,[1] potash predominantly as cainite or carnallite.[2] But the situation with regard to nitrogen was less satisfactory, though it was then rightly recognized as the key nutrient, determining plant growth and setting the level of consumption for the other fertilizers. The best material at the time was Chilean nitrate, since plants take up nitrogen in the form of nitrate. World consumption was then 1·4 million tons.[3] Supplies of nitrate were limited, extraction methods wasteful, and it was also required in large quantities for the manufacture of sulphuric and nitric acids. No wonder then that pessimists and even experts voiced concern lest the growth of the world's population should exceed that of food supplies. Crookes and the neo-Malthusians contributed in a way to the solution of what came to be known as the Nitrogen Problem by urging chemists to find a route to nitrogen fixation; this would, of course, provide farmers with an unlimited source of fertilizer nitrogen. While these matters were being ventilated, there was already available to industrial countries an indigenous source of nitrogen in the form of sulphate of ammonia, obtained in Britain mainly from gas works and in Germany from recovery ovens. Greater use of recovery ovens by the steel industry was to lead shortly to greatly increased supplies of sulphate of ammonia; in 1900 world production was about half a million tons.[4] The drawback to sulphate of ammonia was that repeated use caused a rise in the acidity of the soil, so necessitating the application of lime. Thus the predicted shortage of nitrates and the problems associated with sulphate of ammonia (both greatly exaggerated as events were to show) played their part in stimulating research.

The other mainstream of the new applied chemistry had its source in biochemistry, and in this case pathology was the driving force. As with nitrogenous fertilizers, the situation presented a challenge. Public health had made, at least in Western countries, prodigious advances during the nineteenth century. Indeed the progress made was one of the factors that prompted Crookes to make his depressing forecasts, for the prevention of disease had as its corollary the feeding of more mouths. The improvement in public health, embedded in sanitary legislation, was maintained by novel applica-

[1] World production in 1900 amounted to 4·6 and 1·6 million tons respectively. *Annu. Lambert*, pp. 298, 319.

[2] German production of all potash salts was 3 million tons in 1900; Krische, op. cit., vol. 1, p. 135.

[3] *Annu. Lambert*, p. 311. Some guano was still being used, but the quantity was only about 50 000 tons.

[4] The figures vary. Kubierschky, K. Die künstlichen Düngemittel und ihre Bedeutung. *Z. angew. Chem.* **26** (1913), pt. 1, 727 mentions 0·5 million tons, *Annu. Lambert* p. 134, 0·4 million tons.

tions of industrial chemicals, such as the use of carbolic as disinfectant in hospitals and institutions, the employment of bleaching powder during epidemics and, in the early 1900s, the introduction of chlorine for water treatment. But while preventive medicine was progressing, advances in the treatment of disease were much slower, and as for veterinary medicine, it was only just beginning in the 1880s.

Most progress was made in the treatment of infectious diseases after the work of Pasteur, Koch, and Ehrlich. Pharmacology advanced more slowly along the lines marked out by organic chemistry in general, and it is therefore not surprising that many of the early remedies were the by-product of dyestuffs research, notably the antipyretics and the analgesics. Among the former, antipyrine and phenacetin were discovered in the 1880s, among the latter was aspirin, produced on a commercial scale from the late 1890s onwards. These medicaments were at first made only by dyestuffs firms, probably because the intermediates were most readily available to them. Meanwhile another group of firms, the manufacturing chemists, specialized in the preparation or compounding of the active ingredients of vegetable substances, notably the alkaloids. At the end of the century their contribution to medicine was still much more significant than that of the man-made drugs of the organic chemical industry.

By 1900 many general preparations were available, but specific remedies to strike at a particular disease were rare: quinine against malaria and emetine against dysentery spring to mind, but chemotherapy was still in its infancy.

3

ACADEMIC CHEMISTRY

THERE is always debate on the function and dissemination of science. At the turn of the century two aspects of special interest to us came to the fore: one was the increasing fragmentation of chemistry, illustrated by the formation of specialist societies and the weakening of some of the older learned societies: the other centred on the German system of academic chemistry, its merits, and its adoption elsewhere.

We may conveniently begin by looking at chemical societies. The Paris Exhibition, at which chemistry occupied an important place, also provided the occasion for holding the fourth International Congress of Pure and Applied Chemistry. International meetings of chemists had taken place as far back as 1859, when about 140 went to Karlsruhe to discuss chemical nomenclature, without however reconciling their differences of opinion so that the French went one way and the Anglo-German contingent another. The next gathering was not held for thirty years, but the 1889 Congress again discussed a standardized nomenclature, and this time with more success. Subsequently the American Chemical Society took the lead in convening a meeting, the World's Congress of Chemists, at Chicago, timed to coincide with the exhibition of 1893. From this sprang the concept of regular congresses of pure and applied chemistry.

The first, at Brussels in 1894, had only four sections, concerned respectively with the chemistry of sugar, agricultural chemistry, nutrition, and public health. The second was held in Paris (1896), the third in Vienna (1898), the fourth (1900) was again in Paris. Some 1750 chemists registered, about three-fifths of them from France; there were large groups from Germany and the United States. While the number of sections increased as an arithmetic progression, that of papers was subject to different laws altogether and at the fifth Congress in Berlin (1903) 500 were submitted. Initially these congresses had done much useful work settling definitions, standardizing methods of analysis, and agreeing on samples. Later however, they became veritable chemical jamborees: at the seventh Congress in London (1909) there were over 3000 registrations, though actual attendance was much lower and the foreign chemists outnumbered the British.[1] It became obvious that with such large gatherings only careful organization could keep the entire show going, and their chief value lay in the personal contacts made and the discussion of specialist papers. These were supplemented by broad reviews of the state of industrial chemistry, at which the Germans excelled. In 1912, at the eighth

[1] *Z. angew. Chem.* note, **24** (1911), no. 38, 1793–4.

Congress in New York, they expressed much self-satisfaction and one chemist observed that, next to English, German was the most common language.[1]

The mass of material prepared for these occasions was often of minor interest and of little original value. Yet the need for disseminating the progress of modern chemistry was urgent, and led to the formation of many specialist societies. Their survival to the present day testifies not only to their tenacity, but also to their usefulness. The creation of such new societies did not go unchallenged: resistance came mainly from the older bodies, which as time went on became increasingly concerned with pure chemistry. The Chemical Society of London was the oldest, dating back to 1841. Half a century later it had 1751 members and by 1911 there were about 3100.[2] Next in order of seniority came the Société Chimique de France. Its origins went back to 1857; two years later the Société Chimique de Paris was officially incorporated and in 1906 the present title was adopted. By 1912 it had about a thousand members.[3]

In 1867 A. W. von Hofmann, a Fellow of the Chemical Society of London, founded the Deutsche Chemische Gesellschaft in its image. It had about 3300 members in 1912. Its main care was pure chemistry, but the Gesellschaft collaborated with the other chemical associations founded in the last quarter of the nineteenth century. It was not only god-parent or at least friend to these bodies, but also played a quite exceptional role in the dissemination of scientific knowledge. Apart from publishing the famous *Berichte*, it assumed the publication of the *Chemisches Zentralblatt*, a weekly journal of abstracts, which after the First World War was extended to include chemical engineering. It also took over from Beilstein, who was unable to continue the vast task of collation, the job of editing the *Handbuch der organischen Chemie*, which to the present day remains an essential source-book for the organic chemist.[4]

The American Chemical Society (A.C.S.), now the largest if not the most influential in the world, dates back to 1876, but until 1889 it was mainly a local society with members residing in New York. It was reorganized in 1889–91, and then grew rapidly either by taking over other local societies or by establishing new branches. By 1896 it had 1000 members and in 1912 5600.[5] The A.C.S., unlike its German sister, was apt to look on other societies

[1] Rassow, B. Was haben wir Chemiker in Amerika gelernt? *Z. angew. Chem.* **26** (1913), pt. 1, 705–8.

[2] Anon. *The jubilee of the Chemical Society* (1896), p. 186; Browne, C. A. and Weeks, M. E. *A history of the American Chemical Society* (1952), p. 99.

[3] Delaby, R. (Ed.) *Centenaire de la Société Chimique de France* (1957), pp. 6–8.

[4] Beilstein and one assistant working in St. Petersburg prepared the first three editions: they listed 15 000 compounds in the first edition, 1880–2, which ran to 2201 pages. The third edition (1892–4), together with its supplement (1899–1906), aggregated 11 400 pages and mentioned about 150 000 compounds. The supplement was the first editing job of the Deutsche Chemische Gesellschaft, which published the fourth edition in 26 volumes between 1918 and 1938.

Richter, F. (Ed.) *75 Jahre Beilstein's Handbuch der organischen Chemie* (1957), pp. 49–52; Jacobson, P. Die Adolf Baeyer Gesellschaft zur Förderung der chemischen Literatur. *Z. angew. Chem.* **33** (20 July 1920), pt. 1, 177–8.

[5] Browne and Weeks, op. cit., pp. 37–9, 99.

as intruders and therefore formed not only geographical sections, but also semi-autonomous divisions for the major branches of chemistry. The first of these, founded in 1908, was the Division of Industrial Chemists and Chemical Engineers. This may have been intended as a counter-move to the American Institute of Chemical Engineers, a small body, with rigid qualifications for admission, which gave its members a cachet not enjoyed by those of the A.C.S.[1]

All these societies, with the partial exception of the A.C.S., represented predominantly academic chemists, and they focused on pure chemistry. Industrial chemists and the growing number of those working in public or local service had to look elsewhere to advance their knowledge and safeguard their professional status.

In Britain the Society of Chemical Industry, founded in 1891, had many practical concerns apart from technological progress; examples are the interest it took in the Alkali Acts, patent-law reform, water pollution, technical education, and so on. Membership grew very quickly, probably because local sections were formed from the start. Initially there were about 300 members; they numbered 1400 in 1889/90 and almost 4000 by the early 1900s. The Society spread outside the United Kingdom, and, in particular, formed an American Section in 1894. This had 36 members at the start, and grew so quickly that twelve years later there were 1500 members of whom one-half also belonged to the A.C.S.[2] They kept in touch through the Society's journal, which was competently edited, and which besides reprints of lectures on major advances in industrial chemistry, regularly published abstracts and news of the sections' activities.

The Institute of Chemistry was set up in 1877. The founders' purpose was to create a body that would give chemists professional standing, in a manner similar to the Law Society or the Institution of Mechanical Engineers. Accordingly the Institute required certain qualifications for admission, and so endeavoured to establish minimum standards for chemists who had not been to university. For example, in order to become a Fellow (F.I.C.) the applicant had to be at least twenty-four with three years' training in chemistry and physics followed by another three years as a teacher or industrial chemist: for the Associateship (A.I.C.) the requirements were lower. It was not long before the Institute began to introduce examinations, and these were officially recognized by an agreement in 1899 between the Local Government Board and the Institute specifying the tests that had to be passed in order to qualify as a public analyst. Membership long remained small, but in the 1900s the numbers grew from about 1300 Fellows and Associates in 1902 to almost 5200 a quarter of a century later.[3]

[1] Browne and Weeks, op. cit., pp. 86–8.
[2] Jubilee number of *Chemy Ind.* (July 1931), 14–15, 24–5; Browne and Weeks, op. cit., p. 70.
[3] Pilcher, R. B. *History of the Institute of Chemistry* (1914), pp. 52, 300; Chapman, A. Chaston. *The growth of the profession of chemistry during the past half-century* (1927).

In Germany, the corresponding society occupied a rather more important place. Its beginnings as the Verein analytischer Chemiker in 1877 were inauspicious and it languished for ten years, when it was renamed the Deutsche Gesellschaft für angewandte Chemie. But even in 1895 it had only about 1000 members. The following year the name was changed once more to Verein Deutscher Chemiker (V.D.C.), and now a rapid period of growth began, which brought membership to 2500 in 1901 and almost 4000 in nineteen local sections in 1909. At the latter date it represented about half of all chemists working in Germany. The V.D.C. differed from the Society of Chemical Industry and the Institute of Chemistry in that it combined the interests of industrial and academic chemists. The two were traditionally linked in Germany and both enjoyed considerable prestige. The V.D.C. concerned itself with education, conditions of employment (including holidays), and the development of chemistry generally in Germany. It included members of specialist societies of which several were formed at that time,[1] and maintained links with the Elektrochemische Gesellschaft and the Deutsche Bunsengesellschaft für Angewandte Physikalische Chemie, founded in 1894 and 1902 respectively. All these bodies formed in 1906 the Ausschuss zur Wahrung der gemeinsamen Interessen des Chemikerstandes which, if one may say so, was the trade union of professional chemists and negotiated on their behalf with the government and the employers' association, the Verein zur Wahrung der Interessen der chemischen Industrie (the Chemie-Verein). All these societies had their own journals, but as we have little information about circulation it is not possible to judge their relative importance. The *Zeitschrift für angewandte Chemie* was started as a fortnightly in 1888 by the publishing house of Springer. In 1904 the V.D.C. became sole owner. Circulation rose from about 2000 in 1895 to 3100 at the turn of the century when it became weekly; in 1905 it was 4600 and in 1912 reached 6200. This journal dealt with applied chemistry and technology, leaving prices and commercial news to the *Chemiker Zeitung* founded in 1878.[2] The third major publication was the fortnightly *Chemische Industrie*, the organ of the Chemie-Verein. Its circulation was probably below that of the *Zeitschrift*, but it was competently edited by O. N. Witt (a well known dyestuffs chemist), published excellent statistical series, and carried weight in industrial quarters.

There was nothing quite like this trio outside Germany. The *Chemical Trade Journal* was launched in 1887 and eventually built up a weekly circulation of around 3000 in the 1900s.[3] It was widely read probably because it paid much attention to market intelligence at a time when London was still the centre of the world trade in chemical raw materials and heavy chemicals. The *Journal of the Society of Chemical Industry* was a fortnightly until the end of

[1] For example, societies for chemists in the leather industry, in the foodstuffs and in the paper and wood-pulp industries. Rassow, B. (Ed.) *Geschichte des Vereins Deutscher Chemiker* (1912).

[2] ibid., pp. 36–7.

[3] Private information; the figure seems high. In 1966 it was taken over by its younger rival, *The Chemical Age*, founded in 1919.

1922 when the title was changed to *Chemistry and Industry* and it became a weekly.

In America the *Oil, Paint and Drug Reporter*, first published in 1871, limited itself to recording price changes and trade news. The A.C.S. initially confined itself to the publication of its *Journal*: this was supplemented towards the late 1900s by the appearance of *Chemical Abstracts* and *Industrial and Engineering Chemistry*; the latter was subsequently split into *Chemical and Engineering News* and *Analytical Chemistry*. Initially these journals were sent to every member of the A.C.S. so that in 1910 the print order was at least 5000, and in this respect the A.C.S. played the same role as the V.D.C. in Germany.[1] No other publications were of major significance before the First World War.

Before considering the different systems of chemical education, some general comments are appropriate. There were in the 1900s roughly two main approaches to the teaching of chemistry at university or quasi-university level. There was, on the one hand, the method used in Germany and Switzerland. Its characteristic feature was academic freedom—*Lernfreiheit*—which enabled the student to move from one university to another. The course of study culminated ideally, though not necessarily, in the presentation of a thesis and the award of a doctorate. The staff had freedom to teach—*Lehrfreiheit*—and so could lecture on their own particular subject untrammelled by the restrictions of a rigid curriculum. On the other hand there was the method followed in Anglo-Saxon countries: the student, once having chosen his university, could not change, and the selection of the subjects was governed by the regulations laid down by the university and the guidance of his tutor. The study culminated in the bachelor's degree; the submission of a thesis for the doctorate, which took rather longer, was rare at that time.

There were other important differences between the two approaches. The emphasis on laboratory work was on the whole greater in Germany than in Britain, the United States, or France, and certainly the facilities for this work, such as equipment and materials, were far superior. Another difference was that in Germany, as in other Continental countries, universities were state bodies and their staffs Civil Servants, dependent on the Ministry of Education. By contrast, in Britain their links with the Government were tenuous. Though Continental universities demanded fees, and relied on them to a varying extent, the bulk of their budget was borne by the central Government. The older British and American universities were private or endowed institutions, not in receipt of state assistance; the newer British universities, as well as the American Land Grant and State Colleges, received a measure of public support, the British from local rates, the American from the individual federal states.

A further point was that Germany had a dual system of higher technical education: the universities and the *Technische Hochschulen*; the former specialized in pure (or more precisely organic) chemistry, the latter in applied

[1] Browne and Weeks, op. cit., chap. 18.

UNIVERSITY OF WINNIPEG
LIBRARY
515 PORTAGE AVENUE
WINNIPEG, MAN. R3B 2E9

chemistry. It should be noted that technical colleges or polytechnics existed in other countries, for instance in Switzerland at Zürich, in Scotland at Glasgow, and in the United States at Cambridge (Mass.). But only in Germany did these *Hochschulen* play an important role in contemporary applied chemistry.[1]

There is no satisfactory yardstick for comparing the quality of the German schools of science with those of other countries. We shall see in a later chapter that capable chemists were extremely useful to the German war effort, but equally competent chemists, most of them without the benefit of a German education, skilfully improvised and adapted on the Allied side. Wartime experience provides no conclusive evidence on quality. Nevertheless a discussion of higher chemical education in qualitative terms is of value for the following reason: the progress of applied chemistry between 1890 and 1910 was rapid and was accompanied by an increasing awareness in many countries of the need for well-educated industrial chemists. Similar problems arose in the 1950s and 1960s and again the adequacy and appropriateness of chemistry teaching were called in question.

Since the German system of chemical training was widely praised and occasionally copied, we may start by looking at its drawbacks, leaving the description to a later part of this chapter. At the end of the nineteenth century there was a great deal of correspondence in the learned journals, as well as debate in the Reichstag and in pamphlets, about the weaknesses of the German universities. The allegations covered a wide ground, were well documented, and at times conflicted with each other. There were, for example, complaints about the poor quality of chemists. How, one may ask, can this be reconciled with the boasts of the superiority of German chemistry? But we find C. Duisberg (then a youthful director of the Bayer company) complaining in 1900 about the lack of men with a wide knowledge. He ascribed this to the many doctors of chemistry who had never taken the *Abitur*, the examination marking the end of secondary education.[2] It was certainly quite true that chemistry students (excepting those at *Technische Hochschulen* and those proposing to become food analysts) were not required to pass the *Abitur*. University registrars were insufficiently discriminating in their admission requirements, and only much prodding by outsiders eventually led to the *Abitur* becoming the precondition for university entrance.

A more serious allegation, made by the V.D.C., concerned the universities' extreme preoccupation with organic chemistry. This was probably due to the *Lehrfreiheit* mentioned earlier, and so resulted in distinguished organic

[1] There were nine *Hochschulen* in Germany in 1900. They had been founded as craft or technical schools and had only lately been raised to the status of *Technische Hochschulen* with quasi-university status. They were (the year in parentheses dates the new status): Aachen (1880), Berlin (1882), Braunschweig (1878), Darmstadt (1883), Dresden (1890), Hanover (1880), Karlsruhe (1867), Munich (1877), Stuttgart (1870). Danzig and Breslau were founded just before the Kaiser's war.

[2] Die Abnahme der allgemeinen Bildung bei den Chemiestudierenden. *Z. angew. Chem.* **13**, no. 6, 131–3.

chemists teaching only this branch of the subject. It was said in mitigation that the universities thereby avoided competing with the *Hochschulen.*[1] Be that as it may, for almost twenty years after 1880 there was not a single chair of chemical technology at any of the twenty universities of the Empire; at Berlin alone there was a full-time assistant professor of applied chemistry with his own laboratory. At all the other universities it was taught as a subsidiary subject or by *Privatdozenten* without demonstrations. It was explained that the universities had adopted this policy in order not to compete with the *Hochschulen.* The remedy was simple, but difficult to apply: additional posts had to be created, and the V.D.C. suggested the provision of at least four chemistry chairs at each university, for organic, inorganic, and physical chemistry, and for chemical technology. However even by 1912 very little had been done by the various ministries of education. Although some additional chairs were established, technical chemistry retained a subordinate position at the universities, and was not taught by an *Ordinarius*, that is a full professor.[2] Consequently the doctorate was equated by some with expertise in organic chemistry accompanied by ignorance of other branches of chemistry, not to mention other disciplines. A less extreme and probably more accurate attitude was taken by the V.D.C., which complained to the ministries of education of the different German states about the poor quality of the university students. Thus prodded, the staffs took action and introduced a sort of intermediate examination, known as the *Verbandsexamen.* Together with the *Abitur* it became from the 1900s onwards the essential prerequisite for admission to the doctorate course.

If the universities were allegedly too narrow, the *Technische Hochschulen* by contrast were said to be too broad in their approach, but in other respects they were criticized for aping the universities. The curriculum (there was no *Lernfreiheit*) was certainly directed towards practical matters, but it is not easy to see how the chemistry student was going to benefit from botany, zoology, and drawing.[3] In 1897 a commission consisting of civil servants, industrialists, and university and *Hochschulen* teachers met in Berlin to investigate and report on special examinations for technical chemists. In the published summary the *Hochschulen* were criticized because they examined candidates for their diploma in many subjects, so that the intellectual effort of the student was too dispersed. While no major changes were recommended, the commission agreed that students should be required to conclude the course with a piece of work planned and executed by the candidate.[4]

In Britain and the United States there was also much questioning whether university education, which lasted three and four years respectively, was

[1] Fischer, F. *Das Studium der technischen Chemie . . .* (1897), p. 11.

[2] Rassow, op. cit., pp. 76–8. Halle university was an early exception. In 1899 a laboratory of applied technology was set up which became, in 1902, a separate department.

[3] Fischer, op. cit., pp. 18–24.

[4] Fischer, F. *Chemische Technologie an den Universitäten und Technischen Hochschulen Deutschlands* (1898), pp. 43 ff.

adequate and appropriate to the needs of chemical engineers as distinct from academic chemists. The discussion in the United States need not detain us here, if only because the subject was being ventilated for the first time. But the situation in Britain requires some comment, because technical education had been examined *ad nauseam* for almost a generation.[1] After so much reporting, why was so little done about the teaching of applied chemistry? In the first place the chemical industry, being differently structured from the German, was less in need of professional manpower. Secondly, science teaching, with few exceptions, was not the concern of the Government.[2] (For example, in 1904–5, Government grants to universities for all purposes amounted to £69 000, which may be contrasted with £476 000 paid by the Kingdom of Prussia in 1900–1.[3]) Finally, theoretical and applied physics attracted more interest and talent at Oxford and Cambridge, the intellectual trend-setters, than did chemistry. For instance, the heads of the Cavendish Laboratory at Cambridge during the first fifty years were all physicists: James Clerk Maxwell (1871–9), Lord Rayleigh (1879–84), and Sir J. J. Thomson (1884–1918). Only two places in Britain were able to meet the needs of industrial chemists at the turn of the century, and provide instruction of a quality comparable with that available in Germany: at Manchester University Roscoe had organized a comprehensive four-year course, and at Glasgow, the Technical College had a chair of technical chemistry.

On top of these specific problems there were others applying to English education generally. At one end there was the lack of a state system of secondary education, a defect remedied by the Balfour Act of 1902. At the other there were the immovable rigidities of the prescribed curriculum and the final examination. They were said to give a good general grounding, but to repress the original thinker.

At the beginning of the twentieth century a major step was taken to improve higher technological education, progress being due in a large measure to R. B. (later Lord) Haldane, an alumnus of Göttingen, Sydney Webb, and the then head of the Treasury, Sir Francis Mowat. A Departmental Committee on the Royal College of Science was set up, and its final report in 1906 was a milestone.[4] The Haldane Committee, as it will be referred to here, found that further provision for technological education was essential, that the existing institutions in London were inadequate, that opportunities for research were insufficient because the staff was wholly engaged in teaching, and that the quality of the staff and of the equipment needed improving. The Committee recommended the immediate formation of an institution to

[1] For instance by a Select Committee of Parliament on Scientific Instruction in 1868, by the Royal Commission on Scientific Instruction and the Advancement of Science—the Devonshire Commission—whose reports appeared between 1872 and 1878, and by the Royal Commission on Technical Instruction in 1882–4.

[2] The Royal College of Science and the Royal School of Mines were Government financed. The main purpose of the former was to train science teachers and excise agents.

[3] Cardwell, D. S. L. *The organisation of science in England* (1957), p. 157.

[4] HMSO (1906), Cd 2872.

remedy these defects, and proposed that it be located in South Kensington, where the Royal College of Science and the Central Technical College were already neighbours, though not always on harmonious terms. The outcome was the Imperial College of Science (I.C.S.), established in 1907. Haldane wanted it to be modelled on the German *Technische Hochschule*, and called it hopefully but not euphonically the Charlottenburg of London. The purpose of the College was to provide '. . . the highest specialized instruction . . . especially in its application to industry for which no sufficient provision already exists elsewhere.'[1] It inherited the new laboratories of the Royal College of Science, and its creation marked the Government's firm commitment to the financing of applied science teaching and research at advanced level.

We must now return to Germany and look more closely at its universities and *Hochschulen*, the better to appreciate their influence on industry at home and chemistry teaching abroad.

University education was, and remains, a function of the State, specifically of each state within the Empire. In Prussia, the largest state, universities and *Hochschulen* obtained most of their income from the exchequer[2] and were the responsibility of the Kultusministerium.[3] Its extremely able permanent secretaries, first Althoff and then his successor Schmidt-Ott, were unusually sympathetic to the improvement of higher scientific education. Schmidt-Ott had for some years been in charge of a section with specific responsibility for science and the arts, and in 1908 it became a special department in the ministry.[4]

Higher scientific education was divided into three main groups: the universities, the *Technische Hochschulen* and the *Fachschulen*. We shall not concern ourselves with the third, which in Britain would be called technical colleges or trade schools; German youths in their late 'teens would receive there 2–3 years' special training in tanning, dyeing, brewing, and other skills. Contemporaries thought well of the chemistry classes at the Fachschulen, and Ost, who knew the subject, said that the Fachschulen taught better practical chemistry than the universities.[5]

The universities attracted a much larger number of students than the *Hochschulen*, and their share of the total student population increased slightly during the early 1900s. Detailed figures for the number of chemistry students have not been published: at the universities they formed part of the Faculty

[1] Haldane Committee, report, p. 25.

[2] For example in 1899, 72 per cent of the income of the Prussian universities was obtained from the state.

Rose, F. *Report on chemical instruction in Germany* . . . (1901), pp. 21 ff.

[3] The British equivalent would have been the Board (later Ministry) of Education. In Prussia this government department also dealt with the state church.

[4] Schmidt-Ott, F. *Erlebtes und Erstrebtes* (1952), pp. 107–9.

[5] Die chemische Technologie an den Technischen Hochschulen. *Z. angew. Chem.* **13** (1900), no. 27, pp. 662–3.

of Mathematics and Natural Sciences, at the *Hochschulen* they belonged to the Department of Chemistry and Metallurgy.

The freshman left the strict discipline of the school behind him when he entered the university. He was free to pick his subject, no time limit was set to the duration of his course of study, and he was allowed to move from one university to another. There was one constraint: the candidate for the

TABLE 3.1

Number of students at German universities and Technische Hochschulen[1]

	Winter semester 1899/1900		*Summer semester* 1907		*Winter semester* 1909/10		*Winter semester* 1913/14	
	000	%	000	%	000	%	000	%
Universities								
Matriculated students	33·5	62	46·1	68	52·8	67	59·3	69
Others*	6·7	13	8·2	12	9·9	13	9·3	11
Total	40·2	75	54·3	80	62·7	80	68·6	80
Technische Hochschulen								
Matriculated students	10·9	20	11·1	16	11·3	14	12·8	15
Others*	2·5	5	2·8	4	4·8	6	4·3	5
Total	13·4	25	13·9	20	16·1	20	17·1	20
Total	53·6	100	68·2	100	78·8	100	85·7	100
Index	100		127		145		160	

**Hörer, Hospitanten*, etc.

chemistry Ph.D had first to pass the *Verbandsexamen*. The doctorate examination itself required no broad knowledge of chemistry, but was a test of the student's original work in one particular field: thus research was essential. This system favoured the above-average student by enabling him to carry out original work. (The below-average student either did not complete the course or else transferred to one of the universities—and they existed even in Germany—where originality was not a prerequisite.) The doctorate system produced original chemists, but the procedure necessarily limited the extent of their practical knowledge so that they were ill-suited to enter industry. For most of the nineteenth century this had not mattered too much because chemical manufacturers had looked at the man rather than at his specialized knowledge. But by 1900 they required at least some familiarity with applied

[1] For 1899/1900: Rose, op. cit., pp. 26–7; other years: *Statistisches Jahrbuch für das deutsche Reich*. (The first student statistics to appear in that publication were for the summer semester 1907.)

chemistry. Moreover, the doctorate criteria were not uniform throughout the country and the differences were reflected in the employment prospects: doctors of chemistry from Berlin, Bonn, and Munich were preferred to those from Würzburg or Greifswald.

The *Technische Hochschulen* had been created to train men for industrial appointments. They differed greatly from the universities: there was no *Lernfreiheit*, the curriculum was prescribed and had to be completed in a given time, usually three and a half to four and a half years (7–9 semesters). The *Hochschulen* had their own intermediate examination, the *Diplomvorprüfung*, followed in due course by the *Diplomprüfung* itself. The latter was set and marked by a state examining body, so that the test was uniform throughout the country. One would have thought that this was a decided advantage compared with the varying standards of the university doctorates. But the *Diplom* did not carry the prestige of the doctorate. In a country teeming with doctors, the title *Diplomingenieur* was a social handicap. Prussia took the lead in remedying this state of affairs, and in the 1890s permitted the *Technische Hochschulen* to award the title of *Doktor-Ingenieur*. This certainly raised the esteem in which the graduates were held, but the fact remained that it was still not enough to enable the *Hochschulen* to resume their growth. The Universities were then being reformed and by raising their standards the gap between the institutions widened again.[1]

Foreign students flocked to universities and to *Technische Hochschulen*, because they gave excellent value for money. At the University of Berlin, a chemistry student paid £5 per semester for lectures and a further £6 for the use of the laboratories, and at other universities and *Hochschulen* the fees were about £1–£1. 10*s*. lower. Board and lodging were cheap, and the only big item of expenditure was the fee for the doctorate and the cost of printing the dissertation.[2] As Prussia spent about £40 per student a year on salaries and grants, the foreign students benefited from a large subsidy.[3]

For most of the nineteenth century the foreign student population at German universities was small; the majority were British, American, or Austrian, and collectively they were excellent propagandists for the German university system. But towards the end of the century the situation changed owing to the great influx of Russian students and the decline, absolutely and relatively, of the British and American contingent.[4] This attracted the attention of the V.D.C., which compiled and published statistics on the sub-

[1] The Haldane Committee in order to strengthen the case for the Imperial College drew an idealized picture of the *Hochschulen*, by saying that they 'have drawn nearer and nearer to the university organisation and possessing most of the essential attributes of a university' (pp. 10–11).

[2] The figures were quoted in *Chem. Trade J.* **29** (1901), no. 741, 108 and **29** (1901), no. 742, 132. The cost of a Ph.D. is put at upwards of £40 which even after allowing for the cost of printing and binding the dissertation seems high.

[3] *Z. angew. Chem.* **27** (1914), pt. 1, 604. The figure was mentioned at the 1914 general meeting of the V.D.C. and probably refers to 1913.

[4] The chief reasons for the large number of Russians were the shortage of university places and the limits imposed by Russian universities on the admission of Jews.

ject from 1899 onwards. They show that number of foreigners at universities rose from 1700 in 1899/1900 to nearly 3300 in 1911/12, and their relative importance from 4 to 6 per cent.[1] The Germans were seriously alarmed by the concentration of the foreigners at a few universities, the even larger numbers at the *Technische Hochschulen*, and their tendency to flock to the departments of chemistry, engineering, and medicine, where they represented a sizeable proportion of the total.[2]

Two generations ago the problem of foreign students in Germany was not so dissimilar from that facing British universities in the 1960s. There was the intangible advantage of implanting *Kultur* in receptive minds, and also the practical consideration of selling equipment to German-trained technologists. On the other hand the bunching of foreigners was clearly undesirable, there were language difficulties, and the *Hochschulen* were after all designed for educating Germans. Moreover, the Russian students formed their own clubs and engaged in political agitation, which caused complaints at diplomatic level and at local level resulted in occasional rioting.[3] Eventually restrictions were applied. At first, the admission rules were tightened up, and just before the First World War the Universities introduced a quota system designed to limit the number of Russian students.[4]

The university and *Hochschulen* hierarchy was simple: the professor was head of the department. He taught, examined, administered, and, if there was time, pursued his own research. The professor and the senior lecturer (*ausserordentlicher Professor*) were civil servants, entitled to pensions and enjoying great social prestige. The professor's salary at the beginning of the century was about £325 a year at Berlin, £250 at other Prussian universities and £200–£225 at other state universities: lecturers got about half the salary. The staff at *Technische Hochschulen* received a little less pay than their equivalent university colleagues.[5] In addition some professors were entitled to lecture fees, which could be considerable. Apart from the established grades there were the *Privatdozenten* and the *Assistenten*, whose pay and conditions were scandalously poor and in no way corresponded to the contribution they made to German science. The *Privatdozenten* had taken the examination enabling them to teach at university (*Habilitation*), but had not yet been put

[1] *Z. angew. Chem.* **26** (1913), pt. 1, 758 ff.

[2] In the winter semester of 1906/7 the proportion of foreign students to the total student population (including non-matriculated) at some of the large universities was as follows: Heidelberg 16·1 per cent; Leipzig 14·8 per cent; Jena 14·6 per cent; Berlin 14·5 per cent; Bonn 2·9 per cent; Erlangen 2·6 per cent. At the *Technische Hochschulen* it was as follows: Karlsruhe 39·4 per cent; Darmstadt 38·1 per cent; Dresden 32·6 per cent; Munich 21·9 per cent; Berlin-Charlottenburg 14·5 per cent. These percentages were given in a very hostile speech by L. M. Wohlgemuth, entitled Studium der Ausländer an deutschen Hochschulen and reproduced in *Z. angew. Chem.* **20** (1907), no. 35, 1506–8.

[3] Chaim Weizmann (*Trial and error* (1949), pp. 29–31, 50–51) has some interesting comments on Russian students at Berlin. John Anderson (later Viscount Waverley) who read chemistry at Leipzig in 1903–4 also noted the large number of foreigners (Wheeler-Bennett, J. W. *John Anderson* (1962), p. 14).

[4] *Z. angew. Chem.* **27** (1914), pt. 1, 602–3.

[5] Rose, op. cit., pp. 15–16.

officially on the professorial roll. Their income was derived from occasional lectures, tutorials, or private cramming; some obtained an appointment only late in life. The *Assistenten* had obtained their doctorate and were working towards their *Habilitation.* They were not allowed to teach and officially they were unpaid, but benevolent professors were often able to make some provision for them in their budgets. Their job was to assist the professor, but there was plenty of other work for them: they were often in charge of the laboratories, and generally supervised the students at quantitative and qualitative analysis. In addition the *Assistent* was supposed to get on with his *Habilitation.* Despite all this, professors at the better universities could always rely on a steady supply of such devoted men.

In Germany the ratio of staff to students at Universities and *Technische Hochschulen* was high. At the Charlottenburg Hochschule in Berlin, there were in 1901, 6 professors of chemistry, 6 *ausserordentliche* professors, 12 *Privatdozenten,* and 15 *Assistenten,* a total of 39 for 278 students of chemistry and metallurgy. At the University of Berlin there was a permanent staff of 36 for perhaps 400 chemistry students. In the provinces the ratio was generally lower (except at Strasbourg, which for political reasons was a show place), and at Heidelberg there were only 10 staff to 315 chemistry students in 1898/9.[1]

Staff and students frequently worked in unsatisfactory buildings, and the fine new laboratories of the middle and third quarter of the nineteenth century had become inadequate by the 1900s. Emil Fischer, a peripatetic professor, knew many university laboratories, and his comments on ventilation, lighting, and space were scathing: the chemistry departments at Bonn and Berlin built in the 1860s for the great Hofmann had become too small, at Erlangen the technical equipment was poor, and at Würzburg the construction of a new chemistry block in the 1890s at a cost of £32 000 was long overdue. Fischer visited Bunsen at Heidelberg in 1889 and marvelled how any research or teaching could be done under the conditions obtaining.[2] Improvements came slowly. In 1900 a new chemistry block was opened in Berlin, but the modernization of Liebig's old laboratory at Munich was only completed in 1920.[3] Facilities for teaching applied chemistry were even worse neglected: the first university institute for applied chemistry to be established was that at Leipzig in 1897, and except for a laboratory at Halle (1899) it remained the only one for many years.[4]

Not only the buildings, but also, as we have already seen briefly, the entire system of advanced chemistry teaching was criticized. H. Ost, an authority on applied chemistry, declared in 1900 that the *Technische Hochschulen* were in danger of losing touch with industry. Too much time was spent on theoreti-

[1] Rose op. cit., pp. 21–4; Rose, F. *Supplementary report on chemical instruction . . .* (1902), pp. 3–4, 7. [2] Fischer, E. *Aus meinem Leben* (1922).

[3] Willstätter, R. *Aus meinem Leben* (1949), pp. 278 ff.

[4] Helferich, B. Aus der Geschichte der Chemie der Universität Leipzig. *Chemikerzeitung* **84** (1960), no. 6, 169–72.

cal, mainly organic, chemistry to the neglect of industrial chemistry. The teaching staff used to keep in close touch with industry and had often thoroughly familiarized themselves with one of its branches, but pressure of work now prevented this. Ost also complained that the curriculum was becoming increasingly fragmented: a little was taught on many subjects. He therefore recommended that the first six semesters be spent on general chemistry and the remaining three to four semesters be devoted to specialization in a few fields of applied chemistry. As a result the examination for the *Doktor Ingenieur* at the end of four and a half to five years would be a test of specialized knowledge of a fairly narrow area.[1]

At about the same time criticism of a different kind was levelled against the universities by another eminent chemist, O. N. Witt. He thought that their chemistry departments must adapt themselves to the needs of freshmen who had scarcely any knowledge of chemistry. The raw student had to be taught elementary principles, but Witt questioned whether one and a half years (three semesters) were necessary to teach analytical chemistry. Even worse, little effort was made to teach the manipulative skills so important for experimental techniques. He felt that university chemistry students showed far too little knowledge of the related sciences, notably physics and perhaps also mineralogy and botany. Finally there was too much emphasis on organic chemistry and too little on applied chemistry; the latter subject should be taught in a broad manner and should include simple mechanics and chemical economics.[2]

The debate on the functions and content of chemistry teaching at universities and *Hochschulen* remained inconclusive, but on some points there was agreement, with the result that remedial action was taken. One was to raise the general knowledge of the freshmen, and this was achieved by making the *Abitur*, the equivalent of today's General Certificate of Education at 'A' Level, the essential prerequisite for university admission. Another was to weed out students who did not have the ability to complete the doctorate. This was done by an intermediate examination, the *Verbandsexamen*, which took its name from the Verband der Laboratoriumsvorsteher, an unofficial association of university teachers formed in 1897. By about 1900 most German universities required the candidates for the doctorate to take this examination. These measures had some interesting results: even as late as 1903, 16 per cent of the candidates for the chemistry Ph.D. had not taken the *Abitur*, but in 1906/7 the proportion was a mere 4 per cent. During the same period there was a marked fall in the number of German chemistry doctorates awarded, from 393 to 316, which suggests that standards were tightened up so that the number of worthless doctorates declined.[3]

[1] Ost, H. Die chemische Technologie an den Technischen Hochschulen. *Z. angew. Chem.* **13**, (1900), no. 27, 659 ff.

[2] Witt, O. N. Ueber die Ausbildung der Chemiker für die Technik. *Z. angew. Chem.* **14** (1901), no. 26, 633–40.

[3] Kaselitz, O. Die Vorbildung der Chemiker. *Z. angew. Chem.* **21** (1908), no. 1, 5–7.

The critics were also united on the desirability of research, which was indeed the corner-stone of advanced chemistry education in Germany. But this raised some problems, such as the needs of the professor and the *Assistenten*, the lack of space generally, and the growing demand from industry for chemistry graduates. In 1897 it was estimated that 4000 chemists (excluding university and *Hochschulen* staff) were working in the Reich,[1] and by the eve of the Great War the number had grown to 9000.[2] The big increase in graduate chemists put great pressure on the educational machinery, and accounted for much of the overcrowding in laboratories, the irritation felt towards foreigners, and the difficulties which the post-doctorate men were having with their research work.

Before examining an interesting and important aspect of advanced chemical research, it is necessary to digress briefly and illustrate the growing demand by the German chemical industry for chemistry graduates. From surveys carried out by the V.D.C. in the late 1900s, it would appear that 60 per cent of all doctors of chemistry, but only 30–40 per cent of all *Technische Hochschulen* graduates, were working in large chemical firms, that is companies employing more than twenty chemists. The annual intake of chemists from universities and *Hochschulen* rose from about 300 a year in 1907 and 1908 to about 400 a year in 1912 and 1913. Approximately one-fifth of them were *Assistenten* who had done advanced research. It would appear, though the evidence is not conclusive, that the number of chemists with engineering qualifications grew more rapidly. At any rate there were about 500 of them as against roughly 3100 chemists in the chemical and allied industries at the beginning of 1914.[3] Throughout the period the links between universities and industry were close: *Privatdozenten* at universities were occasionally given financial and other material assistance. Conversely, men who were discharged for reasons beyond their control and who could not, under the terms of their contract, work in other chemical firms, were given compensation and encouraged to return to the universities.

It came to be realized, however, that the rapid changes taking place in pure and physical chemistry required intensive study and research by men who needed to be free from all teaching and administrative duties, and whose conditions of work should counterbalance the attractions of employment in industry. The answer was the research institute. As early as 1900 it was proposed to establish a *Chemische Reichsanstalt* as an independent institution modelled on the Physikalisch Technische-Reichsanstalt.[4] Nothing came of

[1] Fischer, F. *Das Studium der technischen Chemie an den Universitäten und Technischen Hochschulen Deutschlands* (1897), p. 45. Fischer estimated that nearly 1700 chemists were working in the chemical industry.

[2] Annual Report of the V.D.C. quoted in *Z. angew. Chem.* **27** (1914), pt. 1, 598. Roughly two-fifths were working in the chemical industry. It is interesting to note that the proportion was of the same order of magnitude in 1895 as in 1913.

[3] See the Annual Reports of the V.D.C. in *Z. angew. Chem.* **26** (1913), pt. 1, 756–8 and **27** (1914), pt. 1, 597–9.

[4] Fischer, E. and Beckmann, E. *Das Kaiser-Wilhelm-Institut für Chemie* (1913), p. 8.

this project and the only application of the concept in the 1900s was the institute that had been made available to Professor Van't Hoff at Berlin. He had been invited to move from Amsterdam, and accepted on condition that he would be exempt from all teaching duties so that he could devote himself entirely to his laboratory work.[1]

The formation of special research institutes was the result of agitation from different quarters. One combined chemists and leaders of the chemical industry. In 1905 Fischer, Nernst, and Ostwald met Duisberg, Brunck, and other industrialists to discuss once again the need for a *Chemische Reichsanstalt*. But again there was no interest shown by industry, Government, or universities and another three years passed before the Verein Chemische Reichsanstalt was formally constituted and a list of subscriptions opened: the Chemie-Verein headed the list with £19 000 and was followed by Ludwig Mond from England with £10 000, and Krupp, Harries (the rubber chemist), and Ernest Solvay from Brussels with £5000 each.[2] At that point other personalities made their influence felt. In particular Althoff and, after his death in 1908, Schmidt-Ott of the Prussian Ministry of Education felt that the state should actively encourage all sciences; Schmidt-Ott in particular pointed out that the physical and biological sciences were being neglected in Germany. He found the Kaiser receptive to these ideas and William II announced that he would give a site on the royal domain at Dahlem (on the southern outskirts of Berlin) for the building of a research centre.[3] The promoters now enlisted the help of A. von Harnack, an eminent classical scholar, director of the Prussian State Library, and, incidentally, *persona grata* with the Emperor. Harnack was invited to prepare a memorandum on the subject of German science; his report of some fourteen pages contained a brilliant analysis of the structural weaknesses of German science and suggested the remedies. Harnack's proposals were timely and acceptable to William II. The hundredth anniversary of the foundation of the university of Berlin was made the occasion for setting up a special science foundation, the Kaiser-Wilhelm-Gesellschaft (K.W.G.), in January 1911.

The K.W.G. acted as the channel whereby private and public funds could be directed towards the promotion of research in the natural and philosophical sciences. It had an original endowment of about £750 000, then a very large sum, and the promise of substantial annual grants (initially £5000) to pay for the running expenses. The object was to establish a number of institutes, independent of the universities, in which chosen scientists could pursue research into new fields of science. They were freed from teaching duties and given life contracts; indeed, the only condition to which they were subject was that they should not attempt to duplicate the work of the universi-

[1] Van 't Hoff had about six students in his laboratory. F. G. Cottrell, the American chemist who spent about a year with him, reported that the laboratory was open from 9 a.m. to 7 p.m., but that the routine was very elastic.
Cameron, F. T. *Cottrell, Samaritan of science* (1952), pp. 92–4.

[2] Fischer and Beckmann, op. cit., pp. 9–11.

[3] Schmidt-Ott, op. cit., pp. 116–17.

ties. This was designed to avoid the hostility of the universities, which were expected to release some of their best people to the K.W.G. institutes. The stipulation could not always be carried out in practice, but there were few demarcation disputes. In any case these disadvantages were more than balanced by the extremely fruitful cross-fertilization between universities and institutes.[1]

Harnack became the first president of the K.W.G. Contrary to expectation the institutes of biochemistry and physics were not the first to open; this position was claimed by the institutes for chemistry and for physical chemistry, both of which were opened in October 1912.[2] The chemistry institute cost almost £50 000,[3] and was the result of an agreement between the K.W.G. and the Verein Chemische Reischsanstalt.[4] The institute of physical chemistry, which at first did not belong administratively to the K.W.G., was made possible by a grant from the Koppel Foundation. The two institutes had scarcely started work when war broke out, the staff had to abandon pure research, and the buildings became centres for applied research and chemical warfare.[5]

We shall revert to the work of the institutes in a later chapter. Here we may perhaps conveniently sum up their characteristics. The directors and the scientific members (who were both of the same rank[6]) were hand-picked men, a veritable galaxy of Nobel Prize-winners: Haber, Hahn, Neuberg, Willstätter, Meitner, the list lengthens as we move from the 1910s to the 1920s and beyond. Their assistants were *Privatdozenten*. Each institute was left a maximum of autonomy in the choice of work and the selection of collaborators. Though not all the seven institutes that were in existence by 1914 were equipped with equal lavishness, those for chemistry, physical chemistry, and biochemistry were housed in large new buildings, each with many laboratories and designed according to the best current practice.[7] Auxiliary staff such as mechanics, glass-blowers, electricians, secretaries, and librarians were available. In brief, the K.W.G. institutes were the first working examples of the modern type of research centre.

When we turn from Germany to Britain the first impression is one of confusion. In the Kaiser Reich everything is tidy, thoroughly organized, probably riddled with officialdom, but at least understandable. In Britain,

[1] Planck, M. *25 Jahre Kaiser-Wilhelm-Gesellschaft* (1936), vol. 1.

[2] The institute of biochemistry, originally called Institut für experimentelle Therapie, was opened in 1913; A. von Wassermann was its first director. It was hoped to have an institute of physics by 1914, but it was not built until 1936.

[3] *Z. angew. Chem.* **27** (1914), pt. 1, 605.

[4] The Verein was renamed Verein zur Förderung chemischer Forschung in 1913. It made the large funds at its disposal available to the K.W.G. without stipulating that the institute should confine itself to work on applied chemistry.

[5] Planck, op. cit., vol. 1, pp. 53 ff.

[6] In the applied science institutes (of which the first was that for coal research under F. Fischer at Mühlheim) there were heads of departments who carried slightly lower rank than the scientific members.

[7] Fischer and Beckmann, op. cit., pp. 22 ff.

however, the variety bewilders: Oxford and Cambridge independent and aloof, the English universities and university colleges, the polytechnics, the independent technical schools, the Scottish universities, and, to round off the picture, Wales and Ireland. No generalization will suffice, yet to review every institution separately would take too long and the reader is referred to the monographs and the extensive official literature on the subject.[1]

We are primarily concerned here with higher education in chemistry, and immediately encounter statistical difficulties, so that the following numbers of day students attending English universities and university colleges in receipt of a Treasury grant are not very good indicators for our purposes.

TABLE 3.2

Day students at English and Welsh universities and university colleges, selected years, 1891–1914[2]

Year	*Number* (000s)	*Index*
1891–6*	6·4	100
1896–1901*	7·8	122
1901–06*	9·3	145
1908/9†	9·7	151
1911/12†	9·0	140
1913/14†	9·6	149

*Annual average.
†Financial year.

There was little change in the years before the war. From 1911/12 onwards details are available for the various faculties: arts accounted for about 40 per cent, medicine for 25–30 per cent and the sciences for 30–35 per cent. Among the latter, 1700 students were reading pure science, and about 1000 engineering. Applied science was then a new subject, but in the last years before the war it began to attract students: there were 347 in 1911–12 and 459 in 1913/14.[3]

The student population at Oxford and Cambridge grew from about 5900

[1] See in particular Cardwell (1957) and Armytage, W. H. G. *Civic universities* (1955), and the reports of the Royal Commission on Technical Instruction and of the Haldane Committee.

[2] Sources: H.M. Treasury, Report of University Colleges (Gt. Britain) Grant in Aid 1907 PP 1907, **64**, 551 ff., and Board of Education; Report for the Year 1913–14 from those Universities . . . which are in receipt of Grants . . ., PP 1914–16, **19**. The figures exclude Wales during 1891–6, and the London School of Economics, Birkbeck and East London College in 1896–1901 and 1901–06. Oxford and Cambridge are also excluded, though after 1908/9 the Medical Department at Cambridge and the department of Engineering Science at Oxford was included.

[3] Board of Education, report, op. cit. The large increase in the number of technology students may have been due to the development of the Faculty of Technology at Manchester University where the numbers grew from 125 in 1905/6 to 280 in 1913/14.
Charlton, H. B. *Portrait of a university* (1951), pp. 164–6.

in 1900 to 7400 in 1912.[1] There were also the four Scottish universities with 5900 students in the winter semester of 1899/1900, 6456 in 1905/6, and 7474 in 1913/14.[2] There were finally Ireland and the technical colleges, but if we exclude them the total for the British Isles rose from about 21 000 at the turn of the century to about 25 000 by the outbreak of war. This, in absolute terms, is substantially less than the German figure, and in terms of students per head of population much lower. It is also worth noting that while the increase in the number of British students in the period was about 20 per cent, in Germany it was about 60 per cent.

Several characteristics set the British system of higher chemical education sharply apart from that of Germany. Perhaps the most important was inadequate secondary education until the state provided it under the Balfour Act of 1902. This Act further increased the demand for science teachers, who had been recruited in growing numbers from the 1890s onwards for municipal technical schools. Thus the training of science teachers became in the 1900s a factor in university growth.[3] Another important characteristic was the pattern of scientific instruction and its organization. Higher technical education was not controlled by the state and, as far as chemistry was concerned, was shaped by men who had been trained in German schools of science. But the close German links between universities and industry did not exist in England, and the facilities and resources to carry out advanced research for a higher degree were generally lacking. These differences were fundamental: In Germany the Ph.D. marked the culmination of a long university career; in England a short training, into which had to be crammed a lot of elementary education, lasted for three years and was terminated by a diploma or a bachelor's degree. In England an examination on a fairly narrow range of prescribed subjects was the final test and, despite the objections of Kelvin, Roscoe, Playfair, and other scientists, this system had become firmly established even before the end of the nineteenth century. Such an education was necessarily subject to the rigours of a strict curriculum and left no room for *Lehrfreiheit* and *Lernfreiheit*. Since a large minority of the students were intending to become teachers, the course of study tended to be adapted to their needs, rather than to those of the industrial chemist or the potential advanced research student. Chemistry training at English universities consequently resembled that of the old-style *Technische Hochschulen*. But, as we have seen, the latter were themselves changing, and by stressing the need for laboratory work and encouraging research they were following university precedent.

[1] Royal Commission on Oxford and Cambridge, Report, 1922 (Cmd 1588, PP 1922, **10**), pp. 26–7. The figures were obtained by multiplying the number of students matriculating in 1900 and 1912 by three, which gives an approximation of the numbers under instruction in those years.

[2] Between 1899/1900 and 1913/14 the number of students in Scottish science faculties grew from 434 to 1177.

Scottish Universities (Aberdeen, Edinburgh, Glasgow, St. Andrews) were obliged by Act of Parliament to make detailed annual statistical returns. Those for the year mentioned will be found in PP 1900, **66**, PP 1907, **65**, and PP 1914–16, **52**.

[3] Cardwell, op. cit., pp. 195–6.

The grafting of nineteenth-century *Hochschulen* methods on to even older English academic principles was bound to have interesting results. Sir Eric Ashby, reviewing the early years of technological education, concluded that the English system produced 'experts to advise on policy, not leaders to make policy. With modifications, accretions, and improvements, that pattern has persisted ever since.'[1] It may be deplorable that British schools of chemistry failed to produce men with a wider educational background, and particularly with the ability to think creatively. But it is arguable that the greatest need was for Ashby's experts on policy, since the opportunities for the original thinker were no worse than in Germany, and, in physics at any rate, Britain was then rapidly forging ahead. Professor Sir James Dewar, in his presidential address to the British Association in 1902, attributed Britain's weakness in her chemical and dyestuffs industries, and conversely Germany's strength, not to patent or other legislation but to '. . . the want of education among our so-called educated classes and secondarily among the workmen on whom these depend. It is in the abundance of men of ordinary, plodding ability, thoroughly trained and methodically directed, that Germany has so commanding an advantage.'[2] Dewar thought it would take Britain two generations to catch up with the Germans in general training and specialized equipment. He was not so far out, and the delays were due to the attitudes and faulty policies of the 1900s and 1910s.

University education, and indeed research generally, were considered in Britain a matter for private or at best municipal benevolence. Successive governments bowed to pressure and set up commissions, but then refused to implement those recommendations that cost money. Thus it was left to the enterprise of a few individuals to bring about some reforms at the close of the nineteenth century. The 'ginger group' was the National Association for the Promotion of Technical Education, formed in 1887. The moving spirits were T. H. Huxley and Sir Bernard Samuelson, and they were ably supported by the chemist and M.P. Henry Roscoe. They lobbied vigorously for the implementation of the recommendations of the Royal Commission on Technical Education, and their success was reflected in the Technical Education Acts.[3] These empowered local authorities (and after 1891 County Councils) to levy a rate specifically for technical education. In addition the Treasury took the plunge in 1889 and granted £15 000 to university colleges in England.[4] Finally, after some more lobbying, the Local Taxation (Customs and Excise) Act of 1890 provided for grants from the Government to local authorities for financing technical instruction. The funds were obtained by diverting part of the money set aside to compensate publicans for the loss of their licence; one hopes that some of the 'whisky money' helped with the teaching of chemistry.

[1] Technological humanism. *J. Inst. Metals* **85** (1956–7), 464.

[2] Report of the [1902] British Association Meeting (1903), p. 17. [3] The first was passed in 1889.

[4] In the early 1880s small Treasury grants had been made to Welsh Colleges. Armytage, op. cit., pp. 234 ff. and Berdahl, R. O. *British universities and the State* (1959), pp. 51 f.

The cash flow was hardly a flood in the 1900s. The annual grant was increased from £15 000 to £27 000 in 1902 and to £54 000 in 1904. From 1905 onwards it was £100 000 a year until 1912 when it was raised to £150 000 a year. Besides the Treasury Grants, additional disbursements were made by the Board of Education, mainly in connection with Teacher Training programmes, and the total payments made by the Government to English universities rose from £26 000 in the financial year 1897/8 to £69 000 in 1904/5 and £123 000 in 1911/12. At the latter date Prussia was spending £700 000 on her universities and *Technische Hochschulen.*[1]

Private generosity contrasted sharply with official meanness, but created difficulties of another sort: Victorian millionaires might endow university colleges, but the amounts needed were so large that without further help the new institutions were unable to progress and enlarge their scope.[2] Haldane, a strong supporter of the cause of higher education, came to see that '. . . what we needed badly . . . was more universities, and universities of the civic type in different parts of the country'.[3] These municipal colleges frequently took over when private endowments were insufficient, and local pride and support rose when university status was granted. But the change in rank had little permanent effect on finance, and the lack of money was chronic. The income of English universities (excluding Oxbridge) and university colleges rose from £0·54 million in 1908/9 to £0·68 million in 1913/14. The contribution made by local authorities remained at 16 per cent and that by the Exchequer (including grants for teacher training) rose from 27 per cent to 34 per cent, while other sources, such as fees and endowments, fell from 57 per cent to 50 per cent.[4] Manchester, the largest of the civic universities, had an income of only £37 800 in 1900 and £74 800 ten years later. Manchester City Council and the Government made a below-average contribution, so that fees and endowments accounted for 68 per cent of total income in 1910.[5]

As Treasury and municipal grants were relatively small, the proportion of income derived from fees was correspondingly large. Accordingly a university education was expensive. At the Royal College of Science, which received substantial grants, the fees were £30–£40 a year,[6] that is twice to three times those of the *Technische Hochschulen.* The expense must have discouraged some of Dewar's 'men of ordinary plodding ability' from going in for a higher education and led them instead to sit for the simpler tests of the Institute of Chemistry. Much potentially useful manpower was lost in this way.

[1] The English figures exclude the Royal College of Science, Oxford, and Cambridge. Berdahl, op. cit., pp. 53 f. and Cardwell, op. cit., p. 155.

[2] The latter half of the nineteenth century was a period of active college foundation outside London (dates in parentheses refer to granting of university status): 1851 Owens (Manchester University 1903); 1871 Newcastle (later merged with the older Durham University); 1874 Leeds (1904); 1876 Bristol (1909); 1879 Sheffield (1905); 1880 Birmingham (1900); 1881 Liverpool (1903); 1881 Nottingham (1948); 1892 Reading (1926).

[3] *An autobiography* (1929), p. 90.

[4] Reports of the Board of Education for the years mentioned.

[5] Charlton, op. cit., pp. 148–50.

[6] Haldane Committee, Evidence, p. 10 (no. 210).

Another result of the poverty of the universities was that academic salaries could not be compared, except in the faculties of medicine, with what men of ability were able to earn in industry. Hence there were difficulties in staffing departments adequately and paying decent salaries. Senior staff held good appointments; for instance in 1905/6 the Glasgow chair of chemistry paid its holder £1300 a year, F. Soddy, then lecturer in physical chemistry in the university, received £400 a year, and at Oxbridge full-time professors received £600–£800 a year at the outbreak of war as basic salary. But the junior staff fared badly; at Manchester junior assistants and demonstrators received £100 a year, senior assistants £100–£180 and independent lecturers £250–£400 a year.[1] At the same time the social position of the more junior grades was much inferior to that of the *Privatdozenten*, let alone that of the *ausserordentlicher Professor*.

Parsimony also affected scientific equipment and above all research facilities. Here indeed the situation was appalling, from the scientific societies downwards. Dewar looked into the cost of the Royal Institution during the nineteenth century, and discovered that including Faraday's Civil List, the expenditure on the laboratory and salaries came to a total of £119 800, say £1200 a year. This excludes the Davy–Faraday Institute, which Ludwig Mond had endowed, but which was administratively part of the Royal Institution.[2] Government support to the scientific societies was niggardly in the extreme. Those located in Burlington House, which had been built at public expense, received altogether £51 000 for building improvements and £10 500 for maintenance from 1886/7 to 1905/6 inclusive. In addition the Royal Society, one of the Burlington House tenants, received in that period £92 000. The Royal Geographical Society and the Marine Biological Society received £500 and £1 000 a year respectively. The learned societies of Scotland and Ireland shared £2500 a year between them. And that was all.[3]

When the societies had to make do with little official support, the university laboratories could expect no better. Roscoe and Ward had considered 'research as the directive and formative factor in academic development'.[4] The statement expressed a brave attempt to transplant the principles of German *Wissenschaft* to late Victorian Manchester. The high standards could not be long maintained, though Manchester was better in this respect than other universities. Roscoe created a fine school of chemistry complete with laboratories, and for many years Manchester had the only chair of organic chemistry in the British Isles.[5] Nevertheless when Weizmann, in later years an organic chemist of great fame, began working in that department in 1905 he

[1] For details of Scottish salaries see PP 1907, **65**, 546–677; for Manchester see University Colleges (Gt. Britain) Grant in Aid, PP 1907, **64**, 18; for Oxford and Cambridge see report of the Royal Commission on Oxford and Cambridge, p. 189.

[2] Report of the [1902] British Association Meeting, p. 14.

[3] Treasury, Scientific Societies (Government Grants), PP 1906, **65**, 515–18.

[4] Quoted in Charlton, op. cit., p. 55.

[5] Carl Schorlemmer was the first holder of the chair (1874–92), succeeded by W. H. Perkin jr. (1892–1912), A. Lapworth (1913–22), and Robert (later Sir Robert) Robinson (1922–8).

found that the laboratory '. . . was a dingy basement room which had evidently not been used for many months. It was dark, grimy and covered with many layers of dust; the necessary apparatus was there, but a great deal of cleaning and rearranging had to be done . . .'[1]

In southern England considerable efforts were made to improve the laboratory facilities. At Cambridge the chemists had at last obtained their own laboratory in 1887, and this made it possible to abolish college laboratories, which were inadequate and unsatisfactory.[2] But at Oxford, the Clarendon Laboratory, started at about that time, was soon becoming inadequate, and as a result the college laboratories survived till after the First World War, though the Royal Commission thought that they were 'less economical and in some respects less effective than centralized teaching'.[3]

In London a good deal of money was spent on improving the teaching of applied science and in providing the institutions with adequate laboratories. We may conveniently stop here and examine the situation in more detail, since it illustrates the weaknesses and the strengths of chemistry teaching in Britain as a whole.[4] At the opening of the century there were two institutions for higher technological instruction with a practical bias. The Central Technical College, founded by the City Companies, was privately endowed. In the 1870s these Companies began to interest themselves in technological education and in 1880 incorporated the City and Guilds of London Institute for the Advancement of Technical Education, which built new quarters for the Central Technical College in South Kensington between 1881 and 1884. The College consisted of three departments: mechanical engineering, physics (which concerned itself increasingly with practical electrical engineering), and chemistry. Of these the first two, and particularly the second, accounted for the bulk of the student population, while the third, under H. E. Armstrong, one of the country's great educationists, had only 10–20 students.[5]

The Haldane Committee thought the college occupied an intermediate position between the universities and the technical schools of the local authorities.[6] Its narrowly based curriculum was designed to train technicians and terminated in an examination for the diploma. Armstrong's attempts to raise the standard by introducing a four-year course in chemical engineering, though original and valuable in conception, were defeated. Similarly his 'heuristic method' of teaching chemistry by graduated laboratory exercises

[1] *Trial and error*, op. cit., p. 127.

[2] Some years later Cottrell thought that the methods of the chemistry department at Cambridge were so 'cumbersome' that he soon returned to Leipzig to continue under Ostwald.
Cottrell, op. cit., p. 80.

[3] Report, op. cit., pp. 33–4, 117.

[4] The facilities for teaching applied science at advanced level remained unsatisfactory in Scotland until Andrew Carnegie's gifts in the early 1900s brought about an improvement. The Glasgow Technical College (formerly Anderson's College) benefited from the construction of new buildings which began in 1903 and was completed seven years later. It was renamed the Royal Technical College in 1912 and the following year was affiliated to the university. In 1964 it became the University of Strathclyde.
Muir, J. *John Anderson* (1950).

[5] Eyre, J. V. *Henry Edward Armstrong* (1958), pp. 85 ff.

Report, p. 8.

was more suited to the teaching of school-boys than of educating research chemists.[1] He became disappointed in the end, because he had to follow what he considered an unsuitable curriculum, and because the number of students in his department failed to grow. Only the brewers took his men, and this suggests that the demand was for competent analysts and process chemists rather than for chemical innovators.

Similar problems beset the Royal College of Science (R.C.S.), the other of the two institutions. It had been given that title in 1890, before which it had been called the Normal School of Science and Royal School of Mines, thus indicating two of its main activities. It had been created in 1853 by amalgamating the School of Mines with Hofmann's famous Royal College of Chemistry. The R.C.S. was autonomous, and supported itself by fees and grants from the Science and Arts Department, and later from the Board of Education, which paid £18 300 in 1898/9 and £12 000 six years later. By English standards a great deal of capital had been and was being put into the R.C.S. The first or 'old' building at South Kensington cost £65 000, and the new building, including laboratories for physics and chemistry, completed in 1906, cost £260 000.[2] The College had several functions, among them the training of teachers, mining engineers, metallurgists, and Excise Officers. The number of students attending the three-year course (that for Excise Officers lasted for only two) had not changed significantly since the 1870s, when there were 245 in 1873 and 282 four years later. By 1905/6 there were 283.[3] This was discouraging, since teachers were in increasing demand. Another function, inherited from the Royal College of Chemistry, was the encouragement of chemical research and the instruction of industrial chemists. Both were neglected. The available statistics show that the number of chemists completing the diploma course had become a trickle. In the session 1892/3, 35 students in the Royal School of Mines and 23 in the other departments of the R.C.S. completed their studies; among them were only 6 chemists. In the session 1901/2, 59 students completed the course (30 of them in the School of Mines), but there were still only 10 chemists.[4]

The Board of Education was dissatisfied: it was concerned about the lack of science teachers, especially since the Balfour Act, and it was also under pressure to help financially with the training of technologists and to provide facilities for applied research. So a departmental committee was set up 'to inquire into the working of the Royal College of Science [and] to report on

[1] Turner, D. M. *History of science teaching in England* (1927), pp. 142 f. Armstrong stressed that it was essential for students to discover things for themselves. His theory interested teachers and was applied in some schools when laboratory work was introduced for the upper forms. But later, owing to exaggerated stress on laboratory routine, the theory fell into disuse. It was revived again during the 1960s with apparent success.

[2] Haldane Committee, evidence, appendix 4, p. 175, and Annual Reports of the Science and Art Department.

[3] Thomas, D. H. *The development of technical education in England 1851–89* (1940); Haldane Committee, report, p. 5.

[4] Haldane Committee, evidence, p. 19 (no. 298).

any changes which may be desirable'. Haldane was chairman and the conclusions of the committee were short and to the point: further provision for advanced technological education was essential; the existing facilities were inadequate, as were the opportunities for research; the equipment used was not of sufficiently high standard, and the heads of departments should be 'men of the first rank in their profession'.[1] The Committee recommended the creation in South Kensington of an institution of the highest standing, which was to have as its main object the provision of 'the highest specialized instruction . . . especially in its application to industry for which no provision already exists elsewhere'.[2] No agreement was reached as to whether this new institution, the Imperial College of Science (I.C.S.), should have independent status like the *Technische Hochschule* in Berlin. This matter was referred to the Royal Commission on University Education in London, again under Haldane's chairmanship, which recommended in 1913 that the I.C.S. should become part of London University. Another departmental committee was needed in 1924 before the recommendation was implemented in the late 1920s. In the meantime many practical steps were taken. The London County Council and the Board of Education agreed to make available £20 000 a year each to the I.C.S., and the Government undertook to pay the rates and upkeep, and to hand over the R.C.S. building and 4½ acres of Crown Land for additional buildings. It was also decided to merge the Central Technical College with the I.C.S. in 1910; as a result the chemistry department under Armstrong was closed, and only the engineering school remained intact and became the greatly enlarged department of engineering at I.C.S. The new college obtained nearly £200 000 in donations chiefly from Alfred Beit and his company. Thus endowed, the *Technische Hochschule* of South Kensington opened its doors in the academic year 1908/9 to about 700 students, many of them from the R.C.S. and the Central Technical College.[3]

The first fruits of these improvements became noticeable chiefly in engineering and physics. Chemistry was slower to benefit. The evidence collected by the Haldane Committee affords some explanation for this. It would appear that there were, in 1906, 120 places for students in the chemistry laboratories of the R.C.S., and the new building opened in that year had 144 places. There were 15 students to each teacher in the first year, and rather fewer in the second and third years of the course. This compared favourably with good German practice, and indeed some of the witnesses before the Haldane Committee suggested that the best use was not always being made of the available laboratory space. Professor Tilden made this point, but failed to elaborate though he was quite clear in his own mind about the bias of the chemistry course towards teacher-training and the fee-paying metallurgists and mining engineers. These groups accounted for the overwhelming bulk of

[1] Haldane Committee, report, pp. 23–4. [2] Report, p. 25 (para. 88).

[3] First Annual Report of the Governing Body of the I.C.S., Cd 4602, pp. 2, 17; Armytage, op. cit., p. 258 and Eyre, op. cit., p. 171.

the student population. Tilden claimed that in order to attract potential industrial chemists it would be necessary to add a fourth year and create a chair of organic chemistry.[1]

The real trouble was that a career in industry did not seem attractive. Men fresh from the Royal College, were, according to Tilden, 'snapped up' by the local authorities for the new technical colleges and paid a starting salary of £150 a year. The sum seems modest enough, but industry was not paying junior chemists any more, and an educational appointment carried security of tenure. Here was the problem: chemists were not coming forward, because there was insufficient demand for them from industry. (The war created that demand and it was from the universities that the men were drawn to run the chemical side of the war effort.) Britain's chemical industry, as we have seen, was structured differently from that of Germany, so that it required fewer chemists. Nevertheless, even after making due allowance for this element, the discrepancy in the number of working chemists between the two countries was very wide.

In 1902 a committee of the British Association consisting of W. H. Perkin jr., H. E. Armstrong, G. G. Henderson, and G. T. Beilby circularized all members of the Society of Chemical Industry known to be working in industry. About one-half answered, and out of these 502 only 107 were graduates of a British or a foreign university or *Technische Hochschule*. The majority, 395, were non-graduates, most of them men who had not completed their university or technical college education. About 80 had learnt their chemistry at evening classes or in private or industrial laboratories. The lack of qualification among 80 per cent of the chemists can be taken to mean that they were not suitable for advanced research, but were presumably quite competent for routine analysis. The committee obtained 167 returns, roughly a third of the total, from men working in the chemical industry, and it is worth noting that of these only 43 had a degree.[2] Dewar, commenting on these figures, thought that a 'liberal estimate' of the number of trained, though not necessarily qualified chemists in British industry was 1500 and 'at the very outside cannot be put higher than 1500–2000'.[3]

A similar survey was made at about the same time by the Institute of Chemistry at the request of the London County Council. At the end of October 1902 the Institute had 1052 Fellows and Associates. The former, numbering 918, might be considered fully-trained chemists, though only a third had taken a degree at a British or foreign university. The members of the Institute were widely scattered: 30 per cent worked in private practice as analysts or consultants, 27 per cent were in industry, 16 per cent in Government service, and the rest were teachers or unclassified.[4]

If we accept Dewar's highest figure of 2000, we find that Britain had about

[1] Haldane Committee, evidence, pp. 32–3, 125, 127–9.

[2] *Statistics concerning the training of chemists employed in English chemical industries.* Report of the [1902] British Association Meeting (1903), pp. 97–8.

[3] Report of the [1902] British Association Meeting, p. 15.

[4] Pilcher, op. cit., p. 165.

half the number of chemists working in German industry. Unfortunately I have not been able to find a later census of British chemists, so that it is not possible to tell whether their numbers grew as rapidly as in Germany.[1] It seems unlikely, since the available evidence from large firms suggests that their research efforts remained modest throughout the 1900s. Consider Brunner, Mond & Co., if not the largest then certainly the most profitable of British chemical manufacturers, where in 1911 the laboratory 'consisted of one moderate-sized room, with a smaller laboratory for the Research Manager who had arrived there a few years before . . . and had started phase rule studies'.[2] The qualified staff consisted of two men. At the laboratory of the United Alkali Co. Dr. J. B. Raschen, a Ph.D. from Heidelberg, had several graduate assistants. His total staff did not exceed thirty, many of them without any qualification, but whose competence had led to their promotion from the analytical benches to research work.[3] In the dyestuffs trade the situation was no better. Herbert Levinstein said that his father's business at Blackley employed fourteen chemists at the beginning of the century, and he claimed that this was a greater number than in any other chemical works in England.[4] At the nearby Clayton Aniline Co., A. G. Green was research chemist and works manager from 1894 to 1901, when he resigned because he considered the prospects for a British dyestuffs industry hopeless. He was offered and accepted the chair of colour chemistry at Leeds.[5] Arthur George Perkin, the second son of Sir William Perkin, worked as chemist in the alizarin department of Hardman & Holden, and later rose to become manager of the company. He too resigned and moved to Leeds where he became lecturer and then followed Green in his chair.[6] These resignations continued a trend that had begun a generation earlier when Sir William Perkin had left industry to devote himself to private research.[7] The precise effects of this exodus cannot be determined, but it was certainly an unfavourable factor, which was aggravated by the failure of British chemical manufacturers to cultivate contacts with academic chemists. Nevertheless, there were signs of change by the eve of the war: the technical colleges were at last turning out technologists, and these skilled men were being employed in increasing numbers by dyers and colour printers.[8] The students from the Manchester College of Technology had such a high reputation that they were often engaged before they

[1] See p. 48.

[2] Akers, Sir Wallace. The research laboratories of I.C.I. Ltd. *Proc. Journées Internationales de Chimie Industrielle* (1948), p. 349.

[3] I am indebted to Dr. D. W. F. Hardie for this information.

[4] From within the dyestuffs industry. *J. Soc. chem. Ind., Lond.* **50** (1931), 254 T.

[5] *Clayton Aniline Co. Quarterly*, special 75th anniversary issue (summer 1951), pp. 7, 9.

[6] Fargher, R. G. The Perkin family and the textile industry. *Manchester Guardian*, 7 May 1956.

[7] Roscoe, however, was an exception. He left Manchester University to become a director of the Castner–Kellner Alkali Co.

[8] S. H. Higgins, a well-informed observer, thought that the education for textile technologists was better in Britain than in Germany and that training in dyeing techniques was especially progressive. Cf. his *Dyeing in Germany and America* (2nd edn. 1916), pp. 91 f.

had even completed their course. And enquiries made at the end of the Kaiser's war indicated that the universities were also becoming the recruiting ground for industry. For example, the Thomson Committee claimed that, of the 110 Cambridge men who had taken Part II of the Natural Science Tripos since 1900, no less than 80 had gone into the chemical industry.[1]

When we look at the state of chemistry teaching in the United States of the 1900s and 1910s it is important to bear in mind that its characteristics were less important than the results it produced. The reason for this is as follows: when the century opened, the American chemical industry, though huge in scale, was limited in scope. A generation later it was not only the largest, but assuredly the most diversified. American industrial chemists largely wrought this change, and their educational background is therefore highly relevant. The facilities available to them differed from those existent in Germany and Britain, and the broad background merits brief description.

The American system of secondary education was, and remains, free. Such egalitarianism may handicap the able student, but it also produces a large supply of candidates for the universities. Even in 1900 there were scores of them, usually classified into four main groups: the endowed colleges and universities; the state and municipal colleges; the colleges set up under the Land Grant College Act of 1862; the denominational colleges, which specialized in training men for the ministry.

At university or college level the period of teaching normally lasted four years. In the science departments the emphasis was on teaching rather than on research work, and at least one observer considered that the first two years at university should be compared with *gymnasium* work in Europe. The remaining two years were devoted to training for a professional career.[2] Research, pursued critically, was virtually unknown, except at a handful of institutions of which Johns Hopkins in Baltimore was the first and deservedly the most famous. In 1900 the justification for scientific instruction at advanced level in the United States was ideally that of practical service to the community through education. Consequently there was none of that cleavage between university and industry so characteristic of England. Nor was there the pressure to devote oneself to a piece of research, whatever the immediate usefulness of the work itself, which in the early 1900s caused alarm among German chemists. There was another and important consequence: foreign observers noted that pride in the alumni's industrial and political achievements was a characteristic of the university, and that in general the acquisition

[1] The figures, which seem very low, apparently cover the period 1900–14. Report of the Committee . . . into the Position of Natural Science in the Educational System of Great Britain, Cd 9011, p. 71 (para. 170).

[2] Flexner, A. *Universities, American, English, German* (1930), pp. 42 ff. It is fair to say that the curriculum at State and Land Grant Colleges was specifically orientated to applied science, because it was intended to provide the state or the district with technologically trained personnel.

of higher learning, however recondite, was popular. Hence the facilities for pursuing these studies were excellent; for instance the many well-stocked public libraries remained open until late in the evening, and night classes abounded.[1]

American higher education was initially influenced by English precedent and to a lesser, though in science not negligible, extent by Scottish practice. In the nineteenth century however, and particularly in its second half, German influence became a major factor in reshaping and redirecting higher science teaching. It has been estimated that between 1814 and 1915 10 000 Americans went to German universities, the greater part in the fifty years after the Civil War. Over half of these attended the philosophical faculties, at which they would have studied science as well as history and philosophy. Their influence on American scholarship and training methods cannot be overrated.[2] Johns Hopkins, founded in 1876, and the University of Chicago, opened in 1892, were the first big foundations to be modelled on the German system. They were not designed to give a vocational or professional education, but aimed to take the student beyond the graduate level to post-graduate research. Accordingly, advanced research was considered by them from the very start as one of their essential functions. This approach had been tried at other colleges, notably at the University of Michigan in the 1850s and at Clark University (Worcester, Mass.) in the 1880s, but had failed. Johns Hopkins and Chicago, under D. C. Gilman and W. R. Harper respectively, succeeded, partly because of the dominant personalities of their first presidents, and partly because of their large endowments, which made them independent and paid for the equipment. Johns Hopkins bequeathed \$3·5 million (say £0·7 million), while J. D. Rockefeller contributed over \$2 million towards Chicago between 1889 and 1892, and ultimately paid out about \$30 million (£6 million).[3] These universities were remarkable centres of research and quickly became the nursery for eminent university teachers.[4] In this indirect manner the German system was grafted on to the existing Anglo-Saxon pattern of higher education, and it spread rapidly from Baltimore and Chicago.

Statistics of the student population, though available, are somewhat suspect. In 1900 the total number of students enrolled approached 200 000, twelve years later it was 318 000. These figures include the medical and professional schools as well as the preparatory departments of the universities. If these be excluded the numbers fall to about 100 000 and 187 000 respectively. Many of the students were attending extension courses; there was also

[1] Rassow, art. cit., p. 707.

[2] Brubacher, J. S. and Rudy, W. *Higher education in transition* (1958), pp. 171 ff.

[3] ibid., pp. 182 and 360.

[4] Johns Hopkins was outstanding in the very high staff to student ratio (2 : 3 in the very first year), the large proportion of men who went on to postgraduate work (about 25 per cent on average in the period 1897–1900), and the total absence of sports facilities and campus amenities. French, J. C. *A history of the university founded by Johns Hopkins* (1946), pp. 41, 69.

a growing number in graduate departments working for degrees above that of the bachelor: about 5800 in 1900 and 11 600 in 1912.[1]

The number of science students, and specifically of chemistry students, cannot be readily ascertained. A survey of twenty-eight universities and colleges showed that in November 1909 they had almost 90 000 students of which about 13 500 or 15 per cent were enrolled in the departments of engineering, mining, chemistry, and related subjects.[2] A more detailed survey for 1910 reported 174 000 students of which 25 000 (14½ per cent) were in the various science departments, among them 4800 students of chemistry and a further 869 students of chemical engineering.[3] It will be seen from the figures in Table 3.3 that chemical engineering students, despite the rapid growth in their numbers, formed a relatively unimportant group until the war.[4]

TABLE 3.3

Number of students attending engineering courses, selected years 1900 to 1918[5]

	1900	1904	1906	1910	1916	1918
Mechanical engineering	4459	6894	7426	6377	7707	5944
Electrical engineering	2555	4389	5696	5450	7098	5683
Chemical engineering	536	694	1234	869	2774	4548

Though endowments were becoming an important source of university income during the 1900s, fees and assistance from the States or the municipalities contributed 37 per cent of total revenue in 1900 and no less than 50 per cent in 1914. The Federal Government's payments, which nowadays play such a crucial role in university finance, were then of little significance.[6]

It would appear that the study of applied chemistry was more successfully pursued at American colleges than that of pure chemistry. The lack of properly trained teaching staff may have been a reason, but more important was the interest taken in technology: 'In America', wrote de Tocqueville, 'the purely practical part of science is admirably understood, and careful attention is paid to the theoretical portion which is immediately requisite to application. On this head the Americans always display a clear, free, original, and inventive power of mind. But hardly anyone in the United States devotes himself to the essentially theoretical and abstract portion of human knowledge.'[7]

[1] Biennial Survey of Education 1916–18. *Bureau of Education Bulletin No. 90* (1921), vol. 3, pp. 680–4.

[2] *Statistical Abstract of the U.S.* (32nd edn. 1909), p. 112.

[3] *Report of the Commissioner of Education for year ended 30 June 1910* (1911), vol. 2, p. 859.

[4] The fall in numbers in 1910 in Table 3.3 may have been due to more severe matriculation requirements. It is difficult to be positive because figures are not available for 1908, 1909, and 1911–15.

[5] Biennial Survey of Education 1916–18, op. cit., vol. 3, pp. 680 f.

[6] ibid.

[7] de Tocqueville, A. *Democracy in America*, tr. Reeve (1875), vol. 2, p. 36.

De Tocqueville's generalization dates back to the 1830s, but was still applicable to the teaching of chemistry in the early 1900s, with the exception of Johns Hopkins, Chicago, and a few other institutions. But there were signs of change. One was that an increasing number of eminent chemists had obtained their education in their homeland, which would suggest that better teaching staff and facilities were having a beneficial effect. The trend can be seen by examining the qualification of the leading American scientists. Information on these 'starred scientists' was prepared in 1903 and subsequently brought up to date by J. McKeen Cattell, Professor of Psychology and Editor of *Science*. As far as chemists are concerned 175 were starred in 1903, 55 in 1909, 64 in 1920, and 44 in 1926. Between two-thirds and three-quarters of the chemists had taken a Ph.D., and it is interesting to observe that holders of foreign, principally German, doctorates fell from 54 in 1903 to 7 in 1926.[1] There may have been less need for a German qualification because the status of the American-educated chemistry teacher was improving. And we may illustrate this by some examples: there was, and remains, general agreement on the poverty of academics relative to that of business men, but their situation was getting a little better. Before 1900 the ceiling for a professor was about $4000 (£800) a year. Harper at Chicago created a precedent when he announced that he would offer a good teacher $6000–$7000 (£1200–£1400) a year.[2] Harvard also paid well, especially at the senior levels: while the Instructor started at $1200, the Associate Professor (broadly equivalent to the Reader in Britain) ranged from $3500 to $4500, and the professor from $4000 to $5000. The salary levels at such famous institutions had some influence on those of smaller and state universities, which were, in general, substantially lower. They were not dissimilar to the Civil Service rates applicable at the National Bureau of Standards for qualified staff recruited from the universities. In 1906 physicists with some seniority received from $2000 to $2200 a year and Divisional heads from $3500 to $4000 a year.[3] Industrial research laboratories were able to offer salaries about 50 per cent higher.

The facilities for teaching chemistry varied. G. Lunge visited the science departments of several universities in the early 1890s and came to the conclusion that the schools of metallurgy and mechanical engineering had better laboratories and equipment than their European counterparts. But the chemistry departments did not impress him and the equipment, except at the Massachusetts Institute of Technology (M.I.T.), was generally poor. Lunge considered that the laboratory courses were inadequate and fit only to produce analysts. He singled out Johns Hopkins as the only institution with a purely scientific approach and strict matriculation requirements, whereas other institutions lacked both.[4]

[1] Visher, S. S. *Scientists starred 1903–43* (1947), pp. 113, 285.
[2] Mayer, M. *Young man in a hurry* (1957), p. 53; Brubacher and Rudy, op. cit., p. 53.
[3] Cochrane, R. C. *Measures for progress* (1966), p. 98.
[4] Lunge, G. *Die chemische Industrie und die chemisch-technischen Hochschulen in Nord Amerika* (1894), pp. 20–1.

The science buildings at Johns Hopkins dated back to the late 1870s and early 1880s but were still adequate twenty years later. Ira Remsen, the head of the department of chemistry, was an inspired teacher, and there were 150 chemistry students in 1910.[1] The Haldane Committee, some ten years after Lunge, also noted the lavish equipment of the engineering departments and mentioned that this was quite unknown in the United Kingdom. The Committee thought that the rapid progress of the American engineering industries was in some measure due to the influence of university graduates. Johns Hopkins was favourably mentioned, but the committee did not comment on the matriculation requirements beyond saying that they corresponded broadly to those of an English university.[2] There was no reference to chemistry teaching. A little later S. H. Higgins of the Manchester College of Technology visited the United States. He was not greatly impressed by the M.I.T., which he considered was more interested in engineering science than in chemistry.[3]

The situation in the 1900s was characterized by the extensiveness of chemistry teaching, the increasing availability of good equipment, and the absence of study in depth and research. To take only one instance: the first laboratory at the University of Michigan was built in 1844 and had initially 26 work places; in 1901 there were 362 (the Royal College of Science in London had about the same number after 1906). Construction of a new laboratory block was started in 1908; it cost $305 000 (£61 000) and was provided with 950 work places. This was doing things on the scale of the Charlottenburg polytechnic, but there was little chemical work except of the most elementary kind. Indeed until 1905 chemists and pharmacists were in the same department, and only in that year was a separate department of chemical engineering and analytical chemistry established and an additional chair created. In 1908 the two departments of chemistry and of pharmacy had altogether 2599 students of whom just under half attended laboratory courses. But only 28 were doing work at graduate level. In the college year 1915/16, 3497 students were enrolled in the chemistry, chemical engineering, and pharmacy classes. The proportion attending the laboratory courses had grown to two-thirds of the total, but only 119 were graduates engaged on some form of research.[4]

In general the teaching of chemistry at university level was not intended, as in Germany, to culminate in a piece of individual research, but was designed to inculcate efficiently the principles of chemistry. The methods of training were in some respects different from those used in Europe. For example early in the twentieth century the practice was adopted of breaking down the applied chemistry courses into 'unit operations'. The term is said to have been coined by A. D. Little, and meant that processes such as distillation, crystallization, drying, and others were explained in terms of a single opera-

[1] French, op. cit. Remsen was president of Johns Hopkins from 1901 to 1913.

[2] Haldane Committee, report, op. cit., pp. 15, 17.

[3] Op. cit., p. 90.

[4] Campbell, E. D. *History of the chemical laboratory of the University of Michigan* (1916).

tion applicable over a wide area of industrial chemistry and extending to such manufacturing operations as brewing, spirit distilling, sugar refining, etc. It was claimed that this saved teaching time. I do not know how widely this system spread. Certainly it would not create future *Wissenschaftler*, but it was useful for men seeking jobs in industry and fitting them to their tasks.

Nevertheless there was discussion as to whether American universities were turning out properly qualified industrial chemists. The subject was evidently thought sufficiently important for the newly formed American Institute of Chemical Engineers to appoint a committee on chemical engineering education in 1908. The committee reported that the short duration of the university courses was the major problem. A training on the European model had to be crammed into four years, which was inadequate, since half the time had to be spent on completing the schooling of the students, and only two years were left to train at university level. The committee sent out a questionnaire, and the replies showed that the graduates were not suitably equipped for industrial work after four years. Two remedies were suggested: firstly, raising the university and college matriculation requirements so that university training could start at a more advanced level and save time by not teaching elementary science. This began to be done in the years preceding the Great War, with the result that university admissions fell sharply for a time. The second recommendation was the need for post-graduate work leading after three years to a Ph.D. Here too there was an improvement, because the number of students taking higher degrees in chemistry increased considerably. Among the other suggestions was the appointment of professors exclusively for the training of more senior students, the provision of supplementary training to give students some knowledge of related fields, such as mechanical and electrical engineering, the introduction of pilot-plant work, and finally the appointment of 'advisers', that is, specialists in one branch of the chemical industry familiar with the latest developments in that particular branch.[1]

The adoption of these measures was assisted by the development of research centres outside the universities and industry, but in close touch with both. While it would be wrong to consider the United States a pioneer in this field, the role of the consultants and of the research institutes seems to have been particularly important in bringing about an improvement in the quality of chemists and promoting a favourable attitude to chemical research in general. Even before the Great War there were several consultants with considerable influence on the chemical and allied trades. Their work demonstrated that chemical tests and research yielded profits for their clients and so created a climate of opinion favourable to the development of industrial research. Haynes mentions that many college professors of chemistry undertook consultancy work, thus adding to their income.[2] The first of the non-

[1] Frerichs, F. W. Report of the Committee on Chemical Engineering Education. *Trans. Am. Inst. chem. Engrs.* **3** (1910), 122–65.

[2] Op. cit., vol. 1, p. 396.

academic consultants to become famous were A. D. Little and R. B. Griffin, who established their partnership in Boston in 1886. Little, a chemist and an unusual personality, considered that his duties went beyond those of a mere analyst, and he was instrumental in assisting the development of many branches of industry: his early work was concerned with the sulphite paper process, later he became interested in chrome tanning, the Le Sueur electrolytic cell, and aeroplane dopes. As his consultancy business grew it became necessary to divide it into departments, which by 1918 had dealt with 16 287 enquiries. Little, a former student of M.I.T., was instrumental in organizing a course in chemical engineering there, for which the first degrees were awarded in 1889.[1] A. S. Bryant and C. S. Miner established a consultant laboratory in Chicago in 1906. Initially they were primarily concerned with industries processing agricultural products, later they turned to patent matters, and still later to plastics.[2]

Pure and applied research was also encouraged by the various foundations, which around this time were becoming of increasing significance. The Carnegie Institution made large grants to set up research departments at universities and supported individual scientists. About $22 million (£4·4 million) were disbursed for these purposes between 1902 and 1911. Meanwhile the Rockefeller Foundation was also spending lavishly through its General Board of Education where Flexner was able to bring about a major improvement in the quality of American medical schools. The Foundation also financed major campaigns against hook-worm and yellow fever as part of its public health programme.[3] These instances of private generosity were noted in Europe and may have influenced the founders of the Kaiser-Wilhelm-Gesellschaft. There were also developments affecting chemistry specifically that are of interest in this context.

R. K. Duncan, professor of chemistry at Washington and Jefferson College (Washington, Pa.) was the founder of the Industrial Fellowship system. While visiting Europe in 1906 he observed the links between universities and industry, and convinced himself that they were of mutual benefit. The following year he became professor of industrial chemistry at the University of Kansas, and there introduced the first of these Fellowships: manufacturers were invited to endow a Fellowship for a limited period, during which the holder would study the particular problem set by the company endowing the Fellowship. From Kansas, Duncan moved to Pittsburgh in 1910, became professor of industrial chemistry, and met the Mellon brothers, who supported him generously.[4] Duncan re-introduced the Fellowships in 1911, when twenty-three university men were appointed to work in an institute that the Mellons

[1] Little, A. D. *The handwriting on the wall* (1928); Haynes, op. cit., vol. 6, article on A. D. Little.

[2] National Resources Planning Board. *Research—a national resource, pt. 2: industrial research* (1941), pp. 73–4.

[3] Dupree, A. Hunter. *Science in the Federal Government* (1957), pp. 297–8.

[4] The Mellons were the founders of the Gulf Oil Co.

had endowed, and of which he himself became the first director (1913–14). At first, work on industrial problems was carried out at the University of Pittsburgh, but in 1915 the Mellons donated a special building, which, since then, as the Mellon Institute, has become remarkably successful, because it has been filling a need in a novel and practical manner.[1]

An idealistic, but less successful experiment was the Research Corporation. F. G. Cottrell developed the electrostatic dust-precipitating process at the University of California in 1906. The United States royalties were assigned to the university, and in 1911 transferred to the Smithsonian Institution, which, in turn, set up the Research Corporation the following year. The purpose of this non-profit making organization was to help in the development of new inventions, license the patents, and use the royalties therefrom to sponsor research work at universities and to buy other inventions. Cottrell hoped that in this way the work of the university scientists could be turned to practical industrial use without impairing academic independence.[2] These expectations were not fulfilled, because rather less income was received than had been anticipated, and the first significant grant was not made until 1923, when Professor Goddard, the rocket expert, was given $5000 to pursue his researches.[3]

The underlying theme in this brief account of higher chemistry teaching in the United States has been its practical bias. Pure chemistry was not wholly ignored—after all there were luminaries such as Theodore W. Richards and G. N. Lewis, and before them Willard Gibbs, but they were an insignificant minority among the vast numbers of practical chemists that the universities were turning out. By German standards these practical chemists were of low quality, but their training, whatever its shortcomings, was competently technological, and their ranks were stiffened by growing numbers of chemical engineers whose education was modelled on that of the *Technische Hochschulen.* Until 1917 no census was taken of chemists, but in that year the National Research Council made an enquiry and found that about 17 000 were working in the United States.[4] Relative to the total population the number compares not unfavourably with that of Germany on the eve of the war. It is probable that the United States did not suffer from the same lack of competent technologists in the 1920s as Europe, and this may have been one of the key factors in the rapid development of the American chemical industry during the interwar years.

The highlights of chemistry instruction at university level in other countries can be quickly reviewed.

[1] Weidlein, E. R. and Hamor, W. A. *Glances at industrial research* (1936), pp. 18 ff.; National Resources Planning Board, op. cit., pp. 71–2.

[2] National Resources Planning Board, op. cit., p. 70; Cottrell, F. G. The Research Corporation. *J. ind. Engng Chem.* **4** (1912), 864 f.; *Z. angew. Chem.* **26** (1913), pt. 1, pp. 398–400.

[3] Goddard's work attracted renewed attention in the 1950s.

[4] Yerkes, R. M. (Ed.) *The new world of science* (1920), p. 150.

In the eighteenth and early nineteenth century France led the world in the quality of its science teaching, and it may be recalled here that Liebig, the founder of the German tradition of chemical research, was taught by Gay-Lussac at the Arsénal. But as the nineteenth century advanced, so progress ceased, and by 1914 the neglect of scientific instruction was widely considered a national scandal. It would appear that between 1890 and 1910 the proportion of the funds devoted to technical and agricultural instruction in the Ministry of Education's budget remained unchanged at 4 per cent.[1] Such a derisory effort in this key sector contributed much to the technological weakness of French industry.

Contemporaries agreed on the causes, but the remedies were not applied. This was mainly due to the structure of the entire system of education, which was controlled centrally, but remotely, from the Ministry in Paris, and officialdom seemed to be unaware of the profound changes that had taken place in Germany from Liebig onwards. The curriculum of the secondary schools and the regulations of the universities were designed to train lawyers and scholars rather than chemists. In particular the teaching of French and mathematics was given pride of place, and in the competitive examinations to the Ecole Polytechnique and the Ecole Normale, the mathematical tests were of key importance. It is not surprising that these requirements were discouraging to those wishing to specialize in the various branches of chemistry.[2]

An equally serious defect was the neglect of laboratory facilities. Here two factors combined to bring about an intolerable situation. At the universities, as indeed at the *Grandes Ecoles*, the professors continued the tradition of public lectures and private research. The former was a duty, the latter a privilege; and even when the incumbents wished to enlarge their laboratories and adapt them for teaching purposes, the Ministry of Education would refuse to authorize the necessary funds. The other factor was the excessive centralization, which enabled a few chemists to control effectively the curriculum and the staff appointments. These *chimistes officiels* (among them many eminent scientists) were for many years bitterly divided on points of theoretical chemistry, and in particular on the introduction of the chemical nomenclature adopted in Germany and the Anglo-Saxon countries. The quarrel split academic chemists into three factions headed by Sainte-Claire Deville, Berthelot, and Wurtz respectively. We need not go into the details, except to note that the German system was finally adopted, but not before the progress of organic chemistry had suffered great damage. Although the quarrels were eventually composed in the 1880s, their effect on the course of organic chemistry was enduring. By contrast, progress in inorganic chemistry, which had remained unaffected, was not held up.[3]

Dissatisfaction with the condition of scientific education was not polarized

[1] Combe, P. *Niveau de vie et progrès technique en France 1860–1939* (1956), pp. 263–4, 428.
[2] Grandmougin, E. *L'Enseignement de la chimie industrielle en France* (1917), pp. 3–4.
[3] Delaby, op. cit., pp. 84–5.

in official reports. Improvements came as a result of Government action, and were not the outcome of pressure exercised by an informed public. For example, university reform had to wait until the law of 1896 came into force two years later. It resulted in administrative rather than educational improvements. The Ministry of Commerce also bestirred itself and began to take its responsibilities for certain aspects of technological instruction more seriously, though there were delays and occasional clashes with the Ministry of Education. The Government departments might have exerted themselves more had chemical manufacturers shown greater interest in these matters. But the general indifference suggests that too few cared and that the majority were satisfied with the existing system. When, at the beginning of the twentieth century, a group of distinguished educationalists gave a series of lectures on education in a democracy, not one found fault with the method of teaching science. Indeed the one scientist among them, J. Hadamard, considered that simple physical and chemical experiments were adequate for boys in senior forms, and his only regret was that mathematics did not receive more emphasis.[1]

The teaching of chemistry in France was divided between the universities and the institutes attached to them, the *Grandes Ecoles*, and a handful of municipal technical colleges. The last two could not be equated with the German *Technische Hochschulen*, though a few approached that level in chemistry and some surpassed it in engineering. So far as I know no statistics of the number of chemistry students are available.

The universities came under the Ministry of Education, and their budgets were small. For 1906 the proposed expenditure totalled £550 000, of which Paris took £150 000, the provincial universities £300 000, and maintenance generally £100 000.[2] Chemistry was taught at most universities, though the facilities at the Sorbonne were inferior to those at Lyon, Grenoble, and above all Nancy. The reform of 1896 undoubtedly encouraged local initiative, and this resulted in better facilities in some places. At Lyon the first professor of industrial and agricultural chemistry was appointed in 1876.[3] He was Jean Raulin, a pupil of Pasteur, and he took his duties sufficiently seriously to introduce a three-year course (including laboratory work) with financial assistance from the Chamber of Commerce. The Ecole de Chimie Industrielle began with about thirty students in the session 1883/4, and with the ridiculously small budget of 22 000 fr. (less than £1000). A suitable building costing 1·45 million fr. (£60 000) was completed in 1898 with help from industry and wealthy citizens.[4] At Grenoble the university created a special

[1] Croiset, A. (Ed.) *L'Education de la démocratie* (1907). The lectures, in impeccable French, reflect the speakers' mastery of the language. They virtually ignore the question whether the educational system was appropriate to the needs of a modern industrial state.

[2] Haldane Committee, Final Report, p. 20.

[3] This was only the second chair for this branch of chemistry in France. The first had been established somewhat earlier at the Conservatoire des Arts et Métiers.

[4] Matagrin, A. *L'Industrie des produits chimiques et ses travailleurs* (1925), pp. 331–2. Simonin,

Institut Polytechnique in 1902, which had a department of paper industry, and a department of electrochemistry and electrometallurgy. It had the dual function of a research centre for these industries, and for training foremen and technicians, and was intended to form a link between university instruction and the trade school.[1] The most interesting developments were at Nancy, where the scope and equipment and indeed the scale could match those of a chemistry department in a German university or *Hochschule*. The beginnings, however, were modest: in 1883 the Dean of the University, returning from a visit to Germany, called for the establishment of a chemical institute. Eventually some half a million francs (£20 000) were obtained from the state, the Département, the city, and not least local industrialists, among them the alkali makers of Lorraine. A special building in which there was laboratory space for 120 students was opened in 1890, and Professor A. Haller was put in charge.[2] Admission to his three-year course in industrial chemistry was by examination, and at the end of the period there was another examination for the diploma.[3] Haller was a good teacher and an energetic fund-raiser. He obtained 370 000 fr. (£15 000), nearly half of it from Solvay in 1897/8, with which he built laboratories for dyeing and applied electrochemistry. By 1898 Haller's department of industrial chemistry had three professors, an assistant professor, five *chefs de travaux* (probably equivalent to lecturers), six assistants, a mechanic, and a lab. boy. There were seventy-two students, so that the staff to student ratio was very good.[4] The number of students more than doubled in the next fourteen years. A school of brewing technology was added in 1893 and an institute of electrotechnics, the first in France, three years later. The science faculty, which encompassed all three institutes, grew rapidly from 285 students in 1901/2 to 596 in 1904/5.[5]

Technological instruction was administratively outside the scope of the universities. This did not prevent energetic men from providing the necessary courses if, as at Lyon, Grenoble, and Nancy, unofficial backing was obtained, and after the law of 1896 some useful progress was made. But in principle it was the responsibility of the Ministry of Commerce to provide such instruction. In the early 1900s the Ministry became alarmed by the situation, and a bill was introduced in 1905 to provide for better technical education at all levels. But, owing to delays and preoccupation with other matters, no law had been passed by 1914. The political wheels ground slowly, but the Chambre des Députés was at last moved to instruct the secretary of

---

G. (Ed.) *L'Industrie chimique Française*: *Lyon et le Sud-Est* (1959), pp. 19–22. *Paris Exhibition* (1), pp. lxxxvii–viii.

[1] Verlot, C. *Rapport . . . sur . . . l'enseignement technique industriel et commercial* (1912), p. 186; Grandmougin, op. cit., p. 134.

[2] Haller was secretary of the chemical section of the Paris 1900 Exhibition.

[3] Verlot, op. cit., pp. 178–9.

[4] Haller, A. L'Enseignement pratique en général, et en particulier celui de la chimie dans nos universités. *Revue gén. Matièr. Color.* **2** (July 1898), pp. 257 ff.

[5] Haldane Committee, Report, p. 20; Verlot, op. cit., p. 177.

its committee on commerce and industry to write a report on the subject. Verlot's report was damning. Of 3·2 million youths between the age of thirteen and eighteen, a mere 150 000 attended day continuation or evening schools, and the Ministry's proposed expenditure on technical education (including the Conservatoire, the Ecole Centrale des Arts et Métiers, craft schools, scholarships, and subsidies to private schools) was 9 million fr. (£360 000) in 1913, equivalent to 60 fr. a head. In neighbouring Switzerland the Federation and the Cantons managed between them to put up 11 million fr. for the same purposes.[1]

With so little money available it is no wonder that next to nothing was done. The Conservatoire had the only French chair of dyestuffs chemistry, no equipment for laboratory courses, and the work of the professor consisted in delivering a course of lectures to Sorbonne students of chemistry and such members of the public as cared to attend. The lectures were of a general and popular kind and did not represent a course of instruction. This leisurely tradition survived from the nineteenth century, was unaffected by reforms in 1900 and 1901, and continued until the outbreak of war.[2] Elsewhere, the staff at the Ecole Centrale des Arts et Métiers, if they were not teaching arts, confined themselves to the crafts of the engineer. The teachers' training college, the Ecole Normale Supérieure, produced mathematicians rather than science teachers. In 1912 (the date is worth noting) a special school for the latter was established, the Ecole Normale de L'Enseignement Technique, which provided a two-year course. Outside Paris the Ministry of Commerce made grants to the Institut Industriel du Nord at Lille, which provided a two-year course in applied chemistry, and to the Ecole Supérieure de Chimie at Mulhouse, which specialized in dyeing techniques. In both towns local industrialists were largely responsible for creating these opportunities, but we must remember that these colleges were fairly small. The most famous of the municipal colleges was the Ecole de Physique et de Chimie Industrielle, founded in Paris in 1882, which had a three-year course. But in 1910 it had only 94 students, compared with 289 at the municipal school for furniture and engraving and 237 students at the Arts School.[3] The figures suggest that applied chemistry was not a fashionable subject in the capital.

Outside Britain and France the condition of European chemistry teaching requires only very brief mention. But the exceptionally important role of the Swiss Federal Polytechnic at Zurich, henceforth referred to by its German abbreviation of E.T.H., commands attention. Opened in 1855 it was initially

[1] Verlot, op. cit., pp. 243–4.

[2] Grandmougin, op. cit., pp. 47 ff.

[3] Fauque, M. *L'Evolution économique de la grande industrie chimique en France* (1932), p. 162; Verlot, op., cit. p. 151. The Ecole de Physique et Chime now forms part of the Sorbonne. One of its first graduates was G. Claude (1870–1960), the physicist who in his younger years became famous for his process of air liquefaction and a method for synthesizing ammonia, and in old age infamous for collaborating with the Germans during Hitler's war.

modelled on the Ecole Polytechnique in Paris and later was greatly influenced by the German *Technische Hochschulen*. The purposes of the E.T.H. were to give a professional education in civil and mechanical engineering, architecture, chemical engineering, and pharmacy, and to train high school science teachers. Attendance at lectures, mostly in German, was compulsory, discipline was strict and the curriculum rigidly prescribed. At the end of three and a half to four years there was an examination for the diploma. After 1909 students were able to prepare a dissertation for the degree of Ph.D. The buildings, modest at first, were greatly enlarged: the first chemistry laboratory building dated from 1861/2, the second and much larger one from 1886. The E.T.H.'s income was partly derived from fees, but mainly from the Confederation's annual grants: from 1896 the grant was 0·8 million frs. (£32 000) until 1906 when it was raised to 1·1 million fr. (£44 000). In the academic year 1900/1 there were 1004 students and by 1913/14 the number had grown to 1376. The proportion of foreign students in the 1900s fluctuated between 30 per cent and 40 per cent; among them Germans, Austrians, Russians, and Italians predominated.[1]

The chemical department of the E.T.H. was at first divided into two sections, headed respectively by the professor of analytical, organic, and inorganic chemistry, and by the professor of technical chemistry. By 1900 additional chairs had been created, and there were professors of analytical chemistry, general chemistry (organic and inorganic), general chemical technology, applied organic chemistry, and lastly physical chemistry and electrochemistry. Students were thus enabled to specialize to an unusual extent, and it is worth noting that very few *Technische Hochschulen* had so many departments. Each section at the E.T.H. had its own laboratories, but they were not designed as research laboratories.[2] There were 212 students in the various chemistry sections in 1900/1 and 204 ten years later; thereafter the numbers rapidly and inexplicably fell to 128 on the eve of the war.[3]

The remarkable prestige of the E.T.H. in the field of chemistry seems to have been due to a combination of two factors. In the first place the curriculum was designed to produce industrial chemists, not academic scientists, and secondly the full-time students were expected to absorb a large body of knowledge. Willstätter noted that at Zurich the students were earnest and intent on taking copious notes. His own work was demanding: experimental demonstrations were a feature and the staff was expected to lecture six hours a day, six days a week.[4]

The teachers contributed to the great reputation of the chemistry school. Nationality did not matter, what counted was experience and ability. The E.T.H., though not unique in this respect, was willing to recruit from in-

[1] Guggenbühl, G. *et al. Eidgenössische Technische Hochschule 1855–1955* (1955).
[2] Ost, H. Die chemische Technologie an den Technischen Hochschulen. *Z. angew. Chem.* **13** (1900), no. 27, 662.
[3] Guggenbühl, op. cit., pp. 118, 256.
[4] Willstätter, R. *Aus meinem Leben* (1949), pp. 158–60.

dustry as well as universities: Robert Gnehm, who held the chair of applied organic chemistry from 1894 to 1904, had had a distinguished career with the CIBA company, and his successor, the Alsatian E. Grandmougin (1905–8), had come from Hoechst and was subsequently to make a name for himself in France during the First World War by his attempts to reform the teaching of organic chemistry. Their colleague, the Silesian G. Lunge, who held the chair of technical chemistry (later chemical technology) from 1876 to 1907, was considered the greatest authority on the manufacture of inorganic acids and alkalis, and was internationally famous for his textbooks. Willstätter, a future Nobel Prizewinner, came from Baeyer's institute in Munich to teach general chemistry between 1905 and 1911, when he became head of the Kaiser Wilhelm Institut for chemistry in Berlin. Willstätter was succeeded by another German, H. Staudinger, who was also destined to receive the Nobel Prize. Willstätter did his great work on chlorophyll and Staudinger his on ketenes and auto-oxidation at the E.T.H. F. P. Treadwell, an American, was succeeded in the chair of analytical chemistry by his son, and the father's textbook on quantitative and qualitative chemistry went through twenty editions.[1] The professors were well paid, at first 5000 fr. (£200) a year, later 7000–9000 fr. (£360) a year, in addition to which there were various fees; the pay was a considerable attraction.

We have noted that Russian students flocked to German universities as well as to the E.T.H., and it is useful to look briefly at the facilities available to them at home. The Russian university statute of 1884 was reactionary and restricted student activities and also the admission of Jews. But it encouraged the training of technologists, although they had to submit to quasi-military discipline. By 1899 Russia had nine universities[2] with a total student population of 16 500, about half of them in Moscow and St. Petersburg; approximately 3800 were in the natural science departments.[3] The technical schools were inspired by the Ecole Polytechnique, but eventually adopted the system used by the *Technische Hochschulen.* Those at Moscow, St. Petersburg, Kharkov, Riga, Kiev, and Warsaw taught chemistry, and in 1899 had altogether 1300 students of applied chemistry.[4]

With the exception of St. Petersburg, the development of technical instruction dates from the closing years of the nineteenth century and reflects the government's endeavours to provide technologists for the emerging industries. On paper, the education offered was thorough, the course of instruction lasted 4–5 years, but the quality of the staff was often poor. Little

[1] Schwarzenbach, G. W. D. Treadwell. *Neue Zürcher Zeitung* (7 August 1959); Fernausgabe.

[2] Moscow (1755), Dorpat (1802), Kharkov (1809), Kazan (1804), St. Petersburg (1819), Kiev (1833), Odessa (1864), Warsaw (1869), and Tomsk (1888).

[3] de Kovalevsky, M. W. (Ed.) *La Russie à la fin du 19ème siècle* (1900), p. 886. The chapter on higher education was contributed by V. T. Deruginsky. The book was issued on behalf of the Russian commission to the Paris 1900 Exhibition.

[4] ibid., pp. 893–4. In addition to these six (which between them had about 5200 students) there were two schools of mining and the Polytechnic Institute of St. Petersburg, which had about 1800 pupils.

attempt was made at the universities, let alone the technical colleges, to encourage original research among the students. It has been said that Beilstein, who succeeded the great Mendeleev in the chair of chemistry at the St. Petersburg Polytechnic Institute, began the compilation of his famous handbook because he was so disappointed by the quality and work of his students. He turned down several offers of a German professorship, precisely because the system then in force in the Russian capital enabled him to write his book, which in Germany the supervision of research work would have prevented.[1] By contrast V. N. Ipatieff, a self-taught chemist, was not inhibited by the discipline of the Mikhail Artillery Academy, where he was professor of chemistry, from undertaking his studies into the phenomenon of catalysis.[2]

While German methods of instruction were used, there was no tradition of academic research. Professors could indulge in it if they were so minded, and there was no pressure on them to create a tradition of group study in any particular branch of chemistry. Possibly there was no real need for the Russia of those days to follow the German precedent, since the empire's main needs were for civil servants and technologists capable of dealing with the growing industries, rather than for chemists who could widen scientific horizons through research.

[1] Hjelt, E. F. K. Beilstein. *Ber. dt. chem. Ges.* **40** (1907), pt. 4, 5051–4; Richter, F. (Ed.) *75 Jahre Beilstein's Handbuch der organischen Chemie* (1957).

[2] Ipatieff, V. N. *The life of a chemist* (1946), pp. 78 ff.

4

THE NEW CHEMICAL TECHNOLOGIES

THE first thirty years of this century were fertile in chemical inventions. I do not propose to catalogue them all, let alone describe them, but will confine myself to examining the inventions and their consequences in two areas: electrochemistry and nitrogen fixation. The choice can be readily justified because of the role they play in the history of the chemical industry. For example, the new technologies fundamentally affected the supply of a large range of products, and shaped the early years of the then new discipline of chemical engineering. They were to influence materially the location and financial evolution of the chemical industry, and they were major factors in giving to the German and United States industries a decisive lead over those of France and Britain in the inter-war years. Finally, and not least important, electrochemistry brought chemical manufacturers in close touch with electrical engineering and the electricity supply industry. Even more significant were the links formed between producers of fixed nitrogen, the fertilizer industry, and agriculture.

Electrochemistry and nitrogen fixation yielded chemicals which, in most cases, had been obtained by other routes or from natural sources for decades. The products remained unchanged, but the processes were different, and the economic and technological environment wholly novel.

Electrochemistry

In order to broaden the range of their products and lessen the dependence on soda, the firms working Leblanc's alkali process also supplied caustic soda and chlorine. As we shall see, both chemicals could equally be made by electrochemical processes, which yielded products of higher purity. Electrochemistry thus caused the final ruin of the old Leblanc soda industry.

At the turn of the century caustic soda was a key product of the inorganic chemical industry. It was (and remains) an essential material in the production of hard soap. As the soap business grew, so the demand for caustic soda increased. In addition some new uses were found in the 1900s, and by the eve of the Kaiser's war they represented a large additional demand. They included petroleum refining, the preparation of metallic sodium, the treatment of cotton by Mercer's process to give the fibre lustre, and the manufacture of viscose rayon.[1] The co-product of electrolysis was chlorine, which was sold

[1] For every ton of chlorine, about 1·1 ton of caustic soda is made. Until Hitler's war caustic was more important, and there was generally a surplus of chlorine. Since then the picture has been reversed, mainly because of the spectacular development of organic chlorine compounds, notably polyvinyl chloride.

in the form of bleaching powder to the cotton and linen industries. During the 1880s apparatus was designed in Germany to liquefy the gas, and liquid chlorine came to be used extensively in the dyestuffs industry.[1] Modifications of the electrolytic process yielded chlorates, and by the turn of the century potassium chlorate was employed as an explosive, in making matches, and in the preparation of a few dyestuffs. After the Great War potassium chlorate became commercially less important, but the consumption of sodium chlorate as a cheap weed-killer grew rapidly. Another electrolytic alkali was sodium hypochlorite, which had been made by another route since the beginning of the nineteenth century. It was used as a disinfectant, but the real growth of the product dates from its use as a bleaching agent for paper, and it became of importance in Scandinavia.

Metallic sodium had been a laboratory curiosity, until Sainte-Claire Deville in the 1850s prepared it by reducing sodium carbonate with charcoal at a high temperature. He needed sodium to make aluminium, but his process was very expensive, and when, a generation later, the electrolysis of alumina to aluminium was invented, the need for sodium disappeared. But it revived quickly when sodium cyanide was found to be a good flotation agent in processing gold ores (MacArthur and Forrest process). Soon afterwards another new demand was found when sodium peroxide was introduced for straw bleaching, which was the essential first step in the manufacture of the then fashionable boaters. H. Y. Castner, a prolific American inventor, devised in 1890 a process for electrolysing sodium from molten caustic soda, thereby considerably cheapening the preparation of the material, which soon became an important and valuable article of chemical trade: it has been estimated that 1000 tons were made in 1900 and 11 000 tons ten years later.[2] At first sodium was only made by the Aluminium Co. in England[3] and was sold to the Deutsche Gold- & Silber-Scheideanstalt (usually abbreviated to DEGUSSA). This was an old-established firm of assayers, which had more recently begun making cyanides. It now seized the opportunity of abandoning the cumbersome traditional method of preparing cyanides from gas-works residues and leather. DEGUSSA's links with the Aluminium Co. were strengthened when the two companies agreed to build jointly owned factories for making sodium at Niagara Falls in the United States (Niagara Electrochemical Co.) and at Rheinfelden on the German–Swiss border. Their only rival was the Cassel Gold Extracting Co. which used G. T. Beilby's method of preparing cyanides. In 1900 this company agreed to buy all its sodium from the Aluminium Co., so that within about ten years the entire cyanides business had been not only

[1] Baldwin, R. T. History of the chlorine industry. *J. chem. Educ.* **4** (1927), 315–16.

[2] Hardie, D. W. F. *Electrolytic manufacture of chemicals from salt* (1959), pp. 45, 63. The Castner process is now obsolete, having been replaced between the wars by the Downs process in which a fused mixture of salt and calcium chloride is electrolysed.

[3] The company had been set up in 1888 to make aluminium by Deville's process, and used for this purpose Castner's process for making sodium from sodium carbonate and metallic coke patented in 1886. In that year the Hall–Héroult method of electrolysing alumina in fused cryolite made the Deville method obsolete. The Aluminium Co. survived as a producer of sodium.

technologically transformed, but also so organized that it was tightly controlled by a small Anglo-German group.[1]

One totally new material requires mention here: calcium carbide was first prepared by the German chemist Wöhler in 1862. Thirty years later commercial processes for making it were developed simultaneously in France by H. Moissan and in the United States by T. L. Willson. They devised the arc furnace, which found its principal use in the manufacture of carbide, and was also employed for the production of abrasives and refractories. Later, as we shall see, it played an essential role in one of the nitrogen fixation processes. Carbide was the source of acetylene gas, which for a short time enjoyed a great vogue for street and railway lighting, mainly in Central Europe.[2]

The essential prerequisite for the manufacture of all these chemicals was cheap electricity. For industrial purposes the accumulator was useless, and progress was delayed until the dynamo was perfected in the 1870s. The next problem was the source of the electrical energy: initially steam engines driving a dynamo were the rule, but by the turn of the century water power was generally used in the larger electrochemical works. (Germany was the principal exception, because huge lignite deposits provided a very cheap source of thermal energy.) Electrolysis is based on direct current, which at that time necessitated short transmission lines. These electro-technical factors determined the location of the new plants.

In its first years applied electrochemistry was closely connected with the pioneers of electrical engineering[3] and with electricity supply companies, because the steady demand throughout the year of a chemical works had (and still has) obvious attractions to electricity undertakings looking for base loads. For example, the German Allgemeine Elektricitätsgesellschaft (AEG) formed Elektrochemische Werke GmbH in 1893. In France the Soc. des Forces Motrices & Usines de l'Arve took over the business of Corbin & Cie, which made potassium chlorate and 'Cheddite' explosives. Later this electricity supply undertaking became connected with aluminium, calcium carbide, and ferro-alloys.[4] Similarly cyanamide manufacture was, in some countries, closely associated with electrical engineering firms.

The theoretical principles on which electrochemistry was based had been formulated long before cheap power was available. Once the electrical engineering problems had been solved, the industrial development of electrochemistry proceeded simultaneously in several countries. It is difficult to assign to any one person, company, or nation absolute priority. Among the

[1] Anon. *Fifty years of progress: the story of the Castner–Kellner Alkali Co.* (1947), p. 20; *Chem. Ind., Düsseld.* **46** (1923), no. 7, 101–2.

[2] Rose, F. *Report on the rise, progress and present condition of the carbide and acetylene industries in Germany*, HMSO (1900), Cd 353–5.

[3] Among them Le Sueur, Acheson, Gibbs, Downs, Townsend, Oettel, Kiliani, and Rathenau. All these men made major contributions to the design of cells or of electrodes.

[4] It was taken over by the Alais & Camargue company in 1916, and is now part of the Pechiney concern. Gignoux, C. J. *Histoire d'une entreprise française* (1955), pp. 129–30.

pioneering firms, the Griesheim company, after 1898 Chemische Fabrik Griesheim-Elektron, deserves special mention. This German business had begun by making soda in the 1850s. Threatened by the competition of Solvay's ammonia soda, it combined in 1884 with other firms making alkali by Leblanc's process to study electrolysis. For three years they experimented with Breuer's diaphragm cell. The Schuckert company designed a satisfactory dynamo, and they were able to open an electrolytic plant at Griesheim, a suburb of Frankfurt, in 1890. It was decided to treat potassium chloride because caustic potash was the more valuable product. The initial capacity of the little works was 1½ tons a day of caustic potash, and sufficient chlorine to make about 2½ tons of bleaching powder daily.[1]

The study group used diaphragm cells made of specially prepared porous cement; such cells were of simple construction, cheap, gave little trouble, and produced chlorine of good quality. The chief disadvantage was their poor electrical efficiency.[2] The great success of the Griesheim–Breuer cell system was largely due to very competent electrical engineering work and vigorous commercial promotion. In 1892 the study group became the Chemische Fabrik Elektron, bought a site at Bitterfeld and signed a long-term contract with the lignite mine Ilse. The electrolytic plant was opened two years later. It was built to a standard design comprising power plant, cell block, vacuum pans to concentrate the dilute caustic potash to a strength of 50°Bé., and lead chambers for making bleaching powder. The capacity of the unit was 10 tons a day of caustic potash (100 per cent KOH) and about 7½ tons of chlorine; later it was increased to 14 tons a day of caustic potash.[3] The simplicity of the process and of the design led several other firms to apply for a licence. Among the first were Badische (1895) and Consolidirte Alkaliwerke Westeregeln (1896),[4] Lamotte in France (1896), and a Russian company (1898). By 1900 four out of eight German electrolytic plants and two out of four French plants were said to be using the Griesheim diaphragm cell.[5]

In 1898 Griesheim–Elektron leased from the AEG the Elektrochemische Werke, which had plants at Bitterfeld and at Rheinfelden, and thereby became the largest manufacturer of electrolytic alkali in Germany. The deal also included pilot plants for the production of sodium metal and magnesium, and carbide furnaces.[6]

The Griesheim cell was the first commercially successful diaphragm cell. It was followed by about twenty other designs. For example E. A. Le Sueur, who began by designing a mercury cell, developed a diaphragm type and built

[1] Pistor, G. *Hundert Jahre Griesheim* (1958), pp. 30–5.

[2] Hale, A. J. *The applications of electrolysis in chemical industry* (1918), pp. 92–3.

[3] Pistor, op. cit., pp. 38–9.

[4] This was the first instance of a potash mine going into the chemical business. The initia capacity at Westeregeln was 11 tons a day of caustic potash (100 per cent KOH).

Anon. *Fünfzig Jahre Consolidirte Alkaliwerke Westeregeln* (1931), pp. 202–3.

[5] *Paris Exhibition* (1), 245.

[6] Pistor, op. cit., pp. 46–8. Some years later Griesheim became an important producer of magnesium and calcium carbide.

a small works at Rumford, in the state of Maine, in 1893. It was licensed by the Electro-Chemical Co. to several American companies, notably paper-making firms, which needed a cheap supply of chlorine.[1] At about the same time Outhenin and Chalandre in France devised a variant of the Griesheim cell, and the inventors assigned the licensing rights to Société La Volta Lyonnaise, which installed the process in several French and Italian works. In England J. Hargreaves and T. Bird patented yet another diaphragm type and in 1899 formed the Electrolytic Alkali Co. to work it at Middlewich in Cheshire. The works was opened in 1901, and seems to have been the only one to treat the caustic-brine solution with steam and carbon dioxide, thus producing soda ash or sodium bicarbonate instead of caustic soda.[2]

These early diaphragm cells have now been replaced by designs invented in the 1900s. Two American inventors, C. P. Townsend and E. A. Sperry (of gyroscope fame), sold their design to the development company founded by a successful civil engineer, E. H. Hooker. The Townsend or Hooker cell was first used in 1906 in the works of the Hooker Electrochemical Co. at Niagara Falls. Other widely employed diaphragm cells were developed by Billiter, an engineer of the Aschersleben potash mining company (1907), and by A. E. Gibbs working for the Pennsalt company in America (1908).

Cell systems in which mercury is the cathode were invented in the 1890s. The mercury cell has some important advantages over the diaphragm cell, notably good current efficiency, the production of caustic liquor of a higher purity and concentration, and the facility with which large units can be assembled. The chief drawback is the high cost of mercury. The first practical mercury cell was developed by Castner in England and quickly displaced the Griesheim type of cell in that country; it was also used in Germany. Castner's patent was dated 1892 and at about the same time C. Kellner in Austria and Vautin in France also built mercury-type cells.[3] Kellner's cells were apparently never worked on a large scale, though some of the design features are still in general use. Castner assigned his process to the Aluminium Co., while Kellner made over his rights outside Austria to the Solvay company. In 1895 the Aluminium Co. and Solvay agreed to cross-license the Castner and Kellner patents. As a result the mercury cell system was developed more rationally and its commercial significance grew more quickly than that of the various diaphragm types. The Solvay company also secured an important position in electrolytic alkali production and thereby reinforced its role in the soda trade generally. The Castner–Kellner Alkali Co. was founded in 1895, and two years later opened the first big electrolytic alkali plant in Britain at Weston Point near Runcorn.[4] Solvay built electrolytic works at Osternienburg (near Bernburg) in Germany, Jemeppe (Belgium), and Lissichansk (Russia) before

[1] Le Sueur, E. A. Inception and development of electrolytic alkali manufacture. *Trans. electrochem. Soc.* **63** (1933), 188.

[2] Hale, op. cit., pp. 94–5.

[3] Hardie, op. cit., p. 62.

[4] *Fifty years of progress*, op. cit., pp. 5, 26, 29.

1900. All these factories used thermal electricity and the generator at Runcorn, rated at 4000 hp was considered very large by contemporaries.[1] Castner–Kellner licensed the mercury cell to the Mathieson Alkali Co., which opened the first United States plant of this type at Saltville (W. Va.) in 1896, and the following year built a large works at Niagara Falls.[2]

Over a period of years the Solvay company modified Castner's design in some important respects: Castner had devised a rocking motion, which the Solvay engineers replaced by a sloping cell, so that the mercury flowed by gravity and circulation was by means of an archimedean screw. This is now the usual type of mercury cell.

There were yet further variants of electrolytic cells. Oettel devised the so-called bell type cell, which was used by the Oesterreichischer Verein at Aussig, because they were unable to get a licence either from Griesheim or from Kellner.[3] H. H. Dow in the United States designed an extremely simple cell which he used to electrolyse brine with a high bromine, calcium, and magnesium chloride content. Dow's prime object was to recover bromine, and it was several years before he began to electrolyse debrominated brine. In this way he obtained chlorine and started manufacturing bleaching powder, but he made no attempt to recover and purify caustic soda, which was run to waste with the calcium and magnesium impurities. Dow's bromine and chlorine business did not follow the usual pattern of electrolytic works, and it was not until 1908–14 that he and L. E. Ward developed a non-mercury type of cell, which like all the others yielded chlorine as well as caustic soda.[4]

The manufacture of chlorates was based on electrolytic processes patented in England and France in 1887, and used in the latter country from 1889 onwards. The first chlorate cells were of the diaphragm type, but in the 1890s Corbin introduced platinum bipolar electrodes. Potassium chromate was added to accelerate the formation of the chlorate.[5] Corbin's works at Chedde in the French Alps were the largest in France, and in 1900 were reported to be making 4500 tons of chlorates a year, which was considered to be a very large quantity.[6] The only other chlorates process of commercial significance was Hurter's, the chief chemist of the United Alkali Co. His was a diaphragm type cell, which was successfully used by the company at their American works at Bay City (Mich.), and by Griesheim at Rheinfelden.

In the 1890s Kellner constructed the first efficient equipment to make sodium hypochlorite electrolytically. His cells were of the vertical bipolar type, and the chemical was used by paper manufacturers for bleaching purposes. An improved design, known as the Haas–Oettel cell, was patented in 1896, and was licensed by the Oesterreichischer Verein. It had graphite

[1] *Paris Exhibition* (1), 245.
[2] Haynes, W. *American chemical industry* (1954), vol. 1, pp. 277–8.
[3] Hale, op. cit., pp. 110–11.
[4] I am indebted to Dr. D. W. F. Hardie for this information.
[5] Hale, op. cit., p. 125.
[6] *Paris Exhibition* (1), 245, 250.

electrodes, which were cheaper than Kellner's platinum or iridium wire electrodes, and produced hypochlorite continuously, which was an advantage compared with the Kellner process, where recirculation was necessary until the desired concentration had been reached.[1] Nowadays sodium hypochlorite is prepared from the tail gases of chlorine liquefaction plants, and the special cell systems have been abandoned.

This vigorous inventive activity, followed immediately by industrial development, had important consequences. We have repeatedly mentioned the establishment of electrochemical works in Germany, France, Britain, Norway, and the United States. Many were built in other countries in the 1890s and 1900s. Electrochemistry in those two decades had a profound effect on the location of chemical manufacture. The older centres had often been located near raw materials or important markets (Widnes and St. Helens illustrate the former, Basle the latter), but these new enterprises were close to the power supply. This resulted in the building of large plants in central Germany, the French Alps, Norway, and near waterfalls particularly those of the Rhine and the Niagara. In central Germany the works depended on lignite, in England on coal, but elsewhere solid fuel did not become important for electrochemical operations in this or later periods.[2] It was estimated in 1908 that the cost of one horsepower–year was about £6. 9*s*. 0*d*. if coal were used, but only £2. 0*s*. 0*d*. to £3. 10*s*. 0*d*. if water power was the source of energy; for Norway the figure was only £1.[3] For the manufacture of carbide and aluminium the power requirements per ton of product were much higher than for the electrolysis of brine to chlorine, and with the exception of Germany, these operations were based exclusively on hydroelectricity. At first some aluminium works also had carbide furnaces, so that if aluminium production stopped the power units could supply the carbide furnaces. These arrangements were abandoned before the First World War, when Norwegian producers drastically cut the price of carbide.

In the United States the Niagara Falls area quickly became the principal centre of electrochemical operations. J. F. Schoellkopf, an enterprising local business man, owned extensive water rights on the American bank, and his Niagara Falls Power Co. developed an industrial estate there.[4] The first commercial users of hydroelectricity were the Pittsburgh Reduction Co., which made aluminium, Acheson's Carborundum Co., which then produced graphite and electrodes, and a factory which made metallic sodium and cyanides. In the late 1890s the pace quickened and between 1897 and 1900 two carbide works were opened, the Oldbury Electrochemical Co. (the United

[1] *Paris Exhibition* (1), 245, 250; Hale, op. cit., pp. 119–20.

[2] Fuel oil and natural gas began to be used as fuel for power stations in the 1910s and 1920s.

[3] Hale, op. cit., p. 31. The Norwegian figure corresponds to 0·04*d*./kWh which seems very low even for those days.

J. N. Pring, *Some electro-chemical centres* (1908), p. 13, mentions that the Ontario Power Co. charged £3. 10*s*.–£4 per hp–year to large industrial users.

[4] Haynes, op. cit., vol. 3, pp. 17–18. Schoellkopf's two sons owned a dyestuffs factory.

States subsidiary of Albright & Wilson) began the manufacture of phosphorus and potassium chlorate, and the Mathieson Alkali Co. built the first large electrolytic alkali plant in North America. In the next few years other firms settled in the district and by 1906/7, 50 000 kW were used for electrochemical operations.[1]

Outside the United States, the development of the Norwegian water-power resources was to play an important role in the early years of the century. But at first there were some delays because of political obstacles, notably the difficulties encountered by foreign companies wishing to acquire water rights and the uncertainty created by the separatist movement until Norway achieved independence in 1905. The first big electrolytic works was that of the Kellner–Partington Paper Pulp Co. which started making hypochlorite at Sarpsborg in 1896.[2] Several carbide works were built before the turn of the century, and in the aggregate they were large users of power. They comprised Hafslund Carbidfabriken (1898), Meraker Smeltverk (1899), and Notodden Calcium-Carbidfabrik (1900). They were big by contemporary standards, but remained financially unsatisfactory because of the boom-or-bust atmosphere of the early carbide industry. This did not deter later entrepreneurs, for the water-power conditions were exceptionally favourable in Norway: powerful streams flowing evenly, falls or rapids that could be easily dammed, and very low building costs. These characteristics positively invited new construction projects: between 1905 and 1915 several more carbide factories were built, and during this period the first units of the Birkeland–Eyde arc process were installed. By about 1910 the Norwegian factories had a capacity for making 80 000 tons a year of carbide, a figure corresponding to one-third of the European total, but owing to lack of demand, production was only about 55 000 tons.[3]

Electrochemistry also made rapid strides in France and Germany. In the former country the first works were situated in Savoy. They remained small in relation to those of Norway and the United States because the water conditions of the Alpine streams were less favourable and the power companies did not have the resources to finance large plants. The manufacture of chlorates remained the chief electrolytic activity for many years. The production of aluminium and carbide was begun by Soc. Electro-Metallurgique Française at Froges and La Praz, and later at La Saussaz and l'Argentière. But, as elsewhere, carbide production was frequently unprofitable, and the works were converted to the manufacture of ferro-chrome.[4] As for Germany, we have already noted the use of lignite around Bitterfeld; this fuel was also employed at Knapsack near Cologne, where large carbide works were built in the 1900s and drew lignite from nearby deposits. Water power was used

[1] Threlfall, R. E. *The story of 100 years of phosphorus making* (1951), pp. 261–2, 289; Evolution de l'industrie du carbure de calcium; *J. Four élect.* **63** (1959), 259 f. Pring, op. cit., pp. 13–14.

[2] *Paris Exhibition* (1), 252–3; Norges Industri (publ.) *Norwegian hydro-electric industry* (1932), p. 49.

[3] Weber, A. (Ed.) *Über den Standort der Industrien* (1914), p. 92.

[4] Gignoux, op. cit.

on the upper Rhine around Rheinfelden, where AEG built several run-of-the-river power stations between 1895 and 1912. Like Schoellkopf at Niagara, AEG looked for industrial customers with a high load factor. It held shares in the aluminium works at Rheinfelden, and owned two small companies at Rhina, which made ferro-alloys and nitric acid respectively. AEG also supplied power to Griesheim–Elektron's electrolytic alkali plant, the sodium metal works of the DEGUSSA–Aluminium Co. group, and to two factories making abrasives. The rapid development of electrochemical activities in this small district is illustrated by the growth in the number of chemical workers: there were about 520 men in 1899 and more than 1000 on the eve of the Great War.[1]

The fixation of nitrogen

England and all civilised nations stand in deadly peril of not having enough to eat. As mouths multiply, food resources dwindle. Land is a limited quantity, and the land that will grow wheat is absolutely dependent on difficult and capricious phenomena. I am constrained to show that our wheat-producing soil is totally unequal to the strain put upon it . . . it is the chemist who must come to the rescue of the threatened communities. It is through the laboratory that starvation may ultimately be turned to plenty.

The distinguished audience that heard Sir William Crookes make the presidential address to the British Association in 1898 must have been alarmed. Crookes certainly intended to sound a warning, and felt that an appeal to a wider audience than the scientists meeting in Bristol was essential. At the end of a speech that dealt at length with spiritualism and briefly with Freud, he returned to the nitrogen problem: 'There is a gleam of light amid this darkness and despondency'—'the fixation of atmospheric nitrogen'. This, he added presciently, would be 'one of the great discoveries awaiting the ingenuity of chemists . . .'[2]

At that time there were only two major sources of fixed nitrogen: sodium nitrate and sulphate of ammonia. Production of both was rising, but the supply in terms of nitrogen was still small. In addition there was guano, once important, but now declining, sewage, and dung. These last two were of some use despite their low nitrogen content.

Meanwhile demand was rising, and in particular the use of nitrogenous fertilizers was increasing. Farmers in countries with intensive systems of agriculture were manuring with phosphatic and potassic fertilizers, and higher yields, as they knew from experience and Liebig's teaching, required the application of nitrogen. To some extent their needs were met by raising the output of sulphate of ammonia from gas works and coke ovens. But there was an upper limit because coal, on average, contains 1–1½ per cent of fixed nitrogen (not all of which is recoverable in the form of sulphate of ammonia), so that this source of nitrogen depended on the tonnage of coal carbonized.

[1] Horster. *Die Volkswirtschaftliche Bedeutung der chemischen Industrie am Oberrhein* (1922).
[2] Report of the 68th meeting of the British Association, 1899, pp. 4, 18.

The other principal source was Chilean sodium nitrate. At the turn of the century there were some who feared that the entire supply might be cornered by unscrupulous syndicates, or, worse still, cut off by war. Military needs were indeed a justified cause of concern, for nitrogen was becoming the key material for the explosives industry. Nitro-glycerine, dynamite, nitro-toluol, and related compounds required nitric acid for their manufacture, and this, in turn, was obtained from sodium nitrate. Nitric acid could thus become a critical war material. Even in the peaceful years of the late Victorian era a substantial demand for nitric acid was being created by the development of the dyestuffs industry and of explosives for civil engineering purposes. Finally there were some special uses for ammonia, notably as a refrigerant.

In the 1890s two other factors led to the focusing of attention on the problem of nitrogen fixation. One was physical chemistry. This new science concerned itself, among other things, with gas reactions, which led scientists to promising fields for further study. For example there was much speculative enquiry into the 'fixing' of nitrogen with oxygen, so forming oxides of nitrogen and leading to nitric acid, or alternatively nitrogen could be 'fixed' with hydrogen, so leading to ammonia. The other important factor was the progress made in applied mechanical and electrical engineering. Linde invented the air-liquefaction machine in 1895, and so made it possible to obtain nitrogen and oxygen from the atmosphere. Other people followed the lead given by Cavendish in the 1780s: he had shown that the combination of nitrogen and oxygen could be effected by a spark, and a century later sufficiently powerful electrical equipment became available to repeat his experiments on a larger scale. Another interesting development was that by-product hydrogen from the new electrolytic plants was becoming available in large quantities. Finally there was cyanamide, which became commercially practicable once the electric furnace provided the starting material, carbide. This summary points to the following conclusion: the state of chemistry and of engineering had advanced sufficiently by the time Crookes made his famous speech to warrant the industrial development of nitrogen fixation. We must now turn to the three principal fixation methods that were evolved within about ten years of his forecast.

The chemistry of the arc process is simple: at the very high temperature of 2000–3000°C created by the electric arc, oxygen and nitrogen combine to form nitric oxide: $N_2 + O_2 \rightleftarrows 2NO$. The reaction is reversible, and it is essential to cool the gas rapidly to 1000°C, at which temperature the nitric oxide ceases to decompose. When cooled further the gas takes up additional atmospheric oxygen to form nitrogen dioxide, which is passed through an absorption tower, where, in contact with water, it forms dilute nitric acid. The weak acid is concentrated and sold, or is reacted with crushed limestone to form nitrate of lime. This is dried and packed for sale as a fertilizer; and when first marketed contained about 13 per cent N.[1]

[1] Partington, J. R. and Parker, L. H. *The nitrogen industry* (1922), pp. 233–9.

In the wake of Cavendish other scientists had been attracted by this method of nitrogen fixation. In 1859 a Mrs. Lefevbre obtained an English patent for making nitric acid in this way. Crookes himself investigated the arc process in 1892, and a few years later Lord Rayleigh studied it. He and Sir William Ramsay were led in this manner to the discovery of the inert gases. These laboratory studies demonstrated the feasibility of nitrogen fixation, but there were formidable technical obstacles. The conversion efficiency is very low, since only about 3 per cent of the electrical energy is converted into chemical energy to yield nitrogen oxide. Some of the remaining energy can be recovered at the cooling stage and used to raise steam for the concentration of nitric acid and the drying of the nitrate, but by far the greater portion of the power input is lost. Consequently it was essential to have very cheap electricity. Other problems concerned the design of the electrical equipment so that a stable arc was generated, and of the absorbers, which had to be designed for handling a large volume of gases containing only a small proportion of oxides of nitrogen. McDougall and Howell experimented at Manchester in 1899 with a practical process, and Bradley and Lovejoy between 1902 and 1904 at Jersey City and Niagara Falls prepared nitric acid by the arc process. But they only obtained dilute and impure acid, and had to abandon operations.[1] In Germany, BASF began studying the problem of nitrogen fixation from 1897 onwards and employed for this purpose O. Schönherr, who was a chemist, and J. Hessberger, an engineer. They made slow progress, and it was left to two Norwegians, Kristian Birkeland, professor of physics at Oslo, and Sam Eyde, an engineer who was interested in hydro-power projects, to invent and develop the first commercially successful process. They became partners in 1903, and the following year produced small quantities of nitric acid and nitrate of lime in an experimental plant outside Oslo. Birkeland designed the arc furnace in which a powerful magnet imparted to the alternating current the shape of a disc. This circular arc was stable and allowed a large volume of air to be oxidized. The yield was initially much lower than expected: the partners had estimated the energy input at 14·6 kWh per kilo nitric acid (100 per cent HNO_3), but in practice it was 16½–18 kWh, and even 20 kWh.[2] Nevertheless the building of a small works was begun at Notodden in 1904. It was completed the next year, and was reported to be making 1600 tons of nitrate of lime in 1906. A larger works, also at Notodden, was finished in December 1907, but did not start until September 1908.[3]

The work of the German and Norwegian groups naturally attracted attention in scientific quarters. Walther Nernst and Fritz Haber devoted much time to the determination of the thermodynamic equilibrium of oxides of

[1] Waeser, B. *Die Luftstickstoffindustrie* (1st edn. 1922), pp. 445 f.

[2] *La Société Norvégienne de l'Azote 1905–55* (1955), pp. 32–47. (The book is hereinafter referred to as *Norsk Hydro history*.)

[3] Kubierschky, K. Die künstlichen Düngemittel und ihre Bedeutung. *Z. angew. Chem.* **26** (1913), pt. 1, p. 727.

nitrogen. Nernst's studies took place in 1904/5, Haber's from 1907 onwards. While no practical results followed for the arc process, their activities were to have important consequences, for they led to the study of the ammonia equilibrium. And it was the oxides of nitrogen that first brought Haber and Badische together, because the company was prepared to support his interest in nitrogen generally, though reluctant at that time to assist him specifically with his researches into the formation of ammonia from the elements.[1] Badische had financed Schönherr all the time and he patented a process in 1905. It differed from Birkeland's in having a long arc around which the air circulated tangentially, thereby allegedly producing higher yields of oxides of nitrogen. In other respects Schönherr's method differed little from that of the Norwegians. Badische put up a pilot plant at Ludwigshafen where experimental runs were made between 1905 and 1907, but contrary to all laboratory expectations the yields were no better than those obtained by the Norwegians at Notodden. Undeterred, Badische went ahead and built a small plant at Christianssand in Norway in 1907.

Since the Birkeland–Eyde and the Schönherr processes were so similar, a union between the two groups seemed natural. Indeed as early as 1903, O. N. Witt, who knew Birkeland as well as H. v. Brunck, the leading man at Badische, arranged a meeting between them. It was inconclusive, but three years later the negotiations were resumed and led to the formation of a jointly owned nitrates company, called Norsk Hydro, and a power company. Even at this early stage Badische, owing to its great financial strength, was able to prevail on the Norwegians to use the Schönherr furnace.[2] It is more than likely that neither Birkeland nor Eyde were aware at the time that the apparatus had not come up to expectations at Ludwigshafen. The arrangement with Badische left the Norwegians with their nitric acid and nitrate of lime processes. Soon there was friction and ill-feeling. In 1910 it was decided to resolve the problems by comparing both arc systems under stringent test conditions. The trial ended inconclusively and engendered much acrimony. By now the Germans were losing interest in the arc process, because Haber's showed much more promise. Relations between the Norwegians and the Germans worsened and in 1911 Badische sold its shareholding in Norsk Hydro. By pulling out Badische practically wrecked Norsk Hydro, but bequeathed to it the battery of Schönherr arc furnaces.[3]

There were several variants of the Birkeland and Schönherr processes. None were commercially significant, but one of them, invented by I. Moscicki in 1905, is of passing interest. He later became the first president of Poland and in this post contributed significantly to the development of nitrogen fixation in the republic.

[1] Haber, F. *Fünf Vorträge* (1924), pp. 9–11.
[2] Glaser, C. H. v. Brunck. *Ber. dt. chem. Ges.* **46** (1913), pt. 1, 380–1; Bernthsen, A. Über Luftsalpetersäure. *Z. angew. Chem.* **22** (1909), no. 24, 1175.
[3] *Norsk Hydro history*, pp. 49–50.

In practice the arc process was worked on a large scale only in Norway where hydro-power could be obtained so cheaply as to make this method economical. Norsk Hydro, after some early vicissitudes, prospered during the war. But eventually it abandoned the process in the 1920s, owing to the rising cost of electricity, and the greater economy of the Haber–Bosch process.

The cyanamide process, the second of the nitrogen fixation processes, starts from calcium carbide. The material is finely ground and charged into a furnace where, at about 700–1100°C, it combines with nitrogen (obtained by the liquefaction of air) to form calcium cyanamide. This compound is suitable as a fertilizer and contains about 20 per cent N. The development of cyanamide manufacture proceeded along unusual lines, and all the initial work was done in Germany. In 1895 A. Frank, who had taken an active part in founding the German potash industry, started experimenting with the then newly discovered calcium carbide to make cyanides. As we have seen, the demand for cyanides was growing, and Castner and his DEGUSSA–Aluminium Co. associates held the key patents for metallic sodium and sodium cyanide. Carbides were known to provide an alternative route to cyanides. The first trials by Frank and his collaborator N. Caro were made at Hamburg in the works of Dynamit AG. They were unsuccessful.[1] In 1897 they were joined by F. Rothe, who had been working independently on cyanamide.[2] The following year the group discovered that, upon nitrogenation, finely ground barium carbide yielded barium cyanamide, while the same material in coarse lumps when treated in the identical fashion became barium cyanide. With hindsight we may think that the former rather than the latter material would repay further study. Such, however, was not the case at first. DEGUSSA, alerted to the interesting potential of a rival process, combined with Frank, Caro, and Rothe to form the Cyanidgesellschaft in 1899; they were shortly joined by the electrical engineering firm of Siemens & Halske. The object of the company was to make sodium cyanide from barium cyanide, and later from calcium cyanide. Some work was also done on cyanamide: Caro obtained a patent in 1900 for treating cyanamide with steam to produce ammonia. This was the first technical synthesis of the compound. Frank and his son meanwhile demonstrated that cyanamide was a useful fertilizer.[3]

Caro's and Frank's activities were increasingly giving the Cyanidgesellschaft a new direction, and the changing policy was underlined when DEGUSSA left the company in mid-1901. The others now decided to try out the Frank–Caro–Rothe cyanamide process on a large scale. They were pressed for time, for in that year Polzenius made and patented the important discovery that by adding calcium chloride to calcium carbide, the temperature at which the carbide took up nitrogen was cut from 1100°C to 700–800°C.

[1] Grossmann, H. A. Frank zu seinem 80. Geburtstag. *Chem. Ind., Düsseld.* **37** (1914), no. 2, 34.

[2] Rothe later interested himself in the arc process and during the war the AEG and Griesheim built two works to his design: one blew up, the other produced at the rate of a few hundred tons a year. Waeser, op. cit. (2nd edn. 1932), p. 35.

[3] Waeser, op. cit. (1st edn.), pp. 17–18.

Polzenius thereby corrected one of the main defects of the Frank–Caro process, which, owing to its high temperature, caused much trouble to the furnace linings. He secured the backing of the Metallgesellschaft, and a collision between the two groups became inevitable; after many years of quarrelling a compromise was finally reached in 1908, and a cross-licensing arrangement agreed. Meanwhile Metallgesellschaft and the very enterprising potash mining company Consolidirte Alkaliwerke Westeregeln formed the Gesellschaft für Stickstoffdünger in 1904 to make cyanamide by the Polzenius process. The following year the first commercial cyanamide plant was opened at Westeregeln in central Germany.[1] It made a loss. Undeterred the company built a second works at Knapsack near Cologne. As at Westeregeln, lignite was used as fuel for the power station. Knapsack started producing towards the end of 1910, whereupon Westeregeln was closed. By that time the affairs of the Gesellschaft für Stickstoffdünger were in a sorry state: no profit had been made since its foundation and the Metallgesellschaft decided on drastic action. Its subsidiary, the Metallbank, acquired the shares held by Consolidirte Alkaliwerke and reconstructed the business.[2] The trouble at Westeregeln and later at Knapsack was ascribed to marketing difficulties, and with some justification, because the German farmers considered cyanamide to be inferior to sulphate of ammonia. Another problem was the high cost of electricity in the operation.

We must now return to the Cyanidgesellschaft and see what was happening there. The company obtained the backing of the Deutsche Bank, and in order not to aggravate the threatened litigation with Polzenius and his supporters, built a works at Piano d'Orta in Italy. The plant of Stà Italiana per la Fabbricazione di Prodotti Azotati was opened towards the end of 1905 with a capacity of 4000 tons a year of cyanamide. As with later Frank–Caro installations hydro-power was used and this resulted in considerable savings in electricity costs. Siemens & Halske remained partners so that the electrical engineering side was in good hands. In 1908 (after the settlement with Polzenius) the Deutsche Bank stepped in again and helped organize the Bayerische Stickstoffwerke, which began to build a very large factory at Trostberg in Upper Bavaria.

The Cyanidgesellschaft promoted the use of the Frank–Caro process by means of a liberal licensing policy, so that by 1912 eight other firms were using the process and all relied on hydro-power. In that year there were altogether ten cyanamide producers, but only one, the Gesellschaft für Stickstoffdünger, employed Polzenius' method and used solid fuel. The ten companies had an aggregate capacity of 152 000 tons a year of cyanamide.[3] The largest works were at Knapsack and Trostberg in Germany, at Odda in

[1] *Fünfzig Jahre Consolidirte Alkaliwerke Westeregeln* (1931), p. 176; Niebelschütz, W. v. *Knapsack* (1957), p. 18.

[2] Waeser, op. cit., pp. 18 f.; Däbritz, W. *Fünfzig Jahre Metallgesellschaft* (1931), pp. 177–8.

[3] Survey reported in *Z. angew. Chem.* **26** (1913), pt. 3, p. 297. The figure, taken from *Engrais*, is on the high side.

Norway, at Alby in Sweden, near Terni in Italy, and on the Canadian side of Niagara Falls. The last-mentioned is of special interest, since its inception illustrated the close links between hydro-power development and fertilizer manufacture. F. S. Washburn proposed to make nitrate of lime by the arc process at Muscle Shoals (Alabama). The Federal Government was to dam the river and put up a power station. The site was well chosen and much later a nitrogen fixation plant was built there, but for the present the Government refused to build the dam. Upon closer examination the promoters decided that the Birkeland–Eyde process would not be economic under American conditions. So in 1907 they obtained the sole North American rights to the Frank–Caro process, established the American Cyanamid Co. to work it and signed a bulk contract with the Ontario Power Co. for the supply of electricity at low rates. The works, opened in 1909, had an initial capacity of about 4500 tons a year of cyanamide, which was raised to 9000 tons in 1912.[1]

By the eve of the war, the number of cyanamide plants had grown further. In addition to those mentioned above there were works in France, Switzerland, in Dalmatia (then Austrian territory) at Sebenice, and in Japan at Osaka. Others were under construction in Austria, Sweden, Norway, and Italy. This rapid spreading of the cyanamide process is easily explained. The technology was relatively simple, and as it was a batch operation no novel engineering concepts were involved. As a result the capital investment per ton of fixed nitrogen was lower than that for the arc process. Operationally there was the further advantage that less electrical energy was required per ton of nitrogen than by the arc process. The only bulky raw materials were limestone and coal for the calcium carbide unit. Under these circumstances it is not surprising that by 1914 cyanamide was the main source of synthetic nitrogen.

The Haber process, which has displaced the other nitrogen fixation processes, was still largely an unknown factor when the Great War broke out. The synthesis of ammonia from the elements, $3H_2 + N_2 \rightleftarrows 2NH_3$, chemically so simple, represents one of the most important advances in industrial chemistry, and one that has contributed significantly to the shaping of the contemporary structure of the chemical industry. We will observe frequent examples of this contribution in the rest of the book; here we may opportunely summarize the chief reasons. The process, discovered by Fritz Haber and developed industrially by Carl Bosch, was the first application of high pressure synthesis; the technology of ammonia production, appropriately modified, was used later in the synthesis of methanol and the hydrogenation of coal to petroleum. Its influence extends to present-day techniques of oil refining and the use of cracker gases from refining operations for further synthesis. The Haber–Bosch process is capital intensive: it does not require expensive raw materials, but the equipment required for the production of

[1] Haynes, W. On the chemical frontier—the Cyanamid story, unpubl. typescript (1957/8); Warrington, C. J. S. and Nicholls, R. V. V. *A history of chemistry in Canada* (1949), pp. 185 ff.

hydrogen and nitrogen is complex and costly. It is also a process where increases in the scale of manufacture yield great savings and make possible the economic utilization of by-products. The construction of huge plants, designed to handle continuously large volumes of gases at high temperatures and pressures, imposed novel operating practices on works engineers, and contributed to the rapid development of chemical engineering, and also to the production of improved types of steel, and new designs of valves and gas compressing machinery. The confluence of the chemical and engineering sciences dated back to the 1890s, but with the ammonia synthesis the junction of the two became the characteristic feature of modern chemical manufacture. The high capital investment required large resources on the part of the companies working the process. A contemporary authority on the subject claimed that, owing to the high fixed charges, a large throughput was essential, and explained this on the grounds of '... nearly two years' sales being required to turn over the plant investment'.[1] These big works created novel marketing problems: they were designed to operate continuously, but as the bulk of the ammonia was used for fertilizers, demand for which is seasonal, it was necessary to provide large storage facilities and finance the burden of big stocks. Here again only a large company had the necessary resources, and it is hardly surprising that the fear of over-production led towards market control arrangements, not only within a country, but even internationally. Finally, the Haber–Bosch process, being less dependent than either the arc or the cyanamide process on cheap power, made it possible to build nitrogen fixation works in countries or particular districts that had hitherto appeared unsuitable. The process thereby provided nations with the means to attain self-sufficiency in fertilizer and industrial nitrogen, and so answered the threatening question posed by Crookes. After the Kaiser's war factories for the synthesis of ammonia multiplied, and the replication of these plants in Europe, North America, and Japan became perhaps the most important aspect of industrial chemistry in the inter-war years. The details will be described in later chapters, and some attempt made to assess the implications; here it is appropriate to draw an analogy: the impact of the ammonia synthesis may be compared in significance with the adoption of petroleum and natural gas as materials for chemical manufacture in Europe in the 1950s and 1960s.

Despite its apparent simplicity the synthesis of ammonia proved elusive. Priestley first described the constitution of ammonia in 1774, and for the next 100 years it was prepared on a small scale from decomposed animal matter and later from the residual liquors of gas works. By the end of the nineteenth century the former had been virtually abandoned, while the recovery from gas liquors was carried out in several countries on a substantial scale. All

[1] Crane, J. E. Development of the synthetic ammonia industry in the U.S. *J. ind. Engng Chem.* **22** (1930), no. 7, p. 795. Crane was a vice-president of the Du Pont company. Fundamental changes in the preparation of the hydrogen–nitrogen synthesis gas from oil instead of coke have cut the investment per ton per year of nitrogen, but the necessity of operating near capacity remains.

attempts to prepare ammonia from the elements failed. Alternatively there was the indirect approach, based on the observation that certain metals absorb nitrogen when heated and form nitrides, from which ammonia can be obtained by treatment with steam. In 1869 Charles Tellier patented a method for preparing ammonia from iron nitrides, and some years later, in 1880, he formed the Société l'Azote to license his process;[1] this was surely the first business specifically to concern itself with the artificial fixation of nitrogen. His nitride process was unsuccessful, but it did not deter Badische from investigating the subject thoroughly. Otto Serpek, in the years before the war, also studied the preparation of nitrides, in his case those of aluminium, and his process remains of interest today because he attempted to combine the preparation of alumina with that of ammonia, using bauxite, coal, and nitrogen from the air as raw materials. From 1909 onwards Serpek's process was worked on a pilot-plant scale by the Société Générale des Nitrures in Alsace. Later a Norwegian business was formed, but its efforts to work the process failed and it turned to the production of aluminium; only the title, Det Norske Nitrid A/S, now reminds us of the company's first objective.[2]

W. Ostwald was the first to consider ammonia from the point of view of physical chemistry. This was in the last decade of the nineteenth century, and it is interesting to note that C. Bosch, then a young chemist newly recruited by Badische, was given as his first task the re-examination of Ostwald's work. Haber, who was teaching physical chemistry at Karlsruhe, came to the ammonia problem in 1903, when at the request of the brothers Margulies, who wanted to make calcium and other nitrides, he investigated the thermodynamic equilibrium of ammonia. The work of Haber and his collaborator van Oordt in the years 1904–5 was especially significant because they were able to demonstrate that the build-up of ammonia from the elements was possible and required a catalyst. They used iron and other metal catalysts and temperatures of about 1000°C, but no pressure; the yield was very small. Haber also devised a circulating system for the ammonia. This work, which was essentially of a theoretical nature, led Haber to the conclusion that synthesis did not provide a commercial route to ammonia. He thought that manufacture on the industrial scale at such a temperature was impracticable, the more so as the catalysts he used ceased to give worthwhile results above 300°C. He felt that the use of pressure under these conditions was out of the question, and so he abandoned the work on nitrides and turned to the arc process.[3]

W. Nernst became interested in the ammonia equilibrium as a result of his work on the heat theorem that bears his name. In 1906 he examined Haber's findings, disagreed with them and investigated the equilibrium conditions under pressure. At 40–75 atm and about 700°C he obtained traces of ammonia, the yield being lower even than Haber's. Nernst described his work

[1] Mittasch, A. *Geschichte der Ammoniaksynthese* (1951), pp. 35–8.
[2] Waeser, op. cit., pp. 329 ff. [3] Haber, op. cit., pp. 15–16; Mittasch, op. cit., pp. 64, 66.

at a meeting of the Bunsengesellschaft in Hamburg in May 1907, and challenged Haber to repeat the experiments. We may pause here and note that Nernst had thus demonstrated the possibility of synthesis under pressure. It should also be observed that the previous year he had got in touch with Griesheim–Elektron and enlisted their support in his work on the ammonia synthesis. The company assisted these investigations until the winter of 1907/8 when they were stopped because, according to expert opinion, the necessary equipment to sustain such high pressures and temperatures could not be built. This ended the matter as far as Nernst and Griesheim were concerned.[1]

Nernst's intervention confirmed Haber's earlier judgement that the synthesis was not practicable. Still, there was the disagreement about the yield, and Haber took up the challenge. The outcome of his later work, in which he was assisted by R. le Rossignol between 1907 and 1910, was the discovery of the essential features of the ammonia synthesis, and of a practical process. Theirs was a great achievement, and the main principles of their design are still in use: they proposed a pressure of around 200 atm, a temperature of 500–600°C, the circulation of the gases under pressure, continuous removal of the ammonia, and transfer of heat from the hot ammonia leaving the converter to the incoming gases.[2] Haber knew that the industrial success of his process depended on good yields, though the first catalysts that he used were disappointing in this respect. But a yield of up to 8 per cent of ammonia by volume was obtained at 175 atm and 550°C with amorphous osmium. This catalyst was used in the demonstration run of 2 July 1909, which gave the onlookers, among them A. Mittasch of Badische, the first sight of the equipment and an idea of the potentialities of the process. Later that year Haber found that uranium in the form of uranium carbide was cheaper and gave as good results as osmium.[3]

Haber had been helped by Badische since February 1908, but, as we have seen, the company was primarily interested in the arc process and also, even at this late date, in barium cyanide. The demonstration of the process at Karlsruhe in July 1909 changed all that. The company increased its contribution, and Bosch and Mittasch, who were familiar with the principles of the synthesis, were put in charge of the technical development. Between 1909 and the end of 1912 Mittasch and his assistants tried 2500 substances and carried out 6500 experiments, until they found in an activated iron compound the most suitable catalyst. This meticulous work, and the equally thorough patent policy, gave Badische an unassailable position in the domain of ammonia synthesis and a decisive lead over all its rivals.[4]

[1] Coates, J. E. The Haber Memorial Lecture. *Memorial Lectures delivered before the Chemica Society* (1951), vol. 4, pp. 136–7; Pistor, op. cit., pp. 61–2; Mittasch, op. cit., pp. 67–70.

[2] Haber, op. cit., pp. 18–20. The compressor at Karlsruhe had a limit of 200 atm, and this accident determined the maximum pressure.

[3] Mittasch, op. cit., pp. 75–7; Bernthsen, A. Die synthetische Gewinnung des Ammoniaks. *Z. angew. Chem.* **26** (1913), pt. 1, no. 1, 13. There is some disagreement about the catalyst used; I have followed the account in Mittasch.

[4] Mittasch, op. cit., pp. 93 ff.

Bosch was given the task of scaling up Haber's laboratory equipment. Experimental tests on a large rig started at Ludwigshafen in the spring of 1910. They were satisfactory, and this was probably the major factor that led to the decision by Badische to withdraw from the arc process in Norway. Bosch and his team faced two big problems, which they solved brilliantly. Their originality, which made such an outstanding contribution to chemical engineering, was shown by the answer they found to the first of the problems, the design of the converter. Haber's converter was three-quarters of a metre high and a few centimetres wide. Bosch and his engineers built one that stood 1·80 metres in 1910. The next year their tube measured 4 metres and in 1913 it was 8 metres with a width of 45 centimetres. The weight, a mere 300 kilos in 1910, grew successively to 500 kilos, 1 ton (1912), 3.5 tons (1913), and 8·5 tons in the first large scale plant at Oppau in 1913. A few years later converters weighing up to 65 tons were in use.[1] Converters of this size were unknown and required a single forging of high quality steel. Krupps were called in and supplied such a forging. It was then found that hydrogen at high pressure attacks steel and renders it brittle. The ingenious and practical solution to this difficulty was the so-called lined converter: the inner tube of low-carbon steel was not intended to resist the pressure but to absorb the hydrogen. The outer tube of ordinary steel was designed to take the pressure; any hydrogen which had not been absorbed by the inner tube was allowed to escape through small holes drilled through the outer tube.

The other problem facing the chemical engineers at Ludwigshafen was the supply and purification of gases. This necessitates a short digression. The cheap production of hydrogen and nitrogen proved to be the key to the success of the Haber process. Both gases had been obtained for some time on a commercial scale: hydrogen as a by-product of electrolysis, nitrogen by air-liquefaction methods developed by C. v. Linde in 1895 and George Claude a few years later. It will be recalled that a feature of the cyanamide process was the treatment of carbide at a high temperature with nitrogen. Accordingly Frank and Caro got in touch with Linde, who had been working on the liquefaction of coke-oven gas. They combined, and by 1910 had developed a process for liquefying water-gas and removing the constituents (above all hydrogen) by fractional condensation. They extended the process subsequently to the liquefaction of coke-oven gas. The Linde–Frank–Caro process was a much better means of obtaining hydrogen than electrolysis.[2] It was adopted by Bosch for the first of the large ammonia works. For nitrogen Bosch relied (as Frank did) on a Linde air-liquefaction plant. Neither proved entirely satisfactory for very large works, and Bosch decided that some

[1] Hoppmann, H. Die rationelle Gestaltung der chemisch–technischen Produktion, Ph.D. thesis (1934), p. 51; Timm, B. Die Anwendung hoher Drucke in der chemischen Technik. *Die BASF* **17** (1967), no. 1, 23–5.

[2] Rosenberg, E. v. (Ed.) *75 Jahre Linde* (1954), pp. 19, 119. The primary object of the process was to provide the Zeppelins with hydrogen. It was also used by margarine makers for fat hardening by hydrogenation.

cheaper method ought to be found. One of his associates, Wild, studied a process for the catalytic decomposition of water-gas. This enquiry began in 1912, and three years later the Bosch–Wild process (also known as the Bamag process from the engineering firm that supplied the equipment) had been developed. In essence the water-gas was split into hydrogen and carbon monoxide. The latter was treated again with steam, and this shift reaction, as it was called, yielded more hydrogen and carbon dioxide. Nitrogen is, of course, much less critical, since the volume is so much smaller. After further study, the chemists of Badische devised a process for separating nitrogen from the gas obtained when air is blown over coke.

By 1912 the problems of the catalyst and the converter design had been virtually solved; that of synthesis gas remained. The directors of Badische were sufficiently encouraged by the progress made to authorize an increase in the size of the pilot plant, and during the first half of that year ammonia was being synthesized at the rate of one ton a day. A site at Oppau, which adjoins Ludwigshafen to the north, having been bought, it was decided to install there the first large-scale plant. The initial capacity was about 8700 tons of ammonia a year, which was pumped to an adjoining works where sulphate of ammonia was made at the rate of 36 000 tons a year. The works was opened on 9 September 1913, but a little later the directors decided to enlarge it and war-time extensions raised capacity to 60 000 tons of ammonia (equivalent to 250 000 tons of sulphate) by the end of 1915.[1]

From 1910 onwards the German work on the ammonia synthesis was given wide publicity. But, as is often the case, few of the important details were revealed, so that when the war broke out the process was economically speaking an unknown factor. During the war, as will be told later, nitrogen fixation became a major concern in the belligerent countries and numerous cost estimates were made. The most detailed were those of the Nitrogen Products Committee set up by the British Government during the latter part of the war. Their costings were based on such pre-war figures as were available, and showed that nitrate of soda and sulphate of ammonia were by far the cheapest sources of fixed nitrogen. Their costs of production were respectively £10 and £14. 15*s*. 0*d*. per metric ton N. Next came cyanamide with costs ranging from £20. 5*s*. 0*d*. to £23. 16*s*. 0*d*., depending on the cost of power. The cost of making one ton of N by the Haber–Bosch process was estimated at £32. 12*s*. 0*d*., and if the arc process were used the minimum figure was £36. 17*s*. 3*d*.[2]

The report reflected expert opinion and appeared to justify the building of the numerous cyanamide factories during and immediately after the war. Nevertheless, when the Oppau plant could be inspected by the Allies in 1919 it became obvious not only that the process represented a major technical advance, but also that the war-time estimates of production costs were too

[1] *Enquete*, op. cit., pp. 146 ff.
[2] Nitrogen Products Committee, Final Report (1919), Cmd 482, pp. 224 ff.

high. Badische, however, was able to retain control over all the key patents, and even more important, closely guarded the technical know-how and the detailed composition of the catalyst. Their policy encouraged imitation, and during the 1920s several similar processes were developed. Some require mention. The General Chemical Co. in the United States interested itself in the ammonia synthesis as early as 1913. Their process differed little from Haber's, and was used by their successors, Allied Chemical & Dye, after the war. A variant was developed by the Nitrogen Engineering Corp., and during the 1920s was employed not only in the United States, but also in Germany and other European countries.[1] Claude, Fauser, and Casale used different pressures and temperatures in their processes. Claude, one of the most experienced men in gas-reaction chemistry, worked his plant at 900–1000 atm and 600°C, and claimed to get yields of 25 per cent of ammonia.[2] Casale and Fauser operated their plants under less severe conditions. The licensees of Claude and Casale worked predominantly in France and Japan, while Fauser's were especially active in Italy. Finally there was the Mont Cenis process, which took its name from a mine in the Ruhr where it was first worked. Its promoters were a group of mining companies that had been unable to get a licence from Badische. But they knew of the work done by Linde and Claude in the liquefaction of coke-oven gas. This was not such a cheap source of hydrogen as the Bosch–Wild–Bamag process, but the owners of the coke ovens were anxious to diversify, and for many years they had been in the sulphate of ammonia business. So they devised their own process: as Badische prevented them from using pressures above 100 atm, they worked at 90–100 atm and at a temperature of 350–400°C, and got satisfactory results.[3]

None of these processes was very important. Water-gas remained the

TABLE 4.1

Estimated world production of ammonia by process[4]

	1926/7 000 tons N	%*	1928/9 000 tons N	%*
Haber–Bosch	479	88	712	75
Casale	29	5	106	11
Fauser	13	2	42	4
Claude	19	4	47	5
Mont Cenis	—	—	31	3
Others	4	1	15	2
Total	544	100	953	100

*Rounded off

[1] Waeser, op. cit. (2nd edn.), pp. 59–62, 160–1.
[2] Partington, J. R. and Parker, L. H., op. cit., pp. 167–75.
[3] Waeser, op. cit. (2nd edn.), pp. 287–91.
[4] Pollitt, G. P. The synthetic ammonia industry. *Trans. second Wld Pwr Conf.* (1930), vol. 2, p. 151.

principal source of hydrogen, though the use of coke-oven gas by licensees of Claude, Casale, and Mont Cenis led to the establishment of ammonia works alongside coke-oven plants in Germany, Belgium, and Northern France. Few were economic during the slump, but some were still in use until recently in the mining areas of those countries. G. P. Pollitt, who was among those who built the Billingham ammonia works, published some careful estimates of world production of ammonia according to the process used. His figures (summarized in Table 4.1) show that, despite the emergence of rival processes, the Haber–Bosch easily retained its lead.

Pollitt also calculated the total energy requirements of the different nitrogen fixation processes. His figures, expressed in terms of coal (or coal equivalent) per ton of nitrogen, clearly showed why the ammonia synthesis was making all the other processes uneconomic. They were: arc process, 64 tons, cyanamide 12·7 tons, ammonia synthesis 6·3 to 7·3 tons.[1]

At first Badische only made ammonia and sulphate of ammonia, and neglected the production of nitric acid, although as far back as 1839 F. Kuhlmann had demonstrated the feasibility of oxidizing ammonia to nitric acid. For years his work was ignored. Ostwald returned to it in 1900, and the next year carried out the first small-scale oxidation experiments with a platinum catalyst. Griesheim-Elektron also carried out experiments from 1904 onwards, apparently with every success, but once again there were no further developments.[2] It is probable that the traditional method of manufacture from sodium nitrate was so simple, that no one was prepared to take a gamble on the new process, which required ammonia of a high purity. This may explain why only one German firm, Gewerkschaft Lothringen, thought it worthwhile to obtain a licence from Ostwald. From 1908 onwards it oxidized ammonia at the coke-oven plant of Gerthe near Bochum in the Ruhr. The process was also licensed to Nitrogen Products & Carbide Co. Ltd., which, in 1912, built a plant at Vilvorde in Belgium and at Dagenham in England; both obtained ammonia by treating cyanamide with steam. When war broke out, and Germany was cut off from Chile, the works at Gerthe, with a monthly capacity of 150 tons of 36°Bé nitric acid (52·8 per cent HNO_3) was the only factory in the country making this essential material for explosives. Vilvorde was overrun by the invading Germans and the works closed. Dagenham stopped production for lack of cyanamide.[3] Thus it came about that a critical shortage of nitric acid developed unexpectedly in Germany, and later in Britain, although the technical solution was at hand.

Fertilizers

While most of the branches of the chemical industry were affected by electrochemistry and other new chemical technologies, the preparation of fertilizers was scarcely touched by the important changes that were occurring

[1] ibid., pp. 148–9. [2] Pistor, op. cit., p. 61.
[3] Waeser, op. cit. (2nd edn.), pp. 31–2.

in applied chemistry. Consequently the history of fertilizer manufacture stood apart from that of chemicals generally, and what technical development there was, derived from materials handling, which followed a rather different evolution and had little connection with chemicals. Another factor that accounted for the separateness was that the manufacturers were completely tied to and wholly dependent on the fortunes of farmers. Stocks, transport, marketing, credit, and advisory services were all geared to the requirements of this large group of final users, and unlike the rest of the chemical industry there was no product or market diversification. Finally, and perhaps most important of all, manufactured fertilizers meant for many years just a single product: superphosphate. The other materials, such as sodium nitrate or potash, were simply sold as such, or mixed with each other and sold as compounds. Thus the maker of fertilizers was, for long, outside the mainstream of chemical progress, and even if his operations were large, they were, chemically speaking, elementary. Yet these businesses, formed at about the turn of the century in the United States, France, Italy, Japan, and a few other countries, were often the precursors of major companies that were shortly to offer a broad range of products. Taken together, we would certainly be justified in treating these firms, despite their humble beginnings, among the founders of the chemical industry in those countries. The link between the old and the modern fertilizer industry was surely the fixation of nitrogen, which shifted the technical basis from simple crushing and mixing to advanced synthesis, and which transformed the commercial basis from seasonal operations to year-round production and advanced marketing techniques.

The confluence of fertilizer production with the main current of the chemical industry came in the 1920s and 1930s. But even before the Great War fertilizers, whether manufactured or merely mixed, had reached, by contemporary standards, a huge scale. The following figures illustrate the progress made between 1900 and 1913.

TABLE 4.2

Estimated world production of fertilizers
(000 tons)[1]

	1900	1913
Nitrogenous (as N)	270	640
Phosphatic (16% P_2O_5)	1000	2500
Potassic (as K_2O)	230	1000
Total	1500	4140

The volume of nitrogenous fertilizer, though quantitatively the least important of the three, requires first attention.

[1] F.A.O. *Production Yearbook 1960*, 14. The estimated non-fertilizer uses have been excluded, and the figures have been rounded off. Phosphatic fertilizers include Thomas meal.

Liebig's law of the minimum had a wide following at the turn of the century. He had postulated that the plant nutrient in least supply determined the yield of the crop. The great chemist's axiom was vigorously plugged by the owners of potash mines in order to boost their sales at the expense of superphosphate. But in fact it was nitrogen that needed to be increased if the output was to be raised.[1] The use of dung was well established, but the application of commercial fertilizers was restricted by their price, except among horticulturists and wherever intensive farming was practised. Little had been done in the way of scientifically testing the effectiveness of nitrogenous fertilizers by field trials, except at Rothamstead and a very few other research stations. The use of fertilizers on grassland was ignored; in Britain, for example, it did not get under way until the late 1920s.[2]

At the beginning of the century three nitrogenous fertilizers were articles of commerce: guano, sodium nitrate, and sulphate of ammonia. Guano was the least important. It is a natural compound fertilizer, containing calcium phosphate, up to 15 per cent nitrogen in amide form, and a small amount of potash. World production, once large, had declined owing to the gradual exhaustion of the richest deposits on the islands off Peru. In the years before the Great War output fluctuated between a low point of about 40 000 tons and a peak of 133 000 tons. At the most, therefore, the nitrogen content was around 20 000 tons.[3]

From the second half of the nineteenth century onwards sodium nitrate replaced guano. It had been discovered at the beginning of the century, and the enormous deposits had come under the control of Chile after the War of the Pacific against Bolivia (1870–82). The country's finances were largely based on the export tax on sodium nitrate. The nitrate-bearing material, *caliche*, was quarried under arduous conditions and the nitrate extracted by the Shanks process, which was a batch operation on the counter-current principle: the crushed *caliche* was dissolved in water and the sodium nitrate liquor gradually concentrated.[4] From the last tank the strong liquor flowed into crystallization pans and was shovelled from them to drying boards and finally packed into sacks. The method of extraction and purification was inefficient, resulting in the loss of over 50 per cent of the nitrate content of the raw material, and caused heavy expenditure on fuel and transport.[5] This primitive process remained in use for some 75 years. But in the 1920s a method employed for concentrating copper ores was adapted to the treatment of

[1] Later work showed that the amount of nutrient assimilated, rather than the quantity supplied, was the critical factor.

cf. Browne, C. A. *Liebig and the law of the minimum* (1942), pp. 71–82.

[2] Sir Frederick Keeble's book, or perhaps one should say tract, on this subject, *Fertilizers and food production*, was only published in 1932.

[3] *Annu. Lambert*, pp. 160–1; Institut International d'Agriculture. *Yearbook of agricultural statistics* (1922).

[4] Shanks' vats had been designed for the Leblanc process for extracting soda crystals from the crude alkali or 'black ash'.

[5] Cottrell, A. *The manufacture of nitric acid and nitrates* (1923), chap. 1.

nitrates. It was called the Guggenheim process, and the first plant to work it was opened in November 1926: it resulted in the recovery of more than 90 per cent of the available nitrates, and processing costs were almost halved. The process had a major effect not only on the activities in Chile, but indeed on the entire nitrogenous fertilizer industry.

In the years to 1914 the rising demand led to a large increase in the number of producers (*oficinas*). In 1902 there had been 78 and at the end of March 1914 there were 134, a considerable number of them being British- or German-owned.[1] The growth of exports was as follows:

TABLE 4.3

Chilean exports of sodium nitrate
(000 metric tons)[2]

1880	224	1900	1454
1885	436	1905	1650
1890	1063	1910	2336
1895	1239	1913	2738

The statistics for nitrates are only passably reliable. If we assume that all the material contains 15·6 per cent N, Chilean exports represented almost 230 000 tons N in 1900 and about 370 000 tons N ten years later. While a substantial quantity was required for the manufacture of nitric acid and lesser quantities for other nitrogen compounds, agriculture accounted for roughly 75 per cent of total consumption.[3] In 1910, therefore, roughly 280 000 tons of nitrogen in the form of nitrate was used as fertilizer. Farmers preferred nitrate because of its quick action.

Sulphate of ammonia, the third of the nitrogenous fertilizers, has been sold since the middle of the nineteenth century, but its period of rapid growth dates from the closing years. The manufacturing process was simple: ammoniacal liquors, obtained by washing crude coal gas, were distilled, and the vapour absorbed in sulphuric acid, of which about 1·1 ton (as 70 per cent H_2SO_4) was required to make one ton of sulphate. The material was then dried, a stage in the process that was later speeded up by centrifuging, and put into sacks. At first sulphate of ammonia was only available from gas works. Most of them were so small that it was not worth their while to make the fertilizer, and the ammoniacal liquors were usually run to waste. By the 1890s some very large gas works had emerged and the pollution of streams was no longer tolerated. For example the works of the Gas Light & Coke Co. at Beckton made a big tonnage of sulphate and even produced its own sulphuric acid from sulphur recovered in gas making. In Britain 67 per cent of the 180 000 tons of sulphate made in 1895 was obtained from gas works, but by 1913, while the total had grown to 433 000 tons, the importance of gas

[1] James, A. F. B. *Nitrate facts and figures* (1908 and 1914 edns).
[2] Hernandez, R. *El Salitre* (1930), p. 174.
[3] Cottrell, op. cit., p. 36.

works had declined, and their share was only 42 per cent. Coke ovens with plant to recover the by-products contributed 36 per cent, and the remaining production came from shale oil and low temperature carbonization works.[1] In Germany the adoption of recovery ovens was extremely rapid: as early as the mid-1890s they accounted for one-half of the sulphate production, and by 1907 the proportion had become 85 per cent.[2]

If the object had been to maximize the output of sulphate, the best solution would have been the general adoption of low temperature carbonization processes such as Mond's. His yielded about 70 lb of sulphate per ton of coal, while even the most up-to-date recovery ovens produced only 15–37 lb per ton.[3] But the Mond process was of limited application, and high temperature carbonization was the only worthwhile source of ammonia. It needs to be emphasized, however, that ammonia recovery was secondary to efficient metallurgical coke or town-gas manufacture. The former fluctuated with activity in the steel industry, the latter with gas consumption.

For industrial purposes the ammoniacal liquors were purified and concentrated, and sold to makers of dyestuffs, ammonia soda, and other chemicals, as well as for refrigeration. The many agricultural and industrial end uses of by-product sulphate of ammonia and of ammonia itself create statistical difficulties; there are no comprehensive data before about 1900, and even thereafter they are unreliable. However, the figures in Table 4.4 give an indication of the order of magnitude and the ranking of the principal producing countries.

TABLE 4.4

Estimated production of sulphate of ammonia by principal countries[4]
(000 metric tons)

	1900	1905	1910	1913
Germany	108	268	373	549
Britain	217	273	374	440
U.S.A.	n.a.	59	105	177
France	37	45	57	75
Belgium	20	18	36	49
Austria–Hungary	18	20	29	35
Other	24	43	30	76
Total	424	726	1004	1401

If the above quantity is expressed in terms of nitrogen, world production in 1913 was 280 000 tons, which may be compared with 430 000 tons contained

[1] Alkali Inspector, reports for 1895, p. 19 and 1913, p. 15.
[2] Frank, A. The utilisation of atmospheric nitrogen. *Chem. Trade J.* **42** (1908), no. 1100, 581.
[3] Fritsch, J. *The manufacture of chemical manures* (1911), pp. 227–9.
[4] For 1900 and 1905, *Annu. Lambert* (1912 edn.), p. 134; 1910 and 1913, *Yearbook of agricultural statistics* (1923). British figures from annual reports of Alkali inspector.

in nitrates. A larger tonnage could have been obtained if recovery methods had been pushed to their practical limit, but this was unnecessary, for there was no great pressure of demand. Even as late as 1913 there was some resistance among farmers to sulphate of ammonia. They claimed that it was slower acting than sodium nitrate and that it had an acidifying effect on the soil. This was true, but it was not a serious objection. Indeed the loss from leaching was smaller than with nitrate and the acidity could be treated by liming. The more likely reason was the failure, up to 1913, of the makers of sulphate of ammonia to market their fertilizer with sufficient vigour. The nitrate companies were more alert, and they set the price pattern for nitrogenous fertilizers generally, and were able to sell their material at a small premium per unit of nitrogen over sulphate. The disruption of nitrate shipments after August 1914 transformed the situation and sulphate perforce came to be generally accepted.

The supply of nitrogenous fertilizers from the three traditional sources more than doubled between 1900 and 1913, when it reached about 640 000 tons N. To this extent then, Crookes' call had been heeded, but the atmospheric route to nitrogen fixation to which he had pointed was only at the start of its commercial development when war broke out. The cyanamide process alone was being worked on a large scale. The available figures are extremely confusing, because many producers did not say how much they made, but what they could make; the latter was invariably higher, so discouraging competitors and cheering shareholders with the promise of large future profits. Thus the International Yearbook of Agricultural Statistics showed world production of cyanamide at 11 500 tons in 1909, 55 000 tons in 1911, and no less than 186 000 tons in 1913.[1] But these figures are misleading. For example, the combined capacity of the works at Knapsack and Trostberg in Germany was about 50 000 tons in 1913, but they made only 24 000 tons in that year.[2] Similarly the capacity of American Cyanamid in the year to 30 June 1913 was about 29 000 tons, but sales were only 20 600 tons.[3] We have no information on the level of activity in the other factories, but if we assume that all the plants worked at two-thirds of their capacity, then the total output of cyanamide grew from around 7000 tons in 1907 to roughly 120 000 tons in 1913.[4]

Cyanamide was sold before the war with a nitrogen content of 19·5 per cent, so that in 1913 roughly 24 000 tons of nitrogen were made by this route. The small quantity can be readily explained because of the novelty of the product; after all European farmers had been able to buy it in only a few

[1] First edition, 1922.

[2] Waeser, op. cit., 1st edn., p. 19. Waeser is usually reliable, and I have used his figures in preference to the higher ones published by later authors (cf. *Enquete*, pp. 156 and 207).

[3] Haynes, op. cit., Short tons have been converted to long tons.

[4] Kubierschky, K. Die künstlichen Düngemittel und ihre Bedeutung für die Weltwirtschaft. *Z. angew. Chem.* **26** (1913), pt. 1, 727, estimates production at 2200 tons in 1907 and 97 000 tons in 1913. Both are too low.

countries for five or six seasons. Cyanamide was dusty and inconvenient to handle. It caused extra work because it was necessary to cover cyanamide by further ploughing. It was not suitable as a top dressing nor on light soils, but limy soils accepted it readily. Despite the disadvantages and its inferiority compared with nitrates, the new fertilizer was becoming established, and the trend of sales was encouraging. In the United States it was promoted very energetically by American Cyanamid. The company demonstrated that it was suitable as a side dressing in place of nitrates for partly grown tobacco. Later the firm mixed cyanamide with superphosphate and sold the compound, which contained 3 per cent cyanamide, as 'Aero Cyanamid'.[1] In wartime Germany cyanamide was mixed with Thomas meal or potash. Generally speaking the prospects for cyanamide were considered sufficiently promising to attract investors, and numerous factories had been built or were under construction when the war broke out. So far as I know cyanamide was used exclusively as a fertilizer, except in Japan where it became important as a source of ammonia, and on a small scale in Belgium and England for the same purpose.

The arc process was being worked in a large factory by Norsk Hydro in south-east Norway. In the year to 30 June 1914 the company's sales amounted to 115 000 tons, the principal product being calcium nitrate, which was sold under the name of Norgesalpeter. This material when first produced was far too hygroscopic to be of much use to farmers, and in any case its nitrogen content was low. Later the manufacture was altered so as to minimize the first defect and to increase the nitrogen content to 13 per cent. This was still a drawback (only partly compensated by the presence of lime), and the fertilizer took some time to establish itself. By the outbreak of war it was being used extensively in Scandinavia, especially in Denmark. In terms of nitrogen, manufacture was less than 1000 tons in 1908 and still only 3000 tons in 1910, but in the year to 30 June 1913 production rose to 15 000 tons and increased to 17 000 tons in the next financial year.[2]

The synthetic ammonia plant at Oppau, the only one of its kind before the war, though opened in September 1913, did not make any fertilizer that year. During 1914, however, it began to manufacture at its capacity of 36 000 tons a year of sulphate of ammonia.[3]

To sum up: the nitrogen fixation industry was still in its infancy when war broke out. During 1913 it made 40 000 tons of nitrogen, at the most 50 000 tons, out of a total world supply of very roughly 750 000 tons. The new industry's key contribution to the nitrogen problem was still to come. Here it is worth stressing that the tonnage of fertilizer nitrogen more than doubled between 1900 and 1913 because of the increase in Chilean production and the widespread adoption of recovery ovens, coupled with rising demand.

Phosphatic fertilizers in the form of ground bones had been used in the

[1] Haynes, op. cit. The close connexion between American Cyanamid and J. B. Duke was fortunate, for the company's links with the American Tobacco Co. helped sales to the growers.

[2] *Norsk Hydro history*, pp. 417, 459.

[3] Mittasch, op. cit., p. 137.

first half of the nineteenth century, and were found to be particularly effective on root crops, especially on sugar beet. In the 1830s Liebig showed that upon treating ground bones or mineral phosphates with sulphuric acid the resulting product, containing the readily soluble calcium dihydrogen phosphate, was an even better fertilizer as it was assimilated more quickly by the plant. His advice was followed and the European deposits of phosphate were worked from the 1840s onwards. But the manufacture of superphosphate, as the fertilizer is called, only made rapid progress after the discovery of enormous deposits of rock phosphate in the southern states of America after the Civil War. Later big deposits were found in North Africa: the Algerian mines began production in 1889, the Tunisian (discovered in 1873) ten years later. After the Great War phosphate was mined in Morocco, Russia, and the Pacific Island of Nauru.

The manufacture of superphosphate was simple: crushed rock was fine ground in ball mills (later ring mills were employed) and poured into pits or 'dens' where it was treated with sulphuric acid. As a general rule approximately 1 ton of 50°Bé acid (62·5 per cent H_2SO_4) was needed to make about two tons of superphosphate containing 16–18 per cent P_2O_5.[1] The mass was allowed to settle in the den and was then removed with pick and shovel. The design of the first works was very primitive and depended entirely on manual labour; work was very arduous. Soon attempts were made at mechanizing the process: the machinery was tried first in Germany, later in Britain and the United States. Initially modernization comprised the installation of conveyor belts, trolleys, and cable cars. The mechanization of the dens took longer, and here again the Germans were in the lead by introducing the mechanical mixer, a screw-type conveyor in which ground rock and acid were mixed and poured into the pit. Before the war some firms were installing mechanical scrapers, which pulled the superphosphate to the mouth of the den where it was pushed into trolleys; in England the first of these devices was installed in 1910.[2] After leaving the den, the fertilizer was broken up and passed through mechanical screens, after which it was shovelled into sacks. The improvement in handling techniques was accompanied by the introduction of process control and the regular analysis of the raw material and the finished superphosphate.[3]

Mixtures of ground bones or bone meal with guano were sold to farmers at an early date. Mixtures or compounds of superphosphate and potash were occasionally marketed from the 1870s onwards. But the practice of compounding did not become widespread until the turn of the century. Sulphate of ammonia and nitrates were rarely added to the compound, which was found particularly effective on light soils.[4] Some manufacturers prepared triple

[1] Fritsch, op. cit., p. 76.

[2] Gray, A. N. *Phosphates and superphosphates* (1944), pp. 111–15.

[3] Klippert, Die Entwicklung der Technik in der Düngerindustrie . . . bis auf die heutige Zeit. *Z. angew. Chem.* **18** (1904), no. 9, 322–5; Packard, W. G. T. *The history of the fertiliser industry in Britain*, n.d., pp. 14–15; Fritsch, op. cit., pp. 69–71.

[4] Packard, op. cit., pp. 28–31.

superphosphate by treating finely ground phosphate rock with phosphoric acid. It would appear that this fertilizer was not widely used outside Germany before the war.

European farmers used superphosphate as their chief fertilizer, and from the 1890s onwards an enormous new market developed when American cotton and tobacco planters began to employ it. The fertilizer was a cheap, bulk product and its manufacture, even after the introduction of mechanical aides, required much less capital than a works for the fixation of nitrogen. Many firms began making superphosphate in the last twenty years of the nineteenth century, and the following figures indicate the extremely fast development of the industry.

TABLE 4.5

World production of phosphate rock and superphosphate[1]

	Phosphate rock	*Superphosphate*
	(million metric tons)	
1880	0·5	0·9*
1885	1·2	1·6*
1890	1·6	2·0*
1895	2·2	3·5
1900	3·1	4·8
1905	3·8	6·1
1910	5·4	9·5
1913	7·2	11·2

*estimated

By 1910 the United States, France, Germany, and Italy each made more than a million tons a year of superphosphate, and three years later America produced 3¾ million tons, France and Germany almost 2 million tons each. The fertilizer was manufactured on the scale of several hundred thousand tons a year in the United Kingdom, Belgium, and Holland; even Russia made 115 000 tons in 1913. Not only the quantities, but the rates of growth were phenomenal: in Italy, for example, production had been a mere 146 000 tons in 1895, whence it soared to 250 000 tons in 1900 and to a million tons ten years later.[2] The production of superphosphate replaced that of Leblanc soda as the main user of sulphuric acid. And where the Leblanc process had never been established, as in Italy, Spain, and the United States, big sulphuric acid works had to be expressly built for the purposes of the superphosphate factories. Roughly 3½ million tons of acid (as 100 per cent H_2SO_4) were required for this fertilizer alone in 1913. In France three-quarters of the sulphuric acid made went to superphosphate manufacture, and even in Germany, with its highly diversified chemical industry, the proportion was still

[1] Gray, op. cit., pp. 264–5.

[2] Gray, op. cit. *Annu. Lambert*, p. 298.

50 per cent.[1] In countries with a less developed chemical industry sulphuric acid and superphosphate were virtually synonymous.

Thomas meal, obtained by grinding the basic slag of steel works smelting phosphoric ores, provided a supplementary phosphatic fertilizer. It seems to have been first sold for this purpose by Chemische Fabrik Kalk (then trading as Vorster & Grüneberg) in 1885.[2] Other German firms followed and by 1896 sales exceeded the million ton mark; by 1914 they reached 2¼ million in that country alone. During the war Germany was unable to import phosphate rock, but the Lorraine ores were under its control, so that Thomas meal became virtually the sole source of phosphate plant nutrient. France, Belgium, and Luxembourg were also substantial users of basic ores and accordingly produced a large tonnage of Thomas meal. World output grew from about 1·6 million tons in 1900 to 4·2–4·4 million metric tons in 1913.[3] Thomas meal acts more slowly than superphosphate on grain crops, except on acid soils. It was used extensively in vineyards and orchards.

The potash industry was until the Kaiser's war a German monopoly. The huge deposits were first worked at Stassfurt in Prussian Saxony in the late 1850s. Later those in Anhalt and Hanover were developed so quickly that by the early 1900s a big industry had emerged in central Germany. (Potash was also found in Alsace, and the pits began producing just before the war.)

TABLE 4.6

German production of crude potash salts and world consumption of potash, selected years 1895–1913[4]

	German production (million metric tons)	*World agricultural consumption* (000 metric tons K_2O)	*World industrial consumption* (000 metric tons K_2O)	*Total consumption* (000 metric tons K_2O)
1895	1·5	119	51	170
1900	3·0	233	71	304
1905	4·9	407	76	483
1910	8·2	766	92	858
1913	11·6	1004	106	1110

The crude salts contained up to 16 per cent K_2O, and those with a high potash content were simply ground, bagged, and sold. Those with a low content were dissolved and then recrystallized into high strength potassium chloride and potassium sulphate. At first the main demand for potash salts was in the

[1] Hasenclever, M. Zur Lage der chemischen Grossindustrie. *Chem. Ind., Düsseld.* **28** (1905), no. 3, 52 f.; Grandmougin, E. *L'essor des industries chimiques en France* (1917), p. 79.

[2] Anon. *Festschrift zur Feier des 50 jährigen Bestehens der Firma Vorster & Grüneberg* (1908), pp. 9–12.

[3] Kubierschky, art. cit., p. 724; Grandmougin, op. cit., p. 80; *Yearbook of agricultural statistics*, 1922 edition.

[4] Krische, P. *Das Kali* (1923), vol. 1, pp. 134–6 and 184 ff.

manufacture of *Konversionssalpeter*, that is, potassium nitrate made by the double decomposition of potassium chloride with sodium nitrate. But with the decline in the demand for black powder, the consumption of potassium nitrate fell. Its place was taken by other industrial uses, notably by dyestuffs factories and for the manufacture of soft soap. The preparation of caustic potash and metallic potash, to mention only two types of potash for industrial consumption, was entirely in German hands, and was one of the factors that led to the development of electrolytic works in that country. But the growth in the consumption of potash as fertilizer was so rapid that the relative importance of industrial uses fell steadily, until it was less than 10 per cent of the total in 1913.

On the eve of the war German and American farmers, taken together, accounted for three-quarters of the potash fertilizer demand. The Germans used it on their light soils, and mixed it with Thomas meal and superphosphate to provide a compound that became the chief fertilizer for sugar beet and other root crops, notably turnips and swedes.[1] A large tonnage was also bought by Dutch and Belgian farmers for the same application. In the United States tobacco planters (who also cultivated on light soils) found that their yields were significantly higher when using potash, and they began importing steadily growing quantities during the 1900s.

[1] Precht, H. Die Entwicklung der Kaliindustrie, *Z. angew. Chem.* **19** (1906), no. 1, 3.

5

THE GERMAN CHEMICAL INDUSTRY ON THE EVE OF THE GREAT WAR

LEAGUE tables were known and used two generations ago, though the statistics on which they were based were even less reliable than today. Few Germans doubted that in matters chemical they set the pace and that their chemical manufactures were the largest in the world. The value of German chemical production was estimated at £120 million in 1913, and was only exceeded by that of the United States, which, in 1914, was reported to be £170 million. But the American figures were swollen by petroleum refining, seed crushing, and several other non-chemical trades, and if the statistics are adjusted to the European definition, the United States production falls to about £77 million, and Germany retains a clear lead.[1] Its chemical and allied industries employed 115 000 people in 1895 and 180 000 in 1913; at the latter date a further 30 000 were engaged in potash mining and processing.[2]

Germany's foreign trade in chemicals was booming. It accounted for an estimated 28 per cent of world exports of chemicals; the share of the nearest rival, Great Britain, was 16 per cent, and Chile with its huge nitrates business came third with 15 per cent.[3] The German trade statistics were extensively revised in 1906, so that it is difficult and at times impossible to compare the figures for 1913 with those at the beginning of the century. But the figures in Table 5.1 give a fair indication of the importance of chemicals in the country's balance of trade.

In round figures, exports of chemicals trebled between 1900 and 1913, and at the latter date represented a tenth of the country's total. (This proportion was reached in the United Kingdom in the late 1950s.) The two largest sub-groups were chemical raw materials and dyestuffs. Exports of the former consisted of potash and sulphate of ammonia, aggregating almost £4 million in 1913, and also a growing volume of inorganic chemicals, notably alkalis and phosphatic fertilizers. Imports were swollen by about £8½ million of nitrates in 1913. Exports of dyestuffs doubled between 1900 and 1913, and it must be emphasized that the relatively small figure bears no relation to the

[1] League of Nations (Economic and Financial Section). *The chemical industry* (1927), p. 21. The estimates were prepared by Dr. Ungewitter, the secretary general of the chemical industry group in the Association of German Industries. For details of the United States see p. 174.

[2] The numbers are those of the chemical industry as defined by the social security legislation. They comprise not only the chemical and allied industries, but also rubber, wood preserving, and some lesser trades.

[3] League of Nations, op. cit., p. 12. Sodium nitrate was then considered a chemical product not a raw material.

crucial importance of this trade. Finally there were also large exports of intermediates for dyestuffs and other chemicals, and a growing business in such specialities as pharmaceuticals, photochemicals, and synthetic essences.

TABLE 5.1

German foreign trade 1900 and 1913[1]
(£ million)

	1900		1913	
	Imports	*Exports*	*Imports*	*Exports*
Total	292	228	556	509
of which:				
Iron and steel	6	24	5	67
Non-electrical machinery	4	10	4	34
Chemicals and allied products*	13	17	22	50
of which:				
Chemical raw materials	n.a.	n.a.	14	19
Dyestuffs†	0·5	5	0·4	11

*Including coal tar, tar-distillation products, and intermediates returned under headings other than chemical manufactures.
†Excluding organic and inorganic pigments.

In addition to these tangible signs of power, there were others that were not quantifiable, but were nevertheless significant. Perhaps they can best be expressed in terms of the industry having momentum, or, as Rostow put it, 'built-in growth'.[2] The phenomenon was seen most clearly in the organic chemicals sector, where Germany had emerged as the leader in the 1870s, and a generation later was still dominating the business. Elsewhere it was becoming noticeable in the pharmaceuticals industry, and also in that of nitrogenous fertilizers and electro-chemicals. There were many favourable elements supporting this growth: availability of raw materials, good communications, sufficiency of capital both from retained earnings and the banks, the methodical application of inventiveness, and professionally competent manpower, so that advances in chemical knowledge were quickly exploited commercially.

By 1913 the centre of German chemical manufacture was spreading further east. The large works along the Rhine and lower Main, at Ludwigshafen around Frankfurt, and downstream from Cologne, which were already impressive by the turn of the century, continued to expand. But other areas were now becoming prominent: Berlin itself and especially the region to the south-west of the capital. Potash and lignite, having been found, were now worked intensively in the quadrilateral bounded by the line Leipzig–Magdeburg in the east and Cassel–Hanover in the west. Still further east, the development of the upper Silesian coal–steel industry was accompanied by the emergence of several large chemical enterprises.

[1] Statistisches Reichsamt. *Der Aussenhandel Deutschlands* for 1900 and 1913.
[2] cf. his *Stages of Economic Growth* (1960).

Many hundred companies were making chemicals, and most of the larger ones were incorporated with limited liability. Among them was a rather smaller number which, being *Aktiengesellschaften* (as distinct from *Gesellschaften mit beschränkter Haftung*), were obliged by law to issue annual reports. The reports were collated and summarized by the Government and the results published for the first time in 1908.

TABLE 5.2

Financial summary of limited liability companies in the chemical industry[1]

	Years ending 30 June 1908	1910	1913
1. *Capital Structure*			
No. of companies making returns*	137	147	158
	£m	£m	£m
Paid up share capital of those companies	18·0	22·4	25·5
Loan capital (incl. mortgages)	3·5	6·4	7·0
Reserves (excl. pension funds)	6·0	7·0	8·9
2. *Profits*			
No. of companies (included in 1) reporting profits*	117	119	141
	£m	£m	£m
Profits made by those companies	3·7	4·3	5·7
Total dividends paid	2·8	3·3	4·1

*Excluding potash mines, but including manufacturers of explosives, matches, and pharmaceuticals.

The figures show that, on average, the share capital increased from £132 000 per company in 1907/8 to £161 000 in 1912/13, and the average loan capital from £25 000 to £45 000. The increase in reserves was significant. At least four out of five companies declared a profit, and average profits increased during the period from £31 000 to £40 000. The averages conceal important differences between the various sectors of the industry. The largest was the heavy chemicals group, consisting of many small- and medium-sized businesses specializing in alkali and fertilizer manufacture. Potash mining ranked second, and though not strictly a branch of chemical manufacture had several links with it, and so merits attention. The smallest group, but the only sector where individual firms were large and manufacture integrated from raw material to finished product, was that of dyestuffs. (A fourth group consisting of the pharmaceutical businesses would also have been interesting, but, as they were predominantly partnerships or private limited liability companies, no financial statistics were published.) The first analysis by sectors of the chemical industry was issued for companies reporting in the

[1] *Z. angew. Chem.* **24** (1911), no. 48, 2309–12; **27** (1914), pt. 3, 745–9.

year to 30 June 1910. The summarized particulars for that period and for 1912/13 are as follows.

TABLE 5.3

Financial summary of limited liability companies in three sectors of the chemical industry[1]

	Year ending 30 June 1910 Heavy chemicals	Potash mining	Dye-stuffs	*Year ending 30 June 1913* Heavy chemicals	Potash mining	Dye-stuffs
1. *Capital Structure*						
No. of companies making returns	96	30*	22	110	29*	20
	£m	£m	£m	£m	£m	£m
Paid up share capital of those companies	11·8	10·7	7·3	14·6	12·3	7·3
Loan capital (incl. mortgages)	2·6	2·2	3·5	3·0	3·1	3·4
Reserves (excl. pension funds)	2·6	1·8	3·4	4·0	2·7	3·5
2. *Profits*						
No. of companies (included in 1) reporting profits	80	20	16	98	22	19
	£m	£m	£m	£m	£m	£m
Profits made by those companies	1·7	0·7	2·0	2·3	1·1	2·7
Total dividends paid	1·3	0·4	1·5	1·8	0·8	1·8

*Excluding Prussian and Anhalt state mines and the potash works of Deutsche Solvay-Werke.

Note: Manufacturers of explosives, matches, and pharmaceuticals are not included.

The heavy chemicals sector appears, at first sight, to be growing, and overall profits increasing, but for the average firm there was little change in the three years between 1910 and 1913: the share capital per firm remained at about £120 000 and average profits stuck at £21 000. But in the latter year shareholders enjoyed slightly higher dividends. The companies in the potash group were highly capitalized, the average was £420 000 in 1912/13. This can be attributed to expensive shaft-sinking operations and pumping machinery. The year to 30 June 1910 was unprofitable, and only two-thirds of the companies managed to pay a dividend. In 1912/13, however, profits were higher and reflected the boom conditions that followed the new cartel law; the average dividend was well above the 1909/10 level. The dyestuffs sector was in terms of numbers the smallest of the three, but it was without doubt by far the strongest financially. Average share and loan capital were of the same order as in potash mining, but reserves were higher and growing. In 1909/10 three out of four companies made profits and at the eve of the war virtually all reported profits, the average increased from £125 000 to £140 000.

[1] Source: as for Table 5.2.

The three sectors of the industry, taken together, had a paid up share and loan capital plus reserves of roughly £46 million in the earlier period and £54 million in 1912/13. These figures do not assuredly equate the real value of the assets, which was enhanced by the tradition of rapid depreciation and the pursuit of a cautious financial policy. From scattered indications it would appear that the replacement value of the fixed assets was twice, possibly even three times, the amount shown in the published annual accounts. But when speaking of the profitability of the German chemical industry we have to rely on the published data, which make no allowance for undisclosed write-offs or transfers to special reserves. The official statistics show that the published profits as a proportion of share capital and reserves was as follows in 1912/13: heavy chemicals 12·4 per cent, potash mines 7·3 per cent, and dyestuffs companies almost 25 per cent. No further comment is needed.

The production statistics for the inorganic sector of the German chemical industry, though inadequate, give an indication of its size: in 1900 the output of sulphuric acid (measured as 100 per cent H_2SO_4) was 850 000 metric tons, in 1913 it was 1 700 000 tons. Alkali production, excluding caustic soda, grew from 325 000 tons (in terms of 100 per cent sodium carbonate) in 1901 to 460 000 tons on the eve of the war, while that of superphosphate increased from 644 000 tons to 1 860 000 tons between 1900 and 1913 and sulphate of ammonia from 108 000 tons to 550 000 tons in the same period.[1] These large quantities were made by an industry that was technologically advanced. The old lead chamber process was still the usual method for making the weak acid required for fertilizer manufacture, but the modern contact acid process was widely employed to make strong acid for most other purposes. In 1910 one-third of all the acid was made by this method.[2] (Elsewhere the contact acid process was used to only a limited extent, in the United States, Britain, and Russia.) Virtually all the alkali was prepared by the ammonia-soda route. The old Leblanc soda process had never been important in Germany, and by 1910 the only two factories still using it were reported to be making 30 000 tons of alkali in this way. It may be recalled here that the struggle between the two processes had been decided in favour of ammonia soda in the last decade of the nineteenth century without the major upheaval that occurred in Britain. A large volume of nitric acid was produced and it was estimated that in 1910, 150 000 tons of Chile nitrates were used for that purpose. As to bleaching powder, Germany was said to account for half the world production obtained from electrolytic chlorine.[3]

About two dozen manufacturers of inorganic chemicals had been repre-

[1] For sulphuric acid there are official German production statistics. For the other products the *Annu. Lambert* and C. Ungewitter (Ed.) *Chemie in Deutschland* (1938), have been used.

[2] Duisberg, C. Die Wissenschaft und Technik in der chemischen Industrie mit besonderer Berücksichtigung der Teerfarbenindustrie. *Z. angew. Chem.* **25** (1912), no. 1, 6 f.

[3] ibid. About 20 000–25 000 tons electrolytic soda and chlorine were also made by the dyestuffs factories for their own use.

sented at the Paris Exhibition of 1900, and most were still in business during the 1910s. A few examples must suffice: the Rhenania company had several works, of which the one at Stolberg near Aachen was the largest. The business was managed by the Hasenclevers, father and son, and employed about 1300 people, making inorganic acids, bleaching powder, and superphosphate. The firm had a good technical reputation, and was incidentally the last to work the Leblanc process on a commercial scale in Germany. Near Cologne, Chemische Fabrik Kalk specialized in Thomas meal, superphosphate, and ammonia soda, and in 1914 had about 1000 workers and a capital of £300 000.[1] Further south at Wiesbaden-Biebrich the chemical factory of H. Albert, a student of Liebig's, appropriately made superphosphate and Thomas meal. By 1914 the business had altogether fourteen works. The company pioneered the manufacture of triple superphosphate, and independently of Baekeland patented a phenol-formaldehyde resin in 1910, which, under the name of 'Albertol', became well known after the war.[2] The last two companies have retained their identity to the present day (though Albert now belongs to Hoechst), but many others lost theirs as a result of amalgamations. There was, for instance, the Verein Chemischer Fabriken of Mannheim, itself the outcome of a merger, which combined with other firms in the 1920s. In 1914 it had factories at Wohlgelegen, Heilbronn, and Neuschloss, and probably employed 1200–1500 people in the manufacture of acids and alkalis, about 52 000 tons of the latter being made. The firm had developed its own contact acid process, which it had licensed to several other companies, and had designed what was then considered to be a very efficient apparatus for making sodium sulphate and hydrochloric acid.[3] The Deutsche Gold- & Silberscheideanstalt (usually abbreviated to DEGUSSA) of Frankfurt had long outgrown its original assaying business, and specialized in ceramic colours and chemicals for the pottery and chinaware industries. The considerable increase in its capital from £375 000 to £1 million in 1912/13 was mainly to finance the extensions of its interests in sodium metal, which was made in partnership with the Aluminium Co. of Oldbury at Niagara Falls in the United States and at Rheinfelden in southern Germany, and with Soc. d'Electrochimie at Martigny in Switzerland. The connexion with sodium metal stemmed from the company's interest in cyanides, of which it was the largest German producer. During the 1900s the sodium business boomed, as a result of the discovery by one of its chemists, Pfleger, of the use of sodium amide in the synthesis of indigo.[4] Kunheim of Berlin had been started in the 1820s, and by 1913 probably employed over 1000 people. Kunheim was exceptional in the heavy chemicals sector by reason of the great diversity of products made, and also because it was notably unprofitable. There were

[1] Greiling, W. *100 Jahre Chemische Fabrik Kalk* (1958).
[2] Unsigned article on the company in *Chem. Ind., Düsseld.* N.S. **2** (1950), no. 12, 679.
[3] Rüsberg, F. *50 Jahre Kali-Chemie AG* (1949), pp. 33 f.
[4] Anon. DEGUSSA 1873–1923. *Chem. Ind., Düsseld.* **46** (1923), no. 7, 100–3.

several factories, which supplied a comprehensive range of inorganic chemicals, ferro-alloys, coal-tar intermediates, and, in partnership with DEGUSSA, cyanides.[1] Elsewhere in northern and eastern Germany several small soda makers survived from the nineteenth century by favour of the Solvay group. The industrialization of Silesia in the latter part of the century led to the development of several chemical works. At Saarau, Chemische Fabrik Silesia, founded in 1872, employed 1100 people making superphosphate, sodium sulphate, and wood- and coal-tar distillation products. It was reputed to make excellent aniline oil.[2] Further east, Oberschlesische Kokswerke & Chemische Fabriken had a capital of £1¾ million and employed 2200 workers in its four tar-distillation and roofing-paper factories.[3] None of the aforementioned companies could stand comparison with Deutsche Solvay-Werke, which with a capital of £5·6 million was the dominant element in the German soda trade. It probably accounted for two-thirds to three-quarters of the soda ash, and perhaps one-half of the caustic soda made in the Empire. The firm owned the big alkali works at Bernburg in central Germany, several smaller factories in the Reich, potash mines, and a lignite deposit. There were 6400 employees in 1913.[4]

Before leaving the heavy chemicals sector it is necessary to look a little more closely at the Griesheim–Elektron company, because it was the only one in the group to diversify systematically. It was one of the very few firms to develop successfully without starting from or depending on dyestuffs. Griesheim was founded in 1856 to make soda by the Leblanc process, and this remained the principal business until about 1880 when it lost its main customer, Farbwerke Hoechst. It had to find other chemical markets, and so became involved in the development of electrolytic caustic soda and chlorine, of which it became eventually the commercial leader in Germany. Griesheim also dominated the bleaching powder trade by agreements with Rhenania, Verein Chemischer Fabriken Mannheim, Silesia, and others.[5] The company was one of the first to distribute liquid chlorine in pressurized rail tank cars, and from 1905 onwards large sales were made in this way to bleachers who produced their own bleaching solutions. A few years later the practice of chlorinating town water opened a new market.[6] By-product hydrogen from the large electrolytic works at Bitterfeld and Rheinfelden was compressed and used for balloons and for soldering; later Griesheim took part in the development of oxy-hydrogen welding techniques. In 1907 it obtained the German rights to the Claude air-

[1] Spiegel, L. *100 Jahre Kunheim* (1925); passim.
[2] Weber, A. (Ed.) *Über den Standort der Industrien* (1914), pp. 47 ff.
[3] Holländer, H. *Geschichte der Schering AG* (1955), p. 65; *Enquete*, pp. 220–1.
[4] *Enquete*, pp. 314 ff.
[5] During the 1900s Griesheim was sales agent for the chemicals made by some of the potash mines. This agreement ended in 1910 amidst much recrimination. cf. *50 Jahre AG Consolidirte Alkaliwerke Westeregeln* (1931), pp. 212–13.
[6] Pistor, G. *Hundert Jahre Griesheim 1856–1956* (1958), pp. 96–7.

liquefaction process, and together with Linde divided the large German business in industrial gases. (As we saw in the last chapter the company's interest in applied physical chemistry extended to work on the ammonia synthesis from 1905 to 1908.) Electrolytic processes were also applied to a compound of potassium, carnallite, and so to the preparation of magnesium, which began on the scale of 200 tons a month in 1914. Some years earlier the electrochemical activities had been extended to the then novel process of preparing phosphorus in electric furnaces, and it was not long before commercial links were established with the only other European rivals, Coignet in France and Albright & Wilson in England.[1] The manufacture of aniline oil and some intermediates was begun at Griesheim in the 1880s, but for many years the company's organics department was unimportant. In 1900 carbon tetrachloride and monochloracetic acid were added to the selling range; the latter chemical was used in the synthesis of indigo, and production was eventually transferred to a larger plant at Bitterfeld (1902). About 1910 the firm patented a process for making acetic acid from acetylene via acetaldehyde, and by 1913 a pilot plant produced half a ton a day by this route. This first step in aliphatic chemistry was not followed up, because the company did not then have its own supplies of calcium carbide. Other promising developments were the patents in 1912 on polyvinyl chloride and polyvinyl acetate, but the former lay fallow until the 1930s. The latter, however, was made on the scale of 60–70 tons a year during the war for use as coating.[2] In 1905 Griesheim added considerably to the diversity of its activities by the purchase, for about £450 000, of Oehler's dyestuffs and intermediates business at Offenbach, near Frankfurt. It had some association with the pioneering days of organic chemistry, for Hofmann had first made aniline there in 1842. Some sixty years later Oehler's works employed about 550 people and produced 5500 tons of dyestuffs and 16 000 tons of intermediate chemicals. Griesheim enlarged the research section, and in 1911 was able to announce the discovery of the Naphthol AS range of fast dyes.[3] By 1914 Griesheim was among the large German dyestuffs producers, but it is likely that more than half its turnover was still being obtained from the various inorganic chemicals.

The second distinct group of companies in the German chemical industry was that of the fertilizer manufacturers, including the pioneers in the synthetic nitrogen business, and, economically far more significant at that time, the numerous potash mines.

Before the war sulphate of ammonia was by far the most important nitrogenous fertilizer made in the country. In 1913 the coke ovens of the Ruhr were the main source of the 550 000 tons produced; those of Silesia and

[1] ibid, pp. 55, 61–2. [2] ibid, pp. 71 ff.
[3] Entwicklung, Zerstörung und Wiederaufbau der Naphthol-Chemie Offenbach, *Chem. Ind., Düsseld.* N.S. 3 (1951), no. 2, 89 f.

the town-gas works played a subordinate role. The value of this business was estimated at about £3 million in 1906, and £7¾ million seven years later.[1] The fertilizer was sold by the Deutsche Ammoniak Verkaufs-Vereinigung, which was the syndicate of the producers.

On the eve of the Great War synthetic nitrogenous fertilizers were still relatively unimportant. For example the Badische ammonia works at Oppau sold a few hundred tons, and even in 1914 the value of the synthetic ammonia sales was only £125 000, including £50 000 for sulphate of ammonia, which would represent about 4000 tons of sulphate.[2] The cyanamide works, as we saw in the last chapter, were then of greater significance. Bayerische Stickstoffwerke had been founded in 1908 by the Deutsche Bank and the Cyanidgesellschaft (which in turn was owned by the Siemens electrical concern and Kunheim) to work the Frank–Caro calcium cyanamide process. The company had a power station at Trostberg on the Alz, which runs east of Munich near the Austrian border. The cyanamide factory was opened in 1909, but the equipment for the initial capacity of 15 000 tons a year of cyanamide was only installed over the next three years. By 1914 further extensions had been made so that the total capacity was around 25 000 to 30 000 tons a year. Higher figures have been quoted, and were probably attainable if carbide was purchased to augment the works' own supplies. The same owners also operated Ostdeutsche Kalkstickstoffwerke & Chemische Fabriken GmbH near Bromberg. This small factory never operated profitably, but made some carbide, which may have been sold to the sister company at Trostberg.[3] There was a rival business, AG für Stickstoffdünger, which had led a chequered existence. It was the first to make cyanamide in Germany—310 tons in 1905 at Westeregeln—but full-scale operations at the factory at Knapsack (near Cologne) only began in 1910. It is not clear whether the initial capacity was 10 000 tons or 20 000 tons, but by 1913 it seems to have been increased to 25 000 tons a year. About 400 people were employed making cyanamide and also ammonia, which was obtained by treating the material with steam.[4] The first years at Knapsack were unsatisfactory, and in 1911 there was a reorganization, during which Farbwerke Hoechst bought from the Metallbank the shares previously held by Consolidirte Alkaliwerke Westeregeln. (During the war Hoechst bought out the Metallbank and so obtained control.) Hoechst's step-wise acquisition of Knapsack was probably a counter-move to the development of the Haber process by BASF. The investment was, however, much smaller, and Hoechst has retained it to the present day.[5] A third factory at Gross Kayna was established

[1] Bernthsen, A. Über Luftsalpetersäure. *Z. angew. Chem.* **22** (1909), no. 24, 1167–78. The basis for the 1913 figure is the average value of 1·30 M a kilo N.

[2] BASF archives, Geschichte der BASF—Kaufmännische Entwicklung, typescript, n.d., p. 345a.

[3] Waeser, B. *Die Luftstickstoffindustrie* (1st edn., 1922), p. 19; Franck, H. H. *et al. Der Kalkstickstoff in Wissenschaft, Technik und Wirtschaft* (1931), p. 110.

[4] Waeser, op. cit., p. 19; Franck, op. cit., p. 52; Niebelschütz, W. v. *Knapsack* (1957), pp. 65–6.

[5] Niebelschütz, op. cit., p. 66.

by AG für Stickstoffdünger in association with Chemische Werke Albert, Chemische Fabrik Kalk, and Consolidirte Alkaliwerke Westeregeln in 1912.[1] It represented an interesting joint venture by manufacturers of superphosphate, potash, and nitrogenous fertilizer, and was an ambitious undertaking with a planned capacity of 60 000 tons a year of cyanamide. Some sections of the plant were operating by 1913, and the works was completed in 1915, but never functioned properly and was closed during the war.[2]

The total production of cyanamide was said to have increased from 5000 tons in 1909 to 50 000 in 1913 and to 66 000 tons a year later.[3] These figures seem very high and exceed the capacity of the factories. The discrepancy can only be explained on the assumption that carbide was purchased and nitrogenated, and this was in fact done from time to time. Nevertheless it would be prudent to treat the output statistics as optimistic indicators rather than as reliable production returns. Still, the quantity was becoming significant and the cyanamide process was not only working on a large scale, but was also being backed by some very influential German chemical manufacturers.

When we turn to superphosphate we are dealing with an altogether different order of magnitude. In 1900 0·6 million tons were made, in 1913 1·9 million tons, and Germany occupied third place in the world league after the United States and France. Many firms were engaged in this business, and it was estimated that £7½ million were invested in this branch of the industry on the eve of the war, and that annual sales were of the order of £5½–£6 million.[4] In the Rhineland superphosphate was made by companies that had abandoned the Leblanc process and had to find another use for their sulphuric acid made by the lead chamber process. Some of these firms have already been mentioned, and it only remains to add that they obtained their phosphate rock by barge from Rotterdam, where it was transshipped from North America and North Africa. In central Germany the manufacturers started by using local deposits of coprolites, but soon turned to imports. Finally there were factories, some of considerable size, located at the main ports. For example Anglo-Continentale Guano-Werke began by importing Peruvian guano and treating it with sulphuric acid to make a compound fertilizer containing phosphatic and nitrogeneous plant nutrients. Later the company used phosphate rock, and by 1913 owned six factories, employed 1100 people, and had a capital of £1·4 million. In eastern Germany two works, both near Stettin, dominated the superphosphate production in that region. AG der Chemischen Produkten-Fabrik started making alkali in 1857, but around 1900 abandoned the soda trade, and only produced superphosphate; the firm was capitalized at £650 000 and in 1913 employed 900 people. 'Union' Fabrik chemischer Produkte was formed in 1872, and deserves a niche in the history of the industry, for it owned a quarter of the Pacific Phosphate Co., which worked the phosphate deposits in the remote

[1] Greiling, op. cit., p. 56.
[2] Franck, op. cit.; Waeser, op. cit., p. 20.
[3] *Enquete*, p. 156.
[4] *Annu. Lambert*, p. 298 *Enquete*, pp. 74, 241–2.

Pacific islands acquired by Germany in the late 1880s and 1890s. (Nauru, in particular, was becoming an important supplier of phosphate rock in the last years before the war.) 'Union' had a capital of about £850 000 and employed roughly 900 workers.[1] The factories in north Germany had a considerable export business to Scandinavia.

Potash mining, one of Germany's big extractive industries, enjoyed a natural monopoly up to the Great War. Technologically speaking the operations were simple, the most important features being the filtration tanks at the surface and the elaborate pumping machinery underground. Potash mining began in the district to the south and west of Magdeburg and spread later, until it was carried on in central Germany as far west as the line Göttingen–Hanover; there was a recent, but still small, offshoot in Alsace.[2] By 1913 there were 164 pits, compared with 32 in 1905. At the eve of the war about 11·6 million tons of crude potash salts were raised and the value of sales was £10 million. About 30 000 people were employed in the mining and purifying operations.[3]

Potash mining went through some stormy times in the 1900s. The upheavals were of considerable significance, and attracted widespread attention in Germany and among the foreign customers. The troubles stemmed from the marketing organization. Potash sales had long been handled by a syndicate, which was periodically renewed. The effectiveness of the syndicate was weakened when rich potash deposits were discovered in Hanover, Braunschweig, and Anhalt. These small states had less restrictive mining laws than Prussia, with the result that speculators moved in and began to develop the mines at a feverish pace. Soon over-production threatened the syndicate's price structure. The problem was aggravated by the policies of some American fertilizer companies in that period. The Virginia–Carolina Chemical Co. and the American Agricultural Chemical Co., which were large buyers of potash salts, held shares in mining companies, and another American group was reported to have signed a contract below syndicate prices. Schmidtmann, the owner of the Sollstedt and Aschersleben mines, reversed the process, bought shares in the International Agricultural Chemical Co., and contracted to sell it potash at production cost.[4] The Prussian government was alarmed by these disturbing developments, and was in any case an interested party since the Excise owned the largest mines and its share in the syndicate sales

[1] *Enquete*, pp. 271 ff., 280 ff., 285 ff.

[2] Potash was found near Mulhouse in 1904 while drilling for coal and oil. A German group, later taken over by Deutsche Kaliwerke, formed the Amélie mining company which started production in 1910. A French group formed the Mines de Kali de St. Thérèse and began operations three years later, when Alsatian production reached 355 000 tons of crude salts. Amélie was taken over by the French government after the First World War and now forms part of the state-owned fertilizer business.

Gürber, H. Die elsässische Kaliindustrie und das Kaliweltkartell, thesis, (Zurich 1942), pp. 45–7.

[3] Krische, P. *Das Kali* (1923), vol. 1, pp. 136, 257.

[4] Haynes, W. *American chemical industry* (1954), vol. 1, pp. 350–1. The German mines connected with these American firms between them held 13 per cent of the syndicate which operated from 1 January 1905 to 30 June 1909.

amounted to 8 per cent. The government, having failed to reform the industry in 1905, introduced a bill in 1909 which, as the *Reichskaligesetz*, became law on 25 May 1910.

Few pieces of special legislation have had such unsatisfactory results. The law created a compulsory sales cartel, which all mines had to join, and gave the Reichstag authority to fix maximum prices for agricultural potash in Germany. Export prices were the exclusive domain of the cartel, and in principle were to be higher than home prices. The law enraged the foreign buyers, chief among them the Americans, and created economic hostility towards Germany, which became an important factor in the United States after 1914.[1] It also encouraged the search for potash deposits outside the Reich, and during the 1920s these efforts were rewarded and the German monopoly broken. In Germany the law also had important and unfortunate consequences by stipulating that every producer was entitled to a sales quota. It was not a licence to print money, but it gave the go-ahead to an even more frenzied mining boom, for with sales safeguarded by the cartel there were plenty of capitalists eager to sink new mines. On 1 January 1910, 26 pits were under construction, a year later 73, in 1912 102, and at the beginning of 1913 120. The number of mines in operation rose until it reached a peak of 256 in 1915. Meanwhile production per mine fell dramatically from an average of 139 000 tons a year in 1905 to 42 000 tons nine years later.[2] The Government had to step in again, and in 1916 prohibited the sinking of new shafts. This was no solution, and a new potash law was passed in 1919. One might have expected that the failure of the various syndicates and of the 1910 act would have led mine owners to amalgamate and so rationalize their industry, but of this there is only limited evidence. In 1910 half the mines were owned by small firms and half by eight groups, including the Governments of Prussia and Anhalt, Consolidirte Alkaliwerke Westeregeln, Wintershall, Deutsche Kaliwerke, and Deutsche Solvay-Werke.[3] The many new mining operations after 1910 were unaccompanied by any concentration tendencies. To the outsider the remarkable feature is that so little was done to remedy matters. As far as the shareholders were concerned, the main risk was during the building stage. If the money gave out because the expenses had been underestimated, all was lost and millions squandered. But once in operation a quota was assured and a return on capital certain. No wonder then that most of the mine-owners were not interested in chemical manufacture, except for the extraction of bromine and occasionally the preparation of potassium sulphate. Very few had the imagination to diversify into fertilizer manufacture, and when necessity forced them to do so in the 1920s, it was too late, because by then the ammonia plants were in other hands, and they were prevented from making compound fertilizers. Thus the nearsighted

[1] The law also annulled the contracts signed with the American buyers.

[2] Paxman, H. *Die Deutsche Kaliwirtschaft* (1929), pp. 13–14; Krische, op. cit., vol. 1, pp. 135–6.

[3] Paxman, op. cit., pp. 23, 63–4.

attitude of the 1900s permanently weakened a whole sector of the German chemical industry after the war.

The two exceptions to this sorry tale deserve brief mention. Konsolidirtes Salzbergwerk Neu-Stassfurt was formed in 1876. In the 1880s the extraction of bromine and the preparation of sulphates were started, and were followed by the manufacture of caustic potash and bleaching powder. In 1899 Neu-Stassfurt and the Oesterreichischer Verein formed a joint company to build an electrolytic plant at Zscherndorf near Bitterfeld. The works were opened towards the end of 1901 and by 1906 employed 180 people. The alkalis and bleaching powder were sold by Griesheim–Elektron until 1914. Neu-Stassfurt extended its potash operations, but flooding caused heavy losses, and mining ceased after the war; the chemicals business, however, continued to operate profitably.[1] Consolidirte Alkaliwerke Westeregeln, which we have encountered previously, was the other noteworthy company. It was founded in 1881 and remained independent until 1922, when it combined with the Salzdetfurth and Aschersleben potash mines. Its chemical activities comprised for some years shareholdings in cyanamide factories at Westeregeln and later at Knapsack, and also the preparation of electrolytic chemicals. The latter started in 1896, when Westeregeln obtained a licence from Griesheim–Elektron, which, in return, obtained a 25 per cent share in the net profits and handled the sales. The electrolytic plant was opened the following year and by 1906 was producing 4200 tons of caustic potash; later potassium chlorate was also made. The arrangement worked reasonably well for some time, but was terminated in 1911 and Westeregeln henceforth sold its products directly and also was able to broaden its range further.[2]

The dominance of Germany in the dyestuff production and commerce of the entire world is so marked, and inherently of such potential might, that it does not hesitate to make itself felt whenever and wherever an effort is made toward emancipation from its control.[3]

Our purpose is to examine not only how this came about, but also how the business was able to maintain itself so successfully against all possible rivals. There are not many reliable facts about the German dyestuffs industry. According to the 1907 Census of Occupation, 9000 people were employed in the manufacture of aniline dyes and a further 7000 in that of other coal-tar derivatives.[4] The total of 16 000, say one-tenth of the entire chemical industry, certainly understates the relative importance of this branch. We will be on safer ground if we add to dyestuffs manufacture the supporting chemical activities that were so characteristic a feature of the big German works. The

[1] Rüsberg, op. cit., pp. 11 ff.; Anon. *Der Oesterreichische Verein für Chemische und Metallurgische Production* (1906), pp. 51, 75. The company merged with Rhenania–Kunheim in the 1920s, the successor to which now trades as Kali-Chemie.
[2] *Consolidierte Alkaliwerke Westeregeln*, pp. 201 ff.
[3] Norton, T. H. *Dyestuffs for American textile and other industries* (1915), p. 9.
[4] *Statistisches Jahrbuch für das deutsche Reich* **30** (1909), 79.

two were inseparable, and will be so treated in the following pages. On this basis the numbers employed would be at least 30 000 and probably nearer 40 000 by the eve of the war. Dyestuffs production was concentrated in a few companies. Twenty were incorporated with limited liability in 1913, and of this number eight effectively dominated the industry. They were: AG für Anilinfabrikation (AGFA), Badische, Bayer, Cassella, Griesheim–Elektron, Farbwerke Hoechst, Kalle, Weiler–ter Meer.

In 1913 these eight firms and their foreign subsidiaries made about 140 000 tons of dyes out of a world total of around 160 000 tons. Their nearest rivals were the Swiss manufacturers who, between them, made 10 000 tons in Switzerland and elsewhere. Germany herself used about 20 000 tons (including a small volume of imports) and exported roughly 120 000 tons. The trade was valuable and exports amounted to some £11 million. The value of German production was put at roughly £13½ million in 1913, equivalent to three-quarters to four-fifths of the value of world production.[1] Owing to the absence of sales statistics for most of the German firms it is not possible to give a reliable figure for each. The following estimates will, however, serve to show the ranking of the most important.

TABLE 5.4

Estimated dyestuffs sales of German companies in 1913

	(£ million)	
Badische	4·3	(a)
Bayer	3·7	(b)
Hoechst	2·5	(c)
Weiler–ter Meer	0·5	(a)
AGFA, Cassella, Kalle, Griesheim and small firms	2·5	(d)
Total	13·5	

Notes: (a) Company figure; (b) Assumed that half the company's turnover was represented by dyestuffs (excl. inorganic pigments); (c) Last available figure for 1911; this probably understates the value of sales in 1913; (d) Residual figure.

In addition to the sales of finished dyestuffs, all the companies had a large turnover in other products. Badische, for example, sold £1·7 million of intermediates and heavy chemicals; Bayer's trade in pharmaceuticals, photographic film, lithopone, chrome pigments, and other chemicals may have aggregated £3½ million. As early as 1911 Hoechst's sales of pharmaceuticals and veterinary products amounted to almost £0·5 million, and two years later the figure was certainly higher owing to the phenomenal success of the anti-

[1] There are several estimates for production in 1913. Those most commonly accepted are the figures given by T. H. Norton of the USTC in the *Census of dyes 1922* (1923, pp. 123 f.) and by the League of Nations in 1927, op cit., p. 28.

syphilitic compound 'Salvarsan'. Griesheim's big interests in electrolytic alkalis have been previously noted, and the general chemical activities of the smaller firms will be mentioned below. The dyestuffs business was complemented and indeed reinforced by these other activities, and, generally speaking, diversification was fostered throughout the 1900s and was financed out of the profits made from dyes. These profits were still very satisfactory though less so than ten, let alone twenty years previously. The reasons were not only changes in fashions, notably a gradual tendency to switch to lighter colours requiring less dye, but a significant growth of capacity within Germany and also in Switzerland, the United Kingdom, Russia, and the United States. The repeated attempts at cartelization were one response to changing circumstances, another was the formation of groups, a third was diversification.[1] The interest taken by Badische in the synthesis of ammonia and by Bayer in that of rubber were instances of large dyestuffs firms deliberately staking big sums on developments outside their traditional operations, because the rewards appeared to be so much more promising.

We may conveniently start by looking at the common characteristics of the large firms. All had access to a waterway, in particular the Rhine and the Main. This was essential, for they consumed fuels and raw materials on a huge scale. Badische used almost 0·4 million tons of coal and coke in 1914, Hoechst almost 0·3 million tons in 1912. The latter company bought 52 000 tons of pyrites, 38 000 tons of salt, 13 000 tons of tar distillates, and lesser quantities of other commodities.[2] The volume indicates that Hoechst, as well as Badische (and also Bayer), were manufacturers of all types of chemicals on a very large scale. Badische had indeed started that way, Hoechst followed in the 1880s, and Bayer in the late 1890s. By the outbreak of war the daily capacities of their sulphuric acid and chlorine plants testified to the size of these activities: in terms of 100 per cent H_2SO_4, Badische could make 275 tons a day, Bayer 340, and Hoechst 224. For chlorine the tonnage figures were 13, 20, and 4 respectively.[3] They did not go in for coal-tar distillation, and bought benzol and toluol from the Ruhr coke ovens, while for naphthalene they depended in part on imports from Britain. In 1913 (in contrast to the position obtaining since the 1930s) the companies did not make aliphatic chemicals. All required methanol, formaldehyde, acetone, and acetic acid, and, there being no synthetic route, the dyestuffs makers had to depend on the products of wood distillation. In Germany this was controlled by the Verein

[1] Erban, F. *Kartelle und Konventionen in der chemischen Industrie* (1910), tr. anon., p. 33. It is interesting that outright takeovers were rare in the period: apart from Griesheim's purchase of Oehler and Hoechst's of Kalle, the others were of slight importance, e.g. Weiler bought Tillman and joined with ter Meer, and Dahl merged with Wülfing.

[2] Anon. *Farbwerke vorm. Meister Lucius & Brüning*, n.d. (1913?), p. 37; *Die BASF*, n.d. (1922?), p. 25.

[3] Ministry of Munitions, *Report of the British Mission . . . to . . . enemy chemical factories in the occupied zone* . . . (1919), p. 9. Bayer only began making electrolytic chlorine in 1912; the figure for Hoechst is so low because the main plant was at Gersthofen and was not included in the mission's visit.

für chemische Industrie of Mainz, which, even in 1913, had ten factories in Germany, and by Holzverkohlungs-Industrie AG, an associate of DEGUSSA.

To sum up: the dyestuffs companies converted raw materials and purchased chemicals into high-priced finished goods. The element of value added was enormous, and the large cash flow provided the means for further expansion.[1] The depreciation provisions were large, but their full extent cannot be determined from the published accounts. There is some evidence that Badische regularly provided two-thirds to three-quarters of its capital expenditure out of earnings. Whatever the precise figures, the fast write-offs, as we would call them now, were maintained over many years, and resulted in the accumulation of enormous hidden reserves. They are apparent from the large difference between the book value and the original cost of the assets, and are illustrated in the following table.

TABLE 5.5

Book value and original cost of fixed assets of Badische and Hoechst, selected years 1900–13[2]

Year ending 31 December		1900	1904	1908 (£000)	1910	1913
Badische	Book value	1215	1300	1670	1450	2705
	Original cost	3240	4220	5583	6008	8406
Hoechst	Book value	615	955	1130	1155	1465
	Original cost	1855	2410	4400	4740	5701

These financial practices served to strengthen the commercial position which was based on policies systematically pursued over many years. One instance was the invariable practice of patenting all dyestuffs and protecting the manufacturing 'know-how'.[3] (By contrast intermediates were rarely patented and often exported, Switzerland being a large customer.) Another example was the treatment of chemists, who occupied a respected place in the hierarchy and benefited from extensive training, which often included a spell in the dye-house of a friendly customer. The technical service was of a uniformly high standard: the staff consisted of chemists and many of the works had their own dye-houses complete with all the necessary equipment. Dyers were invited to spend some weeks or even longer in the technical

[1] No reliable figures have survived or been published, so that financial analysis is largely guesswork. For example the history of Bayer (1918, pp. 477–9) mentions that in 1906 purchases amounted to £1¼ million and wages came to £0·3 million. Sales were probably of the order of £4–5 million so that there was very roughly £2½–3½ million for salaries, bonuses, depreciation, tax, and profits. As to the latter the published trading profit for that year was £0·85 million.

[2] Based on information from the two companies. The Hoechst figures for 1904 are partly estimated.

[3] It has been said that in 1913 out of the 921 most important dyes, 762 were covered by German patents. Unfortunately neither the tonnage nor the value of the two groups is given. USTC, *Census of dyes* (1922).

application departments to familiarize themselves with the latest colours and techniques. This practice became particularly valuable with the introduction of synthetic indigo and other fast vat dyes, because they presented many novel application problems.[1] The elaborate technical service was supplemented by skilful marketing practices, such as the preparation of splendid and costly pattern books, the care taken with labelling in foreign languages, the network of selling companies dealing with many hundreds of agents, and, not least, careful market research.[2] These characteristics suggest that the leading firms, while possessing a virtual monopoly, took care to maintain it by efficiently exploiting the market for dyestuffs, which they had been largely instrumental in developing in the first place. The purposeful improvement of the technical side, and the energetic, if ruthless commercial attitude, maintained the position of the German industry.

In 1913 dyestuffs manufacture in Germany was in the hands of three groups: the Dreibund of Badische, Bayer, and AGFA; Hoechst and its two satellites Cassella and Kalle; and thirdly 'the independents', of which Griesheim and Weiler-ter Meer alone were significant. In order to complete the picture we might mention that two Swiss companies had factories in the Reich: L. Durand & Huguenin maintained a branch plant at Huningue in Alsace, and Geigy one at Grenzach in Baden. In both cases the reason was lack of space in Basle itself, and not tariff considerations, there being no German duties on Swiss dyestuffs.[3]

The relations between the three groups at the outbreak of war were determined by events that had taken place some ten years previously. C. Duisberg, the young and imaginative director of Bayer, had been impressed by trust building in the United States and canvassed a project to combine all the German dyestuffs and pharmaceutical companies. But H. v. Brunck, the then chairman of Badische, envisaged a union limited to the leading dyestuffs firms. The two men discussed these ideas in the second half of 1903, and at the beginning of 1904 invited the comments of other leading personalities in the industry. The pharmaceutical manufacturers refused to join, and so did Hoechst, whose directors felt that a trust would frustrate inventiveness. Without entirely dismissing these admirable sentiments, it seems more likely that the remarkable success of Hoechst's indigo synthesis led its directors to think that they would do better on their own. Undeterred, the representatives of Bayer and Badische pursued their talks and found that their two firms complemented each other in the range of products made. Bayer was strong in azo-dyes and latterly also in sulphur blacks, and important in the pharmaceuticals trade. Badische specialized in triphenylmethane dyes (the first major

[1] Higgins, S. H. *Dyeing in Germany and America* (2nd edn. 1916), p. 114. Higgins visited Germany in 1905/6, later managed a dyehouse and finally became demonstrator in dyeing at the Manchester College of Technology.

[2] Bayer employed three economists in 1913 in the commercial department.

[3] Jaquet, N. *Die Entwicklung und volkswirtschaftliche Bedeutung der schweizerischen Teerfarbenindustrie* (1922), p. 81.

group of aniline dyes) and in alizarin; more recently it had developed the synthesis of indigo and followed this by the discovery of the first of the anthraquinone vat dyes.[1] These last were, however, posing many difficult technical problems and required complicated and expensive dyeing techniques.[2] It is worth adding that the indigo process used by Badische was more expensive than that of Hoechst.

These difficulties seriously weakened the bargaining position of Badische, because they reduced its profits. The Bayer delegates were equally surprised to see that Badische had a smaller turnover than Bayer. Brunck dismissed these problems as a temporary set-back and insisted that the published accounts, which gave Badische a clear lead, be used for the detailed negotiations. A deadlock threatened, and would have been followed by a break had it not been for the announcement in August 1904 that Hoechst had bought 25 per cent of the Cassella shares, while the latter had obtained 20 per cent of Hoechst's shares. The Bayer people now became worried, because Hoechst was a big producer of azo dyes, as well as a large manufacturer of intermediates, for some of which Cassella henceforth became a captive market. Bayer expected to lose some of its intermediates sales to Cassella and, even more seriously, would now be competing with Cassella, the leading producer of sulphur blacks, backed by Hoechst. Both firms were also likely to become more serious rivals to Bayer in the pharmaceuticals business.

These were alarming prospects, and as Bayer's bargaining position weakened, so that of Badische improved, as a result of signing an agreement on indigo with Hoechst (which effectively shared the market between them) and by drawing closer to AGFA. Duisberg had initially opposed the inclusion of AGFA in the negotiations with Badische, but his colleagues overruled him, and Bayer was now prepared to accept AGFA as a member of the group. The negotiations were resumed, and there were long discussions on profit sharing. The agreement forming an *Interessengemeinschaft* (I.G., literally community of interests) was finally signed on 19 October 1904, to come into effect on 1 January 1906. Badische, Bayer, and AGFA agreed to share profits in the ratio of 43 : 43 : 14, but AGFA was not given a casting vote. The Dreibund had a *Geschäftsführender Ausschuss*, a high level coordinating committee of two representatives from each company. Subcommittees were formed to deal with policy on technical matters, sales, purchases, and patents. In principle the proposal to establish a trust or alternatively to merge the companies completely had been abandoned. The partners, if that is the right

[1] The discovery was made in 1901 by the chemist René Bohn. It was a fine blue dye, and was followed within the year by a yellow ('Flavanthren') and in 1904 by an orange ('Pyranthron'). Later other colours were patented. Bayer's first anthraquinone dye appeared in 1908, those of the other German firms in the last years before the war. In 1922 all agreed to use the trade mark 'Indanthren' originally registered by Badische for these colours. Saftien, K. 50 Jahre Indanthren Farbstoffe. *Die BASF* N.S. **1** (1951), no. 3, 63 ff.

[2] Successful dyeing with these vat colours depended on an efficient reducing agent, and this was sold from 1906 onwards under the name of 'Rongalit'. Dietrich, W. *et al.* Im Dienst der Farbe. *Die BASF* N.S. **6** (1956), no. 6, 245 ff.

description, did not exchange shares. They retained their identities, and made contracts and carried on business as autonomous companies. But soon there was trouble: the affairs of Badische deteriorated, and Bayer and AGFA insisted on improvements. The directors of Badische resented this interference, and personal relations between them and the other directors cooled. While collaboration between the technicians of the three companies was satisfactory, it was non-existent among the commercial staff. The plan for a joint foreign sales organization had to be dropped, because the representatives and the agents refused to cooperate. There was, however, an increase in the trade between the companies, statistics were pooled, agreement was reached on uniform contract conditions, occasionally the three agreed to purchase on joint account, and they bought a coal mine, the Auguste Viktoria pit.[1] Badische, Bayer, and AGFA subscribed in the ratio 43 : 43 : 14 to a half-share in the companies that were formed jointly with Norsk Hydro to work the arc process in Norway. This project was a cause of friction when it was found that the Schönherr process, backed by Badische, did not perform as well as that of Birkeland–Eyde. Neither Bayer nor AGFA were interested in the fixation of nitrogen, and they were content to leave the early development of the Haber process to Badische. In 1911 the Germans withdrew almost completely from the Norsk Hydro venture; Bayer and AGFA thereupon ceased to be connected with nitrogen projects until the start of the large plant at Oppau. All these events were disappointing to one side or the other, and Duisberg was soured by the failure of his grand design: in 1910 he wrote of the Dreibund, that it often seemed to be nothing but an organization for mutually safeguarding profits.[2]

If we look at the trend of the published results we see that Duisberg had cause for dissatisfaction. Bayer was, in effect, supporting Badische. German company accounts of the period are uninformative, but it is apparent from the figures in Table 5.6 opposite that the financial development of Badische and Bayer followed different lines from those anticipated in 1903/4. The results reflected the different tempo at which the two companies were growing, and there is little doubt that, but for the war, a revision of the profit-sharing agreement would have been unavoidable. (The corresponding figures from the accounts of AGFA, which was very much the junior partner in the Dreibund, have not been reproduced in the table, as they do not affect the conclusions.)

The Hoechst group followed a different policy. Cassella, with which Hoechst became closely connected in 1904, was situated not far away, at

[1] It was bought for £850 000 in 1907, and the coal raised increased from 156 000 tons in that year to 715 000 in 1913. Half the production was coked. The mine is still working and now belongs entirely to Badische.

[2] Duisberg, C. Selbsterlebtes und Schlussbetrachtungen, pp. 634–8; written in 1910 and printed as part of the Bayer history, which was never published. Duisberg's comments on the relations between the three firms were more outspoken and interesting on this occasion than the observations on the same subject in his autobiography, written many years later (*Meine Lebenserinnerungen*, 1933) when he was chairman of the IG Farben supervisory board.

Mainkur, a little to the east of Frankfurt. Cassella was a large enterprise, with an issued capital of £1 million in 1908. In that year the two companies tightened their links by a further exchange of shares, and together acquired

TABLE 5.6

Summarized accounts of Bayer and Badische for selected years 1900–13 (£000)

Year ending 31 December		1900	1904	1908	1910	1913
Share capital	Bayer	600	1050	1800	1800	1800
	Badische	1050	1050	1800	1800	1800
Book value of fixed assets	Bayer	435	600	1330	1625	2475
	Badische	1215	1300	1670	1450	2705
Gross profit	Bayer	365	650	920	1160	1480
	Badische	480	545	675	900	1170
Depreciation and to reserves	Bayer	95	215	205	275	655
	Badische	175	230	295	360	565

Note: As far as is known the published accounts were not consolidated. Fixed assets exclude investments in subsidiary and associated companies. The gross profit shown is before taxation, depreciation, and an omnibus item in the accounts called 'general expenses'.

88·8 per cent of the shares in Kalle & Co. This was a relatively small business with an issued capital of only £525 000, but the works were conveniently situated at Biebrich near Frankfurt. The outcome of these transactions was that Hoechst owned 27½ per cent of Cassella's share capital, while Cassella

TABLE 5.7

Summarized accounts of Hoechst for selected years 1900–13 (£000)

Year ending 31 December	1900	1904	1908	1910	1913
Share capital	850	850	1800	1800	1800
Book value of fixed assets	615	955	1130	1155	1465
Gross profit	315	450	690	880	1095
Depreciation	90	100	110	185	275

Note: As far as is known the published accounts were not consolidated. Fixed assets exclude investments in subsidiary and associated companies. Gross profit includes the income received from Cassella and Kalle, and is before tax, depreciation, and an omnibus item called 'general expenses'. The annual report for 1904 is missing, but the figures were available from other sources, except the gross profit, which has been estimated.

held 15·3 per cent of Hoechst's.[1] This was enough to give Hoechst practical control of Cassella and Kalle, and so resulted in the formation of a powerful group of dyestuffs manufacturers, all located in the same district, which acted in concert far more effectively than the various committees of the Dreibund.

[1] Liefmann, R. *Beteiligungs- und Finanzierungsgesellschaften* (1909), p. 244.

The summary figures (see Table 5.7) from Hoechst's annual reports show that it had the same amount of issued share capital as Badische and Bayer. The book value of the fixed assets (excluding Cassella and Kalle, which were simply shown as an investment corresponding to the par value of the shares held) was well below that of the other two companies, but the gross profit, though not growing as rapidly as that of Bayer, advanced more quickly than that of Badische, and by 1913 had almost caught up with it.

Bayer was in 1913 the largest German chemical company. It owned 770 hectares (say 1900 acres), of which the Leverkusen works alone accounted for 370 hectares (900 acres). The firm employed altogether 10 600 people, of whom almost 8000 were working at Leverkusen.[1] The company's original plant at Elberfeld in the Ruhr was devoted mainly to pharmaceuticals, and shortly before the war plans were laid for building a third factory at Dormagen, a short distance from Leverkusen on the left bank of the Rhine. Carl Duisberg, the originator of the Dreibund had, in 1911, become *General Direktor* (that is chief executive), and also chairman of the *Vorstand*, or board of management. He was a chemist by training, but his main interest was management. He ran the business on American lines and his fellow directors were given functional responsibilities; even in those days the company had a full-time director in charge of overseas sales.[2]

One gains the impression when looking at Bayer that its strength lay in commercial development, and that it was short on chemical innovation and chemical engineering. The firm made about 2000 different dyestuffs, the main groups being the azos, alizarin, and sulphur. The manufacture of indigo was started, but before production had got properly under way it was abandoned by agreement with Badische. The output of anthraquinone vat dyes was small because of an agreement with Badische. In dyestuffs generally, Bayer did not show marked originality. The range of pharmaceuticals made was extensive: 'Aspirin', launched in 1898, remained very successful, and was followed in the 1900s by several sedatives, of which the barbiturate 'Veronal' (made under licence from Merck & Co., which had discovered it in 1904) was perhaps the most famous. The company made a large tonnage of inorganic chemicals, and built for this purpose a big contact acid plant at Leverkusen. Bayer actively developed the manufacture of various pigments, and even before the war it was a major factor in the lithopone trade. The company was a latecomer in photochemicals. Its first product in this line was a developer ('Edinol'), discovered accidentally, and launched in 1902. But the business, which was originally carried on in collaboration with the firm of E. Liesegang of Düsseldorf, remained small until it was transferred to Leverkusen in 1907. The production of sensitive paper began in 1906, but as late as 1912 the entire

[1] Pinnow, H. *Werksgeschichte* (1938), p. 141; Farbenfabriken vorm. Friedr. Bayer. *Entwicklung . . . der Firma und ihrer Wohlfahrtseinrichtungen* (1914), pp. 16, 18.

[2] Some of Duisberg's colleagues, among them Krekeler and Mann, became directors of IG Farben in the 1920s and 1930s.

photographic department numbered only eighty-two people. In connexion with the photographic accessories, a great deal of work was done on cellulose acetate non-flammable film, and Bayer appears to have been the first in Germany to sell acetate film ('Cellit') and cellulose acetate lacquers.[1]

As far back as 1906 F. Hofmann began to study the synthesis of rubber at the Elberfeld laboratory, and three years later prepared a rubber-like substance from isoprene. Chemically this was a considerable achievement, but commercially the discovery was of little use, because the synthesis of isoprene was so costly that the end-product could not compete with natural rubber, despite its high price.[2] Hofmann therefore turned to a cheaper raw material: the route started from acetone and proceeded via pinacone to 2,3-dimethyl butadiene (a homologue of isoprene), which was polymerized to 'methyl rubber'. Small-scale production began in 1910, only to be abandoned soon afterwards when the price of natural rubber fell. But sufficient tests had been made to demonstrate that this methyl rubber was an inferior substitute for the natural product. One variant, however, could be treated to yield an excellent hard rubber, and during the war successfully replaced ebonite in battery casings. The synthetic rubber was also used towards the end of the war for other purposes, and some 2500 tons were made in 1917–18. Manufacture was grossly uneconomic and ceased in 1919, but Bayer gained much valuable experience in vulcanization techniques and this led to the manufacture of rubber accelerators and other rubber chemicals in the 1920s.[3]

Badische presented a striking contrast to Bayer, which it approached in turnover, and surpassed in numbers employed. The works were on one site, Ludwigshafen and Oppau, which occupied a long strip of land on the left bank of the Rhine near the town of Mannheim. They covered 272 hectares, and employed about 9240 workers and 2800 foremen and staff in 1913.[4] The company had a long tradition of research: its chemists had pioneered the synthesis of alizarin in Germany, later that of synthetic indigo, and in the 1900s that of the anthraquinone vat colours. Badische had made its process chemicals from the very start, and in the 1890s and later was recognized as one of the leaders in chemical engineering. One of the early achievements was the invention of the extremely important contact acid process by Knietsch and Sapper in the closing years of the nineteenth century, and this was followed about fifteen years later by the industrial development of Haber's ammonia synthesis by Bosch and his team. The Haber–Bosch process, though not yet commercially significant in 1913, played an interesting role in the history of Badische. The fixation of nitrogen involved the solution of many problems in chemical engineering, and the sustained support given to the work by von

[1] Pinnow, op. cit., p. 132; Brand, P. Photographische Erzeugnisse. *Beiträge zur hundertjährigen Firmengeschichte* (1963/4), pp. 172–5.

[2] The large-scale production of isoprene dates from the 1960s.

[3] Schultheis, H. Synthetische Hochpolymere, in *Beiträge . . .*, op. cit., pp. 209–10; Ministry of Munitions, op. cit., pp. 27–8.

[4] Ebert, W. *Die chemische Industrie Deutschlands* (1926), p. 83, and information from company.

Brunck suggests that this was a deliberate effort on the part of the company to take advantage of its engineering skills. Badische became involved in nitrogen fixation from the turn of the century onwards, when Schönherr and Hessberger devised an arc process, which was tried out on a large scale in Norway. The company's annual report estimated that these Norwegian works would cost the Dreibund £1·9 million, so that Badische was committed to spending about £0·8 million on the project. There was also a plan to build another arc plant in Bavaria, but this failed to materialize, and was abandoned in 1910, when the necessary water rights could not be obtained. At the same time the company began the extremely expensive development of the Haber process. All the pilot-plant work was financed entirely by Badische, but the Oppau factory was paid for, if not wholly then in large part, by the Dreibund.[1] Details of the expenditure have not, so far as I know, been published, but certainly exceeded the £1 million that was reputedly spent on the synthesis of indigo, and it was also larger than the cost of the Norwegian enterprise.

To finance these schemes, Badische relied on its traditional dyestuffs business. In 1913 its total sales were about £6 million and included a significant turnover in intermediates (£0·7 million) and inorganic chemicals (almost £1 million). Sales of dyestuffs, at around £4·4 million, were larger than those of the other manufacturers. Exports of all products were about £4·5 million in 1913, and the company claimed to account for one-third of Germany's dyestuffs exports. But the dyestuffs business, the mainstay of the company, was not accompanied by that soaring increase in profitability that distinguished Bayer's operations. Indeed there are some indications that, commercially, Badische was less energetic than its partners in the Dreibund. Its three principal dyestuffs groups were aniline, alizarin, and vat dyes. The first two did not grow very rapidly after 1910; the third did, however, expand extremely fast. Finally, we may observe that the business in inorganic chemicals doubled between 1908 and 1913, and this was due to the development of speciality products, such as reducing agents, and, in 1913, the first small contribution from the synthetic ammonia plant.[2]

[1] It has not been possible to ascertain the extent of the financial assistance given by Bayer and AGFA to Badische. It seems probable that it was in line with the profit-sharing ratios (43 : 43 : 14). The Bayer and AGFA reports for 1913 mention that the increase in their capital during that year was necessary to pay, among other things, for the Oppau plant. It should be remembered that when Badische broke with Norsk Hydro in 1911, Bayer and AGFA also sold their holdings in the Norwegian business.

[2] Badische sales by main product groups were as follows, in £000:

	1908	1910	1913
Aniline dyes	1275	1490	1765
Indigo and anthraquinone vat dyes	1260	1370	1740
Alizarin dyes	615	720	845
Inorganic chemicals	475	625	965
Intermediates	570	670	715
Total	4195	4875	6030

Information from company.

By the eve of the war Hoechst was emerging as a major rival to Badische in vat dyestuffs, but its business in other types of dyes was not growing so rapidly.

TABLE 5.8

Sales of dyes and intermediates by Badische and Hoechst, (£000) 1904, 1908, 1911[1]

	1904		1908		1911	
	BASF	*Hoechst*	*BASF*	*Hoechst*	*BASF*	*Hoechst*
Indigo and other vat dyes	735	380	1260	1075	1480	1065
Other dyes	1495	1030	1890	1035	2330	1375
Intermediates	225	155	570	240	695	340
Total	2455	1565	3720	2350	4505	2780

Hoechst was a much smaller company than Badische, judged by turnover, but it was probably the most diversified of the leaders. At the end of 1913 it employed 8100 workers, 392 foremen, and a commercial and scientific staff of 871, at the main works just to the west of Frankfurt, at Gersthofen (near Augsburg) in southern Germany, where the electrochemical plant was located, and at the Behringwerke near Marburg. Although Hoechst was a large producer of inorganic chemicals, there is no clear evidence of engineering skill or originality as in the case of Badische.

The company's greatest success in the field of dyestuffs was its indigo synthesis, which from about 1904 onwards made a significant contribution to turnover and probably also to profits. While the alizarin dye business was relatively small, the triphenylmethane group occupied an important position. In other sectors, the most notable aspect was chemotherapy, a branch of the pharmaceuticals industry that Hoechst had first developed, and made, as it were, into its own preserve. From 1892 onwards the company supported R. Koch and E. Behring, who discovered a diphtheria serum two years later, and followed this by a large number of bacteriological products, many of them for veterinary use. In the field of human medicine several cocaine substitutes (e.g. 'Novocain') were introduced in the 1900s. But the greatest prewar success resulted from the work of Paul Ehrlich and Sahachiro Hata, and was their discovery of 'Salvarsan' and 'Neosalvarsan'. These organo-arsenic compounds were the first effective remedies against syphilis, and they had a tremendous impact on public health. Hoechst began the commercial production of Salvarsan in 1910, but for some years great manufacturing problems were encountered. The arsenicals are easily contaminated, so that the plant had to work under conditions not previously encountered in large-scale operation, and special machines had to be designed for filling the preparation into vacuum-sealed ampoules. In 1911 the company's entire pharmaceuticals

[1] Information from companies.

business represented about one-eighth of the turnover, but the proportion rose quickly, aided by sales of Salvarsan, which grew from £50 000 in the first year to £150 000 in 1911, and were undoubtedly higher by 1913.[1]

Hoechst's satellites require brief mention: Kalle & Co. employed less than a thousand people, and specialized in dyestuffs. Cassella & Co. dated back to 1807 and began making extracts from dyewoods five years later. By 1900 the company employed just over 2000 people; it differed from the other large dyestuffs manufacturers in that it made no process chemicals and only those intermediates that it had patented.[2] Under the very able direction of A. von Weinberg, Cassella produced excellent direct cotton dyes ('Diamines'), and, as the result of an agreement with Vidal, became the first German company to make direct sulphur black dyes (1895). Other sulphur colours were added to the range in the 1900s. After the exchange of shares with Hoechst, Cassella became its customer for chemicals. The two companies had cross-licensing agreements, and between them were able to supply all the types of dyes then known commercially. Cassella's pharmaceuticals business dated back to 1904, when it began to collaborate with Ehrlich, and resulted in the discovery of bacteriostats sold under the trade names of 'Panflavin' and 'Tripaflavin'. The firm grew considerably in the last years before the war, and by 1913 had about 3000 employees.[3]

The associate of Bayer and Badische in the Dreibund was AGFA, with works in Berlin and at Wolfen, a small town about 25 miles north of Leipzig. The company was not so publicity-minded as the other chemical manufacturers, so that rather less is known of its size and the scope of its activites. At the end of 1913 its share capital was £0·7 million and the book value of its assets was of the same order; gross profits in 1913 were £0·5 million. Employment at the Berlin factory was about 1500. AGFA was well known for its direct cotton dyes (its chemists had discovered the famous 'Congo red' in the 1880s), but the scale of dyestuffs manufacture was small, and remained small after 1904, when all competition with Bayer, its main rival, ceased; a few pharmaceutical specialities, chiefly bromine compounds, were also made. By contrast the business in photochemicals, started in 1888, grew steadily, and by 1914 was said to represent at least 30 per cent of the turnover. Developers and fixers were the initial specialities; later the manufacture of dry plates and film was begun. Cellulose acetate and nitrate became so important that two factories were built at Wolfen, in 1910 and 1914, exclusively for their production.[4]

[1] Farbwerke Meister Lucius & Brüning, op. cit., pp. 30 ff.; Bäumler, E. *Ein Jahrhundert Chemie* (1963), pp. 34 ff.; Vershofen, W. *Wirtschaftsgeschichte der chemisch-pharmazeutischen Industrie* (1958), vol. 3, pp. 104–5; and company information.

[2] *Paris Exhibition* (2), 58–61.

[3] Information from company. Cassella appears in the novel of Sybille Bedford *A Legacy* (1956). One of her characters is Sarah Genz-Kastell who draws her immense wealth from Kastell-Anilin of Frankfurt. Miss Bedford thought of the Gans family, who were Jewish and owned a large stake in Cassella.

[4] Stenger, J. *100 Jahre Photographie und die AGFA* (1939), pp. 24 ff.

The Dreibund and Hoechst with its satellites formed a large group in 1916, from which only the ammonia works of Badische and the cyanamide interests of Hoechst were excluded. The new combine, called Erweiterte IG., will be discussed in greater detail in a later chapter. Here it is sufficient to note that it was joined by the remaining independents, Griesheim–Elektron (which has been previously described) and Chemische Fabriken vorm. Weiler–ter Meer. The latter dated back to the 1860s, and after various mergers emerged as a substantial business with a capital of £0·3 million in 1913, and sales approaching £0·8 million. The works at Uerdingen on the lower Rhine employed about 1500 and specialized in aniline oil (of which about 4000 tons were made for sale under contract to Cassella, AGFA, and Bayer), triphenylmethane and azo-dyes. By agreement with Badische an end was put to competition in the triphenylmethane business. Weiler–ter Meer had its own contact acid plant and in 1914 was manufacturing a camphor substitute and diphenylamine, used as a stabilizer in smokeless powder. The product became an essential war material and the company built a large explosives factory during the war.[1]

Bayer and Hoechst, and to a lesser extent Cassella and AGFA, manufactured pharmaceuticals and proprietary medicines. Except for Bayer, these represented subsidiary activities, but all four companies were associated with the large German pharmaceutical industry. Unfortunately there are no employment and sales figures for the fifteen companies, which, between them, dominated this branch of the chemical industry. Some were descended from eighteenth and early nineteenth century apothecary shops, among them J. D. Riedel, Schering AG, and E. Merck, others had begun as druggists and drysalters, notably C. F. Boehringer, Gehe & Co., and Schimmel & Co. Still others, such as E. de Haën, von Heyden, Th. Schuchardt, and Knoll & Co., had started by making fine chemicals, for which there was a growing demand from laboratories, and later turned to pharmaceuticals.[2]

There was no clear geographical pattern: Schuchardt was at Görlitz in lower Silesia, von Heyden at Radebeul near Dresden, Gehe at Leipzig, Schering in Berlin, de Haën at Seelze near Hanover, Boehringer at Waldhof, a suburb of Mannheim. Merck, the largest of them, had its factory at Darmstadt, and even in 1900 employed 800 workers, 50 pharmacologists and doctors, and a staff of 150; thirteen years later the total number had grown to 2000. The firm specialized in alkaloids and organo-therapeutical compounds.[3] Chemische Fabrik E. de Haën was famous for inorganic fine chemicals, and in 1911 its catalogue listed 10 000 different substances, as well as many pharmaceutical preparations. It remained, like Merck, a partnership for many years, and was not incorporated with limited liability until 1912. In 1923 Riedel and de Haën merged.[4] Schering dated back to 1851, when

[1] Hauser, H. *Geschichte des Werkes Uerdingen . . .* (1956).

[2] Merck, J. H. Entwicklung und Stand der pharmazeutischen Grossindustrie, thesis (1923), p.14.

[3] *Paris Exhibition* (1), 350–3; *Enquete*, p. 458.

[4] Anon. *Hundert Jahre Chemische Fabrik E. de Haën* (1961). In 1955 Riedel de Haën became a subsidiary of Cassella.

E. Schering set up a chemist's shop in Berlin, from which a substantial business developed by the end of the century. Almost at the same time, and independently of Hoechst and Behring, Schering began producing a diphtheria serum (1893), but with less success than its rivals. The manufacture of photochemicals and photographic film began in the 1900s, and quickly became important. The company's chemists synthesized camphor from turpentine (1902), and it was soon produced at the rate of several hundred tons a year as a plasticizer for celluloid. By 1913, 935 workers and a staff of about 300 were employed.[1]

It will be recalled that Duisberg had hoped in 1903/4 to include pharmaceutical manufacturers in his trust. His plan failed, but Merck, Gehe & Co., J. D. Riedel, Knoll & Co., and C. F. Böhringer formed a group of their own in 1905, called Pharma IG. The five businesses retained their identity, but otherwise went far in the direction of a complete merger. For example they agreed to avoid duplication of production, to maintain uniform prices, uphold each other's patents and trade marks, and handle their purchases and export sales on a joint basis. In contrast to the Dreibund, profit sharing was only adopted to a limited extent. There was not much overlap between the members of the Pharma IG. and the dyestuffs manufacturers, and the contacts between the two groups were not close. Bayer and Merck, however, had agreements covering the business in sedatives,[2] and there may have been some other arrangements.

Germany had a virtual monopoly in the manufacture of synthetic essences. The firm of Haarmann & Reimer had developed these specialities in the nineteenth century, and others had been drawn to this small but profitable business. In 1913 the four leading firms employed altogether about 500 people, and the value of their turnover was estimated at just over £1 million.[3]

[1] *Enquete*, p. 433; Holländer, H. *Geschichte der Schering AG* (1955).

[2] Waller, P. *Probleme der deutschen chemischen Industrie* (1928), pp. 81–2; Schmitt, R. Die pharmazeutische Industrie und ihre Stellung in der Weltwirtschaft, thesis (1932), pp. 113–14.

[3] Ausschuss zur Untersuchung . . . der deutschen Wirtschaft, *Die deutsche Seifen- und Parfümerieindustrie* (1931), p. 47.

6

GERMANY'S RIVALS

Britain

BRITAIN was Germany's main commercial rival, but her chemical manufactures, which had once been the dominant factor in world trade, were marking time in the years to 1914. The evidence for this is not precise and the available statistics are often unreliable, but the principal indicators all point in the same direction.

In 1907 the first Census of Production was taken. It was followed by the less detailed Census of 1912, but, owing to the war, the full results were not published. At the earlier date the net output of British industry was £712 million, and almost 7 million workers were employed. Against this broad background we must place the chemical industry, which for census purposes included the drug and perfumery trades, but excluded fertilizer, glue, and disinfectant manufacture. The net output of the industry thus defined was £9·6 million in 1907, and about 52 000 people were employed. Five years later the figures were £11·3 million and 61 000.[1] Obviously the chemical industry was only making a small contribution to the economy as a whole, and certainly did not play the same role as that of Germany. It is also disappointing to observe that between 1907 and 1912 the increase in net output was about 18 per cent, say 3·4 per cent a year, which, even allowing for the limitations of the Census, is unimpressive growth.

The statisticians estimated that in 1907 the value of the industry's sales, after eliminating double-counting as far as possible, was £18–22 million, plus a further £6–7 million for fertilizers, glues, and disinfectants.[2] The total turnover was therefore, at the very highest, of the order of £30 million. By comparison, the sales of the German chemical industry were estimated at £120 million in 1913.

Recent work on foreign trade in chemicals helps to explain why contemporaries were so worried by the trend and by the better performance of the German manufacturers. In terms of volume and at base year prices Britain's chemical exports between 1900 and 1914 increased at only half the rate of exports as a whole. Conversely imports of chemicals grew at twice the rate of all imports.[3] The details of such calculations may be open to argument, but

[1] Final Report of the Third Census of Production (1924), *The Chemical and Allied Trades* (1931), pp. 22 f.

[2] Census of Production 1907, Preliminary Tables HMSO (1910), Cd 5005, p. 544.

[3] Richardson, H. W. Chemicals, in Aldcroft, D. H. (Ed.) *The development of British industry and foreign competition 1875–1914* (1968), pp. 289 ff.

the conclusions confirm the rather disappointing performance recorded by the published trade figures at current prices. Thus the value of chemical and dye exports between 1907 and 1913 rose by only a fifth, which is certainly not very impressive. Such an increase would do little to give a boost to the production of the industry as a whole.

TABLE 6.1

United Kingdom exports of manufactures[1]
(£ million)

	1907	1908	1912	1913
All manufactures	342·0	297·0	385·0	411·4
of which:				
Cotton manufactures	110·4	95·1	122·2	127·2
Iron and steel	46·6	37·4	48·6	54·3
Electrical goods and apparatus	2·5	1·9	4·3	5·4
Chemicals and dyes	8·8	8·5	10·2	10·6
Fertilizers	4·0	3·9	5·3	5·7

Note: chemicals and dyes exclude medicines, drugs, medicinal preparations, paints, pigments, colours, and methylated spirits.

The effect of the recession in world trade in 1908 is apparent, but the recovery of the chemical trades compares unfavourably with that of electrical goods. The overseas sales of many industrial chemicals changed little in the pre-war years, and the increase in value in the above figures reflects in the main

TABLE 6.2

United Kingdom imports of wholly or mainly manufactured articles[2]
(£ million)

	1907	1908	1912	1913
All articles wholly or mainly manufactured	154·9	143·1	185·5	193·6
Of which:				
Chemicals and allied products, and fertilizers	14·3	13·2	15·5	16·2
Including:				
Chemical manufactures	4·1	3·4	4·3	4·5
Drugs and medical preparations	1·8	1·4	2·2	2·0
Dyestuffs	2·1	2·1	2·5	2·5
Fertilizers	2·7	3·0	3·0	3·3

Note: fertilizers include basic slag, bones, guano, sodium nitrate, superphosphate, and phosphate rock. All were classified by the Customs as raw materials, but have been shown here to complete the picture.

1 Annual statement of the trade of the United Kingdom with foreign countries and British possessions 1910, HMSO (1911), Cd 5699; and 1913, HMSO (1914), Cd 7401.

2 Source: as for Table 6.1.

the larger shipments of sulphate of ammonia, soda ash, and copper sulphate. This would suggest that the range of products made was not being broadened, and the pattern of imports confirms the impression. Foreign chemicals may be imported for final use (e.g. dyestuffs), or as intermediate products requiring further treatment. If the industry as a whole had been growing vigorously between 1907 and 1913 one could reasonably have expected a significant increase in imports caused by larger requirements for intermediates. But again the figures, summarized in Table 6.2, do not bear this out.

The failure to grow at a higher rate was due to the lack of competitiveness abroad, and to the slowing down in the development of the main consuming industries at home. For example soap production, which largely governed the output of caustic soda, reached 353 000 tons in 1907, and then 'remained pretty well stationary' until the war.[1] Similarly the manufacture of superphosphate, which had grown rapidly in the early 1900s, later stuck at around 0·8 million tons, and this affected the production of sulphuric acid. It was estimated that the demand for acid was increasing by only 2–3 per cent a year before the war.[2] Nor were conditions among the other major users of industrial chemicals particularly favourable. Finally, the technological changes that led to new products and so gave rise to additional requirements for process chemicals—examples were mercerized cotton and viscose rayon—were only beginning to make themselves felt.

In 1913 as in 1900 the principal branch of the British chemical industry was the soda business, or more precisely the manufacture of a wide range of alkalis by the ammonia–soda, Leblanc, and electrolytic processes. There are no comprehensive statistics, but it has been reported that the United Kingdom consumption of alkalis increased at an average rate of 3½ per cent a year before the war and in 1913 was of the order of 400 000 tons. Exports fluctuated, but usually exceeded 200 000 tons annually and in 1913 reached a peak of 272 000 tons. Imports were small. Production may therefore be estimated at around 700 000 tons and possibly as high as 750 000 tons.[3] Brunner, Mond & Co. effectively dominated the business from its four ammonia-soda works in Cheshire at Winnington, Middlewich, Sandbach, and Lostock, the last being the site of the Bowman, Thompson factory, which had been bought in about 1907.[4] In the financial year 1913/14 Brunner, Mond produced 670 000 tons of alkali. Soda ash was the main product (over half a million tons being made), but soda crystals, bicarbonate, and caustic soda were also supplied.[5] The survival of the Leblanc process had been the *raison d'être* of

[1] Wilson, C. *The history of Unilever* (1954) vol. 1, p. 116.

[2] *Chem. Trade J.* **64** (1919), no. 1658, 185, referring to the statistical supplement of the Departmental Committee on Sulphuric Acid.

[3] Steel, L. Forty years on. *I.C.I. Magazine* **41** (1963), no. 310, 8 ff. The Census gives 0·7 and 0·95 million tons for 'soda compounds' in 1907 and 1912 respectively. These quantities are inflated by some double counting.

[4] Irvine, A. S. *A history of the alkali division*, n.d. (1959?), pp. 22 f.

[5] Information from Mr. W. J. Reader. Production figures for the other companies have either disappeared or are manifestly incomplete.

the United Alkali Co., but by the eve of the war this large firm had been forced to accept change. It still made some alkali by the old route at Widnes, but most of its output came from the new ammonia-soda works at Fleetwood. The company's total production was well below the 100 000 ton mark and probably did not exceed 50 000 tons.[1] The third manufacturer was the Castner–Kellner Co., founded in 1895 to make caustic soda and chlorine by electrolysis. Its main works were at Weston Point near Runcorn, opposite Widnes. It is unlikely that it made more than about 20 000 tons a year of caustic soda.

An account of the structure of the British soda trade may conveniently begin with a reference to the links and agreements between the manufacturers. For instance, Castner–Kellner had obtained from Solvay & Cie the right to use the Kellner process. Solvay, a Belgian firm, had a substantial holding in Castner–Kellner and also in Brunner, Mond, so that we may reasonably assume that the relations between the two British companies were friendly if not close. They also had made an agreement with the United Alkali Co. It seems that in the early 1890s Brunner, Mond and United Alkali had reached an understanding on prices and markets for soda ash; caustic soda and some other products were included a few years later. Representatives of the two firms occasionally met those of Castner–Kellner and Chance & Hunt of Oldbury to discuss prices and export policy. There were also agreements not to poach each other's workers, and though the companies did not pay the same wages, they seem to have acted in concert when making changes in the rates of pay.[2]

The main purpose of the various arrangements and agreements was the prevention of price wars at home and the maintenance of a common front in foreign markets. Exports of soda ash, amounting to 156 000 tons worth £560 000 in 1913 were, in this respect, less critical than those of caustic soda, for which the figures were 75 000 tons and £720 000. The trade statistics show that between 1907 and 1913 shipments of soda ash to Japan and to British Dominions and Colonies almost doubled from 42 000 tons to 81 000 tons, and this increase represented three-quarters of the improvement in exports. The growth of sales to these particular countries suggests that there were some international agreements, which Solvay may have helped to bring about. But caustic soda presented different problems. There had been agreements for many years whereby the leading manufacturers undertook to maintain price lists and commission terms; sometimes there were also limits on the annual tonnage delivered. In the 1900s the English Caustic Soda Makers' Association was active for some years and included the United Alkali Co., Brunner, Mond, Castner–Kellner, Chance & Hunt, and Joseph Crosfield & Sons Ltd.

[1] Alkali sales in the home market rose gradually and by 1911 reached 35 000 tons of ammonia soda and caustic soda. In addition there was some export business but the quantities are not known. Carl Duisberg in a lecture given in 1911 (Die Wissenschaft und Technik in der chemischen Industrie. *Z. angew. Chem.* **25** (1912), no. 1, 6) alleged that 120 000 tons of alkali were still being made by the Leblanc process, Even if sodium sulphate is included, this is much too high.

[2] I am indebted to Dr. D. W. F. Hardie for this information.

The Association's job was to shore up the export business, which in 1910 accounted for half the United Kingdom production. American tariffs effectively excluded British-made chemicals. If the price were cut sufficiently, say by about a fifth, the United States tariff could be overcome, but at such a low level Brunner, Mond alone could work at a profit.[1] With the United States closed to foreign producers, British manufacturers had to struggle for a share in the other markets of the world. Here the problem was that American firms could and did cut prices and so outsold the British. No agreement could be reached between the parties because the Americans feared the Anti-Trust Act. Trade with Europe presented other problems: in the first place there were the Castner–Kellner licensees controlled by Solvay, who would not tolerate competition. And in the second place the British apprehended retaliation by other manufacturers of electrolytic alkali, who threatened to dump bleaching powder on the British market. This had occurred earlier in the century. Griesheim–Elektron (though not a Castner–Kellner licensee) was the leader of the German syndicate, and had undersold the United Alkali Co. first in Europe and then in England, until United Alkali had capitulated and reduced its sales of caustic and bleaching powder to Europe. (The *Chemical Trade Journal* referred to these events in 1900–4 as the 'Bleaching Powder War'.) The British exporters of caustic soda were thus left with secure markets only in southern Europe, Brazil and Japan, which between them took around 35 per cent of exports, and with the colonies, which accounted for about one-fifth. The proportions hardly altered after 1907, and, as the consumption in these countries grew slowly, it is not surprising that caustic soda shipments moved narrowly between 73 000 and 78 000 tons a year between 1907 and 1913. The performance of exports would certainly have been worse if the Colonies and Japan had not been, as it were, reserved for British firms.

These alliances and understandings extended beyond the manufacturers to their suppliers and customers. Among the former the Salt Union played a key role, for it controlled over 60 per cent of the country's salt production.[2] But Brunner, Mond had their own brine fields, and the two companies agreed that the Salt Union would discourage other firms from making alkalis in Cheshire, while Brunner, Mond promised not to assist anyone intending to produce salt.[3] As for the customers, a Government committee quoted an allegation of the soapmakers that 'neither of the two alkali companies was desirous of supplying a soapmaker who had previously been supplied by the other alkali company'.[4] The soda makers denied this, but 5-, 10-, and even

[1] Musson, A. E. *Enterprise in soap and chemicals* (1965), pp. 236 ff.; Notes of Meeting of English Caustic Soda Makers' Association, 15 December 1910, in Chance and Hunt Archives.

[2] Standing Committee on Trusts . . . the Existence of a Ring in the Salt Trade, 1920 (Cmd 832), p. 4.

[3] Calvert, A. F. *Salt in Cheshire* (1915), pp. 562 ff. By 1913 the agreement had become meaningess because the Salt Union was almost bankrupt, and Brunner, Mond stronger than ever.

[4] Report on the soap industry (1921), Cmd 1126, p. 15.

25-year contracts were usually made with soap works before the war, and in this way the customer was effectively tied up. Brunner, Mond were more successful at this sort of bargaining than United Alkali, because they were more powerful. But it was to cause them much trouble when they were confronted by an equally powerful customer in the person of Hesketh Lever.

The various national and international agreements helped to stabilize the alkali business. With demand at home rising by about $3\frac{1}{2}$ per cent annually and exports growing slowly, this was perhaps the best solution to a difficult trading situation. But the outcome was that the largest sector of the British chemical industry was going through a long period of semi-stagnation. Old-fashioned businesses, unable to modernize, suffered from these unfavourable conditions, and among them the United Alkali Co. was hit most severely.

This giant among chemical manufacturers had been formed by merging the leading Leblanc soda makers in 1890. Some thirty years later the company was said to retain 'the bulk of the heavy chemical trade of the country, manufactured by methods probably the most advanced in the world'.[1] The statement reflected post-war optimism, but in the years leading to 1914 the United Alkali Co. was declining, and the capital had to be written down in 1913. The following figures tell their own story, that of the shareholders is equally dismal: their £10 shares had been without dividends for years, and at the beginning of 1913 were quoted at £2. 6*s*. 0*d*. A year later, after the capital reconstruction, the new £1 shares were traded at 10*s*. 9*d*.!

TABLE 6.3

United Alkali Co.: summarized balance sheet and profit figures selected years 1900–13

	Years ending 31 December (£*million*)			
	1900	1910	1912	1913
Fixed assets after depreciation	7·6	8·3	8·2	5·7
Net assets	8·4	9·4	9·3	6·8
Represented by:				
Share capital (ordinary and preferred)	5·7	5·8	5·8	3·4
Loan capital and mortgages	2·5	3·0	2·9	2·9
Reserves	0·2	0·6	0·6	0·5
Gross profit from trading	0·4	0·5	0·5	0·4
as percentage of net assets	4·8	5·3	5·0	5·9
Net profit	0·2	0·3	0·2	0·2
as percentage of net assets	2·3	3·2	2·2	2·9

Note: Reserves include the balance of the Profit and Loss account after payment of the interim dividend. Gross profit is before rents, rates, and transfer to Debenture Redemption Fund. A dividend of 2 per cent was paid in 1910.

[1] Anon. *Centenary of the alkali industry 1823–1923* (1923), p. 7.

In retrospect it is easy to see what had gone wrong. In 1890 the Leblanc process was moribund, but the formation of United Alkali delayed the end. In 1906 this cumbrous and out-of-date concern, though reorganized, still had 47 factories, mostly in Widnes and St. Helens, employed 12 000 people, and claimed to be the largest producer of chlorine and sulphuric acid in the world. It owned salt deposits and coke ovens, and, in Spain, pyrites mines. During the 1900s the company was still working the Leblanc process, but also operated an ammonia-soda factory at Fleetwood. The directors tried to develop the by-products and co-products of the Leblanc soda business and a great many chemicals were made, among them nitric acid, cyanides (about 1000 tons a year), copper sulphate, coal-tar distillates, monochlorobenzene, chloroform, acetone (from imported acetate of lime), carbon disulphide, soap, sulphate of ammonia, and, until about 1906, superphosphate, after which fertilizer production declined.[1] At Bay City in the United States a subsidiary made chlorates from 1898 onwards.

This miscellaneous collection of products had little in common except that it could be made by others at greater profit. The directors, descendants of the great Victorian soda makers, were imbued with the traditions of the Leblanc trade, and unable or unwilling to carry out the drastic changes that were overdue. For example, little was done to improve fuel efficiency, though coal was becoming dearer, and only in 1914 were the shareholders told that steps were being taken to reduce the wasteful use of fuel. But far more serious was the technological short-sightedness of the leading men. The company's chief chemists, F. Hurter, and after him J. B. Raschen, were competent men, but they underrated the potential of electrochemistry, which their directors also failed to perceive. Hurter turned down the offer of a Castner licence from the Aluminium Co., an event which led directly to the formation of the Castner–Kellner Alkali Co. The directors eventually decided to build an electrolytic works. This was in 1912, but owing to the war the Gibbs cell plant was not opened until ten years later. The Leblanc process itself was retained, though overdue for scrapping, in order to make alkali and provide chlorine for bleaching powder and the newly introduced organic compounds.[2] With so much old equipment around and profits stationary there was no hope of far-reaching modernization. As a result United Alkali was in decline. One sign of the trend was the run-down of labour. By the outbreak of war the number had fallen by roughly a third to about 8000–8500. Widnes suffered most: its fortunes had been linked with the rise of the Leblanc soda trade, and in 1894 United Alkali had employed 5760 in that chemical town; by 1913 there were 3500. Other jobs were not easy to find in Widnes, because the detritus of the old works did not attract new industries.[3] In nearby St. Helens there had been

[1] Anon. *The struggle for supremacy* (1907); Hardie, D. W. F. *A history of the chemical industry in Widnes* (1950), pp. 154–5.

[2] Hardie, op. cit.; *Centenary of the alkali industry*, op. cit. The production of soda by the Leblanc process came to an end about 1921.

[3] Diggle, G. E. *A history of Widnes* (1961), p. 136; Annual Report of United Alkali for 1914.

eighteen chemical works in 1868, when the Leblanc boom was approaching its peak, but by 1912 only nine were left, of which United Alkali owned six. It had carried out the closures in St. Helens with some vigour, and the redundant labour was in many cases able to move to Pilkington's and to Beecham's, to glass and to patent medicines, which were prospering. In 1911 United Alkali's soap works at Runcorn was sold to Levers, which closed it the following year. There were also large factories on Tyneside at Gateshead and Hebburn, which were still making soda compounds in 1914, and the works at Fleetwood was making progress.

Brunner, Mond & Co., the traditional rival, was smaller but more profitable. The accounts, summarized below, cannot be readily compared with those of United Alkali, but the shareholders had no cause for complaint. The ordinary £1 shares stood at around £5 for many years regardless of bonus and rights issues. At the beginning of 1914 the Stock Exchange valued the business at £13·6 million, and a few years earlier Brunner, Mond & Co. had come after J. & P. Coats, Imperial Tobacco, and Guinness when ranked according to gross profits of British companies.[1]

TABLE 6.4

Brunner, Mond & Co., summarized balance sheet and profit figures selected years 1900–14

	Years ending 31 March (£ *million*)			
	1900	1910	1912	1914
Fixed assets after depreciation	2·0	2·9	3·1	3·8
Net assets	2·7	4·2	4·9	6·0
Represented by:				
Share capital (ordinary and preferred)	1·7	2·8	3·5	4·0
Reserves	1·0	1·4	1·4	2·0
Gross profit	0·4	0·8	0·9	0·8
Net Profit	0·4	0·7	0·8	0·8
as percentage of net assets	14·8	16·7	16·3	13·3

Note: Reserves include the balance of the Profit and Loss account after payment of the interim dividend. Gross profit is after depreciation and transfer to reserves. Net profit is after salaries and taxes, but before directors' fees.

Brunner, Mond's satisfactory trading record was due in some measure to able management, but to an even greater extent to the superiority of the Solvay process as a method of making alkali. The close links with Solvays were mutually beneficial since there were extensive technical exchanges and commercial understandings, which were reinforced by the presence of A. Solvay and G. Querton on the board. By the eve of the war the company

[1] Hart, P. E. Business concentration in the U.K. ***Jl. R.** statist. Soc.* **123** (1960), pt. 1, 56–7. In 1924 Brunner, Mond dropped to sixth place after Levers and Courtaulds.

had eliminated most rivals, and gone into the soap business through the purchase of Joseph Crosfield & Sons Ltd. Crosfields was a large buyer of soda ash, from which it prepared caustic soda and sodium silicate. Lever Brothers also bought alkali from Brunner, Mond, though on less favourable terms, and in 1911 they acquired land and salt deposits at Lymm in Cheshire. Brunner, Mond feared, not without some justification, that Levers would go into the soda business. They reacted swiftly and by an exchange of shares bought out Crosfield as well as another soap business, Gossages of Widnes. But Brunner, Mond obtained little benefit from the move, since both companies were already customers, while relations with Hesketh Lever and his associates deteriorated. Lever retaliated by taking over the remaining large soap makers. At the beginning of 1913 Brunner, Mond and their two soap companies were facing the largest British soap maker, from whom they wanted a new long-term contract for alkali. A compromise was the only solution. Brunner, Mond agreed not to extend their soap interests and to operate Crosfields and Gossages for the benefit of Lever Brothers as well as of themselves. In return Lever signed a ten-year contract on preferential terms and sold them the land at Lymm. The settlement nevertheless created a new problem: the Crosfield management, who detested Lever, considered that they had been ill-used by Brunner, Mond, and indeed they had been pawns in a struggle of big business. Brunner, Mond also became dissatisfied and after further litigation agreed in 1917 to sell half their shares in Crosfield–Gossage to Lever Brothers, but the transaction was not carried out. Two years later, however, the entire holding was sold, and Brunner, Mond could console themselves with a profit of about £1·8 million on the deal.[1]

The intricate negotiations and the disputes indicate Brunner, Mond's preoccupation with a market that was only growing slowly. While United Alkali looked for a solution to this problem by diversifying, but failed, Brunner, Mond found the answer by extending their hold over the soda trade so that, but for United Alkali, they had the business to themselves: during and immediately after the war they bought Electro-Bleach & By-Products of Middlewich (the successors to the Electrolytic Alkali Co.) and the Ammonia Soda Co. of Plumley; the Castner–Kellner Alkali Co., in which Brunner, Mond already held a minority interest, was an acquisition of an altogether different size, with large electrolytic works at Runcorn and on Tyneside.

Cheshire and South Lancashire remained the centres of the British heavy chemical industry. There were less important clusters of factories along the Tyne, and in London, Manchester, and Glasgow. In the Midlands, Chance & Hunt of Oldbury made a range of inorganic chemicals, and owned salt works in Staffordshire and a sulphuric acid plant in Monmouthshire. They and Albright & Wilson, another Oldbury firm, owned British Cyanides. Albright

[1] Musson, op. cit., pp. 240 ff. and Wilson, op. cit., vol. 1, pp. 126 ff. have detailed descriptions of this complicated business. For the sake of simplicity I have omitted the litigation over the fat hardening (hydrogenation) process, which further exacerbated relations between the two firms.

& Wilson's main business was phosphorus making and their principal product was white phosphorus for the match industry.[1] The company used the electric furnace process at Oldbury and the Niagara Falls (N.Y.) factory, which had been opened in 1897. From 1903 onwards Albright & Wilson also controlled the Electric Reduction Co. at Buckingham (P.Q., Canada), and taken together, its North American enterprises were probably more important than those at Oldbury.

Connexions with foreign manufacturers extended beyond the soda trade. For example 25 per cent of the shares of the British Oxygen Co. were held by Ges. für Linde's Eismaschinen, which owned Linde's patents on industrial gases.[2] Even more important was Nobel's Explosives Ltd., which were by far the largest producers of explosives in the British Isles. They had bought up many rivals, including Prentice's New Explosives Co. of Stowmarket. This was then the leading producer of gun cotton in the British Isles, and was developing new uses for nitrocellulose in varnishes. Nobel's formed part of the Nobel Dynamite Trust Co., a concern with extensive interests in Germany, and on friendly terms with the Soc. Centrale de Dynamite, a holding company with investments in France, Spain, Switzerland, and Italy.[3]

The manufacture of fertilizers, which in many other countries represented a growth point for the chemical industry generally, developed unevenly in Britain. The sulphate of ammonia business was making satisfactory progress: production rose from about 370 000 tons in 1910 to 430 000 tons three years later, and the proportion contributed by coke ovens increased from 27 per cent to 31 per cent of the total. Three-quarters were exported: Japan and Spain were the largest single markets and between them accounted for over two-fifths of overseas sales. At a time when chemical exports were growing slowly, those of sulphate were bounding ahead, and rose from £3·4 million in 1910 to £4·4 million in 1913. By contrast the position with regard to superphosphate was unsatisfactory and this reacted on sulphuric acid. A Government Committee reported

> . . . that the difficulties were due to a variety of causes of which one of the most important was the smallness of the scale of production in most cases, in consequence of the limited home demand . . . the productive power of the manufacturers was substantially in excess of the amount which could be disposed of in the home market whilst export was restricted by natural circumstances.[4]

While the conditions for the manufacture of superphosphate were not in general more favourable on the Continent, the deterioration of the balance of

[1] *I.C.I. Magazine*, December 1951, pp. 376–7; Threlfall, R. E. *The story of 100 years of phosphorus making* (1951), pp. 123, 147.

[2] Monopolies and Restrictive Practices Commission, *Report on the supply of certain industrial and medical gases* (1956), pp. 9 ff.

[3] Trotter, R. (Ed.) *The history of Nobel's Explosives Co. Ltd.* (1938), pp. 99 ff.

[4] Committee on Commercial and Industrial Policy after the War, Final Report, 1918 (Cd 9035), p. 20.

trade was exceptional. The Committee's reference to 'natural circumstances' as restricting exports is obscure, unless shipments by the sea route were meant. But this was no obstacle to imports, which doubled in a few years.

TABLE 6.5

Great Britain and Ireland: superphosphate trade selected years 1907–13[1] (000 tons)

	1907	1910	1912	1913
Estimated production	845	722	801	810
+Imports	11	30	50	64
−Exports	115	153	89	63
=Apparent consumption	741	599	762	811

The fertilizer trades (which for Census purposes included glue and disinfectants) were large in that they employed about 12 500 people in 1907, but their backwardness is illustrated by the limited extent to which they used mechanical aids: they had a mere 22 000 hp to handle 1·3 to 1·6 million tons of products.[2] The business was seasonal, labour intensive, and insufficiently profitable. At the outbreak of war superphosphate manufacture represented another sector of the British chemical industry where change was overdue.

The pre-war history of the British dyestuffs industry is a sorry tale. An account of the situation is made difficult by the post-war recriminations and the special pleadings of those who, after 1918, painted a black picture of the position before 1914 in the hope that the comparison would redound to their credit. The result is to cause confusion and to create the general impression that before the war the British makers of dyestuffs, despite valiant efforts, were being driven against the wall by the ruthless and unscrupulous Germans. The story, however, is not so simple.

In 1907, 7000 tons of dyes valued at £373 000 were made in the United Kingdom; the average value was £53 a ton. Six years later production was estimated at 4100 tons; imports were 18 300 tons and exports 2400 tons, so that the apparent consumption was 20 000 tons.[3] British production was equivalent to about a fifth of the country's requirements for dyestuffs, and consisted predominantly of the simpler and cheaper colours. This is borne out by the export statistics, which show that the average f.o.b. value was between £70 and £80 a ton. The average value of imports was significantly higher.

[1] Gray, A. N. *Phosphates and superphosphates* (1930 edn.), p. 159.

[2] Census of Production, 1907, pp. 543, 558. The tonnage is an estimate by the Census office and is free from double counting.

[3] Census of Production, 1907; Report of the Dyestuffs Industry Development Committee, 1930 (Cmd 3658), p. 10.

TABLE 6.6

United Kingdom imports of coal tar dyestuffs, selected years 1907–13[1]

	1907		1910		1912		1913	
	000 *tons*	£000	000 *tons*	£000	000 *tons*	£000	000 *tons*	£000
Alizarin and anthracene	3·1	269	3·1	283	3·1	264	3·0	272
Aniline	10·9	1271	12·8	1453	14·2	1464	14·1	1543
Indigo	2·1	158	1·4	101	1·4	91	1·2	77
Total	16·1	1698	17·3	1837	18·7	1819	18·3	1892

It is apparent from the above statistics that imports of aniline dyes rose appreciably between 1907 and 1913, thereby reflecting the inadequacy of British production. By contrast imports of indigo were halved, because the Germans considered it prudent to comply with the Patent Act of 1907 and began making the dye from imported intermediates in England. This was apparently the only occasion when the Act achieved its objective as far as chemical products were concerned. The Act had been steered through Parliament by Lloyd George when President of the Board of Trade. A significant innovation was the provision in Clause 27 whereby a foreign patent taken out in the United Kingdom had to be worked there. The patentee had to work his patent to an 'adequate' extent, and if he failed to do so had to give a satisfactory account of his reasons. Failure to comply with Clause 27 entailed the revocation of the patent. An important and novel concept was that the patentee had to prove, if called upon, that there was no abuse of monopoly, 'the abuse being in certain circumstances presumed, unless disproved'.[2]

For a short time the supporters of Patent Law reform were jubilant, and there were high hopes of a new attitude by foreign chemical companies to manufacture in Britain. They were confirmed when Badische and Hoechst announced plans for the building of dyestuffs factories. Badische took the Act sufficiently seriously to assign 270 of its British patents to the newly formed subsidiary, Mersey Chemical Co. Nearly one-fifth of the parent company's patents were thus transferred, including most of the synthetic ammonia patents. Further transfers of patents from Badische to Mersey Chemical ceased in 1910.[3] By then it was clear that the Act was not working as intended. It had come into force on 1 January 1908, but almost two years later only 69 applications for revocation had been made, and of these 15 had resulted in revocation of the foreign patent.[4] The Germans were aware that

[1] Source: as for Table 6.1.
[2] R.P.C. **26** (1909), p. 240, line 7. Judgement by Mr. Justice Parker in the High Court.
[3] Zimmermann, P. S. *Patentwesen in der Chemie* (1965), p. 121.
[4] *Chem. Trade J.* **45** (1909), no. 1175, 498.

Clause 27 aroused little interest, and the outcome of the 'Eternit' case showed them that it was in practice ineffective.

One C. W. Zerenner applied under Section 27 for a revocation of the foreign patent on 'Eternit' asbestos cement on the ground of non-working. The Comptroller General of Patents heard the case in November and December 1908 and granted the application. The patentee Hatschek, and his licensee the Eternit Co., appealed, but lost in the High Court in March 1909. This settled the matter as far as the 'Eternit' patent was concerned, but in his judgement Mr. Justice Parker gave a new interpretation to Section 27. In the first place he considered it unfair to put the patentee on the defence, and placed the onus on the applicant to show abuse of the monopoly by reason of not working the patent in Britain. In short the applicant must first present his evidence, and then the patentee should prove his defence. And in the second place the judge upon examining the precise meaning of the clause decided that its object was 'to secure fair play between foreign industries and the industries of this country, and not to secure to the latter, during the period of the Patent, a protection which they would not enjoy if no Patent had been granted'.[1] His first point certainly did not reflect the intention of Parliament and his second was extraordinarily naive. The judge by drawing the teeth of Section 27 strengthened the inclinations of those who preferred agreements to competition. Under the changed circumstances there seemed henceforth little point in applying for revocation. Even if it were granted, would the foreign patentee under the new conditions pass on 'know-how'? And what if the British firm having been granted the revocation had to endure retaliatory action by the patentee in other countries? Such considerations probably had a greater practical effect than Mr. Justice Parker's decision in the 'Eternit' case. The Act of 1907 was a brief interlude, and did not really change the position which was summarized some years later by the Birchenough Committee:

> ... It has been ascertained that a considerable part, probably the greater part, of our pre-war production was not competitive, but was tolerated by the Germans upon conditions, which practically made British dye manufacturers tributary dependents of the gigantic German industry.[2]

There are several instances of agreements that enabled British manufacturers to survive, among them one for aniline oil and another for the important group of sulphur black dyes. The British Alizarine Co., the sole United Kingdom producer of that dye, was a member of the international alizarin convention. The manufacture of indigo was blocked by Hoechst, which made it at Ellesmere Port and thereby secured its British patent and monopoly. As for the other types of dyes made by British firms, it was reported that the Germans could have driven them out 'at any time had they

[1] R.P.C. **26** (1909), pp. 1–11, 228–48.

[2] Report of the Departmental Committee appointed by the Board of Trade to consider the position of the textile trades after the War, HMSO (1918), Cd 9070, p. 107.

chosen to do so'.[1] This may have been an exaggeration, but generally speaking the situation was bleak. A tariff would not have helped the British producers because they lacked patents, know-how, and, above all, research and enterprise. Their small size was not at the root of the problem, for the Swiss manufacturers and the lesser German firms were no larger, but they had what the British lacked: competent management and a market effectively regulated by conventions and cartels.[2]

Levinstein Ltd. was the largest British firm.[3] The works at Blackley outside Manchester had initially specialized in intermediates, but the business, which was thriving at one time, suffered when the Germans reduced their purchases after 1908. It was said that Levinstein's prices and quality were unsatisfactory, but these were mere pretexts. The real cause was retaliation by the German buyers against Ivan Levinstein for having played such an active role in Patent Law Reform and the Act of 1907. The loss of the intermediates trade caused Levinstein to develop the manufacture of dyestuffs, and by 1914 over 150 different colours were made. Attempts to broaden the product range led to patent litigation with the Germans, but he won at least one case against AGFA over sulphur blacks.[4] But the range of dyes remained incomplete, and the situation was extremely difficult because Levinstein had to import some of the essential intermediates from German firms that were also clients of his chief rivals. The favourable verdict in the sulphur black litigation cannot have endeared him to AGFA and Read Holliday who between them had supplied almost the entire British market until then. By the eve of the war, when Herbert Levinstein had taken the place of his father, the company could make 5000 tons of dyes a year, which put it among the large producers, but for commercial reasons the works were operating well below capacity.[5] Levinstein blamed the dyers and their customers (such as the War Office, which specified German dyes for khaki), but it is possible that his production methods were inefficient.

Read Holliday & Sons Ltd. were large makers of inorganic acids, intermediates, and a few dyestuffs. In 1901, on the death of Robert Holliday, Joseph (later Sir Joseph) Turner and L. B. (later Major) Holliday became joint managers. They were extremely energetic and under their direction the business prospered and paid regular dividends. Hollidays specialized in sulphur black for which they obtained monochlorobenzene from United Alkali. The works at Turnbridge near Huddersfield was sufficiently competitive to interest the Cassella Co., which made an abortive take-over bid for them in

[1] Report on Dyes and Dyestuffs, HMSO, (1921), Cmd 1370, p. 4.

[2] Higgins, S. H. *Dyeing in Germany and America* (2nd edn., 1916), p. 101.

[3] From 1890 to 1895 AGFA and Bayer between them held two-thirds of the shares. Upon the formation of Levinstein Ltd. they became minority holders and in 1897 withdrew completely. Information from Mr. W. J. Reader.

[4] *Chem. Trade J.* **54** (1914) no. 1397, 263.

[5] Levinstein, H. From within the dyestuffs industry. *J. Soc. chem. Ind., Lond.* **50** (1931), transactions and communications, pp. 253T–256T.

1913/14. The company was the only British dyestuffs firm to have a sales office in Germany.[1] We may note here, because it was to become an important factor during and after the war, that Holliday's managers did not get on with Levinstein's, and that personal and business relations between them were deteriorating.

The United Turkey Red Co., the Calico Printers' Association, and several other dyers owned the British Alizarine Co., which had a substantial factory at Silvertown in London. The company benefited from membership of the international alizarin convention and from having a captive market for its product. The other half dozen or so British firms in the dyestuffs business did not have either (or both) of these advantages and so had a struggle to maintain themselves.

The foreign-owned businesses were small: the Clayton Aniline Co. of Manchester, at one time a public company, had had a chequered career. A. G. Green, was works manager until 1901, when he left in disgust to become professor of tinctorial chemistry at Leeds; later Chaim Weizmann was a consultant. But these eminent chemists were unable to preserve its independence. In 1909 CIBA of Basle made the first bid, and two years later bought the entire business for over £160 000. Dr. C. Dreyfus, who had founded the company, retired in 1913, and the new owners were then able to reorganize the business, a job which took them about two years. In 1913 Clayton Aniline employed 244 workers and a staff of 56 and its main products were aniline oil, trinitrotoluene and some intermediates. The bulk was made for export to the parent company in Basle, which thereby became less dependent on purchases in Germany and the goodwill of the German suppliers.[2]

The Ellesmere Port factory of Hoechst was built in 1907/8 at a cost of £30 000–£35 000. It was a very small plant and initially the capacity was 500–600 kilos a day of synthetic indigo from imported phenylglycine. The works was enlarged a little later and was fully working from January 1910 until the outbreak of war, when production fell off quickly because imports of the key intermediate ceased.[3] The Mersey Chemical Co. was set up by the Badische–Bayer–AGFA Dreibund on 24 acres of land bought from the Port Sunlight Estate. The factory was working in 1909 making reducing agents and pigment dyes for Badische.[4]

These foreign enterprises were of no great significance at the outbreak of

[1] Whittaker, C. M. Some early stages in the renaissance of the British dyemaking industry . . . *J. Soc. Dyers Colour.* **72** (1956), no. 12, 558–61.

[2] C.A.C. Quarterly: Special Anniversary Issue—Annals of C.A.C. Summer 1951, pp. 5 ff.; Gesellschaft für chemische Industrie 1884–1934 n.d., passim; *Ciba Blätter* **15** (1957), no. 152, 4–5. Jaquet (*Die Entwicklung und volkswirtschaftliche Bedeutung der schweizerischen Teerfarbenindustrie* (1922), pp. 80–1) thinks that the 1907 Patent Act was a contributory factor in leading to CIBA's decision. But, as I have explained, the Act's ineffectiveness after the Parker judgement was known on the Continent and it is likely that other factors determined CIBA's decision.

[3] I am indebted to Dr. W. H. Cliffe of I.C.I.'s Dyestuffs Division for this information.

[4] *Chem. Trade J.* **52** (1908), no. 1087, 272; Zimmermann, op. cit., p. 123.

war and except for Clayton Aniline their role in the wartime dyestuffs crisis was minimal. The lack of German dyes was felt quickly because stocks were low; fortunately a good deal of spare capacity was available.

The manufacture of dyestuffs was probably of lesser importance than that of medicines, including the lucrative patent medicine business. Exports of pharmaceuticals and proprietary preparations rose from £1·7 million to £2·4 million between 1907 and 1913; imports in that year were £2 million. Most of the firms were small, and the only amalgamation of any note was the merger in 1908 of several wholesale druggists to form British Drug Houses Ltd. (now part of the Glaxo Group). The role of foreign manufacturers was still negligible: the American firm of Parke, Davis set up a London branch in 1891, and Hoffmann–La Roche of Basle established a subsidiary at Welwyn in 1909.[1] Among the manufacturers of fine chemicals, W. J. Bush, founded in 1851, was perhaps the most active. It specialized in flavouring essences and in 1911 was the first to make synthetic vanillin in Britain. When war broke out Bush began making aspirin, which had hitherto been obtained solely from Bayer.[2]

It is not possible to compare the British and German chemical industries in 1913. They were different not only in size and diversity of products, but also in the outlook of their leaders and their attitude to technological change. We have examined the main sectors in some detail so as to highlight the contrast. One final example must suffice to indicate the difference in what was then the entirely novel business of plastics manufacture. In 1913 the only commercially significant plastic was celluloid. It was made in France by seven factories, in Germany by six, in the United States by three, and in Japan by two, but in Britain by only one company, the British Xylonite Co.[3]

France

In 1913 the country's wealth stemmed principally from its agricultural products, vineyards, and textiles. Exports of silk goods represented the largest single group of manufactures, and the French chemical industry was oriented towards the particular needs of agriculture and textiles. It was smaller than that of Germany, Britain, and the United States, and unlike those countries retained a surprisingly narrow product range until the outbreak of war. Moreover the technical outlook of the leading firms was, with few exceptions, unprogressive. Nevertheless production of chemicals was not stagnating, and exports were growing rather more rapidly than imports.

The vigorous development of external trade points to considerable internal activity, but the contemporary evidence is, to say the least, unsatisfactory. The problem is not only the lack of statistics, but also the tendency of French

[1] Duckworth, A. *Rise of the pharmaceutical industry*, reprint from the centenary number of *Chemist Drugg*. November 1959.

[2] Since 1961 Bush have formed part of Albright & Wilson.

[3] Departmental Committee on Celluloid, minutes and evidence (1913, Cd 7159), p. 108. Evidence of C. P. Merriam, managing director of British Xylonite.

TABLE 6.7

French exports and imports of chemicals, selected years 1907–13[1]

	Exports (£ million)				*Imports* (£ million)			
	1907	1910	1912	1913	1907	1910	1912	1913
Chemicals	5·3	6·3	8·1	8·5	6·6	6·3	9·3	9·6
Colours	0·5	0·5	0·4	0·4	0·4	0·4	0·4	0·5
Total	5·8	6·8	8·5	8·9	7·0	6·7	9·7	10·1
Index	100	118	146	153	100	97	138	144

Note: Chemicals include superphosphate and other fertilizers; colours include vegetable and coal-tar dyes.

observers to ignore chemical manufacture altogether, or alternatively to refer in general terms to its shortcomings *vis-à-vis* those of Germany. The figures mentioned in the report of the Paris 1900 Exhibition are often reproduced without acknowledgement, and, with little regard to accuracy, are used to describe 'the pre-war' situation. But in some areas of chemical manufacture we can measure the growth since 1900. Readers of Sir John Clapham's history[2] will look in vain for a reference to superphosphate, though there are a few to 'manures'. Yet it was superphosphate, more than any other product, which gave impetus to chemical manufacture in the 1900s. This reflected, at one remove, the modernization of French agriculture, which was proceeding vigorously in the decade before the war. There were also important secondary effects: superphosphate is simple to make, but expensive to transport, and the high cost of freight gave the manufacturers local or regional monopolies. Scores of small factories were established on the coast or by navigable rivers. They caused the chemical industry's scattered appearance and its characteristic structure of many small factories.

In 1906 the French chemical industry comprised about 500 works and 37 000 employees; only nine factories employed more than 500 people each. The 'average' plant, if there was such a thing, had about 75 workers.[3] No statistics appear to exist for 1913. In Germany and Britain chemical manufacture tended to be concentrated in certain districts, but such localization was less marked in France, except possibly in the northern suburbs of Paris, near the Belgian border, in Lorraine, and around Lyon. The war was to effect a major change in this pattern, because the factories in the occupied territory around Lille, or close to the battle lines, as in Lorraine, were unusable and new works had to be hurriedly built in the south. The country's water-power resources in the Alps and Dauphiné were exploited to only a small extent for chemical production before 1914. Of the 90 000 hp of hydro-electricity in the Alps on the eve of the war, only 12 000 were used for chlorine and caustic

[1] George, F. *Le rénovation de l'industrie chimique française* (1919), pp. 19, 21.
[2] *The economic development of France and Germany 1815–1914* (4th edn., 1936).
[3] Grandmougin, E. *L'Essor des industries chimiques en France* (1919), p. 212.

soda manufacture, and a further 18 000 for cyanamide and chlorates.[1] The greater part was used in the manufacture of calcium carbide and alloys.

The narrow product base gave the French chemical industry a lop-sided appearance. Products of significance in other countries were neglected in France. Examples are the cyanides, phosphorus, sulphate of ammonia, and electrolytic alkalis (except chlorates), among the inorganic chemicals. It is difficult to explain these instances of arrested development. France was not worse placed than Britain and Germany for the manufacture of cyanides, yet failed to make them. The manufacture of phosphorus was decidedly backward: the sole producer, Coignet & Cie of Lyon, made it as a by-product of its glue and gelatine business. Coignet's raw material was bones and the firm did not install an electric furnace until 1925.[2] The production of sulphate of ammonia was small and did not grow very quickly. This can be attributed to the belated installation of recovery ovens. The first were built in 1905, and the efficient recovery of benzene from coke-oven tar only began in 1908.[3] These and other examples show no consistent pattern and we can only speculate on the causes for this state of affairs: technical backwardness and lack of adaptability were among the factors, so was the small size of some of the consuming industries, notably dyestuffs; it is possible also that insufficient capital was attracted into some of the newer branches of chemistry, because the risk was judged too high. In this last respect one is tempted to make comparisons with the United Alkali Co. But the analogy would be misleading, for most French chemical enterprises appear to have been profitable, perhaps because the owners of the small factories were careful men. To this general rule there were two exceptions, both applying to ventures outside metropolitan France: the first was the rapid and costly development of the North African phosphate deposits, the second was the successful rescue of Norsk Hydro, after the departure of the Germans, by the Banque de Paris & des Pays-Bas. As far as dyestuffs were concerned, the patent law and the tariff combined to create a situation in which a large tonnage was made almost entirely from imported intermediates. The absence of facilities for making them was the most conspicuous example of uneven development affecting directly many important sectors of the industry, and indirectly, but importantly, academic chemistry.

The principal statistics of the French superphosphate trade are summarized opposite. The estimated value of superphosphate production in 1913 was £4·2 million, equivalent to very roughly one-quarter of the industry's entire turnover. The dependence of the sulphuric acid business on the fertilizer trade is therefore not surprising, and the consumption of acid for dyestuffs, copper sulphate, and other uses was relatively unimportant. Owing to the close links with

[1] Baud, P. *Les industries chimiques régionales de la France* (1922), p. 128.

[2] Simonin, G. (Ed.) *L'industrie chimique française—Lyon et le Sud Est* (1959), pp. 42–3.

[3] Bedhome, R., L'evolution de la carbochimie dans le Bassin du Nord et du Pas-de-Calais. *Chim. Ind.* **87** (1962), no. 4, 435–6.

superphosphate manufacture, acid plants were usually on the same site. In the fertilizer year 1913/14 about 90 superphosphate factories were working in France, nearly half of them in the north and near Paris. Production was generally seasonal, and the average factory made about 21 000 tons a year. Those in the south-west, however, were rather larger, possibly because they were of more recent date, and their yearly output averaged 33 000 tons. There were 87 sulphuric acid plants with a total capacity of 1 350 000 tons of 53° Bé acid, (which would correspond to 0·9 million tons of 100 per cent H_2SO_4,) and they produced almost 1·2 million tons in 1913/14.[1]

TABLE 6.8

French superphosphate trade, selected years 1900–13[2]
(000 metric tons)

	1900	1907	1910	1912	1913
Production	615	1633	1634	1950	1920
+Imports	n.a.	66	133	89	101
−Exports	n.a.	198	258	170	145
=Apparent consumption	615	1501	1509	1869	1876

Two firms in the superphosphate trade owned several fertilizer works and, as they also had considerable interests in other chemical manufactures, merit a short description. Cie des Glaces & Produits Chimiques de St. Gobain, Chauny & Cirey—for short St. Gobain—dated back to 1665, to Colbert and Mercantilism. In the nineteenth century it had pioneered the Leblanc soda process and its improvements in France, and in 1872 had taken over Soc. Perret-Olivier with extensive chemical factories in Lyon and Provence. It was that very rare thing in France, a veritable industrial giant, owning over 40 works with 20 000 employees and 300 chemists and engineers. Many were employed in the large glass works, but even in 1900 the chemical operations consisted of 14 plants and 6600 people.[3] By 1913 there were 21 chemical works including pyrites mines, phosphate-rock deposits at Gafsa in Tunisia, and limestone quarries in metropolitan France. St. Gobain's development is illustrated by the growth of its sulphuric acid and superphosphate production between 1900 and 1913. The former rose from 465 000 to 650 000 tons (as $62\frac{1}{2}$ per cent H_2SO_4), the latter from 500 000 to 900 000 metric tons. By the eve of the war the company's share of French output was about 41 per cent in respect of the former and 47 per cent of the latter.[4] Such concentration in the hands of a single firm was unique in the French fertilizer industry and indeed exceptional in Europe. During the 1900s the company built or bought

[1] Lambert, M. *Les engrais chimiques et produits chimiques utiles à l'agriculture* (1917), pp. 28 ff.
[2] Gray, op. cit., p. 147.
[3] Matagrin, A. *L'industrie des produits chimiques et ses travailleurs* (1925), pp. 276–8; *Paris Exhibition* (1), 39.
[4] Matagrin, op. cit., pp. 276–8.

several superphosphate factories, notably at Ivry (near Paris), Poitiers, and Toulouse, and in general held a strong position in the centre and south-west. In 1914 it owned the second largest ammonia-soda works in France; it had been opened at Varangéville near Nancy in 1906, had a capacity of 100 000 tons a year, and was designed to replace the last Leblanc soda factory at Marseille l'Estaque.[1] The chemical business was profitable, not only because the company had a captive market for much of its soda, but also because it was the leading voice in the French superphosphate syndicate. It also participated actively in the affairs of the Soc. des Agriculteurs de France, an old-established pressure group of the farmers, whose president was the Marquis de Voguë, who, by a happy coincidence, was also head of St. Gobain. There were of course, some who disapproved of such close ties and claimed that the Société was more concerned with keeping up the price of superphosphate than with supporting the agricultural syndicates in their policy of bulk buying on a co-operative basis.[2] The war caused a decline in the relative importance of St. Gobain and weakened its position in the fertilizer trade.

Etablissements Kuhlmann was the other large company, and unlike St. Gobain it was purely a chemical firm, owning outright five factories in northern France and employing about 2000 workers.[3] Superphosphate, copper sulphate, and acid were the main products; it also operated a small contact acid plant, the only one in France. The tonnage of fertilizers made (predominantly superphosphate) soared from a mere 55 000 in 1900 to 130 000 in 1908, and to 220 000 in 1913–14. Kuhlmann's net profits reached a peak of £90 000 in the financial year 1911/12, but fell back to £60 000 two years later.[4] During the war its factories were in the battle zone or in German occupied territory.

A few of the smaller companies deserve brief mention. Ets. Malétra, founded in 1808 at Petit Quévilly near Rouen, was employing about 1000 people and making 25 000–30 000 tons a year of superphosphate, as well as some sodium compounds.[5] Cie Bordelaise de Produits Chimiques had been started at Bordeaux in 1891 to make superphosphate, and from 1911 onwards had a second works at Rouen. Other companies in the south-west were Soc. des Phosphates Tunisiens (renamed Pierrefitte in 1924) and La Cornubia, which used to distribute imported sulphate of copper, and from 1906 onwards began making it at Bordeaux; it belonged to the Dennis family of Macclesfield.[6] These firms, like so many others, were of limited industrial significance.

[1] Choffel, J. *Saint Gobain* (1960), pp. 74, 80; Baud, P. *L'Industrie Chimique en France* (1932), pp. 262–3.

[2] The French fertilizer trade. *Chem. Trade J.* **56** (1910), no. 1188, 219.

[3] Lambert, op. cit., p. 29. Kuhlmann held a half share in another factory and also owned a works at Rieme in Belgium, which employed 400.

[4] Anon. *Les Etablissements Kuhlmann 1825–1925* (1925). In 1966 Kuhlmann combined with Ugine and Soc. des Produits Azotés and now trades as Ugine–Kuhlmann SA.

[5] *Paris Exhibition* (1), 46.

[6] Anon. Les industries chimiques Normandes. *Ind. Chim.* **47** (1960), no. 515, 181–6; Gallo, A. Les industries chimiques de l'Aquitaine et des Pyrénées. *Ind. Chim.* **48** (1961), no. 523, 42–50.

They enjoyed a powerful position in the local market, protected by the high cost of transporting superphosphate over long distances. They were little touched by technical change, traded satisfactorily, and, being run on frugal lines, were not greatly troubled by the vigorous expansion of St. Gobain. Some of the companies, principally those in the south, also made copper sulphate. This material in the form of a solution with lime was called 'Bordeaux mixture', and used extensively after the mid-1880s to spray vineyards against mildew, from which the rootstocks imported from the United States to replace those killed by the phylloxera suffered. By 1913 copper sulphate had become one of the major French chemical products. Production was estimated at around 26 000 tons, and was supplemented by imports of 19 000 tons from the United Kingdom.[1]

French consumption of potassic and nitrogenous fertilizers was relatively small. About 42 000 tons of the former were used in 1913, principally in the beetroot-growing area of the North.[2] Imports of Chile nitrates in that year were 314 000 tons and of artificial nitrates from Norway 10 000 tons, equivalent to altogether 50 000 tons N.[3] The production of sulphate of ammonia was growing, and consumption reached 97 000 tons in 1913, but this was relatively small when compared with Belgium, which used 50 000 tons.[4] In addition about 12 400 tons of cyanamide were made by Soc. des Produits Azotés at Notre Dame de Briancon in Savoy.[5]

The country of Nicolas Leblanc had entirely given up his process at the outbreak of war. The figures for alkali production vary considerably. One authority gives an estimate of 625 000 tons for 1912, and Baud mentions 640 000 tons of soda ash and crystals, and a further 40 000 tons of caustic soda in 1913; these figures are certainly swollen by double-counting.[6] A volume of the order of 400 000 tons seems altogether more plausible.[7] Such an amount would in any case mark a substantial increase over the progress achieved in the 1900s: 245 000 tons in 1903, 312 000 tons in 1908, and 317 000 tons in 1910.[8] The growth in production compares favourably with that recorded in Britain, and is attributable to the rising consumption of soap and

[1] Grandmougin, op. cit., p. 125; George, op. cit., pp. 54–5.

[2] Krische, P. *Das Kali* (1923), vol. 1, p. 156. By comparison Holland used 44 000 tons in 1913.

[3] Ogburn, W. F., and Jaffé, W. *The economic development of post-war France* (1929), p. 432, George, op. cit., pp. 50–1, 85–6.

[4] Grossmann, H. and Weicksel, P. *Die Stickstoffindustrie der Welt* (1930), pp. 91, 127.

[5] Grossmann and Weicksel, op. cit., pp. 125–7; Grandmougin, op. cit., p. 87 gives the production as 7500 tons.

[6] Fisch, R. Les industries chimiques de la région Lyonnaise, thesis (1923), p. 91; Baud, op. cit., pp. 267–8.

[7] L'Association Nationale d'Expansion Economique, *Index to French production* (1920 edn.), p. 353. The statistics quoted in this publication have been used by other writers on the French chemical industry. Separate figures are given for sodium carbonate, crystal soda, and caustic soda, but it is not stated whether the last-mentioned includes electrolytic caustic.

[8] ibid. The figures refer only to sodium carbonate. Up to 1908, and possibly even later, the production of electrolytic caustic was not sufficiently important to affect the total to any significant extent.

higher exports. In 1913, 82 000 tons of soda were sold abroad, chiefly to Holland, Belgium, and Switzerland.

Solvay & Cie were the chief producers and accounted for over two-thirds of the total French output. Their main works was at Dombasle near Nancy and drew its salt from local deposits. A second, but smaller, works at Salin-de-Giraud in the Camargue used solar salt.[1] Solvay's rivals, if such a word is appropriate in this well-regulated trade, were St. Gobain's subsidiary Soudières de la Meurthe at Varangéville, also in the Dombasle area, and a smaller company, Marcheville, Daguin & Cie at la Madeleine, between Nancy and Varangéville. In 1899 it made only 20 000 tons.[2] During the war the country's alkali works in Lorraine were close to the battle area and the greatly increased demand for caustic soda and chlorine necessitated the building of new factories elsewhere. They were located in the Alps and Dauphiné, although most of the best sites were already bespoke for the production of carbide, aluminium, and ferro-alloys. Nevertheless the water-power resources had not been fully developed before the war (except perhaps on the river Arc), partly because the conditions were less favourable than in Norway. Owing to lack of capital, small run-of-the-river power plants had been built instead of dams.[3] As a result operations were frequently seasonal and confined to spring and early summer. This did not matter too much with chlorates, but was a serious drawback for the more costly caustic soda and chlorine installations.

Though the electrolytic alkali industry hardly deserved that name and was certainly smaller than that of Britain and Germany, it had a complicated structure, which must be briefly described on account of the considerable post-war development. Solvay, though holding the Continental rights to the Castner–Kellner process, did not own electrolytic works in France. There were several producers of caustic soda and chlorine; all worked on a small scale, even St. Gobain, which, for some years, was one of the very few to use the Hargreaves–Bird system. The extent of the chlorine business may be judged from the output of bleaching powder, which was a mere 45 000 tons in 1913 and reflected the small size of the cotton and linen industries.[4]

Most of the producers of chlorates dated back to the 1890s. One of them, Soc. de Produits Chimiques & d'Explosifs Bergés, Corbin & Cie was reported to be making 3000 tons annually at the turn of the century.[5] Its works were

[1] Solvay was even in 1900 one of the biggest French chemical manufacturers and employed about 2000 people. The number was probably larger in 1913. *Paris Exhibition* (1), 53.

[2] Hoff, L. Le centenaire des Soudières Réunies. *Chim. Ind.* **73** (1955), no. 5, 998–1001. In 1948 Marchéville-Daguin acquired Soudières de la Meurthe, and the new company in which St. Gobain retained a stake, was called Soudières Réunies la Madeleine-Varangéville.

[3] In 1906 for instance there were five aluminium works and one chlorate factory in the valley of the Arc but between them they had only 22 000 kW available. Pring, J. N. *Some electro-chemical centres* (1908), pp. 97–8.

[4] L'Association Nationale d'Expansion Economique, op. cit., p. 357. For comparison Britain exported 37 000 tons of bleaching powder in 1913.

[5] *Paris Exhibition* (1), 27.

at Chedde in Haute Savoie, and gave their name to 'cheddite' explosives, which were used extensively for civil engineering purposes on the Continent.[1] The company later came under the control of Soc. de l'Arve, an enterprise specializing in the financing of large users of electricity and the building of power stations. By 1914 Soc. de l'Arve had been taken over by another holding and financing company, which during the war sold the various chemical and metallurgical subsidiaries to Cie des Produits Chimiques Alais, Froges & Camargue, which later changed its name to Cie Pechiney. A. R. Pechiney, the founder of this enterprise, had been interested in aluminium and chemical manufacture, but his successor, A. Badin, decided to put the emphasis on the former. On the eve of the war the company owned four aluminium smelters, a works near Avignon for the production of copper sulphate, and the original factory at Salindres, which made a number of inorganic compounds, chief among them the sulphates of aluminium and copper. Finally the company owned extensive solar salt works in the Camargue, which supplied Solvay under long-term contract.[2]

French aluminium production was about 15 000 tons before the war, and there were at least six factories making the metal. The structure of the carbide industry was equally fragmented: there were nine works with a total capacity of 50 000 tons, but in 1913 they were making perhaps half the tonnage.

The dyestuffs business was also divided into numerous small units. It centred on Lyon and Paris and consisted of ten companies: five German, one Swiss, and four French-owned. Of the French, only one was making intermediates as well as finished dyes.[3] In 1913 France was reported to be producing about 8000 tons of dyestuffs, importing 2000 and exporting 1000, so that the apparent consumption was 9000 tons, of which azo-dyes accounted for roughly one-half. The value of the business was about £1 million, but French-owned companies held merely 10–15 per cent of the market.[4] The tonnage made in France was larger than in Britain, but the volume is misleading, and we are justified in generalizing that the French industry specialized in making dyes from imported intermediates, and to a large extent was doing no more than diluting or repacking foreign colours. There was no infrastructure for the preparation of intermediates, and hence virtually no manufacture of organic chemicals, nor of the connected inorganics such as nitric acid and the different oxidizing and reducing agents. The reasons for this state of affairs were the patent legislation and the tariff. The former effectively stifled all innovation because it protected the end product and so stopped the adoption of alternative routes. French patent legislation had a working

[1] Cheddite contains about 80 per cent potassium chlorate, castor oil, and mononitro-naphthalene; the last named improves the stability of the chlorate. cf. Marshall, A. *Explosives* (1917), vol. 1, pp. 380 f.

[2] Gignoux, C. J. *Histoire d'une entreprise française* (1955), pp. 109–13.

[3] Ogburn and Jaffé, op. cit., p. 463.

[4] League of Nations Economic and Financial Section, *The chemical industry* (1927), p. 28; Grandmougin, op. cit., pp. 163 ff.; L'Association Nationale d'Expansion Economique, op. cit., p. 365.

clause, but this could be readily satisfied by making just a few tons, and in any case German firms were exempt from the provision under the Franco–German commercial agreement of 1903. The tariff reinforced this nonsensical situation. Intermediates and dyestuffs were subject to different rates of specific duties, the former being chargeable at 15 fr. per 100 kilos, the latter at 100 fr. per 100 kilos. Accordingly there was every incentive to bring in intermediates, and as the two classes were ill-defined it was possible, and indeed seems to have often been the case, that dyestuffs were passed off as intermediates.[1] There was no French pressure group demanding a change in the patent law and a revision of the tariff structure; indeed the question of an *ad valorem* duty on dyestuffs was apparently first raised during the war. There is a simple explanation for such indifference: the principal enterprises were foreign-owned, and the French manufacturers were completely dependent on the supply of intermediates from abroad and therefore quite unable to move on their own. The only French company of any size was the S.A. des Matières Colorantes & Produits Chimiques de St. Denis. It had been making alizarin for some years, but not having a captive market like the British Alizarine Co., had to abandon this line, and by 1913 was specializing in azo- and sulphur dyes, of which it made perhaps 800 tons. The firm also manufactured some intermediates, superphosphate, varnishes, lacquers, and enamels, and employed about 300 people.[2]

The German dye makers set up branch factories in France at an early date, sometimes buying the business of a French producer for this purpose. Cassella, for example, bought the Manufacture Lyonnaise de Matières Colorantes at Gerland in 1885, and AGFA controlled the business of Picard & Cie.[3] Hoechst formed the Cie Parisienne des Couleurs d'Aniline in 1881, and two years later bought Coupier's aniline works at Tremblay (near Creil); in 1913 150 workers, 15 chemists and engineers, and 60 salesmen were employed. Among the main products was synthetic indigo, of which roughly 450 tons were made from imported intermediates.[4] The large number of salesmen suggests that the main purpose was to 'finish' and market imported dyes rather than to make them. Badische, however, built its own factory at Neuville sur Saône (near Lyon), and the other German companies followed its example. Thus Bayer established a small business at Flers near Roubaix in 1882 to work some of its patents and in 1904 a peak of 50 tons of dyes was reached. But when the working clause lapsed a decline began, and by 1908 a mere 8 tons were made. The bulk of the turnover consisted of selling intermediates to French firms.[5] CIBA was established in the St. Fons quarter of

[1] Higgins, op. cit., pp. 128–9; Grandmougin, op. cit., p. 154.

[2] Gérard, J. (Ed.) *Dix ans d'efforts scientifiques et industriels* (1926), vol. 2, pp. 2781–2.

[3] Simonin *et al.*, op. cit., p. 46.

[4] Anon. *Farbwerke vorm. Meister Lucius & Brüning 1863–1913*, n.d., p. 40. Grandmougin, op. cit., p. 166.

[5] Burgdorf, K. Geschichte der Fabrik Flers, in *Geschichte und Entwicklung der . . . Fr. Bayer . . .* (1918), pp. 258–61.

Lyon. In 1913 the Swiss company employed 68 workers and a staff of 26. The French subsidiary was a relatively unimportant venture by comparison with the factories at Manchester and in Russian Poland, which were employing about 300 people each.[1]

In other branches of the organic chemicals industry French firms flourished and widened the range of their products. In particular the perfumes and essences of Provence maintained themselves successfully against the competition of German-made synthetics and the manufacture of tanning extracts from chestnuts was a major activity. The Lumière brothers, who are probably better known as pioneers of the cinema, made photographic supplies and developers at Lyon, and Poulenc Frères at Ivry-Port and Montreuil near Paris specialized in fine chemicals and laboratory supplies.[2] Soc. des Produits Chimiques Gillet & Fils, the predecessors of Progil, had close links with the leather and silk trades. They began by making dyeing and tanning extracts and charcoal, later they produced carbon disulphide, which at one time was used against the phylloxera, but increasingly found its main market in the manufacture of viscose fibre. The company benefited greatly from the practice of charging silk, and supplied stannic chloride and other materials to the silk dyers and finishers of Lyon.[3] By far the largest of these firms was the Soc. des Usines du Rhône, which even in 1900 was employing 700 workers and 27 chemists at St. Fons. It traced its descent from one of the first French makers of aniline dyes, and by 1913 had become the leading French firm in the organic chemicals industry. Under the direction of J. Koetschet and N. Grillet it specialized in pharmaceuticals, photochemicals, essences (vanillin and coumarin), and cellulose acetate film.[4] Its indigo synthesis, though practicable, was not commercially viable. It was the first in France to synthesize phenol.[5]

Artificial fibres and plastics were rapidly developing sectors of French chemical manufacture. Chardonnet had pioneered artificial silk, and his factory at Besançon had been making nitro-cellulose fibre for many years. But on the eve of the war it was taken over and converted to the viscose process by the Soc. Française de la Viscose which already owned a factory near Rouen. There were also factories at Givet and at Izieux making cuprammonium fibre. In 1910 these enterprises formed a joint selling organization, the Comptoir des Textiles Artificiels, which handled the country's entire production of about 2000 tons of cellulosic fibres in 1913.[6]

[1] Gesellschaft für Chemische Industrie, op. cit.

[2] Löffl, V. K. *Die Chemische Industrie Frankreichs* (1917), pp. 186–7, 202–3. Lumière is now controlled by CIBA, while Poulenc merged with Usines du Rhône in 1928 to form Rhône-Poulenc.

[3] Simonin *et al.*, op. cit., pp. 54–5. The addition or 'charging' of silk with chemicals or other substances lowered its price and created new markets for the fibre from which Lyon in particular benefited.

[4] Baud, *Les industries chimiques régionales de la France*, op. cit., p. 144; *Paris Exhibition* (1), 26 ff.

[5] Simonin *et al.*, op. cit., pp. 47–8.

[6] Ward-Jackson, C. H. *A history of Courtaulds* (1941), p. 38. Grandmougin, op. cit., p. 187

As regards plastics, that is celluloid, France was well behind Germany and the United States, with an output of about 3600 tons in 1913.[1] Three companies handled the bulk of the business: Soc. Française du Celluloïd had been founded by the Hyatt brothers at Stains near St. Denis, and was the first to produce plastics in Europe. The Soc. Génerale pour la Fabrication des Matières Plastiques was set up at La Rivière-Saint-Sauveur (Calvados) by the Centrale de Dynamite. More recently, in 1906, the third firm Soc. Industrielle du Celluloïd had started factories at Villetaneuse (Seine) and Chauffry (Seine & Marne).[2] The celluloid was worked into articles in Paris and especially at Oyonnax. There was a tradition of woodworking in that little Jura town during the winter season, and from the turn of the century onwards the men specialized in making combs, umbrella handles, and *bibelots*.[3] Although celluloid is now a very minor article of commerce, Oyonnax remains a centre of plastics fabricating to the present day.

In 1913 Belgium was the largest single customer of the French chemical industry and imported £2·4 million worth of goods, equivalent to 27 per cent of all French chemical exports. Belgium was dependent on imports because its own chemical manufactures were small and dispersed. A survey carried out in 1905 showed that there were 224 firms with altogether 256 factories and 10 000 workers in the chemical trades.[4] Superphosphate was the principal product made, and a few years before the war was manufactured by 33 factories, half of which were also making their own sulphuric acid. Production was about 450 000 tons a year and consumption 160 000–170 000 tons in 1911–13. Belgium was a big exporter of superphosphate, principally to Holland and Germany.[5] The largest group in this branch was the Union Commerciale des Fabricants Belges de Produits Chimiques, a federation established by six companies in 1889 to pool sales of superphosphate and sulphuric acid. In the 1910s the members were also said to be dominating the manufacture of sodium sulphate (an important product owing to the country's large glass industry) and hydrochloric acid.[6]

Solvay & Cie represented the manufacture of alkali, and its factories at Couillet and Jemeppe, which relied on imported rock salt, employed about 500 people in 1905. Annual production was about 30 000 tons of soda ash and

mentions 3000 tons, but this is high. It is worth noting that Courtaulds, then the only producer of viscose in Britain, made about 1400 tons (Coleman, D. C. *Courtaulds* (1969), vol. 2, p. 127).

[1] Baud, op. cit., p. 128. Grandmougin's figure (op. cit., p. 189) is only 2500 tons, but this may be for the unplasticized product.

[2] The last two companies combined in 1924 to form the Soc. Industrielle des Matières Plastiques, which, after further mergers, became part of the Nobel–Bozel group (1947).

[3] Baud, *Les industries chimiques . . . régionales*, op. cit., p. 128.

[4] Lucion, R. *La grosse industrie chimique en Belgique* (1908), p. 2. The author was head of research of Solvay.

[5] Gray, op. cit., p. 129.

[6] Norton, T. H. The chemical industries of Belgium. *Chem. Trade J.* **52** (1913), no. 1339, pp. 51–2; *Paris Exhibition* (1), 93–4.

a few thousand tons of caustic soda in the 1900s.[1] The Belgian factories were the oldest in the group, but by 1913 were among the less important of the family concern. Solvay owned outright factories at Dombasle and Salin-de-Giraud in France, Saarralbe (then in German Lorraine), and Torrelavega in Spain. Between them they made 375 000 tons of alkali in 1910, mainly in France. There were subsidiary and associated companies in Germany, Russia, the United States, and Austria, and, if their annual make is added to the Franco-Belgian factories, and if Brunner, Mond & Co. are included, the Solvay group made altogether 1·9 million tons of alkali and employed 25 000 people. This private partnership directly and indirectly was certainly the most international of all the chemical enterprises and, measured by the yardstick of employees and volume of output, the largest in the pre-war world. It was managed from Brussels by *gérants commanditaires*, all related to the founders of the business, who collectively had unlimited responsibility in this partnership.[2]

Belgium made no organic chemicals and dyestuffs, and the products of its coal-tar distillation works were exported to Germany.[3]

Switzerland

Chemical manufacture in Switzerland had some unique characteristics. The country lacked most of the essential raw materials such as coal, brimstone or pyrites, phosphates, and nitrates. There were, however, considerable salt deposits in the cantons of Basel Land, Aargau, and Valais. Such was the particularism of the inhabitants that the deposits in German-speaking Switzerland were not worked efficiently until after 1909, when the Vereinigte Schweizerische Rheinsalinen was established. It is worth mentioning that, in this land of free enterprise, salt mining came, after much petty bickering, under cantonal ownership, and also that salt for chemical use was dutiable until the war.[4] Under such circumstances the building of a soda works was impracticable, the more so as Solvay could supply the country from a factory at Wyhlen in Baden just across the border. Nor were inorganic chemicals made to any significant extent before the war. The factory census of 1905 recorded 205 establishments and 2203 people working in the inorganic chemical and fertilizer trades.[5] Among the chemical minnows there were a few bigger fish: Chemische Fabrik Uetikon near Zurich and Chemische Fabrik Schweizerhall near Basle specialized in acids, there were works at Turgi and Monthey producing electrolytic alkalis, and a French-owned factory at Martigny had

[1] Ministère de l'Industrie et du Travail—Monographies Industrielles, *Fabrication des produits chimiques* (1905), p. 116.

[2] Solvay & Cie, *Cérémonie de commémoration du 50e anniversaire de la fondation* (1913), p. 10; illustrated brochure of Solvay & Cie., n.d. (c. 1956), pp. 24–5.

[3] Ministère du Travail, op. cit., pp. 285 ff.

[4] Waldmeyer, E. Die schweizerische Salz- und Soda Industrie, thesis (1928), p. 136. The English salt excise ended on 1 January 1825!

[5] Neumann, H. and Weckerle, E. (Eds.) *Die Schweizerische Industrie und ihre Arbeiter: Papierfabrikation–Chemische Industrie* (1947), p. 41.

a capacity of 7500 tons of cyanamide a year.[1] This, it would appear, was the sum total of the Swiss heavy chemical activities; the term industry would be misleading. During the war the reliance on imported chemicals caused increasing difficulties, until the large users combined to set up an acid factory and a tar distillery.

These large users were the companies making dyestuffs and other organic chemicals, and, then as now, they were the dominant element in the Swiss chemical industry. They were: CIBA (the abbreviated title of Gesellschaft für Chemische Industrie), J. R. Geigy AG, Chemische Fabrik vorm. Sandoz, and L. Durand & Huguenin. In addition F. Hoffmann–La Roche & Co. made pharmaceuticals. All five had their works in Basle.

By 1913 the reasons for establishing dyestuffs factories in that town fifty years earlier were no longer apparent. Effluent disposal into the Rhine, once easy, was now beset with restrictions; patent considerations had lost much of their force with the introduction of a patent law in 1907; and silk dyeing had become less important. German labour, paid less than the locals and not a charge on the rates, had contributed to the early development of the dyestuffs factories, and even in 1911 foreign workers still represented 51 per cent of the labour force.[2] But the employers were anticipating a gradual decline, because they established branch factories in Alsace and Baden, where relatively cheap labour was more readily found. The works in Basle with their branch factories just across the border and subsidiaries further afield gave the dyestuffs firms a special European characteristic well before the war.

TABLE 6.9

Estimated employment of Basle chemical firms in 1913[3]

	Basle	*Rest of Switzerland*	*Abroad*	*Total*
CIBA	1600	500	800	2900
Hoffmann–La Roche	550*	—	400*	950
Geigy	400*	—	300*	700
Sandoz	340	—	negl.	340
Durand & Huguenin	100*	—	100*	200
Total	2990	500	1600	5090

*Estimated

Individually the four dyestuffs manufacturers were small, even collectively they appeared insignificant alongside the German Big Three. In 1913 they

[1] Franck, H. H. *et al. Der Kalkstickstoff in Wissenschaft, Technik und Wirtschaft* (1931), p. 114.

[2] Jaquet, N. *Die Entwicklung und volkswirtschaftliche Bedeutung der schweizerischen Teerfarbenindustrie* (1922), pp. 114–15.

[3] Die Basler Chemische Industrie Kollektiv-Gruppe. *Schweizerische Landesausstellung Bern* (1914), pp. 3–6; Horster, *Die volkswirtschaftliche Bedeutung der chemischen Industrie am Oberrhein* (1922), pp. 60, 64.

exported 9000 tons of dyes valued at £1·1 million. With tanning extracts and pharmaceuticals the total rises to around £1·35 million.[1] Before the war chemical exports were still playing a minor role in Swiss exports as a whole, of which the largest constituents were embroideries (£8·4 million), watches (£7·5 million), and silk goods (£5·6 million). Since then the order has changed profoundly, and chemical products have been among the chief exports for many years. The background to this transformation requires comment.

In 1913 (as forty or fifty years later) the bulk of production was exported, and it is a remarkable fact that the Swiss firms grew and prospered while the British and French were languishing. How did the Swiss manage it? In the first place they did not make some of the commercially important tonnage dyestuffs such as alizarin or Congo red. Synthetic indigo and vat dyes which were important in pre-war Germany were only produced by CIBA, and that on a small scale. In the second place the Swiss firms were less interested in introducing novelties over a wide range of colours than in matching the better quality German dyes. Thirdly, a considerable effort was directed to maintaining or improving quality, and up to the Great War there was relatively little fundamental research. Finally, most of the firms had a few specialities in which they were technically among the world leaders. The outcome is reflected in the average values of Swiss and German dyestuffs: both fell in the 1900s, but in 1913 Swiss exports averaged £130 a ton, while the German figure was £100 a ton.[2]

The Swiss manufacturers could not have survived merely by making dyes as well as or slightly better than the Germans. More was needed to ensure prosperity, and it would appear that the Germans, who accounted for roughly nine-tenths of world trade, did not wish to squeeze out the Swiss. They were not threatening rivals and indeed their elimination by one or the other of the German groups might have exacerbated relations and upset the *status quo* between them. It is possible too that the Germans would have found it uneconomic to extend their range so as to match every Swiss speciality, which was in most cases made on a small scale. There were also practical business reasons that enhanced the value of the Swiss manufacturers to the Germans. For example in 1913 the former bought more than four-fifths of their raw materials and intermediate products from the latter, who took care not to jeopardize the trade. Accordingly they even tolerated the import of Swiss dyes, valued at £240 000 in 1913, and put up with the commercial treaty that exempted Swiss firms from the compulsory working clause in the patent law. The relations between the dyestuffs manufacturers of the two countries were further strengthened by various agreements and understandings. One can surmise their existence between CIBA and Badische, there was a profit-sharing pool between Geigy, Bayer, and Badische for a

[1] Kollektiv-Gruppe, op. cit., p. 8; League of Nations, op. cit., p. 50.

[2] League of Nations, op. cit., p. 50. According to Jaquet (p. 106) the average values in 1910 for Switzerland and Germany were £148 and £124 a ton respectively.

black dye, and an agreement covering tanning extracts linked Geigy with German and Austrian firms.[1]

One further factor, which cannot be quantified, but which was certainly beneficial to the Swiss, remains to be noted: at a time of rising economic nationalism (which often reflected the jingoistic posturings of the politicians), the products of neutral manufacturers probably benefited from the political innocuousness of the country of origin.

CIBA was easily the largest of the Swiss companies, and even in 1909 had a turnover of three-quarters of a million pounds.[2] The previous year it had bought Basler Chemische Fabriken and so become the owner of an electrolytic alkali works at Monthey in the Valais. The factory was extended and was also equipped to make about 200 tons a year of indigo.[3] CIBA's main factory was at Kleinbasel on the right bank of the Rhine. Besides a substantial quantity of sulphur- and azo-dyes, it was the only Swiss firm to make anthraquinone vat dyes. CIBA began the manufacture of pharmaceuticals at an early date, and in the 1900s their preparation had become so important that it was transferred to a separate section. Various ethical products were made, and also 'Vioform', a very successful antiseptic for internal use.[4] The accounts of Swiss companies were particularly uninformative in those days, and CIBA's were no exception. Net assets rose from £0·2 to £0·8 million between 1900 and 1913, and the trading profit from £44 000 to £120 000.

The other firms require only passing mention. Geigy was still making a line of dyewoods and tanning extracts, and the exports of these two, valued at almost £100 000, were of the same order of magnitude as the firm's dyestuffs sales from the Basle factory. At the eve of the war Geigy specialized in wool dyes, having failed to develop an economic indigo synthesis. Sandoz and Durand & Huguenin were very small businesses in those days, the turnover of the former being about £116 000 in 1910, that of the latter even less.[5]

Other European countries

In Austria–Hungary, Italy, Spain, Scandinavia, and the Russian Empire there was no chemical industry with distinct characteristics and structure. In some of the countries chemical manufacture of sorts had long been established, for example in Bohemia for about two centuries. But nowhere had the scale become significant, and the product range was very narrow. In all the countries chemicals played a negligible role in the economy, and home production, though protected by tariffs, remained small. There were, however,

[1] Bürgin, A. *Geschichte des Geigy Unternehmens 1758–1939* (1958), p. 253.

[2] CIBA 75th Annual Report, p. 13.

[3] CIBA. *Cinquantenaire de l'Usine de Monthey*, n.d. (1958?), pp. 38, 54. The indigo process used had been developed by Basler Chemische Fabriken but some parts were probably of German origin.

[4] Kollektiv-Gruppe, op. cit.; CIBA, *Histoire et Développement de la Société . . . 1864–1926*, n.d., pp. 9, 65 f.

[5] Bürgin, op. cit., p. 244; Anon. *Sandoz 1886–1961—75 years of research and enterprise* (1961), p. 136.

some interesting regional variations. In the Mediterranean countries superphosphate was the key product that led to the development of the industry, while in Scandinavia the starting point was the introduction of electrochemistry. In Russia neither electrochemicals nor fertilizers were particularly important, but alkalis provided the foundation for the expansion of chemical manufacture generally.

In 1910 the factory census of the Hapsburg Empire recorded 2475 works and about 70 000 employees in the chemical and allied trades. According to an estimate made a year later, there were about 13 000 people in the chemical trades proper, excluding oils and fats, soap, toilet articles, and petroleum products. The number is small in absolute terms, and insignificant by comparison with the textile and engineering industries. Austria was entirely dependent on Germany as far as dyestuffs and pharmaceuticals were concerned, and to a large extent for a great many other chemicals. Imports of chemicals were valued at £5·2 million in 1912, while exports were substantially higher at £7·7 million and reflected the growing business in raw materials for the chemical industry, notably wood-distillation products, and in particular methanol and acetate of lime. Germany took 40 per cent of the exports, the other neighbouring countries most of the remainder.[1]

A few chemical firms in Bohemia and Vienna represented the greater part of the production of heavy chemicals. On the eve of the war 300 000–350 000 tons of sulphuric acid, about 400 000 tons of superphosphate, and 100 000 tons of soda were made.[2] The largest enterprise was the Oesterreichischer Verein für Chemische & Metallurgische Production AG, founded at Aussig (now Usti) in 1856, which employed about 3400 people in 1906 at Aussig, Kralup, and in subsidiary works. In 1911 it took over a rival, the Erste Oesterreichische Fabrik für Sodaproduktion Miller & Hochstetter at Hruschau. Even in the early 1900s the Verein had a turnover of £0·9 million and made a large range of inorganic chemicals, including sodium sulphate and copper sulphate. It took advantage of the increase in the tariff on dyestuffs and in 1905 started making aniline dyes, to which alizarin was added later.[3] The Verein and the Solvay group were the joint owners of the country's main soda factories. The first was opened in 1882 at Podgorze (near Cracow) in Galicia, in the neighbourhood of the Wilica salt mines. Other plants were built during the following quarter-century at Ebensee in the Austrian Salzkammergut, Lukavacs in Bosnia (1899/1900), Nestomitz near Aussig (1906/7), and Maros-Ujvàr in Hungarian Transylvania. The location of the works was dictated by the availability of salt and limestone, and with the exception of Ebensee and Nestomitz they were small and supplied a limited market.

[1] Glaser, J. *Die chemische Industrie Oesterreichs und ihre Entwicklung* (1918), pp. 13, 25–6, 81.

[2] Glaser, op. cit.; Gray, op. cit., p. 125.

[3] Anon. *Der Oesterreichische Verein für Chemische und Metallurgische Production 1856–1906* (1906), pp. 60 ff. After the Great War the company was renamed Aussiger Verein; it came under the control of IG Farben in 1939 and is now nationalized.

In Vienna the main enterprises were Nobel's explosives factory and Wagenmann & Seybel, which were old-established manufacturers of heavy chemicals.

Chemical production was reported to be flourishing in Bohemia before the war, probably on account of the brisk activity in the glass and fertilizer trades.[1] But it was less vigorous in Austria, because of the slow growth of the superphosphate business, and more generally that of the main chemical-using industries.[2]

The situation in Austria may be contrasted with the vigorous development in Italy, illustrated by the following statistics.

TABLE 6.10

Italian production of chemicals, 1906 and 1913[3]
(000 metric tons)

	1906	1913
Superphosphate	512	972
Sulphuric acid (96% H_2SO_4)	243	430
Copper sulphate	13*	44
Calcium carbide	28	49
Cyanamide	1	15

*1900

The role played by superphosphate requires no emphasis, though the period of most rapid growth ended in 1910. The production of copper sulphate increased steadily, and by the eve of the war Italy had become the leading continental manufacturer.[4] These heavy chemicals were made in numerous small factories in Lombardy or in the main ports. They got their brimstone from Sicily and phosphate rock from Tunisia. There was no indigenous manufacture of soda ash and soda crystals, and the textile and other users depended on imports from France and Britain. Solvay decided in 1913 to build an ammonia-soda plant at Rosignano (near Leghorn), but the war intervened and it was not completed until six years later.[5] Electrolytic caustic soda was made on a small scale by Stà Italiana di Elettrochimica at Bussi, in central Italy, from 1902 onwards, and by Stà Elettrica & Elettrochimica del Caffaro at Brescia after 1905. Production was estimated at 13 000 tons in 1913.[6] The co-product chlorine was used to make bleaching powder and copper oxychloride for spraying vineyards.

[1] Schicht, H. The development and present situation of the chemical industry in Czechoslovakia. *J. Soc. chem. Ind., Lond.* **50** (1931), 276T–277T.

[2] cf. Benedikt, H. *Die Wirtschaftliche Entwicklung in der Franz Joseph-Zeit* (1958). Industry generally was booming in Hungary because of official assistance, but there was virtually no chemical manufacture.

[3] Parravano, N. (Ed.) *La chimica in Italia* (1938), p. 9; Gray, op. cit., p. 173.

[4] *Chem. Ind., Düsseld.* **28** (1905), no. 10, 311; Fenoglio, G. *Die volkswirtschaftlichen Folgen des Weltkrieges in Italien* (2nd edn. 1915), p. 24.

[5] Parravano, op. cit., p. 171.

[6] Morandi, L. Aspects et perspectives de l'industrie chimique Italienne. *Chim. Ind.* **63** (1950), no. 1, 79. Parravano (op. cit., p. 9) quotes 6600 tons, but this may refer to only one company.

The production of calcium carbide was started by the Terni company in Umbria in 1898. Although the country's hydro-electric resources were less favourable than Norway's, they were rapidly developed by companies that combined electricity supply with the manufacture of carbide. Calcium carbide was sold for acetylene lighting, but from 1905 onwards an increasing proportion was used for cyanamide. In that year the Cyanidgesellschaft established the Stà Italiana per le Fabricazzione di Prodotti Azotati at Piano d'Orta in the Apennines. It was the first commercial-size cyanamide factory and was extended some years later so that it had a capacity of 14 000 tons a year by 1913. The Terni company also built a works at Collestate (Umbria) after 1908, and Montecatini (then a small business making acid and superphosphate) established a factory at St. Marcel in the Aosta valley three years later. All the companies were working well below capacity in 1913 when they made about 15 000 tons of cyanamide, equivalent to 3000 tons of nitrogen.[1] This was insufficient to meet the country's needs, and Chilean nitrates and sulphate of ammonia had to be imported.

These activities represented the bulk of Italy's chemical manufactures. For completeness' sake one should add that the preparation of citric acid attained considerable importance. In Piedmont there were several firms, of which Lepetit, Dollfuss & Gansser (now the Ledoga Co., an associate of Dow Chemical) was the most important, specializing in chestnut extracts for use in the silk industry and in tanning. Carlo Erba's pharmaceutical company in Milan was the only one of any significance.

Geographical and economic conditions prevented the development of a diversified chemical industry in Scandinavia. But by 1913 some substantial enterprises had emerged,[2] and, more important, the exploitation of the water-power resources directed production into certain specific areas. In this connexion the situation in Sweden need not detain us,[3] but the electrochemical boom in Norway deserves attention.

On the eve of the war there were at least five big companies in Norway using hydro-electricity for the production of chemicals. Since the country lacked a capital market, most of these firms were established by foreign investors, and the largest were foreign-controlled or owned. The Kellner–Partington Paper Pulp Co., then a British enterprise, now a branch of A/S Borregaard, had been established in 1896 to make wood pulp and hypochlorite at Sarpsborg in south-eastern Norway. Some twenty years later the company employed about 1000 people mainly in the wood pulp business, and in the calcium carbide, ferro-silicon, and electrolytic sections. A/S Hafslund, which specialized in carbide and ferro-alloys, was controlled by a Franco–Swiss group, and in 1912 the Pechiney company became the principal share-

[1] Franck *et al.*, op. cit., pp. 87–8.

[2] Among them Nobel's explosives factory near Stockholm and Stockholms Superfosfat Fabriks (now Fosfatbolaget AB), which dominated the dynamite and fertilizer trades respectively. The Norwegian businesses described in the text were even larger.

[3] Sweden produced about 10 000 tons of cyanamide and also some potassium chlorate.

holder in Det Norske Nitrid A/S. This firm had been established to make alumina and aluminium nitrides by Serpek's process, but, having failed in both, was now considering the conversion of the equipment to the production of aluminium by the usual Hall–Héroult electrolytic method.[1] Cyanamide was made by North Western Cyanamide Co., the operating subsidiary of Nitrogen Products & Carbide, a British company. It used the Frank–Caro process, at Alby near Sundsvall in Northern Sweden, and at Odda on the Hardanger Fjord in Western Norway. At Alby 9500 tons and at Odda 70 000 tons of cyanamide were made.[2] The Norwegian material was exported, some of it to England, where the parent company owned a small works at Dagenham for making ammonia by reacting cyanamide with steam. The profits of North Western Cyanamide were unsatisfactory because the business was dependent on imported coal and coke for the manufacture of carbide. Just before the war the business suffered from the fall in the price of carbide and disappointing sales of cyanamide, and during the war the high cost of coke caused more difficulties. The company sold the works at Alby in 1918, and liquidated that at Odda two years later.

By far the largest enterprise, and one commanding respect in the chemical world generally, was Norsk Hydro Kvaelstof AS, for short Norsk Hydro. Its foundation has been mentioned in an earlier chapter,[3] and it will be appropriate to describe here the position after the withdrawal of Badische from the venture. The Banque de Paris & des Pays-Bas took over after the departure of the Germans, and in 1913 the company's share and loan capital amounted to the large sum of £7 million. This reflected the heavy capital expenditure spread over several years on building two big power stations (their capacity was 130 MW in 1913 and 266 MW in 1916), fertilizer works at Rjukan and Notodden in southern Norway, as well as ancillary services including a 30-mile railway and a train ferry. The company employed 2000–3000 people. The directors were able to claim that technically and financially the arc process was a success, and certainly the turnover was considerable. In the year to 30 June 1914 Norsk Hydro sold 115 000 tons of chemicals, mostly calcium nitrate or 'Norgesalpeter'. On sales of almost £1·6 million the net profit was £0·3 million.[4]

At the outbreak of war Norsk Hydro had to manœuvre diplomatically. In the year 1913/14 it had made about 17 000 tons of fixed nitrogen in the form of nitrates, nitric acid, and some other nitrogen compounds, and was therefore one of the world's largest producers of these essential materials for the production of high explosives. French shareholders had a clear majority and were represented on the board. The Germans retained a small stake in the

[1] Norton, T. H. The chemical industries of Norway. *Chem. Trade J.* **52** (1913), no. 1348, 311; (1913), no. 1349, 339–40; Norges Industri (publ.), *Norwegian hydro electric industry* (1932), pp. 55 ff.; Gignoux, op. cit., pp. 99, 114–15, 131–2.

[2] Franck *et al.*, op. cit., pp. 99, 110. Waeser, op. cit. (2nd edn.) p. 106, gives the capacity at Odda as 52 000 tons a year.

[3] See Chapter 4, p. 87.

[4] *Norsk Hydro history* (1955).

firm and had two directors, among them the redoubtable Duisberg of Bayer; they were also completing the supply of some key equipment. Sam Eyde had returned as chief executive in 1911 and his pro-Allied attitude was well known. A tug of war between the two parties began in August 1914 and continued for the next six months; at first the Germans benefited because Eyde and the French decided not to stop exports to Germany until all the machines had been delivered. But from 1915 onwards the Norwegians were able to dispense with German materials, and shipments fell to a trickle, while sales to France increased quickly, and that country obtained a large share of its nitrogen supplies from Norway for the rest of the war.

The early years of chemical manufacture in Czarist Russia are remarkably well documented. Contemporaries described at length the rapid expansion which began at the turn of the century, was briefly interrupted by the Revolution in 1905, and resumed vigorously in the last pre-war years. In Russia proper alkalis and dyestuffs benefited most from the chemical boom, while in Poland and the Baltic provinces the manufacture of superphosphate suddenly became a substantial business. It would be inappropriate to speak of a Russian chemical industry, but by 1914 there were in that vast country several centres of production where a narrow range of products was made on a large scale. The situation did not resemble that obtaining in Britain, Germany, or even France. The nearest parallel to Russia were the chemical manufactures of the United States, and both countries showed interesting similarities in the structures of their foreign trade: they were large importers of chemicals, but did not export them to any significant extent, but for raw materials such as wood distillation products.

When we seek to document these changes we encounter no lack of statistics, but on the contrary a wealth of confusing and conflicting numbers. Contemporaneous writers picked what best suited their purpose, while post-war authors comparing 1913 with the 1920s were often careless in defining the territory of the empire, sometimes including, sometimes excluding, the regions lost after 1918. Finally Czarists no less than Communists manipulated figures. Subject to these qualifications, this much can be said with fair confidence: in 1913 chemical manufacture in Russia employed 33 000 workers, including those in the big explosives and ammunition trades; a year later there were 40 000. As far as one can judge this was little different from the numbers employed in 1900, which, according to Haller, were estimated at 35 000.[1] The value of production of the main branches, including wood-distillation products, cosmetics, paints, and medicines, was about £6·3 million in 1900 and £11 million in 1912. The heavy chemical sector represented about 40 per cent

[1] Strumilin, S. Khimicheskaya Promyshlennost S.S.R. *Problem ekonomiki* (1935), no. 2, 118 ff.; *Paris Exhibition* (1), lxiv–v. According to Strumilin there were only 15 000 workers in the chemical trades in 1900. He excludes ammunition and explosives, which employed 10 000–12 000, but even so the total is below Haller's figure. The discrepancy cannot be explained.

of the total, and wood-distillation products recorded the most rapid increase.[1] The figures suggest that productivity increased, and this is likely despite complaints about the poor quality of the equipment and the high prices of chemicals; even the simpler ones were sometimes a third or a half above the levels ruling in Britain or Germany. The enormous distances and poor communications were partly responsible for the high prices.[2] The quality of the foremen and the managers was also said to be uneven; at first many Germans had been employed, but after 1900 their numbers diminished and the easy informality of their Russian replacements often degenerated into sheer inefficiency. The explanation here was the different discipline and attitude to factory work of the German and Russian staff, and perhaps also the latter's response to the climate.[3]

The figures in the following table show the relative importance of the main centres of production. As the criteria for measuring this have not been given the numbers should be treated as rough orders of magnitude rather than precise indicators.

TABLE 6.11

Regional distribution of chemical manufacture in Russia in 1912[4]

	Czarist Empire			*Post-war boundaries*		
Main regions	*All chemicals* %	*Heavy chemicals* %	*Organic chemicals* %	*All chemicals* %	*Heavy chemicals* %	*Organic chemicals* %
North-west	13	17	9	21	21	21
Moscow region	19	11	28	30	14	63
Ukraine	16	26	5	26	33	10
Baltic and Poland	37	22	56	—	—	—
Others	15	24	2	23	32	6
Total	100	100	100	100	100	100

Poland, the Baltic towns, and the Moscow region were the main centres, and the loss of the first two after the war therefore deprived Russia of a large share of its chemical capacity. Certain manufactures were very highly concentrated. For example the principal sulphuric acid works were located near Moscow and Warsaw, and also close to the then Russo-German border in Polish Silesia. In the Ukraine, which was the heart of Russia's coal–steel industry, there were only two provinces (each the size of several English counties) in which chemical manufacture was carried on to any significant

[1] Nekrassov, N. N. *Ekonomika khimicheskoi promyshlennosti* (1957), p. 28.

[2] Pantyuchoff, N. Ten years of Russian chemical industry 1898–1908. *Chem. Trade J.* **66** (1910), no. 1182, 63–5.

[3] Hauffe, M. Die chemische Grossindustrie in Russland geschildert auf Grund eigener Erfahrungen und Beobachtungen. *Z. angew. Chem.* **13** (1900), no. 1, 26–30.

[4] Ioffe, Y. A., Kozlov, A. I., and Alman, G. L. *Ekonomika i planirouaniya khimicheskoi promyshlennosti* (1940), pp. 20, 24.

extent: north-east of Kiev in the province of Chernigov fertilizers and wood-distillation products were important, while in Ekaterinoslav (Dnepopetrovsk) province five factories with 2425 workers testified to the importance of the soda trade that was established there.[1] The Leblanc process was no longer used except to make sodium sulphate and hydrochloric acid. Electrolytic alkalis were commercially unimportant. Most of the soda was made by the ammonia–soda process. Output increased significantly in the years before the war, and by 1914 the country had become virtually self-sufficient. Given the size and population of Russia the production of alkali was relatively small.

TABLE 6.12

Russian production of alkali, selected years 1908–13[2]

Year	*All alkalis* 000 tons	*Caustic soda* 000 tons
1908	117	40
1910	132	48
1912	164	49
1913	160	55

Manufacture was primarily in the hands of two firms, of which Liubimov, Solvay| & Co. a joint Russian–Belgian company, was easily the more important. Its works were at Beresniki in the Urals and at Lissichansk in the Ukraine. About 1500 people were employed in the 1900s, and the figure was certainly larger by 1913 owing to the extensions at Lissichansk. Liubimov, Solvay accounted for about four-fifths of the soda ash and the greater part of the caustic soda made in Russia. It was profitable and paid regular dividends. The other firm was the result of a merger of three small foreign controlled and unsuccessful businesses. Their factories were at Slaviansk, in the province of Kharkov, and at Zombkowice in south-west Poland.[3]

In 1913 there were said to be thirty-four sulphuric acid works with a combined capacity of 213 000 tons. They made 165 000 tons (as 100 per cent H_2SO_4) in that year, mainly from imported pyrites and brimstone.[4] In Russia

[1] Dyes, W. A. Produktion und Rentabiltät der Russischen Chemischen Industrie. *Chem. Ind., Düsseld.* **28** (1905), no. 11, 339–40; Busemann, M. (Ed.) *Die wirtschaftliche Bedeutung der Ukraine* (1918), p. 80; Ioffe *et al.*, op. cit., p. 66.

[2] For 1908 the source is Pantiuchov, art. cit., pp. 149–50, and for the other years Nutter, G. W. *et al. Growth of industrial production in Russia* (1962), pp. 412, 423. The figures accord reasonably well with those quoted in *Z. angew. Chem.* **26** (1913), pt. 3, 303, and *Chem. Ind., Düsseld.* **37** (1914), no. 13, 389.

[3] Dyes, art. cit., pp. 406 ff.

[4] Information from Mr. Kaufmann of the National Bureau of Economic Research. The tonnage made is equivalent to 260 000 tons of 50° Bé acid (62 per cent H_2SO_4). His figure lies midway between that of his collaborators (see Nutter *et al.*, op. cit., pp. 412 and 424) who estimated production at 292 000 tons for the Czarist Empire and at 226 000 tons for that made in the post 1918 territory. The strength is not given, but was probably around 50° Bé.

proper the acid was used for general chemical purposes and also for oil refining, and the two main producers were the Tentelev and Ushkov companies. The former owned large works in St. Petersburg, the latter, an old-established firm, had several factories in central Russia.[1] Tentelev, and even more so Ushkov, specialized in sodium sulphate, hydrochloric acid, and bleaching powder. The first of these was then employed as a cheap substitute for soda ash, and as such retained a market long after it had ceased to be important in Western Europe. The co-product hydrochloric acid was also significant because it was treated by the Weldon process to yield chlorine. This was the source of bleaching powder and zinc chloride, the latter being used at the time in place of creosote (which was scarce) for timber preservation, and the railways were large buyers.

Superphosphate was a minor outlet for sulphuric acid in central Russia and the Ukraine, but its manufacture developed rapidly in Poland and the Baltic ports. Altogether 115 000 tons of the fertilizer were made in 1913 from imported phosphate rock, and almost three-fifths of the total was produced in these border regions. It is interesting to note that agriculture had developed sufficiently in the Baltic (and probably also in the region around St. Petersburg) not only to absorb all the Russian-made superphosphate, but also to require upwards of 200 000 tons of imported material.[2]

In the last years before the war the manufacture, generally on a small scale, of many new products was begun. The list includes calcium carbide, acetylene, liquid oxygen, hydrogen peroxide, carbon disulphide, and viscose rayon. Wood-distillation products for export became commercially important, and the Schering company, which was German-owned, was one of the large producers, owning a factory at Wydriza in White Russia, which employed several hundred people.[3]

The most interesting and significant pre-war development was in coal-tar products and dyestuffs. It was directly linked with the introduction of recovery ovens in the Ukrainian steel industry. The first of these ovens were built about 1910, and by the eve of the war there were 848 in the Don basin making almost 40 000 tons of crude tar.[4] Simultaneously sulphate of ammonia became available, and production jumped from nothing to 14 000 tons, between 1910 and 1913, and the following year reached 17 000 tons.[5] As far as dyestuffs are

[1] Tentelev were technically progressive and had developed (some said imitated) a contact process for making oleum. It was profitable; Ushkov was not. The Ushkov family were related by marriage to the Liubimovs. The industrial activities of these two families extend over two generations and possibly longer, and would repay study.

[2] Kats, A. M. *Voprosy Ekonomiki Osnovnoi Khimicheskoi Promyshlennosti SSR* (1939), p. 83. Nutter *et al.*, op. cit., pp. 412, 424. Gray, op. cit., p. 191, states that Russian production of superphosphate was 225 000 tons.

[3] Holländer, H. *Geschichte der Schering AG* (1955), p. 37.

[4] The source is W. Pochitonov, in a lecture given to the Russian Chemical Society on 24 September 1914 on the state of the Russian chemical industry. The lecture is quoted in Hesse, A., and Grossmann H. (Eds.) *England's Handelskrieg und die chemische Industrie* (1915), p. 389. The production of crude steel doubled between 1906 and 1913, when it reached 5 million tons.

[5] Institut International d'Agriculture, *Yearbook of agricultural statistics* (1923).

concerned the story begins in 1876 when the Russian tariff on colouring materials was amended to encourage indigenous production. In 1900 the Quartermaster General first allowed the use of coal-tar dyes for dyeing army cloth. The changes in the regulations created more favourable conditions, but the absence of Russian-produced coal-tar distillates until just before the war delayed the emergence of integrated manufacture. When war broke out the situation resembled that of France. Russia used about 7000 tons of dyes, of which roughly 4000 were made from Russian or imported intermediates and 3000 were imported from Germany and to a small extent from Switzerland.[1] German firms dominated the dyestuffs business, and their works were of considerable size. Hoechst, for instance, made alizarin and aniline dyes at Moscow, where it employed just over 500 people in 1913.[2] Cassella and Kalle subordinated their activities to that of the senior partner. The Dreibund followed a similar policy. Badische and AGFA owned small factories, and allowed Bayer to expand theirs. The Bayer factory near Moscow grew quickly and by 1908 made just over 1400 tons from imported and local materials, and in addition distributed 2000 tons of German dyes. The company had a laboratory and a dye-house besides the factory and packing plant, and employed several hundred people, among them chemists and technical service staff.[3] The Swiss firms operated on a smaller scale. Geigy was connected with businesses at Sokolniki, near Moscow, and at Libau, which made tanning extracts and sulphur dyes and also sold the dyes made by the parent company in Basle and by Durand & Huguenin.[4] CIBA owned a half share in the Pabianizer AG für Chemische Industrie, situated near the textile centre of Lodz in Poland. In 1913 this company employed 300 people and specialized in khaki colours for wool and cotton.[5]

The manufacture of and trade in dyestuffs was profitable. The high tariff was indeed an obstacle, but it could be circumvented by shipping the dyes in bulk as concentrates, and diluting and repacking in Russia. Large stocks had to be kept, but the high margin on sales or agency commissions amply covered the expense. These stocks were extremely useful after the outbreak of war, and certainly played their part, along with skilful improvisation, in making the dyestuffs problem less acute in Russia than in Britain or France.

The United States of America

By any standards the American chemical industry was enormous. Including the allied trades, the value of production was estimated at $840 million (say

[1] League of Nations, op. cit., pp. 49, 52.

[2] *Farbwerke vorm. Meister Lucius & Brüning 1863–1913*, op. cit., p. 40.

[3] Blank, A., and Löw, W. Geschichte der Fabrik . . . in Moskau; *Geschichte und Entwicklung der Farbenfabriken vorm. Fr. Bayer*, op. cit., pp. 270 ff. The factory was among the first in the world to use a diesel engine for electricity generation.

[4] Bürgin, op. cit., p. 186.

[5] Jaquet, op. cit., p. 77; Ges. für Chemische Industrie, Basel 1884–1934, op. cit.

£170 million) in 1914, that is about 40 per cent larger than that of Germany.[1] On a more narrow definition of chemical manufacture, which, by excluding ammunition, carbon black, petroleum refining, etc., corresponds better with that of Europe, the output was valued at \$378 million (about £77 million) thereby placing the industry mid-way between that of Germany and Britain.

TABLE 6.13

Employees and value of production of the United States chemical industry in 1914[2]

Sector	*Employees* (000)	*Value of production* \$ million	*Net output* \$ million
Explosives (excluding ammunition)	6	42	16
Fertilizers	23	153	45
Sulphuric, nitric, and other acids	3	15	8
Wood distillation and charcoal	3	10	4
Chemicals not elsewhere specified	32	158	69
Total	67	378	142

By comparison with other technologically new industries, chemical manufacture was relatively small: in 1914 the automobile industry employed 79 000 people and had a net output of \$211 million, and electrical engineering 118 000 and \$180 million. Nevertheless the chemical industry as a whole was growing fast, and the range of acids, alkalis, and fertilizers made was being enlarged, so that by 1914 the country had become effectively self-sufficient in these products. In that year the United States occupied first place in the world league for superphosphate, sulphuric acid, and sodium compounds. The country had not yet become a significant exporter of chemicals, excepting raw materials such as sulphur and wood-distillation products. It was a big importer of potash and nitrates, neither of which was then indigenously available, and also of dyestuffs, ammonium compounds, and cyanides.

Tariff policy fostered and maintained the development of inorganic chemicals. On account of the great distances, their manufacture was concentrated in certain areas: New England, the Philadelphia region, Niagara Falls, and Wyandotte near Detroit were the most important. Even in 1914 the industry was still characterized by the predominance of small works, and the average chemical factory employed about 80 people. Apart from Du Pont in explosives, General Chemical in inorganic chemicals, and a few of the specialist superphosphate producers, there were no firms in the United States

[1] *Enquete*, p. 8. The rate of exchange used is \$4.90 = £1. The 'allied trades' include ammunition, carbon black, patent medicines, paint, polishes, petroleum refining, soap, seed crushing, tanning materials, and town gas.

[2] Statistical Abstract of the U.S. (1925).

playing even remotely a role resembling that of the German dyestuffs companies, or the British soda makers. As a rule the chemical enterprises were still owned by the founders and their local associates, and unlike their European counterparts very few had their shares quoted on the stock exchange.[1] Individualism flourished, and when there was excess capacity any tendencies towards market sharing and agreements were checked by the then politically fashionable trust busting. The Sherman Act (1890) was vigorously applied under Theodore Roosevelt and Wilson, and the Clayton Act (1914) reinforced the earlier legislation by prohibiting price discrimination, interlocking directorates, and the acquisition of shares by one company in another if the effect was to lessen competition. The Federal Trade Commission was set up under the Act to investigate and report on infringements.

The manufacture of fertilizers represented the largest single sector of the American chemical industry. Production was concentrated in a few big firms, while compounding was fragmented among a large number of small enterprises. The cotton growers represented about half the consumption of fertilizers,[2] and production was therefore located mainly south of the Mason–Dixon line, and more specifically in South Carolina, Florida, and Tennessee. Output grew from roughly 1·8 million tons of superphosphate in 1905 to 2·9 million tons in 1913, equivalent to about one-quarter of world production.[3] But since 1911 the trade had been in the doldrums. The other fertilizers played a subordinate role, except that the consumption of potash, entirely imported from Germany, doubled between 1905 and 1913.[4]

At the eve of the war three companies had emerged as the leaders of the superphosphate producers. The oldest was the Virginia–Carolina Chemical Co. founded in 1895. Its chequered career ended ingloriously in 1924 in bankruptcy, from which it was reconstructed, without being any more profitable. Its rival, the American Agricultural Chemical Co., started in 1899 and fared rather better. The third, the International Agricultural Chemical Co., was the outcome of an amalgamation in 1909 of several phosphate-rock mines, and had links with the German potash industry through the Schmidtmann mining group, which was for some years a notorious price cutter.[5] The

[1] In 1910 only two chemical firms were among the 25 largest U.S. companies measured by total assets. They were Du Pont and Virginia-Carolina Chemical Co. with total assets of $75 and $72 million respectively. They must be compared with the four leaders: U.S. Steel ($1822 million) Standard Oil ($372 million in 1906), American Tobacco ($286 million), International Harvester ($173 million). Their names succinctly reflect the nature of U.S. industry at the time.

[2] Haynes, W. *American chemical industry* (1945), vol. 3, p. 180.

[3] Grey, op. cit. (1930 edn.), pp. 249–51. Figures are in long tons.

[4] It has been estimated that in 1913 the United States used, in terms of plant food, 510 000 tons of P_2O_5 mainly as superphosphate and bone meal, 218 000 tons of K_2O and 154 000 tons N, mainly in the form of nitrates. Federal Trade Commission, *Report on the fertilizer industry* (1950), p. 159.

[5] Two of the big firms survived precariously until the 1960s when they were swallowed by oil companies that were extending their fertilizer activities. Thus Mobil Oil took over Virginia Carolina in 1963 and Continental Oil acquired American Agricultural a year later. International Agricultural Chemical has been renamed International Minerals & Chemical Corp. and is still trading.

three companies, taken together, accounted for 55 per cent of the phosphate-rock production in Tennessee and 96 per cent of that in Florida. The first two each sold more than 0·9 million tons of fertilizers, and together with the five next largest handled 3·6 million tons of manufactured fertilizers, that is roughly three-fifths of United States consumption.[1] This degree of concentration is the more remarkable when we note that not all the manufacturers made their own sulphuric acid: International Agricultural purchased theirs from Tennessee Copper, others bought from firms located on the Atlantic seaboard which specialized in making acid from imported pyrites.[2] One other aspect requires comment: the big manufacturers competed with scores of small compounders, which bought superphosphate and potash, did their own mixing, and supplied a relatively small area. These independents did not face the same distribution and stock problems as the big producers, and, not being burdened by heavy debt charges, they tended to be more profitable.

American farmers used a substantial, but not precisely measurable, tonnage of organic manures such as dried blood, fish meal, and, of course, dung. They also bought Chilean nitrates. But sulphate of ammonia, which was then an important fertilizer in Europe, played a minor role in the United States until the last two or three years before the war, and few attempts had been made to sell it systematically. By contrast, the new methods of nitrogen fixation attracted attention. The American Cyanamid Co., which worked the Frank–Caro process on the Canadian side of Niagara Falls, was an important element in the American fertilizer industry. Though the capital, partly raised in Britain, was quite small—\$5·8 million (about £1·2 million)—the size of the plant was steadily being increased. In the twelve months to June 1914 about 33 000 tons of cyanamide were made, and later in that year the capacity was enlarged to 57 000 tons. At this point J. B. Duke, later president of the American Tobacco Co., become financially connected with the company. He owned phosphate-rock deposits, which were to be used for making ammonium phosphate, but this ambitious scheme, which was ahead of its time, was replaced by a much simpler one, in which a small amount of cyanamide was mixed with superphosphate.[3] In 1914 American Cyanamid was the only enterprise of its kind in North America, and without Duke's support might have failed, because its capacity greatly exceeded output. The company provides an example of the American attitude to the size of chemical works, which was, in short, to build a large plant and then find a market for the products.

The alkali trade also followed this policy, and as a result was facing a difficult trading situation by 1914, when the United States made 835 000 tons of

[1] Haynes, op. cit., vol. 2, pp. 185, 187; vol. 3, pp. 172–3.

[2] In 1913 only 3 per cent of all the sulphuric acid made in the United States was prepared from brimstone, and the huge sulphur deposits in the south were at the beginning of their development. Haynes, W. *Brimstone, the stone that burns* (1959), p. 91.

[3] Haynes, W. On the chemical frontier—the Cyanamid story, typescript dated 1957/8.

soda ash and 260 000 tons of caustic soda.[1] The main problem was excess capacity: tariff changes in the late 1890s had given American soda makers adequate protection, and several big ammonia-soda factories were built. There was not, as in Britain, a struggle between rival technologies, but unlike Britain and the rest of Europe, no single firm had the same power as Solvay, which through subsidiaries and agreements was able to discipline the leading manufacturers and even exercise control over electrolytic caustic. Another difference from Europe was the absence of a dyestuffs industry with its considerable demand for alkalis.

There were in 1913 five companies making ammonia soda: Solvay Process with three factories was the largest; Columbia Chemical and Diamond Alkali were linked with glass makers; Michigan Alkali and Mathieson Alkali were perhaps a little smaller than the others and were not growing rapidly. Michigan Alkali at Wyandotte on the Detroit River owned limestone quarries and coal mines, while Mathieson had works at Saltville (W.Va.) for making ammonia soda and was extending its electrolytic alkali plant at Niagara Falls, which used Castner's cell.[2] These five competed with each other, and also with the fast growing electrolytic caustic soda manufacturers. Michigan Alkali exported soda and so aggravated the difficulties that were then facing British producers. In 1910 John Brock of the United Alkali Co. visited the United States in the hope of reaching an understanding on soda exports, but found the Americans unreceptive. They did not want to attract the attention of the Department of Justice, which at that time was extremely active in fighting restraint of trade and was prosecuting Du Pont and Standard Oil.[3] The Brock mission failed, and excess capacity and falling prices plagued the American soda trade until the war.

The difficulties of the ammonia-soda makers were aggravated by the rising trend of electrolytic caustic soda. Production rose quickly in the 1900s, and the companies established at Niagara Falls benefited from favourable electricity tariffs; they included Mathieson (which pioneered the sale of liquid chlorine in the United States), Hooker Electrochemical, which had started in 1906,[4] and Niagara Alkali Co., which was founded in 1909. Elsewhere there were several substantial manufacturers, notably the old-established Pennsylvania Salt Mfg. Co. of Natrona (Pa.), which owned a diaphragm-cell plant at Wyandotte. The company's sales were about $8 million (£1·6 million) in 1913, and consisted of various inorganic acids besides alkalis.[5] The Warner Chemical Co. of Carteret (N.J.) was roughly of the same size, and from the

[1] Census figures quoted by Haynes in *American chemical industry*, vol. 3, p. 430. Caustic lye sold to soapmakers is excluded.

[2] Pound, A. *Salt of the earth* (1940); Anon. (Mathieson Alkali Works) *Fifty years of chemical progress* (1942), pp. 19 f. The company now forms part of Olin Mathieson Chemical Co.

[3] Notes of meeting on 15 December 1910 of the English Caustic Soda Makers Association in Chance & Hunt archives; Haynes, op. cit., vol. 3, p. 5.

[4] Anon. *Hooker Electrochemical after 25 years* (1929), p. 5. Hooker was taken over by Occidental Petroleum Co. in 1968.

[5] Leavitt, R. K. *Prologue to Tomorrow* (1950), pp. 53–9.

outset made several chlorine compounds, including carbon tetrachloride.[1] Among the small firms only one name need be retained: the Dow Chemical Co. of Midland (Mich.) was then a very modest enterprise with sales of \$1·1 million (£0·2 million). It specialized in bromine, bleaching powder, and carbon tetrachloride. Its founder, H. H. Dow, was extremely energetic and added other chlorine derivatives to the range, but the phenomenal growth that propelled Dow Chemical into the front rank of the American chemical industry was still many years hence.[2]

The general chemicals business, that is the preparation of inorganic acids and salts, was handled by numerous firms. Trade was less buoyant in the last years before the war because the pattern of consumption was changing: for example sulphuric acid had been used in the treatment of paraffin, and, while the oil lamp was unchallenged, sales to the petroleum refiners had boomed. Now, however, gas and especially electricity were replacing the oil lamp and the demand from the refineries ceased to grow at spectacular rates. There were, of course, other large customers, but steel-making and superphosphate were subject to cyclical fluctuations, which reacted on the profitability of the chemical manufacturers. Few had the resources and the managerial ability to diversify their operations successfully. Among the exceptions was the General Chemical Co., a merger that fulfilled the expectations of its founders: it had been set up in 1899 by amalgamating twelve small manufacturers. By 1912 its principal works was at Marcus Hook (Pa.), and its chief products sulphuric acid, oleum,[3] sodium sulphate, and various sodium compounds (excluding, however, soda ash and caustic soda). General Chemical, in association with other firms, extended its interests before the war by starting the manufacture of aniline oil and simple intermediates. But at first this branch, trading as Benzol Products, was a failure. Nevertheless General Chemical was a fairly profitable business earning around 8–9 per cent on assets in the period 1900–13 (with the exception of the slump year of 1908), and its considerable size is indicated by the large figure of \$31·8 million (£6·5 million) for its assets.[4]

The other makers of heavy chemicals need not long detain us. The Grasselli company produced sulphuric acid, fertilizers, and sodium silicate in New Jersey and at Birmingham (Ala.), the Merrimac Co., then the leading chemical enterprise in New England, made alumina, insecticides, and sulphuric acid.[5]

[1] The founder was L. C. Warner (1841–1925) who, having made a fortune out of corsets, turned to chemical manufacture late in life. The corset business bears his name to the present day, while the chemical company eventually became part of the F.M.C. Corp. For details of Warner see Haynes, W. *Chemical pioneers* (1939).

[2] Gerstacker, C. A. The Dow story. *Armed forces chem. J.* (July/August 1956), reprint (no page number) and anuual reports of company.

[3] The company had designed its own contact acid process and after litigation with Badische had reached agreement on an exchange of information. Lunge, G. Zur Geschichte des Kontaktverfahrens .. . in den Vereinigten Staaten. *Z. angew Chem.* **23** (1910), no. 16, 720 f.

[4] Anon. *The General Chemical Co. after 20 years* (1919); annual reports of the company.

[5] Haynes, *Chemical pioneers*, op. cit. Grasselli was later bought by Du Pont, and Merrimac by Monsanto Chemical.

Subsidiaries of foreign companies played a subordinate role. The United Alkali Co. owned a chlorate works at Bay City (Mich.), and Albright & Wilson made phosphorus and sodium chlorate at Niagara Falls. The Solvays held shares in the Solvay Process Co., but the extent of their control has not been revealed. Finally and importantly, the Deutsche Gold- und Silberscheideanstalt (DEGUSSA) controlled Roessler & Hasslacher. We have encountered DEGUSSA before, and its North American interests were diverse: Roessler & Hasslacher together with the Germans owned the Niagara Electrochemical Co., which used the Castner process for making metallic sodium and cyanides. Later it added hydrogen peroxides and perborates. At Perth, Amboy, Roessler & Hasslacher prepared formaldehyde from methanol, and was one of the suppliers to General Bakelite's new phenol-formaldehyde resin business.[1]

The American textile industry accounted for one-sixth of world dyestuffs consumption, and its requirements were estimated at 24 000 tons in 1913. A mere 3000 tons were made indigenously.[2] This enormous gap between consumption and home production was due to several causes. In the first place the United States lacked what is now called infrastructure, and in this connexion refers to adequate supplies of coal-tar distillates and of simple intermediates. In the last years before the war Benzol Products attempted to remedy this lack and prepared, among other things, aniline oil and nitrobenzene. But the company's operations were uneconomic and its contribution towards the establishment of dyestuffs manufacture unimpressive.[3] A second obstacle was the existing patent legislation. As the product was patentable, alternative routes to that end-product could not be developed. (A similar situation had stopped the development of the French dyestuffs industry half a century before.) Moreover there was no working clause in the United States patent legislation. These problems might have been overcome if there had been sufficient interest in the establishment of dyestuffs manufacture. But there was little evidence of this, and the tariff policy reflected official and legislative indifference. Furthermore, chemical businessmen, of whom few were chemists, were in general more concerned with large tonnage products than with applied research and technical service: quick returns were expected and investment in projects likely to mature slowly was avoided. So far as I know American chemists made no contribution to dyestuffs chemistry before the war, nor were staff and men predisposed for the painstaking work that was a prerequisite for the successful manufacture of dyestuffs.

One of the main reasons why dyestuffs production could not flourish was the detrimental effect of United States tariffs.

[1] For DEGUSSA see p. 113; Haynes, op. cit., vol. 3, p. 18, and summary of the Report of the Alien Property Custodian in *J. ind. Engng Chem.* (April 1919), pp. 356–7.

[2] League of Nations, op. cit., pp. 28, 52, but according to the Review of the Dyestuffs Industry prepared by the British Dyestuffs Corp. in 1927, U.S. consumption was 26 000 tons in 1913. The difference in the figures does not affect the argument.

[3] Haynes, op. cit., vol. 3, p. 213.

TABLE 6.14

United States tariffs on dyestuffs and intermediates[1]
(Rates are in per cent *ad valorem* unless otherwise stated)

	1897	1909	1913
Aniline and toluidine	20	free	10
Alizarin, natural and synthetic	30	free	free
Other intermediates	20	20	15
Indigo, synthetic	¾ ¢/lb	¾–10 ¢/lb*	free
Other dyestuffs	30	30	30

*According to concentration.

It is very difficult to see any coherent policy in these changes. The American legislators, like the Russians and the French, imposed lower duties on intermediates than on finished dyes, but later changed their minds and exempted aniline and toluidine, as well as alizarin, from duty in 1909. All three were easy to make. Four years later the duty on synthetic indigo was removed and that on intermediates cut by a quarter, an action guaranteed to hurt American producers. These manipulations made no sense, and one can only explain them on the ground that the users of dyestuffs and the importers carried great weight in Washington, and were able to persuade Congress that an indigenous dyestuffs industry was unnecessary, because the Germans could and would supply the entire range regularly and cheaply. Events were shortly to prove them wrong.

The value of dyestuffs imports in the year to 30 June 1913 was \$9·1 million (say £1·9 million); a year earlier the figure had almost reached \$10 million, and it would appear that the level of activity in the textile industry rather than tariff changes accounted for most of the annual fluctuations.[2] As for intermediates, the duties were not sufficiently high to protect Benzol Products, nor were they sufficiently low to enable United States firms to make dyes from imported intermediates and compete with the German products. In short the margin between the tariff on intermediates and finished dyes was insufficient, and the only party to benefit from this policy was the Germans. They owned or controlled seven companies in the United States, but of these only Bayer & Co. at Rensselaer (N.Y.) and its subsidiary the Hudson River Aniline & Color Works at Albany made dyestuffs (the latter also prepared aspirin and phenacetin). The other companies operated primarily as repackers and distributors of German-made products.[3] The Germans were brisk salesmen and did not shrink from unscrupulous marketing methods.[4] Their American

[1] Haynes, op. cit., vol. 1, pp. 431 ff.; vol. 2, pp. 332 ff.

[2] Norton, T. H. *Dyestuffs for American textile and other industries* (U.S. Dept. of Commerce, 1915), p. 17.

[3] Alien Property Custodian report, art. cit., p. 357.

[4] They cut the price of aniline oil as soon as Benzol Products put theirs on the market. In 1912 a salesman for Bayer was tried for corrupt practices, and at about the same time Badische, Bayer, and Hoechst were sued for infringement of the Anti-Trust Act. They lost, and Hoechst was fined \$40 000. Alien Property Custodian report, art. cit., p. 358.

rivals complained of unfair competition, but we may note in passing that the same accusations were levelled against the American oil and sulphur companies in Europe.

Only two American firms require mention in connexion with dyestuffs. Schoellkopf Aniline & Chemical Co. made some intermediates and perhaps 1000–1500 tons of dyes. The lesser enterprise was that of Heller & Merz of Newark, which specialized in paper dyes. On the outbreak of war in Europe, prices soared in the United States, and the textile industry lived from hand to mouth until large-scale production got under way in 1917.

The other branches of applied organic chemistry reflected more vigorous activity. Patent medicines were being supplemented by ethical products, and both were growing fast, production being valued at $102 million (£21 million) in 1914. Firms that have since become famous, such as Eli Lilly, Abbott Laboratories, G. D. Searle, Upjohn, and a few others, were then beginning to place more emphasis on medicinal preparations.

Wood distillation represented another sector of the industry that was developing rapidly. It was being carried out on a very large scale in the southern pine forests. The United States was probably the largest producer of methanol in the world, and the operations also yielded in 1914 74 000 tons of acetate of lime, which represented then the only source of acetone. There was virtually no manufacture of specialized organic chemical products, with the notable exception of the small firm of Monsanto Chemical. The assets of this business were a mere $0·4 million in 1913. It made vanillin, caffeine, phenacetin, phenolphthalein, and a few other sophisticated chemicals, which competed directly with those imported from Germany. J. F. Queeny, who had founded the company a few years earlier, was assisted by chemists from Switzerland and was willing to risk a fight with the Germans. His gamble was justified many years later when his company ranked among the six largest in the United States chemical industry.[1]

The history of the explosives industry is outside the scope of this book and accordingly the nineteenth century growth of the Du Pont company has been omitted. But in the last years before the war this large producer of powder and dynamite began extending its interests to other areas. The changes came to fruition in the 1920s and 1930s and were so important that the antecedents require detailed treatment. From 1909 onwards Pierre Samuel du Pont was effectively in charge of the company and had to implement the division or dissolution that the Court ordered when the Department of Justice won the anti-trust case against Du Pont.[2] The court's final decree of 1912 set out the

[1] Baumann, E. Obituary of G. F. du Bois, *Neue Zürcher Zeitung* (for. edn., 15 November 1953); Haynes, *Chemical Pioneers*, op. cit., pp. 235 f. and information from Mr. F. J. Curtis, Treasurer's Dept., Monsanto Chemical Co., St. Louis.

[2] Du Pont agreed that it had a monopoly of the military smokeless powder trade. When the dissolution was ordered the Secretaries of State for the Army and Navy pointed out that this would endanger national safety. Despite widespread protests, the company was allowed to retain the military business in its entirety. The Sherman act had thus been set aside on the ground of national interest. This decision caused ill-feeling against the company in Congress and even in the

manner of the dissolution, and so led to the establishment of two new explosives companies, which began to trade in 1913 as the Hercules Powder Co. and the Atlas Powder Co.[1]

The judgement still left Du Pont as the largest United States manufacturer of powder and dynamite, and at the end of 1913 it employed about 5000 people. The hiving off of some of the properties to Hercules and Atlas hardly made a dent in Du Pont's total assets, which stood at \$86 million (£17·6 million) on 31 December 1912, the eve of the dissolution, and at \$74·8 million (£15·3 million) a year later. But sales fell smartly from \$36·5 million in 1912 to \$26·7 million in 1913; net earnings, however, fell less than proportionally from \$6·9 million to \$5·3 million.[2]

At the beginning of the century the old patriarchal administration gave way to an 'Executive Committee' consisting of younger members of the du Pont family, and a few outsiders. They set up a Development Department, which controlled the Experimental Station, that is to say the research laboratory.[3] The new department soon began to concern itself with the effect of the decline in Government orders for smokeless powder, which raised the problem of spare capacity in the nitro-cellulose plants.

In the 1900s a rapidly growing use for nitro-cellulose was to plasticize it with camphor and sell the material as celluloid to the fancy goods trade; latterly a new market had been found for the transparent sheet as side screens for motor cars. In 1913 the United States production of unwrought celluloid was estimated at 3700 tons. Nitro-cellulose was also employed as a lacquer for brass and other metal ware, and as a coating for fabric, which as 'leather-cloth' was widely used in book-binding. (Chardonnet's artificial silk was based on nitro-cellulose, but was being rapidly displaced by viscose.) The Du Pont company was unfamiliar with these products, but in 1908 it had set up a small committee to report on the idle nitro-cellulose capacity and to recommend what steps should be taken '. . . in the direction of developing further uses for guncotton or any of the other products of our smokeless powder plants'.[4] The committee in its report recommended that the best way of achieving this objective was to buy firms that used nitro-cellulose. It rejected the development by Du Pont itself of new uses, because the company's research facilities were inadequate. Indeed the Chemical Department, which was to carry out

---

Government. Du Pont Smokeless Powder Dept. *A history of the Du Pont Company's relations with the U.S. Government 1802–1927*, pp. 69–72.

[1] Hercules powder was rather larger than Atlas and had about 1000 workers and several factories. Under the final decree both companies were allowed to use Du Pont's engineering, purchasing, and chemical departments until 1917. The ordinary shares of both companies were passed on by Du Pont to its shareholders, and the loan stock was sold to raise cash. Anon. *Hercules Powder Co.—a history* (1947), pp. 1, 10–11; Annual reports of Du Pont for 1912 and 1913.

[2] Annual Reports of Du Pont. The company gave particulars of gross receipts which are close enough for our purposes to sales.

[3] James, M. *Alfred I. du Pont* (1941), p. 125; Dutton, W. S. *Du Pont* (3rd edn. 1951), p. 185.

[4] Dutton, op. cit., pp. 186, 296–7; D.P. 69 (note on Executive Committee Meeting of 16 December 1908). A company bought by Du Pont in 1905 made metal lacquers on a small scale.

this kind of research and development work, was not created until 1911. The advice of the committee was sound, because Du Pont had considerable administrative and financial expertise, which had proved itself in earlier take-overs of explosives manufacturers. These firms had been bought by issuing loan stock in exchange for their assets, and in this way Du Pont's equity was not diluted and control remained with the du Pont family. There was a further advantage: the acquisitions would provide the company not only with a captive market for nitro-cellulose, but would also secure at the same time the marketing skill for selling the products. The Development Department, in endorsing the recommendations of the committee, was thus using a well-tried tactic in a novel context, and, in a phrase which has since become classic in business history, was 'buying the customers along with the business'.[1]

Once the objective had been defined and strategy decided, action followed swiftly: in 1909 a pilot plant to make artificial leather was established at Oakland (Cal.), and negotiations were started with the Chardonnet company in France, but were abandoned when it was realized that viscose rayon had greater prospects.[2] The next year the first major acquisition outside the explosives business was made by the purchase of the Fabrikoid Co. of Newburgh (N.Y.) for about \$1·2 million. Fabrikoid specialized in bookcloth and employed fewer than 100 people. Its sales were a mere \$0·7 million in 1911, but had doubled by 1913, when the firm was renamed Du Pont Fabrikoid.[3] The Development Department, which was responsible for the execution of the diversification policy, next turned to celluloid. There were then four manufacturers in the United States, of which the Arlington Co. was the largest. Its assets at the beginning of 1914 amounted to \$5·7 million (almost £1·2 million). Arlington made celluloid and used 40–50 per cent of the make for fabricating into various plastic semi-manufactures. It was bought in 1915.[4] The following year the Fairfield Rubber Co., a producer of leather-cloth, was purchased, and was followed shortly by Harrison Brothers, makers of paints and varnishes.

None of these acquisitions on its own represented a significant contribution to Du Pont's sales and profits. But they illustrate the company's systematic and purposive approach to diversification and forward integration. We will leave to a later chapter the description of how the process transformed Du Pont, and the example it set to others.

[1] GTX 108, Development Department report, 17 April 1914.

[2] DP 72, Annual report of the Development Department, 9 February 1910; DP 87, memorandum from Development Department on General Utilization of Plants after the War, 15 May 1916.

[3] GTX 103, Letter from R. R. M. Carpenter to Executive Committee, 16 March 1912; GTX 104, memorandum by Development Committee on Pluviusin Co. of 11 October 1912; GTX 105, memorandum by R. R. M. Carpenter, 10 February 1913. GTX 106, Short History of the Fabrikoid Industry by R. R. M. Carpenter, 4 November 1921. Fabrikoid used the patents and know-how of the British Winterbottom Book Cloth Co. Ltd. which had a small shareholding in the American Company.

[4] GTX 108 and 110, Development Department memoranda of 17 April 1914 and 12 September 1915.

7

THE GREAT WAR

WAR is a technological forcing house, and so it is not surprising that applied chemistry was profoundly influenced by the events of 1914–18. The effects were specific as well as general, and the former are particularly relevant to our story, because of the changes they wrought in the orientation and structure of the chemical industry in the principal countries. Among the Allied powers the manufacture of organic chemicals, exemplified by dyestuffs and their intermediates, was still embryonic in 1914. But within four years the United States and Britain, and to a lesser extent France, Italy, and Japan, created their own dyestuffs industries and related activities. On the German side the situation was totally different: organic chemicals presented no problems, so that they can be safely ignored in this chapter, nor was the adaptation of dyestuffs works to military needs particularly difficult. The crucial problem resulted from the cessation of imports, which caused an acute shortage of nitrogen. Thus a study of the interaction of war and chemical manufacture must take account of the contrasting experience among the belligerents, dyestuffs on one side, nitrogen on the other. The solutions that were evolved to these specific problems had wider and more general effects on chemical manufacture as a whole.

Dyestuffs

> No direct, concrete attempt was ever made to focus national thought in the U.S. upon the problem [of dyestuffs manufacture], to enlist the cooperation and combined efforts of producers of crude products, actual or potential manufacturers of finished dyes, the many trades consuming dyestuffs in quantity . . . and the general industrial world.[1]

This was written in 1915 by T. H. Norton, a career Civil Servant, member of the Chemical Society of London, and one-time American consul at Chemnitz in Saxony. What he wrote of the United States applied equally to Britain and to France, and reflected the shock caused by the war, when the Allies and neutrals (conspicuously the Americans and the Italians) discovered how dependent they had been on German dyestuffs.

In 1914 the United States was, after China, the largest user of dyestuffs. There were then two American-owned producers of dyestuffs, which together

[1] Norton, T. H. *Dyestuffs for American textile and other industries* (U.S. Dept. of Commerce, Special Agents Series No. 96, 1915), p. 51.

with the German-owned firms made about 3000 tons, or roughly one-eighth of the country's requirements.[1] In August–September 1914 it became obvious that the output of intermediates was totally inadequate, and that synthetic indigo, vat dyes, and certain types of alizarin were not made at all. In short the United States appeared to be utterly dependent on imports. Fortunately the Germans had no intention of disturbing the flow of trade. Supplies were routed via Rotterdam, and the proceeds from these sales were used to buy cotton. The American subsidiaries of the German dye makers put up their prices, strengthened their grip on the market, and extended their manufacturing activities. Bayer, which traded as the Hudson River Aniline & Color Works, raised its output, and H. A. Metz, the Hoechst representative in the United States, was instructed to make dyes and for this purpose set up the Consolidated Color Co. The Cassella company also established a factory under the name of Century Color Co.[2] The American-owned firms, notably Schoellkopf Aniline & Chemical Co., W. Beckers Aniline Co., and Benzol Products Co., benefited from the higher prices and enlarged the production of intermediates as soon as coal-tar distillates became available from newly-completed recovery ovens. Here a problem arose, because the tar distillers were giving priority to the urgent and very valuable Allied orders for toluol and phenol, and initially were little concerned with the supply of raw materials for the manufacture of dyestuffs.[3]

The American Chemical Society addressed itself to the question of dyestuffs supplies, and on 9 October 1914 set up a Chemical and Dyestuffs Committee, which included academics as well as businessmen. Among the latter was Metz, who represented, as it were, the views of the German chemical industry in the United States. In November 1914 the Committee agreed that the producers of dyestuffs needed more protection than they had been given by the Underwood Tariff of the previous year. It is indicative of the lack of pressure and probably also of the prevailing climate of opinion that there was no demand for more direct Government intervention. The Committee did not press for the compulsory licensing of German patents until 1917.[4]

By the end of the year prices were rising fast, and on 1 January 1915 the Senate passed a resolution directing the Secretary of Commerce to report on the dyestuffs situation. Before the report (from which the above quotation has been taken) was published, Britain tightened the blockade, and the last German shipment of dyes arrived in the United States on 19 March. A period of acute shortages and soaring prices now began. Stocks became fabulously valuable overnight, vegetable dyes and dyewoods were used, and costly efforts were made to secure synthetic dyes from merchants in Mexico and China.

[1] ibid., p. 8; League of Nations, *The chemical industry* (1927), pp. 28, 52.
[2] Alien Property Custodian report, quoted in *J. ind. Engng Chem.* (April 1919), p. 357.
[3] Higgins, S. H. *Dyeing in Germany and America* (2nd edn., 1916), p. 126.
[4] Stone, I. F. *The aniline color, dyestuff and chemical conditions from 1 August 1914 to 1 April 1917* (1917), p. 125. Stone was President of the National Aniline Co. and a member of the American Chemical Society's Committee.

Few textile mills closed for lack of dyes, but most of them avoided closure by only a very narrow margin; the quality of the dye-work deteriorated.[1]

Although the users were becoming increasingly alarmed, little was done during 1915 towards the creation of an American dyestuffs industry, chiefly because the tariff question remained unresolved. There was, however, an increase in production, though in some cases it was found that German patents blocked progress.[2] The German-owned companies benefited most, for they had no problems with patents and were able to take full advantage of the readier availability of intermediates. In these early years of the war their penetration of the United States chemical industry was pursued vigorously and systematically, and operations were often masked, so that later it was sometimes difficult to take over German property.[3] Towards the end of 1915 Du Pont was approached by American Badische with a proposal to make intermediates which would be exchanged in Europe for dyes. It was said that Lord Moulton, who was then in charge of the Explosives Supply Department in London, was agreeable to importing such materials and exporting British-made dyes. The incident is bizarre, and the story odd, for Moulton would hardly need a German company as intermediary between Du Pont and the British firms. Perhaps American Badische was willing to release some of its patents for that purpose, and it would certainly benefit if it were able to resell the British dyes.[4] Be that as it may, Du Pont refused to follow up the suggestion on the grounds that it lacked experience, had no manufacturing capacity to spare and was afraid that if it put up the additional equipment it would be left with excess plant after the war! The company's Executive Committee resolved that 'we are not prepared to commit ourselves at this time to go into the manufacture of intermediates for the production of chemical dyes'.[5] A few months later the decision was reversed. An agreement was made with Levinstein Ltd. and Du Pont thereby obtained much valuable know-how, particularly on the indigo synthesis. Moreover, Du Pont secured the exclusive right to sell Levinstein's dyes in the United States, Mexico, and Latin America. Levinsteins were also satisfied because the link with Du Pont strengthened their hand against their rivals British Dyes, and also brought in £25 000 a year for know-how. Later, however, there was to be considerable trouble between the two firms.[6] Du Pont now went ahead, and in February

[1] Evidence of Dr. Grinnell Jones (a chemical expert) to the Tariff Commission at Hearings of the House of Representatives Committee on Ways and Means on H.R. 2706 and 6495, *Dyestuffs* (1920), p. 9.

[2] There were however some notable exceptions. For example the chemists at Dow devised their own, extremely successful, synthesis of phenol.

[3] Alien Property Custodian report, art. cit., p. 352.

[4] The go-between was M. R. Poucher who later joined Du Pont and became head of their dyestuffs sales organization.

[5] DP 85 and DP 86, memo by W. S. Carpenter to the Executive Committee, 29 December 1915, and minute of that Committee on the same day.

[6] Bolton, E. K., Du Pont Research, reprint of the Perkin Lecture to the American Soc. of Chemical Industry on 5 January 1945, p. 8; *Chem. Trade J.* **64** (1919), 1673, p. 531 and **66** (1920), 1721, p. 640. It is possible that the link between the two companies was made by Harry McGowan

1917 began to build a large dyestuffs factory at Deepwater Point (N.J.). The directors earmarked \$7 million (say £1·4 million) for the project. By 1922 \$22 million had been spent, but the venture was unprofitable for many years.[1]

Several smaller companies began making dyes during 1916, but their initial efforts were on a very modest scale. For example the Calco Chemical Co., established at Bound Brook in 1915 to make intermediates and dyes, employed only sixteen people a year later.[2] Dow Chemical, Ault & Wiborg, Newport Chemical, and a few others also started dyestuffs production around that time. The most important move towards the establishment of a large group in the organic chemical business came in May 1917 when W. Beckers Aniline, Standard Aniline, Schoellkopf Aniline & Chemical, National Aniline, and parts of Benzol Products, Semet-Solvay, the Barrett Co., and the General Chemical Co. combined to form the National Aniline & Chemical Co. This became the largest United States producer of dyes, and after the war was merged with other companies into Allied Chemical. Within a short time of its formation the new company was reported to be making 60 different intermediates, 150 colours and almost 16 000 tons a year of dyestuffs. If reliable, the figure would correspond to two-thirds of the country's pre-war consumption.[3]

All these efforts owed nothing to the Government. They owed a good deal to the improvement in the supply of tar distillates, and also to the willingness of banks and others to advance the necessary funds. Certainly the business was lucrative in 1915 and 1916, but the effect of larger production made itself felt from the beginning of 1917 onwards; for instance beta-naphthol had been about \$2/lb in the latter half of 1915, but a year later had dropped to \$1·01/lb, and in the second half of 1917 was quoted at 61 ¢. Aniline oil, another important material, fell as sharply.[4]

From the end of 1915 onwards the outlook for protection improved. In December of that year a bill was introduced in the House of Representatives that embodied most of the recommendations made by the American Chemical Society's committee as far back as November 1914. The bill failed to pass in Spring 1916, but the users of dyestuffs now sided with the protectionists, and, as a result of their combined efforts, the dyemakers secured some benefit under the Emergency Tariff Act of September 1916. The act, while not unduly severe by American standards, gave the United States dyestuffs manufacturers a substantial measure of protection for five years.[5] The legislation was criticized after the war, but for the present, under the shadow of deteriorating

---

who was then on the board of Levinstein Ltd. and as chairman of Nobel's Explosives was familiar with the leading personalities of Du Pont.

[1] Dutton, W. S. *Du Pont* (3rd edn. 1951), pp. 288–9.

[2] McMurray, J. H., The History of Calco, unpubl. typescript (1943).

[3] Haynes, *American Chemical Industry* (1945), vol. 3, p. 242.

[4] Written evidence by J. M. Matthews of the American Dyes Institute at the Hearings of the House . . . Committee on Ways and Means, op. cit., pp. 184–5.

[5] Haynes, op. cit., vol. 3, pp. 250 ff.

relations with Germany, it provided manufacturers with a sufficient incentive.

The output of dyestuffs rose significantly from the second half of 1916 onwards, and the shipments from Germany by the submarines *Deutschland* and *Bremen* in July and December of that year, though sensational, were of little practical importance.[1] After the declaration of war in April 1917 there was much confusion over the sequestration of German property, and an extraordinary state of affairs came to light when it was discovered that German-controlled companies had obtained valuable war-materials and diverted them profitably to non-military uses! The Alien Property Custodian took over and sold the Century Color Co. to National Aniline & Chemical, and Bayer's dyestuffs factory went to the Grasselli Chemical Co.[2]

Reliable figures measuring the progress made between 1915 and 1917 are not available. In the latter year the United States organic chemicals industry, then comprising essentially the production of dyes and intermediates, employed almost 20 000 people; in 1918, 27 000 were working in that branch.[3] The country's dyestuffs capacity sufficed to meet most of the users' requirements, but was inadequate as to range of products and quality. Although manufacture was carried on by dozens of firms, many of them inefficient, two large companies, Du Pont and National Aniline, emerged as leaders, and dominated the field for many years.

In Britain the dyestuffs question was beset from the start by problems of a different nature, and particularly by the antagonism of producers and users of dyes. The outbreak of war found the dyestuffs manufacturers entirely unprepared. United Kingdom production in 1913 was estimated at around 4000 tons, equivalent to roughly one-fifth of consumption.[4] In general British-made dyes were of the simpler types; for the more sophisticated products and many intermediates the country was dependent on Germany, and to a lesser extent on Switzerland. Stocks were low in the summer of 1914 and shortages were felt at once. For sulphur black the scarcity was so great that customers came in their own carts to take away unground, untested material in lumps, as it was knocked out of the drying trays.[5] The scarcity of basic blues and violets was equally serious.

The Government acted quickly, and by the end of August a committee had been appointed to advise on the replacement of chemicals hitherto imported from Germany. The use of German patents was authorized, and the export

[1] Each submarine was loaded with dyes and pharmaceuticals amounting to several hundred tons. Badische sent about 60 tons of dyes on each voyage. Their value was 1·2 million M f.o.b. and the New York office was able to dispose of them for 5 million M, corresponding to £60 000 and £250 000 at pre-war rates of exchange; (Voigtländer-Tetzner, W. and Pinnow, H. Geschichte der BASF: Kaufmännische Entwicklung typescript, n.d. (ca. 1938–9), p. 309.

[2] Alien Property Custodian report, art. cit., pp. 359 f.

[3] USTC, *Census of dyes and coal tar chemicals 1917* (1918), p. 37; ibid., *Census . . . 1918* (1919) p. 54.

[4] Report of the Dyestuffs Industry Development Committee (1930), HMSO, Cmd 3658, pp. 10 ff.

[5] Whittaker, C. M. Some early stages in the renaissance of the British dyemaking industry . . . *J. Soc. Dyers Colour* **72** (1956), 560.

of dyestuffs was prohibited with effect from 8 September 1914.[1] These measures gave little relief, and the worst difficulties were at first caused by the lack of certain inorganic chemicals and intermediates. For instance there were only three firms equipped to make oleum: Nobel's Explosives and Clayton Aniline made it for their own use, and both were immediately ordered to increase output and earmark it for the explosives factories; Spencer, Chapman & Messel made oleum for sale, but their capacity was too small. Oleum had therefore to be ordered in the United States, but owing to faulty containers there was much waste. Early in 1915 the situation became so difficult that on Lord Moulton's advice the Advisory Committee on the Supply of Sulphuric Acid for the Production of Munitions was set up. Despite its long title it produced quick results, and by 1916 supplies of oleum were sufficient. The pre-war consumption of concentrated sulphuric acid and oleum had been small, but in 1917/18 the dyestuffs firms alone were using 30 000 tons a year.[2]

Trinitro-phenol and trinitro-toluol were required at once for explosives, and this uncovered a shortage of phenol and toluol. Read Holliday restarted making synthetic phenol and allowed others to use their process, while the Clayton Aniline Co., which had made 200 tons annually of TNT before the war, raised production to 3000 tons by 1916. Some other important organic chemicals with a dual purpose had not been made on any scale before the war; dimethylaniline was suddenly and urgently needed for detonators, while salicylic acid became important as a mordant for khaki on wool.[3]

The position became so difficult that the Board of Trade had to authorize imports of German dyestuffs through neutral countries. When the Germans found out, they proposed to barter dyes for rubber and cotton. It was a thoroughly distasteful business and was stopped as soon as possible.[4] But the Board of Trade was spurred on to conclude negotiations with the Swiss manufacturers for an arrangement whereby exports of intermediates were allowed in return for dyestuffs. The Anglo-Swiss agreement had important consequences. On the British side it encouraged the diversification of intermediates production, which was an essential prerequisite to dyestuffs production generally, and indeed the first step towards the creation of an organic chemicals industry. It also secured, for the duration, the supply of more complex dyes. As for the Swiss, they obtained the valuable British market for their specialities and freed them from dependence on German intermediates, which even before the war had been found inconvenient.

The Government did not propose to leave matters there. In contrast to the *laissez-faire* attitude of the American administration, it intended to support actively the development of a large British-owned dyestuffs industry. In October and November 1914 the Board of Trade was in touch with producers

[1] Higgins, op. cit., p. 119.

[2] *Chem. Trade J.* **57** (1915), no. 1480, 326; (1915), no. 1481, 339–40; **62** (1918), no. 1607, 203–7.

[3] Whittaker, art. cit., pp. 557, 560; *Clayton Aniline Quarterly*, Special 75th Anniversary Issue, Summer 1951, pp. 11–12.

[4] Report on Dyes and Dyestuffs (1921), HMSO, Cmd 1370, p. 5.

and users on this question, and W. Runciman, the President of the Board of Trade, spoke about dyestuffs in the House of Commons on 23 November, and 'of a combined national effort on a scale which requires and justifies an exceptional measure of State encouragement'.[1] It was certainly unusual for a Minister, and in a Liberal Government at that, to think in those terms. But soon it transpired that however desirable such a 'combined national effort' might be, there were differing opinions on how to set about it. *The Times* printed some correspondence which was scarcely constructive: Professors Armstrong, Ramsay, and Pope, who could be counted among the qualified public, wrung their hands at the absence of a dyestuffs industry, but criticized Runciman's project because it lacked scientific direction.

Undeterred by the academic critics, the Board of Trade enlarged on its plans, which were at first limited to financial support. The immediate object was to increase the manufacture of intermediates, and broaden the range for munitions, as well as for the Swiss. The users of dyestuffs were of a different mind. They were accustomed to handling German dyes and were not interested in intermediates. For this reason they were unwilling to back any Government scheme, and their sole concern was to obtain the pre-war range at the lowest price. Their attitude, though inexcusable, is explained by their utter ignorance of dyestuffs technology. The Chambers of Commerce of Lancashire and Yorkshire also had their say: they echoed the opinions of the users and for good measure were openly suspicious of Government intervention and the possibility of a Government-sponsored dyestuffs monopoly. Finally there were the makers of dyes. Only Read Holliday and Levinstein Ltd. need concern us here. The two companies were rivals and their approach to dye-making differed. Read Holliday were fairly important in intermediates and before the war had had reasonably good relations with the Germans. They were particularly strong in sulphur black and, working from this base, were able to offer a simple range of dyes by the end of 1914. Production was quickly increased and for a time early in 1915 the factory was turning out colouring materials at the rate of 1300 tons a month, but their quality was inferior.[2] Levinstein Ltd. had a good reputation for research, and their works were large by contemporary standards, with a capacity of 5000 tons a year. But they lacked facilities for making intermediates, and for this reason encountered serious problems after the outbreak of war, so that the factory was working below capacity. Herbert Levinstein, the managing director, was a pugnacious man and on bad terms not only with the Germans but also, and more importantly, with the Read Holliday people.

Faced with differences of opinion and delays engendered by quarrels, the Government decided to act. Early in 1915 it chose Read Holliday to spearhead the development of dyestuffs manufacture. In retrospect the Board of Trade was undoubtedly right and rendered the country a great service by

[1] Hansard (House of Commons), 5th ser. **68** (23 November 1914), cols. 759–62.
[2] Whittaker, art. cit., p. 560.

selecting the Huddersfield firm. The history of British dyestuffs during the war thus became to a large extent the history of Read Holliday, or more precisely of British Dyes Ltd., which took their place. The Commons discussed the formation of the proposed company, but the debate followed party lines and was not constructive. The question of future protection arose: the Conservatives were prepared to concede it after the war, but the Liberals remained opposed, and the Government refused to commit itself. Runciman took the opportunity of criticizing the users, but his remarks had no effect beyond increasing their distrust.[1] Neither the Calico Printers' Association nor the Bradford Dyers' Association, two of the chief dye users, were prepared to give a lead by underwriting a sizeable portion of the proposed share issue. The negotiations with them and others delayed the issue of the prospectus until the beginning of March. One of its significant features was a sort of cooperative principle: as long as dyes were scarce they would 'be primarily appropriated for supplying the wants of users who are subscribers'.[2] The other was the extent of Government support: the original capital of £1 018 000 consisted of £854 550 in mortgage debentures, all held by the Treasury, and £163 450 in ordinary £1 shares of which 5*s*. was paid up, held mainly by the users. The Government was entitled to appoint two directors with absolute veto powers over questions of supply or price discrimination.[3] The first response to the prospectus was disappointing, and *The Times* positively discouraging:

> But those actually connected with the industry—from whom alone, except for anyone else who subscribes from purely patriotic motives, applications for shares are likely to be made—must be fully acquainted with the controversies which have arisen over the scheme, and they will put their money into it, if at all, with their eyes open.[4]

More pressure was exerted on the big dyestuffs users, and eventually the Calico Printers' Association took up 5000 shares, which considering its size was a trifling investment.[5] The Bradford Dyers' Association followed suit.

So British Dyes Ltd. got under way. The works at Huddersfield were taken over at the beginning of May 1915 for about £422 000. Most of the staff remained at their posts, and the manager, Joseph Turner, was paid the then considerable salary of over £5000 a year.[6] Extensions were put in hand at once, and building continued throughout 1915 and 1916, and most of 1917. The grid lay-out used at Leverkusen was followed as far as the site allowed, and by 1918 the works covered almost 500 acres and comprised 12 oleum units

[1] Hansard (House of Commons), 5th ser. **70** (22 February 1915), cols. 91–5, 102–9, 110–14. There was another debate on 11 March.

[2] Prospectus in *The Times* 6 March 1915, p. 12. The arrangement resembled that for British Alizarine which was controlled by the big users.

[3] Prospectus and *Chem. Trade J.* **57** (1915), no. 1469, 57. [4] *The Times*, 8 March 1915, p. 13.

[5] L. B. Lee at the Annual General Meeting of the Calico Printers' Association, *Chem. Trade J.* **57** (1915), no. 1479, 301.

[6] Hansard (House of Commons), 5th ser. **70** (11 March 1915), cols. 1640–99; *Chem. Trade J.* **57** (1915) no. 1469, 57.

each rated at 10 tons a day, a nitric acid plant, ice works, producer-gas unit, power station, and two laboratories.[1] Construction interfered with manufacture and it was difficult to extend the range beyond sulphur- and basic dyes, which had been made by the predecessor company, until the middle of 1916. But the effort put into the intermediates section had quicker results, and the variety as well as the volume increased markedly in the first twelve months.[2] In October 1917 the managing director, J. Falconer, claimed that his company made sufficient intermediates for its own needs and those of the Swiss.[3] During 1917/18 production of dyestuffs increased further and British Dyes were then probably accounting for two-fifths by value of British consumption.[4]

Despite these not inconsiderable achievements, criticism continued unabated throughout the war. It came from those who thought that the company was not sufficiently science-minded. But this was untrue, for eminent chemists served on the company's Advisory Council, which did much to foster university research in organic chemistry; there were also scientists on the Technical Committee, which played an active role in the management of operations. Another group of critics consisted of the dye users. It is difficult to sympathize with their attitude, because the supply problems were enormous and it may be doubted whether anyone at that time knew how to operate a large integrated dyeworks. As late as 1919 a temperate critic such as A. Hoegger of the British Cotton & Wool Dyers Association could still maintain that the activities of the British dyemakers 'on the whole, have been rather disappointing' and that fast colours were still inadequate.[5]

In one respect, however, the critics were justified: dye making was to be not only a national effort, but also a combined effort, yet there was the minimum contact between British Dyes Ltd. and Levinstein Ltd., until the Government brought about their amalgamation in 1918.[6] Personal antagonism, which ultimately led to open hostility between Joseph Turner and Herbert Levinstein, was one element, another was Levinsteins' more efficient production methods, which gave them a competitive advantage, and thus sharpened the conflict.

Before the war Levinstein Ltd. made about 700 tons of intermediates, and by its end produced 7000 tons. The figures for the production of dyestuffs are not available, but it is likely that they were of the same order of magnitude in 1918.[7] Levinstein Ltd. certainly benefited from the fact that their works

[1] *Chem. Trade J.* **62** (1918), no. 1620, 463.

[2] *Chem. Trade J.* **59** (1916), no. 1527, 179–80 (summary of 1st Annual Report of British Dyes).

[3] Report of the annual general meeting of British Dyes Ltd. on 31 October 1917, summarized in *Chem. Trade J.* **61** (1917), no. 1589, pp. 367–8.

[4] So far as I know the quantity made has not been published. It was probably a little larger than Levinstein. These two companies between them represented three-quarters of the United Kingdom production according to the report on Dyes and Dyestuffs, op. cit., p. 8.

[5] Annual General Meeting of the Association, summarized in *Chem. Trade J.* **64** (1919), no. 1671, 480.

[6] See the next chapter.

[7] Report of the last Annual General Meeting of Levinstein Ltd., summarized in *Chem. Trade J.*

were not operating at capacity in 1913–14, and extensions appear to have interfered relatively little with output. There was a fairly large staff of chemists, so that immediate advantage could be taken of the German patents. The company enlarged the intermediates section in 1914–16, and then cast about for an opportunity to extend the scope of its activities. Here it had some luck. In 1916 the small business of Claus & Rée in Manchester and the erstwhile Hoechst indigo factory at Ellesmere Port were bought. The latter, at £70 000, was a bargain from the Board of Trade; Professors Tilden and Pope had advised H.M.G. that Levinstein Ltd. 'were better qualified to carry on this particular manufacture' than their rivals, and so to the disgust of British Dyes, the Board of Trade did not call for tenders when selling the Hoechst factory. The works had been closed owing to lack of intermediates, but Levinsteins managed to restart them quickly, and this gave them a lead in the indigo business. They also got the Hoechst patent for 'Novocain', and began making this drug at Ellesmere Port.[1]

Levinstein's annual reports and the company's statements were invariably optimistic, and the directors possessed ample self-confidence. This may have given some people the impression that Levinsteins were able, unaided, to solve Britain's dyestuffs problems. But in the absence of statistics caution is indicated, and one must remember the successful efforts made by British Dyes and the valuable contribution of the many smaller firms. Among them a few had begun as tar distilleries, as for instance Brotherton & Co., which in 1917 bought the Mersey Chemical Co., previously owned by Badische, and began making dyes and auxiliaries.[2] The Clayton Aniline Co. greatly enlarged its intermediates plant. L. B. Holliday joined the forces after British Dyes had taken over the family business, but he was recalled from the army in 1916, and opened works at Deighton near Huddersfield to make picric acid, and in 1918 began making dyes; his firm, L. B. Holliday & Sons Ltd., survives to the present day.[3] The British Alizarine Co. continued with its specialities at Silvertown. Not one of these firms was able to reproduce the German anthraquinone vat dyes, and so James Morton, whose furnishing-fabric business depended on them, decided he would have a try. He got British Alizarine to make the intermediate for him, and towards the end of March 1915 produced small batches of vat dyes corresponding to 'Indanthren' yellow and blue. More equipment was obtained and installed near Morton's printing and dye works at Carlisle. The Solway Dyes Co., though small, successfully manufactured these special dyes throughout the war, and became famous for them in the 1920s.[4]

64 (1919), no. 1651, 32. As it was the last meeting, it is possible that the chairman was carried away into giving a rather glowing picture of the progress made.

[1] *Chem. Trade J.* **59** (1916), no. 1534, 329; no. 1535, 352–3, 358; **61** (1917), no. 1589, 367; **65** (1919), no. 1676, 13.

[2] I am indebted to Dr. W. H. Cliffe of I.C.I.'s Dyestuffs Division for this information.

[3] Miall, S. *A history of the British chemical industry* (1931), p. 95.

[4] Morton, J. *History of the development of fast dyeing and dyes*, reprint of a lecture delivered before the Royal Society of Arts, priv. pub. (1929).

The situation in France was unusual. In 1913 the country required about 9000 tons of dyestuffs, of which roughly 1000 tons were made by French-owned works, 7000 tons were produced by the branch factories of Swiss and German firms from imported intermediates or semi-finished dyes, and 2000 tons were imported. In October 1914, the Government having created a special department to co-ordinate and develop the chemical industry, it was found that there were substantial stocks. While there was no immediate supply problem, the outlook for indigenous production was unpromising.[1] In the first place France did not have a big tar-distilling industry, and most of the northern plants were lost in August and September 1914, and remained under enemy occupation for the next four years. And secondly no systematic effort was made at first to raise the output of coal-tar benzene and toluene in the remainder of the country.[2] It was necessary therefore to import from Britain and later also the United States, and distillation plant to extract benzene and toluene from Borneo crude oil was installed at an oil refinery near Marseilles. During 1916 the supply problems were solved and the production of coal-tar aromatics became adequate for the explosives and dyestuffs factories. Intermediates had been made before the war by the French-owned S.A. des Matières Colorantes & Produits Chimiques de St. Denis. The works was extended and the factories of Hoechst at Creil and Badische at Villeneuve near Lyon were requisitioned and suitably adapted. Some of the intermediates made were delivered to the Swiss in exchange for dyes, and this arrangement (which followed the earlier British one) became effective towards the end of 1916. Swiss-made dyes contributed importantly to French requirements in 1917 and 1918.

There remained the problem of indigenous dyestuffs manufacture. Progress was slower than in Britain, partly owing to the large stocks, which seemed to lessen the urgency, and partly on account of the delays over coal-tar distillation. Cie Française de Produits Chimiques & Matières Colorantes de St. Clair du Rhône and the Soc. de Produits Chimiques & Colorants Français at Villeneuve-St. George were established to make dyes, but their production, added to that of the St. Denis factory and the former German works, was still considered inadequate. The Government felt, as in Britain, that a national effort was called for. It was prepared to help, but effective measures were delayed until 1916: in February, the Government set up an inter-ministerial council for dyes and in April the Syndicat National des Matières Colorantes was formed by a group of industrial enterprises, among them Kuhlmann. The Government assisted by offering two munitions factories that had suitable site services, and also a requisitioned factory at Creil. G. Patard, a senior official from the Service des Poudres, was seconded to become the managing director of the Syndicat, and the Ministry of Education

[1] Kap-Herr, J. de. Les matières colorantes in Gérard, J. (Ed.) *Dix ans d'efforts scientifiques et industriels* (1926), vol. 1, pp. 1311–12.

[2] A law making the recovery of coal-tar aromatics mandatory was belatedly passed in November 1915; Fisch, R. Les industries chimiques de la Region Lyonnaise, Ph.D. thesis (1923), p. 27.

released Professors Grandmougin and Noelting from teaching duties to take charge of the scientific side. In return the Government reserved itself the right to determine the production programme, and was entitled to one-quarter of the distributable profits above 5 per cent.[1] These preliminaries culminated at the end of February 1917 in the formation of the Cie Nationale de Matières Colorantes (CNMC) with a capital of fr. 40 million (say £1·6 million). The Banque de Paris & des Pays-Bas took charge of the financial side, and the Board contained representatives from industry, including Kuhlmann. The chairman was R. Masse of the French Gas Industry Federation. Technical links were established with British Dyes. CNMC with its 2000 workers and 100 chemists was large by contemporary standards.[2] Another company that was partly financed by dyers was the Soc. de Produits Chimiques & Colorants Français, which started at Villeneuve-St. George in the early months of 1918.

The contribution of the new firms, the sequestrated German factories, and CIBA's works were useful. In 1917 a little under 2000 tons were made, chiefly at Creil. That factory had to be evacuated for a time during the German spring offensive of 1918 and so production was interrupted. Output recovered later in the year and during 1919 3000 tons were made, a creditable achievement representing three times the pre-war volume of wholly-indigenously prepared colours.[3] Although more was done in Britain and the United States, the French measures had some important and permanent results: in the first place the manufacture of coal-tar derivatives and of intermediates became firmly established on a substantial scale, secondly the scope of French dyestuffs manufacture was enlarged, and the two steps were decisive in the gradual development of self-sufficiency in this key sector.

Before the war Italy did not make dyestuffs. Imports from Germany continued, though in smaller quantities, until December 1914, when they fell to a trickle, and they ceased altogether on the outbreak of war with the Central Powers in spring 1915. For the next two years all supplies had to be bought from Switzerland and the Allies. Levinsteins helped establish the Industria Nazionale Colori di Anilina in December 1916, and some smaller works began making simple dyes in 1917 and 1918. By the end of the war, production was reported to have reached 2000 tons a year, principally of sulphur dyes for cheap, dark-coloured cotton goods.[4]

Russia, like France, entered the war with large stocks of dyestuffs, and the supplies were stretched further by improvisation and making do with vegetable colours. The chief problems were the lack of organization and the

[1] Anon. Les Etablissements Kuhlmann 1825–1925 (1925), pp. 96–7; *Z. angew. Chem.* **31** (1918), pt. 3, 147, 445; *Chem. Trade J.* **59** (1916), no. 1543, 528.

[2] Matagrin, A. *L'Industrie des produits chimiques et ses travailleurs* (1925), pp. 64–5.

[3] Gérard, J. (Ed.), op. cit., vol. 1, p. 1313; vol. 2, p. 2765; Les Etablissements Kuhlmann, op. cit., pp. 96–7; L'Association Nationale d'Expansion Economique, *Index to French production* (1920 edn.), pp. 365–6.

[4] Fenoglio, G. *Die volkswirtschaftlichen Folgen des Weltkrieges in Italien* (1915), pp. 26–7; *Chem. Trade J.* **60** (1917), no. 1546, 11; **66** (1920), no. 1716, 469–71.

vast distances. The first explains why the manufacture of coal-tar dyes, which had been started a few years earlier by foreign companies, was not quickly increased. In 1915, however, there were some developments: several companies in the chemical and cotton industries set up the Russian Dye Co. in a requisitioned factory. Swiss chemists, possibly those previously employed by CIBA at Pabianice, were put in charge of the technical side. According to Ipatieff, who was closely connected with the chemical industry throughout the war, the new company did not fulfil the expectations of its founders.[1] This may not have been due to technical faults or organizational incompetence, but to shortages caused by the enormous transport difficulties. An account of the problems connected with benzene and toluene will serve as illustration. The production of coal-tar distillates was rapidly stepped up in the Ukrainian steel centres. But as there was insufficient rectifying plant locally, the crude benzol and toluol had to be sent all the way from the southern Ukraine to St. Petersburg, where Tentelev had the necessary equipment. From Tentelev the benzene was sent to Moscow where phenol was made by the sulphonation process, while the toluene went to a TNT factory at Samara (Kuibyshev). The phenol eventually returned to the Ukraine, where it was converted to picric acid in a factory at Slaviansk.[2] Under these circumstances prices were unrealistic and interruptions to freight trains could (and did) upset the entire programme. Despite such obstacles, considerable plant extensions, together with various expedients and luck, gave some hope that in the course of 1916 the country would have a sufficiency of organic and inorganic chemicals to meet the demand for explosives and for dyes. So far as I know the extent to which these objectives were achieved has not been published. In any event the growing disorganization of the railways increasingly frustrated these efforts, and in 1918 the Civil War divided the country into several regions between which communications ceased. The Germans and Austrians marched into the Ukraine in Spring 1918, and by the summer of the year they more or less effectively controlled the south-eastern steel works, thereby cutting the supplies of tar distillates to the dyestuffs works in the Moscow region.

The Japanese Government assisted with the formation of dyestuffs factories. Legislation was passed in 1915 granting subsidies to firms owned or controlled by nationals who proposed to make dyes, and several businesses were formed.[3] The Tokyo Gas Co. supplied the intermediates, and towards the end of 1915 the first simple aniline colours began to appear. Within the year they were supplemented by sulphur black, and during 1917–18 the range was further extended. By the end of the war the manufacturing capacity was 5000 tons a year, which corresponded to consumption in 1913.[4] This, if true,

[1] Ipatieff, V. N. *The life of a chemist* (1946), pp. 374–5; Higgins, op. cit., p. 128.
[2] Ipatieff, op. cit., p. 207. At a later date rectifying plant was built in the Ukraine.
[3] USTC, *Census of dyes and coal tar chemicals, 1919* (1921), p. 56.
[4] Yamasaki, K. and Ogawa, G. *The effect of the World War upon the commerce and industry of Japan* (1929), pp. 278 ff.

would be a considerable achievement, but it is more likely that the volume represented the potential capacity for making fairly simple coal-tar dyes. There was not enough know-how to make the more complex dyestuffs, and the manufacture of intermediates and of process chemicals (notably oleum) was insufficiently developed to sustain a sophisticated dyestuffs technology.

Throughout the war the Swiss firms played an absolutely crucial role. In the early weeks there were great difficulties because German supplies were delayed and after mid-September 1914 were subject to export licensing. The works in Basle cut production and sold from stock. They also made frantic efforts to buy intermediates in England and to enlarge the capacity of their subsidiaries. But this took some time, and Geigy, which had a factory just across the border in Grenzach, fared worst of all, because the Germans refused access to the works until December 1914. For lack of German materials the output of the parent works fell from almost 700 tons in 1914 to just under 500 tons in 1915.[1] The other firms also had serious problems, and the agreement with the Board of Trade was essential to them. The exchange of English intermediates for Swiss dyes began in the first half of 1915, but British Dyes Ltd. was not used as the authorized distributor.[2] The arrangement with the British left other customers dissatisfied, and in September 1915 the Swiss Government was forced to introduce some controls, which were reinforced a few weeks later by export licensing.[3]

The outcome of these measures was a recovery in exports of dyes. The value of the trade rose from 29 million fr. (£1·16 million) in 1913 to 31·2 million fr. in 1915, and then soared to 58 million fr. in 1916 and to about 98 million fr. (nearly £4 million) annually in the next two years. Germany, which had taken 20 per cent, ceased to be a buyer, and Britain became the principal customer, her share of Swiss exports increasing from 12 per cent in 1913 to a peak of 49 per cent in 1915, whence it fell gradually to about 35 per cent. Swiss exports to France increased until they accounted for 26 per cent of the total in 1918. During the four critical years 1915–18, the United Kingdom, France, and the United States between them took 73–77 per cent of Swiss dye exports.[4]

During the war, the Swiss replaced the Germans as the leading exporters of dyestuffs. It was a profitable business and permanently strengthened their trading situation. At the same time the exchange of intermediates for dyes freed them from dependence on Germany, which in 1913 had supplied 86½ per cent of the raw materials and intermediates. By the end of the war, two-thirds of those supplies came from the United Kingdom. Finally the emphasis

[1] Heitz, W. *Erinnerungen . . .*, n.d. (1937?), pp. 73, 96.

[2] *Chem. Trade J.* **57** (1915), no. 1482, 366. A committee of dye users was formed to handle the allocation of Swiss dyes.

[3] Jaquet, N. *Die Entwicklung und volkswirtschaftliche Bedeutung der schweizerischen Teerfarbenindustrie* (1922), pp. 63–4.

[4] ibid. It is worth observing that dyestuffs were exported in a highly concentrated form during the war, so that the 1914 volume of 9600 tons cannot be compared with the 5900 tons exported four years later.

on special dyes, which was noted already before the war, became even more pronounced. This characteristic benefited the Swiss manufacturers after the war, when, in the 1920s, the capacity for making ordinary dyestuffs was found to exceed greatly the demand for them, while the business in speciality dyes maintained itself better.

Nitrogen

Before the war 80–90 per cent of all the nitrogen was used in agriculture in the form of nitrate, sulphate, or cyanamide. The quantity was around 650 000 tons N, compared with 80 000, or at the most 160 000 tons N used industrially.[1] During the war the situation was reversed: industrial nitrogen, specifically nitric acid, was absolutely essential for the manufacture of explosives. The quantities required for this purpose surpassed all expectations: in Britain, for example, the monthly production (in terms of 100 per cent HNO_3) climbed from less than 1000 tons before the war to 4300 tons in July 1915 and to 20 700 tons at the beginning of 1916. From this peak it fell back to an average of about 17 000 tons in 1917.[2] The acid was required for the preparation of organic nitrogen compounds, notably picric acid and TNT, Cordite and other smokeless powders made from gun cotton and nitroglycerine, and also for ammonium nitrate. The output of these explosives was enormous, and in Britain alone amounted to almost 0·7 million tons from October 1915 to December 1918.[3] Given the scale of these requirements, the nitrogen problem as it developed after August 1914 was not simply a choice between guns or grain, but between guns or defeat. Looking back, it is remarkable that despite all the warnings of an impending shortage, no practical steps had been taken to safeguard nitrogen supplies for military purposes. But few expected the war to last long, and fewer still had anticipated the size of the combined fertilizer and explosives demand. It quickly became a critical factor in all belligerent countries, and in Germany remained the most crucial military–technical problem.

In 1913 the Reich needed about 225 000 tons N of which roughly one-half was imported, predominantly from Chile.[4] In August 1914 imports stopped abruptly, and the mobilization immediately reduced the output of by-product ammonia. Stocks were seasonally low, and no attempt had been made to build up a strategic reserve. It was reported later that Germany had a stock-pile of 450 000 tons of nitrates in the summer of 1914,[5] but the available evidence

[1] Pollitt, G. P. The synthetic ammonia industry *Trans. of the 2nd Wld Pwr Conf.* (1930), vol. 2, p. 147. He estimated world nitrogen production at 782 000 tons N in 1913. The quantity used industrially has been estimated by others at 10–20 per cent of the total.

[2] *History of the Ministry of Munitions*, pt. 4, Materials for the explosives industry (1921), p. 38.

[3] Ministry of Munitions, Review of munitions output 1914–18; n.d. (1919?), pp. 81–5. Production of cordite by the Admiralty is excluded; figures for the period August 1914 to September 1915 are not available.

[4] Waeser, B. *Die Luftstickstoffindustrie* (2nd edn. 1932), p. 5.

[5] Haynes, op. cit., vol. 2, p. 57.

does not support the allegation. German imports of nitrates rose until 1911, and then declined slightly. The greater part of the imports was received in the period January–April in time for the late winter–early spring fertilizer season.[1] There were no signs of a last minute flurry of precautionary purchases, because prices at Chilean ports were lower in 1914 than in 1913.[2] So, in the absence of special reserves, the military planners had to rely on importers' and merchants' stocks, which amounted to 45 000 and 55 000 tons of nitrates respectively; to this we must add about 550 000 tons a year of sulphate of ammonia and 25 000 tons a year of artificial nitrates from Norway.[3] The advance into Belgium led to the capture of nitrate stocks in Antwerp. At the time their importance was greatly inflated, but the quantity probably did not exceed 20 000 tons.[4]

In round figures the Germans had 130 000–140 000 tons N at their disposal in the form of stocks and of more or less firm annual supplies. This would not last long under severe battle conditions, and certainly was inadequate to ensure victory in a long drawn-out war. Imports, except from Norway, were ruled out. Consequently the nitrogen fixation capacity in the Reich became of the utmost importance. Sulphate of ammonia obtained as a by-product from coking operations was easily the leading indigenous source and in 1913 provided almost 110 000 tons N. There were two cyanamide works, which at full output could supply 10 000 tons N a year, and the small synthetic ammonia plant of Badische. The figures (see Table 7.1) show how the Germans compensated for the decline in by-product output by increasing that from other processes.

The most immediately urgent problem was the shortage of nitric acid. It was caused by inadequate capacity and failure to husband the available supplies of nitrates. The alternative route to nitric acid was by the catalytic oxidation of ammonia. The process was an old one, but had been improved some ten years previously by the German chemist Ostwald. Only one German company, the Gewerkschaft Lothringen at Gerthe in the Ruhr, used it, and

[1] The statistics of German nitrate imports are as follows (000 m.t.):

	1911	1912	1913	1914
January–June	460	543	507	590
July–December	360	270	267	not pub.
Total	820	813	774	590 incompl.

1913 was considered a poor year for nitrates generally.

[2] The quarterly fluctuations in shillings and pence per Spanish quintal (101·4 lb) f.o.b. Chilean ports were as follows: 30 June 1913: 7/9½; 30 September 1913: 7/9¾; 31 December 1913: 7/11; 31 March 1914: 7/10½; 30 June 1914: 7/6½; 30 September 1914: 7/9. James, A. F. B. *Nitrate facts and figures* (1915).

[3] Ungewitter, C. (Ed.) *Ausgewählte Kapitel aus der chemisch-industriellen Wirtschaftspolitik* (1927), p. 336.

[4] E. Castelein, the president of the Antwerp Chamber of Commerce, in his list of the German requisitions valued the nitrate stocks at over 4 million fr. (£160 000). This would be equivalent at pre-war prices to 15 000–17 000 tons. cf. *Annuaire de la Chambre de Commerce d'Anvers de 1871 à 1926* (1927), p. 70.

TABLE 7.1

German production of nitrogen 1913–19[1]
(000 metric tons N)

	1913	1914	1915	1916	1917	1918	1919
By-product sulphate	108	95	69	68	57	54	55
Cyanamide	10	13	15	34	37	36	43
Ammonia (Haber–Bosch)	<1	6	12	43	75	95	63
Total	119	114	96	145	169	185	161

Note: The figures for cyanamide in 1913 and 1914 may be too high as the plants were not working at rated capacity.

the capacity of its plant was 150 tons a month of 36° Bé acid. The military and the Raw Materials Board were unaware of the general situation and of the existence of this small oxidation plant. As early as September 1914 Emil Fischer and Fritz Haber, both academic chemists with good contacts in industry, drew attention to the impending crisis, and on 1 October, Fischer sent a detailed report to the War Ministry on the subject.[2] It was followed by prompt action, and within weeks the factory at Gerthe was enlarged and others built; Badische developed a cheaper catalyst and started an ammonia oxidation unit in Spring 1915. All the plants (except Oppau) used ammonia obtained from by-product operations, which yielded a weak nitric acid suitable for the manufacture of artificial sodium nitrate, from which strong nitric acid was prepared by the traditional route. During the winter of 1914/15 about 5000 tons a month of nitrate were made in this round-about and expensive way.[3] But staving off a crisis in one area uncovered an even greater problem in another. In the first months of 1915 the supply of fertilizer nitrogen became inadequate, because of the diversion of coke-oven ammoniacal liquors to the oxidation plants and so to nitric acid. This time no quick solution was in sight, but the Government hoped to provide enough for agriculture by the end of that year. In order to implement the policy it decided in March 1915 to support the development of cyanamide manufacture.[4]

The farmers did not like this. In the first place it provided no relief for the Spring dressing, which was their immediate concern. In the second place many were unfamiliar with cyanamide, and when they used it complained that because it was like dust it was difficult to apply. The Prussian Ministry of Agriculture endeavoured to overcome the objections by offering a prize for the best instruction leaflet and another for an efficient cyanamide spreader. The second prize was substantial: 10 000 M, and a further 10 000 M if the

[1] *Enquete*, pp. 154, 156; Mittasch, A. *Geschichte der Ammoniaksynthese* (1951), p. 182.

[2] Forster, M. O. Emil Fischer Memorial Lecture. *Memorial Lectures delivered before the Chemical Society* (1933), vol. 3, p. 41.

[3] Haber, F. *Aus Leben und Beruf* (1927), p. 14.

[4] At the same time the production of potash was increased. German consumption rose from 0·5 to 0·9 million tons K_2O between 1913 and 1918. Krische, P. *Das Kali* (1923), vol. 1, p. 190.

machine went into mass production.[1] But by the end of the year a satisfactory spreader had not been developed. Despite the complaints, the Government's policy was not unreasonable, because the cyanamide process had been working satisfactorily for some years and there was sufficient technical expertise. It must be remembered that outside Badische, relatively little was known of the economics of the Haber-Bosch process.

The Government decided to support Bayerische Stickstoffwerke in the building of two large cyanamide plants. Bayerische already operated a works at Trostberg in upper Bavaria, which had a capacity of 25 000–30 000 tons of cyanamide in 1913. It was impossible to enlarge Trostberg quickly because of the delay in increasing the size of the hydro-electric power station. The planners therefore decided to rely on thermal electricity for the new works, the so-called 'Reichswerke'. The first was located at Piesteritz near Wittenberg on the Elbe, the second at Chorzow near Kattowitz in Upper Silesia. Piesteritz obtained electricity from a lignite-fuelled power station near Bitterfeld, Chorzow used coal. Building proceeded quickly and the first batch of cyanamide was made at Piesteritz in December 1915. The works was designed to produce 150 000 tons of cyanamide, and Chorzow a further 110 000 tons. Together with extensions at Trostberg, which added a further 50 000 tons a year, the three works could produce 340 000 tons a year, say 66 000 tons N.[2] But owing to constant trouble with the electricity supply, the factories never operated at more than about 40 per cent of their capacity.[3] The initial arrangements between Bayerische Stickstoffwerke and the Reich were altered during 1915, so that Piesteritz and Chorzow became wholly state-owned, but operated by Stickstoffwerke.

The Reich also helped the AG für Stickstoffdünger with a loan of £750 000 to increase the capacity of the cyanamide works at Knapsack near Cologne. The extension from about 25 000 to 100 000 tons a year began in 1915 and was completed in 1916. A subsidiary company, the Mitteldeutsche Stickstoffwerke at Gross Kayna near Merseburg was formed in 1912, but the works failed to operate satisfactorily and made no worthwhile contribution to the supply of cyanamide.[4]

The combined capacity of the German cyanamide works was roughly 0·45 million tons, and in 1917–18 their output was approximately 0·2 million tons a year. (The factories also made calcium carbide for lighting, welding, and some other purposes.) At their war-time peak the cyanamide works supplied 36 000–37 000 tons N, which corresponded to roughly one-fifth of the pre-war fertilizer consumption.[5] The amount was, of course, totally inadequate to maintain food production. Hence it was necessary to divert

[1] *Chem. Ind., Düsseld.* **38** (1915), no. 9/10, 203–4.
[2] Franck, H. H. *et al. Der Kalkstickstoff in Wissenschaft, Technik und Wirtschaft* (1931), p. 55.
[3] *Enquete*, p. 156.
[4] Niebelschütz, W. v. *Knapsack* (1957), p. 67. Knapsack was also working below capacity.
[5] Waeser, op. cit., pp. 24–5; Matignon, C. L'Azote, see Gérard, J. (Ed.), op. cit., vol. 1, pp. 401–2. I have excluded carbide facilities for uses other than fertilizers.

sulphate of ammonia from industry to agriculture, and in this way provide altogether two-fifths to one-half of the fertilizer nitrogen requirements. The transfer of by-product sulphate of ammonia to agriculture became possible once the production of synthetic ammonia increased and could materially contribute to industrial requirements. In this manner the development of the Haber–Bosch process between 1915 and 1918 enabled Germany to continue the war at home and at the front.

Badische opened the Oppau works in September 1913. It had a capacity of 36 000 tons of sulphate of ammonia, corresponding to 7200 tons of nitrogen. During the next two years the capacity of the plant was raised by stages to about 50 000 tons.[1] In 1914 production, in terms of nitrogen, was 6000 tons, in 1915 it rose to 12 000, and the next year reached 43 000 tons.[2] The relatively small increase in 1915 was due to the failure to appreciate the role of the process under wartime conditions, and the consequent delay in placing orders for equipment, particularly reactors. Oppau had been designed for the production of sulphate of ammonia, but it was decided late in 1914 to concentrate on technical nitrogen compounds, chiefly ammonia liquor and nitrate of soda. The decision was undoubtedly correct in that it released cyanamide, and, to an increasing extent, by-product sulphate for agriculture. As a result Badische virtually opted out of the fertilizer business for the duration.[3]

Towards the end of 1915 the Oppau plant was working up to its full capacity, but the shortage of ammonia continued, and additional capacity was essential if the military requirements were to be met without encroaching on the already insufficient supplies earmarked for agriculture. Oppau itself had been attacked by planes, and though the damage was slight by World War II standards, it was imprudent to expose this most vital factory to greater danger by enlarging the target area still further. In any event the resources of Badische were strained and it could not extend the works on its own. At this stage, we have not the precise date, the Government decided to transfer its main financial support from the cyanamide to the Haber–Bosch process and in January 1916 instructed Badische to build a second plant. The terms of the loan to the company were signed on 10 April 1916 during the battle for Verdun, in which more explosives were used than ever before. The Government initially wished to obtain a holding in the new works, but Badische resisted and in the end merely undertook to supply nitrogen at a fixed price.[4]

The site of the works was Leuna, near Merseburg, about 20 miles west of Leipzig. It covered 600 hectares (roughly 2½ square miles) and was close to large lignite deposits in the Geiseltal. Work on constructing a branch railway

[1] *Enquete*, pp. 146 ff.

[2] Mittasch, op. cit., p. 182.

[3] From 1 January 1915 to 11 November 1918 the company's sales of nitrogen compounds amounted to 414 million M (almost £21 million). This included 24·5 million M (£1·2 million) of fertilizer nitrogen. Geschichte der BASF, op. cit., pp. 345 f.

[4] Waller, P. Probleme der deutschen chemischen Industrie, Ph.D. thesis (1928), p. 172.

line began on 28 April 1916, and a month later a start was made on the foundations. Construction was pressed forward despite a wet summer and the severe winter of 1916/17. The first reactor came into operation on 27 April 1917. During construction the plans were altered: in spring 1916 the initial capacity had been set at 30 000 tons N a year, but this was raised in June to 75 000 tons. The Hindenburg programme later that year provided for a second stage, and work on an additional group of reactors with a capacity of 55 000 tons N began in April 1917 and was completed in the late summer of 1918. In June 1918 a third stage of a further 75 000 tons N was ordered, but lack of materials delayed completion until 1920. By the end of the war the Oppau and Leuna factories had annual capacities of 50 000 and 130 000 tons N respectively. Their output was rather smaller, but during the whole of 1918 they made 95 000 tons of nitrogen; this represented more than two-fifths of the pre-war consumption of all forms of nitrogen.[1] From 1916 onwards both factories specialized in the manufacture of ammonia liquor, and indeed this was the sole product made at Leuna. Ludwigshafen and other factories converted the ammonia to nitric acid, which in turn was used to make TNT and ammonium nitrate. Details of the cost of this huge building project have not, as far as I know, been published. But the company's balance sheet provides some indication, though the inflation contributed importantly to the soaring increase in current liabilities from 19 million M at the end of 1915 to 462 million M (£23 million at the pre-war rate of exchange) three years later.

The company rounded off the operations at Leuna by buying lignite mines, and also a deposit of anhydrite near Niedersachsenwerfen.[2] The production of sulphate of ammonia by the anhydrite process (essentially the reaction of coke and ammonia with crushed anhydrite in a rotating kiln) had been developed at Oppau during the war. The company decided to make the fertilizer at Leuna and building work began in summer 1917, but the project did not have a high priority and owing to lack of materials was not completed until May 1919. By the end of that year the plant could make 100 tons a day of the fertilizer by the new process.

In 1917 and 1918 ammonia made by the Haber–Bosch process represented 45 and 50 per cent respectively of the country's production of nitrogen compounds. Without this contribution, the war would undoubtedly have ended sooner. But this was not the end of the story: at the time of the Armistice, Germany was approaching self-sufficiency in nitrogen, though lack of fuel and materials were soon to cut production to a fraction of the potential. The return of normal conditions some years later caused a decisive change in the fertilizer nitrogen position of the country, and was to have far-reaching effects throughout the 1920s. Here we may note that the implications of these changes

[1] Greiling, W. 25 Jahre Leuna 1918–1943, typescript (n.d.). Shortly before the Armistice the daily make at Leuna was 250 tons of ammonia, equivalent to an annual rate of 73 000 tons N. Interruptions to the electricity supply and lack of spares prevented the achievement of the planned daily make of about 440 tons of ammonia.

[2] Greiling, op. cit., pp. 51–2.

were perceived even in 1918, and that Badische by reason of owning the process was becoming the most important company in the German chemical industry.

On the Allied side the nitrogen problem was intermittently acute, and, unlike the Central Powers, the demands of agriculture and armaments could usually be reconciled, the more so as the North American grain crops were less dependent on nitrogenous fertilizers than the German. Britain, the leading exporter of sulphate of ammonia, was able to supply its Allies, chiefly France, which lost many of its coke-oven plants early in the war. To the Allies, in short, the problem was not so much one of supplies, but of access to supplies in neutral countries, particularly Chilean nitrates and nitrogenous compounds from Scandinavia. Additionally, steps were taken in the United States and Britain to establish the manufacture of cyanamide and of synthetic ammonia, while in France the existing cyanamide works were enlarged.

The production and exports of Chilean nitrates fell steeply upon the outbreak of war.[1] Germany had been a very large customer, and shipments to other countries were attacked by von Spee's raiding squadron. With his defeat in December 1914 at the battle of the Falkland Islands, trade recovered. Exports had been only 1·8 million tons in 1914, but next year 2·0 million tons were shipped, and in 1916 the figure was 3·0 million tons. But submarine warfare threatened supplies once again in 1917, and led to the urgent consideration of other sources of fixed nitrogen.[2]

The Scandinavian producers were well placed to help, but Norsk Hydro, though French-controlled, still had to work off a contract made with the Germans before the war. The company completed the installation of German equipment during 1915, after which it became possible not only to increase output, but also eliminate sales to Germany. Production in the year to 30 June 1915 was 17 000 tons N, a year later it was 24 000 tons, and in 1916/17 it rose to a peak of 40 000 tons. (In 1917/18, however, output fell back to 36 000 tons.) Sales to France and other Allied countries rose from one-third to two-thirds of the total make.[3] Nitrogen Products & Carbide, the other large Scandinavian manufacturer, was British-controlled, and used the cyanamide process at Alby in Sweden and Odda in Norway. The works were not enlarged during the war owing to lack of materials and difficulties with the

[1] The following figures show the extent of the fall (million long t):

	Jan.–July		*Aug.–Dec.*		*Total*	
	1913	1914	1913	1914	1913	1914
Production	1·6	1·8	1·7	0·7	3·3	2·5
Exports	1·4	1·4	1·3	0·4	2·7	1·8

Rowe, L. S. *The early effects of the European War upon the finance, commerce, and industry of Chile* (1918), p. 51.

[2] Chilean exports dipped to 2·8 million tons in 1917 and were 2·9 million in 1918.

[3] *Norsk Hydro history*, pp. 415, 417.

supply of coke. The combined capacity of the two factories was about 80 000 tons of cyanamide (say 15 000–16 000 tons N).[1] The owners were hoping to use a large proportion of the cyanamide for the preparation of ammonia and so of nitric acid by Ostwald's process (of which they were licensees). But the plants erected for this purpose in England at Dagenham and Trafford Park did not fulfil expectations and were adapted to other uses. The cyanamide was sold for fertilizer use, and Odda also made a large tonnage of calcium carbide.

The American contribution to the French and British nitrogen requirements was indirect, and consisted of large shipments of explosives and smokeless powder. But the rising Allied contracts and the country's own rearmament programme caused a re-appraisal of the United States nitrogen resources. Chile was the principal supplier, and, apart from shipping problems, prices were increasing. Accordingly the emphasis shifted to indigenous supplies, and in the first place to sulphate of ammonia. In 1913 this had provided 36 000 tons of N. Within a few months of the outbreak of war many recovery ovens were installed in place of the wasteful beehive ovens, and production rose quickly from 47 000 tons N in 1915 to 61 000 tons in 1917 and over 70 000 tons a year in 1918 and 1919.[2]

The development of nitrogen fixation in the United States was marked by delays, unusual muddle, and vast expense that yielded negligible results. It will be recalled that the Government had not intervened in the dyestuffs crisis except to provide some tariff protection, but by contrast it was deeply involved from the outset in the nitrogen projects. Section 124 of the National Defense Act of June 1916 authorized the President to investigate various methods of nitrogen fixation.[3] Although much information was available on the cyanamide process, and the Haber–Bosch process had proved itself, the President's advisers took until May 1917 to make up their minds. They eventually recommended building a works, the so-called No. 1 Plant, which was to use de Jahn's modification of the Haber process. This had been developed by the General Chemical Co. over a period of several years and was now to be operated on the modest scale of 27 tons a day of ammonia, say 7000–8000 tons a year of nitrogen. The site was at Sheffield (Ala.) and building work was completed in September 1918. But the plant apparently did not become operational, and was scrapped after the Armistice, so that the process never proved itself.[4] The President's advisers also proposed to employ Bucher's method for making ammonia from sodium cyanide. This extraordinary process had been tried out at the works of the Mathieson Alkali Co., but had been found expensive as well as useless.[5]

[1] Franck *et al.*, op. cit., pp. 99, 110; *Chem. Trade J.* **58** (1916), no. 1493, 5, 10.

[2] Pollitt, art. cit., p. 147.

[3] Haynes, op. cit., vol. 2, p. 93.

[4] Anon. *The General Chemical Co. after 20 years* (1919), pp. 61–2; Haynes, op. cit., vol. 2, pp. 103–4.

[5] Anon. *Fifty years of chemical progress* (Mathieson Alkali Works) (1942), p. 29. Mathieson built an ammonia plant in 1926.

In the autumn of 1917 wiser counsel began to prevail, and it was belatedly decided to rely on the cyanamide process. The delay has not been satisfactorily explained, and may have been caused by distrust of American Cyanamid, although it was then, without doubt, the most experienced United States Company in the field of nitrogen fixation.[1] At any rate it was awarded the contract to build the No. 2 Plant at Muscle Shoals (Te.) on 4 December 1917. The capacity of the plant was fixed at 540 long tons a day of cyanamide, say 190 000 tons a year or about 38 000 tons N. The cyanamide was to be converted to ammonia, some of which was oxidized to nitric acid, which was reacted with the rest of the ammonia to yield ammonium nitrate.[2]

An immense amount of energy was devoted to the building work, and as an achievement it ranks with Leuna. The first batch of carbide was made on 26 October 1918, that is less than a year from the start of construction. A fortnight later the war was over, and the works had not even been run in properly when it was hurriedly closed. It has been said that Muscle Shoals produced '. . . only a sufficient quantity to make it possible for the construction engineers to deliver the plant to the United States Government and to receive payment'.[3] The whole enterprise was to cause endless argument in the post-war decade, and the power plant and dam eventually became part of the Tennessee Valley Authority's operations. Two other cyanamide works, No. 3 and No. 4, which were being built when the war ended, remained uncompleted. These four projects, excluding power stations and dams, cost the Government $107 million (£22 million). Despite the huge expenditure, the United States failed to solve the problem of nitrogen fixation during the war, and the first successful manufacture of synthetic ammonia did not take place until 1921.

The supply of indigenously-made nitrogen compounds fell sharply in France in 1914 and did not recover until after the war. The production of sulphate of ammonia from coke ovens was 9300 tons in 1917, or about a quarter of the pre-war figure. Gas works maintained their output and made 19 000 tons, but the recovery from sewage plants, once important, became negligible.[4] So France had to depend on British sulphate of ammonia, Norwegian nitrates, and larger imports from Chile. The Government also decided to support the construction of cyanamide works. The factory at N.D. de Briançon, which had been producing cyanamide for some years, was enlarged to 30 000 tons a year, and a second works to make 25 000 tons a year was hastily completed during 1914/15 at Bellegarde near the Swiss border. Several others were authorized, but owing to various delays were not finished until after the war. All were small and so contributed to the fragmentation of

[1] The company had been making cyanamide on a large scale on the Canadian side of Niagara Falls for several years. In 1914–15 the capacity of its works was raised to 57 000 tons a year, and in the 12 months to 30 July 1916, 45 000 tons were actually made. American Cyanamid also sold nitric acid made by Ostwald's process.

[2] Haynes, op. cit., vol. 2, pp. 104 ff.

[3] Brand, C. J. Some fertilizer history connected with World War I. *Agric. Hist.* **19** (1945), no. 2, 104 f.

[4] Fontaine, A. *L'Industrie Française pendant la guerre* (1925), p. 251

nitrogen production that was so characteristic a feature of the French chemical industry generally in the inter-war years. The cyanamide was used to make ammonia and nitric acid in the works of the Service des Poudres.[1] The Government also financed the construction of an arc plant at Soulom in the Hautes Pyrénées. Norsk Hydro was put in charge of the project in 1915/16, and Soulom made a few thousand tons a year of ammonium nitrate; after the war, nitric acid was produced. The works was small and did not have its own power station, so that when the contract with the electricity supply company terminated in 1921, the business was closed and the site sold.[2] The cyanamide works, the arc plant, and the connected nitric acid and nitrate factories are said to have cost the French Government £15 million.[3] The figure seems exaggerated, and for our purposes it is sufficient to note that the various nitrogen fixation works made only a relatively small contribution to the problem during the war.

In Britain the large town-gas industry together with the coke ovens produced the equivalent of about 90 000 tons a year of nitrogen. This was adequate, particularly when steps were taken to introduce export licensing for sulphate of ammonia and to raise imports of Scandinavian cyanamide and Chilean nitrates. An unexpected difficulty was the lack of equipment to make nitric acid on a sufficiently large scale, and this created a bottleneck in the supply of ammonium nitrate. Fortunately some work on producing this key compound by another route had been done before the war by the chemists of Brunner, Mond & Co. During the war, F. A. Freeth and his colleagues developed the process, which derived from the well-tried principles of ammonia-soda manufacture. In this way, starting from ammonium compounds and Chilean nitrates, ammonium nitrate was produced in large quantities at the Sandbach works of Brunner, Mond and Government factories specially built for this purpose. Between the end of 1915 and the Armistice, 280 000 tons were made without the use of nitric acid.[4] The shortage of this acid remained a problem nevertheless. Nitrogen Products & Carbide intended making it from ammonia derived from cyanamide, but never managed to attain volume production. Nobel's Explosives, Chance & Hunt, and British Cyanides next tried their hand, and eventually devised an oxidation process, modelled on Ostwald's, to work on gas works ammonia. They built factories at Ardeer and Oldbury which began making nitric acid in January 1918. But the operations were expensive, and the acid weak, so that the factories were closed as soon as the most urgent need had passed.[5] In

[1] Sevin, R. L'usine electrochimique de Bellegarde. *J. Four élect.* **63** (1959), no. 4, 163–4. Franck, op. cit., pp. 80 f.

[2] *Norsk Hydro history*, pp. 180 ff. The site was acquired by the predecessor of the Pierrefitte group to make fertilizers.

[3] *Chem. Trade J.* **65** (1919), no. 1695, 547 (report of the Annual General Meeting of Nitrogen Products and Carbide).

[4] Freeth, F. A. Explosives for the First World War. *New Scient.* **23** (1964), no. 402, 274 ff.; Ministry of Munitions, op. cit., pp. 81 ff.

[5] Miles, F. D. *A History of research in the Nobel Division of I.C.I.* (1955), pp. 80–1.

Britain, the traditional route to nitric acid from nitrates and sulphuric acid prevailed during the war and continued to be used until the late 1920s.

The country's nitrogen supply problems were solved on the basis of the customary raw-materials, processed by novel methods. But the submarine campaign and its threat to nitrate imports caused the Ministry of Munitions to form the British Nitrogen Products Committee in June 1916. This large and influential committee concerned itself not only with the most efficient use of ammonium sulphate and nitrate, but also with the investigation of new processes. One of its duties was to advise which of these processes was to be operated on an industrial scale, and unlike its American counterpart, the committee made up its mind very quickly: early in 1917 it recommended the building of a plant for making 50 000 tons a year of cyanamide and the construction of a pilot plant for the Haber process. The cyanamide project was eventually dropped, but work on the Haber process continued.[1] The various sub-committees of the main Committee studied such aspects as power supplies and gasification processes. At the beginning of 1918, the Committee recommended building a plant for the manufacture of ammonia by the Haber process, and the Government offered a site for this at Billingham on the north bank of the river Tees. When the war ended, no fixed nitrogen had been made by the Haber process, but the Government was still firmly committed to it, although soon afterwards it decided to end its involvement by selling the site to Brunner, Mond, which later built a large works at Billingham.[2]

Interaction of war and chemical technology

The development of dyestuffs manufacture in Britain, the United States, and France, and the fixation of atmospheric nitrogen in Germany were the most spectacular manifestations of applied chemistry during the war. In both cases neither the chemistry nor the end products were novel, and generally speaking it was the rapidity of the introduction and the large scale of the technology, rather than inventiveness, which characterized the war-time chemical industry. This applies not only to the products we have considered so far, but also to most organic and inorganic chemicals. Examples abound, and two important ones will illustrate the point: chemical warfare because of its effect on the chlor-alkali sector of the industry, and acetone because it marked the first stage in the later growth of aliphatic chemistry. Both were supported from the outset by Government departments, and so serve to highlight another important change in the history of the industry.

Chemical warfare, that is the use of poison gas, flame-throwers, and smoke generators, together with protective measures, conspicuously gas-masks, began in the closing months of 1914. In most countries academic chemists

[1] *Chem. trade J.* **61** (1917), no. 1593, 459–60.
[2] Parke, V. E. *Billingham—The first ten years* (1957), pp. 4–6.

played an active role, and the pioneering work was carried out in university laboratories and research institutes.[1]

The first major gas attack of the war was made by the Germans on the Western Front near Ypres on 22 April 1915. It involved the discharge from cylinders of a large volume of liquid chlorine, and while technically successful was of little consequence, for the Germans failed to exploit the surprise element. The emission of chlorine clouds was gradually superseded by gas shells. They made it easier to pinpoint and saturate targets, and reduced the danger of blow-back. By the end of the war over one-quarter of the German shells were filled with gas warfare materials, and on the Allied side the proportion may have been of the same order.[2] From 1916 onwards, chlorine was being replaced by phosgene (carbonyl chloride, also called 'green cross'), which seems to have been first employed by the French at the opening of the battle for Verdun in February 1916.[3] The Germans copied them on 22 June 1916, as a prelude to their last and biggest attack on the fortress. It yielded them no worthwhile gains.[4] In the following months the horrors of gas warfare mounted: 'Vincennite', a compound of hydrocyanic acid, tin chloride, arsenic chloride, and chloroform, was introduced on the Somme in June 1916, and the following year the Germans first used 'mustard gas', 'yellow cross', or 'Yperite' (dichlorodiethyl sulphide), which differed from the others in that it also attacked the skin by raising blisters.

As a method of warfare gas is condemned, and the efforts to justify it on the ground that escape is possible, are mere casuistry and conveniently ignore the fate of those who, wounded or shell-shocked, are unable to get away and die in agony. But as a material of war, gas had an important effect on the chemical industry. With rare exceptions all poison gases were based on chlorine or on chlorinated hydrocarbons, and because a single attack might require several hundred tons, the scale of manufacture had necessarily to be considerable. This created few problems in Germany, where the dyestuffs industry had used appreciable quantities of liquid chlorine and even phosgene before the war. Production of chemical warfare agents was therefore handed over to the dyestuffs factories, particularly Bayer and Hoechst; Badische concentrated on phosgene.[5] By contrast France was entirely unprepared in this respect, and immediately after the Ypres attack steps were taken to establish a chemical warfare department. In September 1915 it was renamed Service du Matériel Chimique and made directly answerable to the energetic Albert

[1] In Germany the work was directed by Haber, and his institute for physical chemistry became the centre of chemical warfare research. His role was appreciated in unlikely quarters: 'It would be difficult to exaggerate Germany's debt to Haber, yet he had never attained a higher rank than that of captain. His country, however, made proper use of him. Had he been born an Englishman, he would certainly have attained a higher rank, but almost as certainly he would not have been properly utilized until it was nearly too late.' *Chem. Trade J.* **65** (1919), no. 1677, 37.

[2] Haber, F. *Fünf Vorträge* (*Zur Geschichte des Gaskrieges*) (1924), pp. 77 ff.

[3] Moureu, C. *La chimie et la guerre* (1920), p. 63.

[4] For a brilliant description see A. Horne, *The price of glory* (1962), pp. 284 ff.

[5] Ministry of Munitions, *Report of the British Mission . . . to . . . enemy chemical factories in the occupied zone . . .* (1919), pp. 11–12.

Thomas, who was then Under-secretary of State for Artillery and Munitions.[1] Orders were issued to import chlorine, at first from Britain and Italy, later from the United States. It was also decided to build immediately seven electrolytic works for making chlorine. The initial daily capacity was to be merely 30 tons, but by 1917 it had been raised to 50 tons.[2] Six of these plants were located in the Alps or the foothills of the Alps and, with the exception of St. Auban, they were, even by contemporary standards, pretty small, possibly because of the lack of hydro-power and the delay in getting additional capacity into operation.[3] Salt was obtained from the Mediterranean by solar evaporation. These seven chlorine plants, together with four built by the Service des Poudres, produced altogether 24 000 tons of chlorine during the war, half of it as liquid, half as chloride of lime. All were operated by the Service du Matériel Chimique, but the actual preparation of the poison gas was carried out by a few chemical firms, notably Usines de Laire at Pont de Claix, Air Liquide, and Poulenc Frères.

During 1916 the French began making sufficient chlorine to retaliate, and as their production increased they assumed responsibility for supplying their Allies. They made 16 000 tons of phosgene by 1918, and a large tonnage of other gases and smoke compounds.[4] The Service du Matériel Chimique was also in charge of the defensive measures, and designed various types of gas-masks, some of which were produced in astronomical quantities.[5]

Liquid chlorine plants were built in several other countries. In Russia, after the Germans had used gas on that front in May 1915, a factory was quickly put up at Slaviansk. Additional electrolytic capacity was installed in the Donets region, in Finland, and a large works was under construction at Samara at the time of the Armistice. Phosgene was first made about 1916.[6] In Britain and the United States the manufacture of liquid chlorine was developed less in response to the needs of gas warfare, than to meet the requirements of dyestuffs manufacture.

These war-time factories were built when chlorine was more important than its co-product caustic soda. In this way the output of electrolytic caustic increased, and when war ended there was considerable excess capacity. This state of affairs obtained throughout the 1920s, with the result that the price

[1] Moureu, op. cit., pp. 32–3. Similar organizations were created by the other Allies. The United States had no Chemical Warfare Service when it entered the war. In March 1917 the laboratory of the Bureau of Mines at the American University at Washington became the centre of chemical warfare research. The Chemical Warfare Service was formally constituted in July 1918. Yerkes, R. M. (Ed.) *The new world of science* (1920), pp. 150–1.

[2] Moureu, op. cit., pp. 74–5.

[3] At St. Auban 6400 tons of chlorine were made between 1916 and 1918. Gignoux, C. J. *Histoire d'une entreprise Française* (1955), pp. 143–4.

[4] According to Moureu (pp. 77 f.) the amount came to 12 600 tons. The British forces were supplied with phosgene by Usines de Laire from a works specially built at Calais. It obtained liquid chlorine from England.

[5] The visor of the M 2 mask, of which 30 million were made, was of cellophane. This seems to have been the first use of regenerated cellulose in sheet form. Celluloid was rejected because of its flammability.

[6] Ipatieff, op. cit., pp. 212–15.

of caustic soda was depressed, and it was difficult to find profitable markets for the chlorine.

Our second example to illustrate the interaction of war and chemical manufacture is acetone. Up to 1914 acetone had been principally used as a gelatinizing agent to incorporate gun cotton into nitroglycerine. This essential stage in the manufacture of smokeless powder required substantial quantities of the chemical, about 50 per cent by weight of the gun cotton. Recovery of the acetone was not particularly efficient, but in those days there were no supply problems, for it was produced on a large scale in the United States and the Austrian Empire from a by-product of wood distillation. About 1 ton of acetone was obtained from about 80–100 tons of birch, beech, or maple.[1]

At the outbreak of war the demand for smokeless powder soared and with it the need for acetone. Although recovery methods were improved and the wastage cut down, the supply situation became difficult. The solution to the problem was found in the preparation of acetone by fermentation methods. Such methods, based on specially cultivated bacteria in place of the traditional yeast, had been developed just before the war as a route to the production of hydrocarbon compounds from which synthetic rubber could be prepared. The synthesis of rubber had attracted considerable attention on account of the very high price of natural rubber, and between 1900 and 1914 500 patents were taken out (mainly in Germany and Britain) for the manufacture of elastomers from isoprene, butadiene, or methylisoprene (2,3 methylbutadiene). Isoprene rubbers appeared particularly promising and were indeed successfully vulcanized in Germany before 1914. Various materials were used as a source of isoprene: Bayer obtained it from acetone; Staudinger and Engler, with remarkable foresight, drew attention to the by-products of oil cracking;[2] Tilden proposed turpentine; Badische and Strange & Graham started from iso-amyl alcohol derived from fusel oil.[3] A. Fernbach at the Institut Pasteur and Chaim Weizmann at Manchester were involved in the work on fusel oil, and they were able to increase the yield of iso-amyl alcohol. But in connexion with our story the more important discovery was incidental to the main task; it was made by Weizmann and was a bacterium that fermented grain to acetone and butanol. Weizmann's fermentation process yielded about 37 tons of solvents for a hundred tons of grain, and the solvent mixture consisted of roughly one part of acetone for every two of butanol.[4] When the price of rubber fell, all work on fermentation ended. After the out-

[1] Marshall, A. *Explosives* (1917), vol. 1, pp. 304, 307, 340.

[2] They proposed to convert mixed pentanes to trimethyl ethylene from which isoprene was to be prepared. It may be recalled that this was one of Staudinger's first major investigations. Later he was to be called 'the father of high polymers'. For a contemporary account see Holt, A. Neuere Arbeiten auf dem Kautschukgebiet. *Z. angew. Chem.* **27** (1914), pt. 1, 153 ff.

[3] Strange and Graham were consultants connected with the Synthetic Products Co., which from the beginning of 1912 onwards made iso-amyl alcohol from potatoes at Kings Lynn. The factory operated during part of the war on grain. Nathan, Sir Frederic, The Manufacture of acetone. *J. Soc. chem. Ind., Lond.* **38** (1919), no. 14, 271T–273T.

[4] Fernbach later claimed to have preceded Weizmann, but it appears that this method was only worked in France, and that on a modest scale from 1917 onwards. Nathan, art. cit.

break of war, Weizmann offered his process to the Admiralty. He was ignored. But during the early months of 1915 the Admiralty had a change of mind and became interested. After these initial, and in retrospect incomprehensible delays, Weizmann's fermentation route was developed energetically.[1] Experiments at the Lister Institute in London and at Nicholson's gin distillery in Bow were successful, an acetone plant was built at the Royal Naval Cordite Factory, and six distilleries were converted for the same purpose. Grain, and, increasingly, damaged rice, were used as raw materials.

By 1916 the Weizmann process was making an essential contribution to the British war effort, and he was rewarded by the Balfour Declaration of 1917.[2] But the shortage of grain prevented the large-scale utilization of the process in wartime Britain, and the operations were therefore transferred to North America. In 1916 British Acetones Ltd. was formed in Toronto. With technical help from Gooderham & Worts Distillery, a large fermentation works was built and went into production in January 1917. At its peak it made 200 tons a month of acetone. At the beginning of 1918 the British Mission bought a distillery at Terre Haute (Indiana), had it converted to the Weizmann process, and began deliveries of acetone in May of that year.[3] Meanwhile the old route to acetone from wood distillation via acetate of lime continued to be worked, and in the United States the Hercules Powder Co. made it in this way and also from giant kelp on a very large scale.[4]

In Germany the acetone situation was at first less acute, though there too fermentation processes were adopted to augment supplies available from wood distillation. There was, however, another and most significant development. Before the war Hoechst and a smaller company, Alexander Wacker, patented a method for making acetone from acetic acid.[5] The process was first used during the war on a commercial scale by Hoechst at Knapsack and Wacker at Burghausen in Bavaria. The Germans had patented the process abroad, and the patents were sequestrated and assigned to Canada Electro Products Co. of Montreal, a subsidiary of Shawinigan Water & Power. The company completed the building of a synthetic acetone plant in November 1916, and a few weeks later production began. In 1917, 1200 tons of synthetic acetone were sent to Britain by Canada Electro Products, but later, when British Acetones were able to supply more cheaply, production ceased. However, the first stage of manufacture, the preparation of synthetic acetic acid

[1] Weizmann, Ch. *Trial and Error* (1949), pp. 218–20. Weizmann says that it was only in Spring 1916 that he was invited to meet Sir Frederic Nathan, the then head of the Admiralty Powder Department. Nathan in his lecture (see footnote 3, p. 211 mentions that it was a year earlier. This seems to be the more likely date.

[2] Weizmann turned down offers of honours or financial rewards, and years later his life-long work for Zionism culminated in his becoming the first president of the new state of Israel.

[3] Nathan, art. cit., pp. 272T f.; Haynes, op. cit., vol. 3, pp. 151–2. The works at Terre Haute were bought after the war by the Commercial Solvents Corporation, for many years one of the leading firms in this branch of chemical manufacture.

[4] Anon. *A history* (1947), p. 2.

[5] This was obtained by the catalytic oxidation of acetaldehyde, which in turn was made by hydrating acetylene.

from acetylene, continued, and altogether 10 000 tons were made before the end of the war.[1] The carbide route to acetylene and acetic acid was not practicable in Britain, but an alternative approach was worked in which fermentation alcohol was dehydrogenated to acetaldehyde, from which acetic acid was made. This was to become an important process in aliphatic chemistry, but during the war only Joseph Crosfield appear to have employed it.[2]

One might have thought that all these processes would meet every conceivable need for acetone with the utmost flexibility. But the story was not yet over. In the latter part of the war two new and unexpected uses emerged. The lesser was confined to Germany and was connected, once again, with the manufacture of synthetic rubber. Bayer at Leverkusen, on orders from the Government, began the complicated synthesis from acetone, which was reduced to pinacol, from which, in turn, dimethylbutadiene was obtained. This compound was polymerized with a metallic alkali to form an ebonite-type material, suitable for battery cases. The addition of toluidine and dimethylaniline to dimethylbutadiene yielded a vulcanizable rubber, inferior to the natural product, but nevertheless of use to the Germans. Output, at its peak, was 150 tons a month, equivalent to about 10 per cent of the country's wartime rubber consumption.[3] For our purposes it is sufficient to note that this was the first large-scale use of synthetic rubber, ante-dating by some twenty years the self-sufficiency policy of Hitler's Germany, and that it was of some significance during the Great War. The rubber was extremely expensive, and the British Chemical Mission was told shortly after the Armistice that the process did not offer 'a reasonable prospect of being a commercial success'.[4]

The same might have been said (and perhaps was), of the second and much more important of the new uses for acetone, namely as a solvent of cellulose acetate for aircraft dope. Special varnishes of different sorts had been employed for aircraft before the war, but it seems that a cellulose acetate coating to protect the fabric wings of planes was unknown. Before 1914 cellulose acetate was made chiefly in the form of non-flammable cine film and to a lesser extent as sheet by Bayer and AGFA in Germany, Usines du Rhône in France, and the Dreyfus brothers in Basle.[5] The quantity was small, and the film, though safe, was much more expensive than celluloid, while the sheet was only used for luxury packaging and display purposes. Upon dissolving cellulose acetate in acetone, a varnish or dope of excellent properties is

[1] Warrington, C. J. S. and Nicholls, R. V. V. *A history of chemistry in Canada* (1949) pp. 173–5.

[2] Musson, A. E. *Enterprise in soap and chemicals* (1965), p. 272.

[3] Duisberg, C. Lecture at 24th general meeting of the Deutsche Bunsen-Gesellschaft. *Z. angew. Chem.* **31** (1918), pt. 3, p. 243.

[4] *Report of the British Chemical Mission on the chemical factories in the occupied area of Germany* (1919), p. 100.

[5] Urquhart, A. R. Cellulose derivative rayons, past, present, and future. *Development of some man-made fibres* (1952), p. 14.

obtained. Between 1916 and 1918 the aircraft factories used very large quantities of cellulose acetate: in France the amount was 6000 tons, and in Britain and the United States the quantities are not likely to have been smaller.[1] In this way there emerged a big demand for acetone, especially as these was no efficient solvent-recovery process at the time, and also for acetic acid.

The additional and combined requirements for these chemicals raised fresh problems, and much money was spent in providing the necessary manufacturing facilities.[2] After the war, the demand for acetic acid and acetone (and the co-product butanol) fell steeply and much plant became idle. It was fortunate therefore that within a few years a new use was found for acetic acid and butanol: butyl acetate was (and some say remains) the best solvent for nitro-cellulose lacquers, which became in the 1920s the preferred and indeed principal coating for motor-car bodies. The business in acetic acid and butanol suddenly boomed. The change in the output of a single United States manufacturer illustrates the remarkable transformation. In 1923, Commercial Solvents made just over 2000 tons of butanol, the next year almost 6500 tons.[3] The dates may be taken as the beginning of the aliphatic solvents industry in the United States, which for roughly a quarter of a century was to be based predominantly on fermentation processes. In Europe the development of these solvents was slower, the scale smaller, and their effect on chemical manufacture less conspicuous. In Germany, moreover, the acetylene route to solvents continued to play a more important role than fermentation.

Conclusion

The wartime increase in chemical manufacture and the structural changes this entailed in the principal countries were far-reaching. Large-scale production of chemicals was occasionally met with in the nineteenth century and became more frequent in the 1900s, but in the years 1914–18 it was a common enough phenomenon. Size had many aspects. We have already instanced the German nitrogen fixation works. There are just as appropriate examples to be found in other countries. In October 1914, Du Pont, then the principal manufacturer of smokeless powder in the United States, had an annual capacity of 4500 tons. Three years later its factories at Carney's Point, Haskell, and Parlin had a combined capacity of 130 000 tons, and a few

[1] Clement, R. and Rivière, C. Les matières plastiques, in Gérard, J. (Ed.), op. cit., vol. 1, pp. 851–3. The authors estimated that on average 120 kilos of dope were used per plane. The French built about 50 000 planes during the war, the British just under 55 000.

[2] In Britain the Dreyfus brothers tendered successfully for the cellulose acetate contract, and in 1915 set up the British Cellulose & Chemical Manufacturing Co. at Spondon (near Derby) to supply it. Quantity production was delayed until July 1917. The factory ultimately cost £3 million, most of it provided by the Government. The business was the predecessor of British Celanese, which during the 1950s became part of Courtaulds Ltd. For an account of the difficulties and the enquiry into the delay see *Chem. Trade J.* **63** (1918), no. 1628, 81; **65** (1919) no. 1683, 197–200.

[3] USTC, *Census 1924* (1925), p. 123.

months later the completion of extensions brought the figure to 200 000 tons a year.[1] In Britain, Levinstein Ltd. and Read Holliday Ltd. between them made a few thousand tons of dyes before the war; in 1918, combined under the name of British Dyestuffs Corporation, their output was around 16 000 tons. In France, the Kuhlmann company had been a substantial producer of sulphuric acid and superphosphate in the area of Lille. When the German advance halted at last in September 1914, the firm was left with only one small works, Amiens, in Allied hands. So it had to start from scratch: it built a factory for the Government at Port de Bouc (near Marseilles), bought glue works at Aubervilliers and Nevers, and fertilizer plants at Bordeaux, Nantes, and Petit Quevilly (near Rouen); it also acquired Soc. Lambert-Rivière, the leading French chemical merchant, and was one of the founders of the Cie Nationale de Matières Colorantes.[2] When the war ended Kuhlmann was a different company, not only much larger than in 1914, but also with factories in many parts of the country. The experience of the company was repeated many times, with the result that most problems caused by enemy action were eventually solved. For example, the French production of sulphuric acid in the year to 30 June 1914 had been about 1·2 million tons (as 53° Bé acid). Heavy losses were sustained, but new plant was hastily built and by the end of 1916 the manufacturing capacity was reported to be 1·6 million tons (as 53° Bé acid, equivalent to about 1·05 million tons as 100 per cent H_2SO_4). The figure was well above pre-war level, and no less than a third was located in the Midi and the South-East.[3]

The war-time changes were perhaps even more far-reaching in Japan and Russia. In the former we have previously noted the emergence of dyestuffs manufacture. Similarly the soda trade was transformed. Up to 1914 the country had relied on imports and on some small Leblanc works. During the war the production of electrolytic alkali was begun, and by 1918 about 10 000 tons a year of caustic soda were made in this way.[4] The Russian chemical industry developed rapidly under the stress of war: it employed 33 000 workers in 1913, 54 000 in 1915, and a peak of 117 000 in 1917.[5] Building materials were scarce and engineering supplies subject to long delays, but the Russians who replaced the German engineers and chemists were remarkably ingenious and achieved much between 1914 and 1917. For example, liquid chlorine was made on a substantial scale, and the compressors required for this purpose were designed and built entirely in Russia. Another considerable success was the evacuation of factories and their move to safe areas. In this way enough equipment was salvaged to begin the manufacture of sulphuric acid and superphosphate in the region of Nijni Novgorod

[1] Dutton, op. cit., pp. 223, 227.
[2] Les Etablissements Kuhlmann, op. cit., pp. 27 ff.
[3] Lambert, M. *Les engrais chimiques et produits chimiques utiles à l'agriculture* (1917), p. 51.
[4] Yamasaki and Ogawa, op. cit., pp. 281–4.
[5] The figures include persons employed in the manufacture of explosives. Strumilin, S. Khimicheskaya promyshlennost SSR. *Problem ekonomiki* (1935), no. 2, 118 f.

(Gorki), and of chlorine at Samara; the peak of production was reached in 1916.[1]

The structural changes resulting from the emergency measures created the framework of the contemporary chemical industry, with which we have become familiar because we are aware of its principal characteristics: the large factories, the big concerns that control them, and their enormous capital requirements. But the uniformity of this simplified and generalized picture masks some significant and special characteristics. In Germany the relative importance of the dyestuffs and pharmaceutical sectors of the industry diminished during the war. The very opposite was the case in Britain, the United States, and France. Japan and Russia were brought into the mainstream of industrial chemical developments, though in the latter country 1917 marked the end of a most promising phase.

Another feature of the wartime chemical industry was the speed with which firms diversified their activities. Activities that might have been spread over five or ten years, were now telescoped into two or three. A small firm such as Dow extended its range to relatively advanced products such as aspirin and acetic anhydride, a large firm such as Du Pont began making stabilizers for explosives, which it had previously imported. In Switzerland, diversification took a different direction, caused by the shortage of inorganic process chemicals, which had hitherto been imported. The leading Swiss makers of dyes combined to build a plant for making contact acid, and together with the Cantons established a factory for making soda.[2] Even in Germany, where the range of chemicals had been wider than in any other country, it was necessary to adapt, extend, and improvise. For example the supplies of imported pyrites from Scandinavia and the Balkans were inadequate, and so the low-grade indigenous deposits were worked, the Silesian zinc smelters enlarged, and sulphur recovery from gas-making residuals improved; the last-named source alone yielded 26 000 tons of sulphur in 1917/18. When animal and vegetable fats for soap making became scarce, the production of glycerine fell and improvisation was essential: the substitute obtained by the fermentation of sugar residues was inferior to the genuine article, but adequate, and about 41 000 tons of this *Ersatz* material were made in the latter part of the war.[3] The dyestuffs companies, which used to make inorganic chemicals for captive use, were now obliged to increase the output of sulphuric and nitric acids, and also of chlorine for the explosives and chemical warfare programmes. At the end of the war their equipment for making these chemicals far exceeded their foreseeable needs. It was inevitable therefore that in order to keep the works

[1] Ioffe, Ya. A., Kozlov, A. I., and Alman, G. L. *Ekonomika i planirouaniya khimicheskoi promyshlennosti* (1940), pp. 25 f.; *Russkaia promyshlennost v. 1921 godu i ee perspketivy* (1922), pp. 244–5; Ipatieff, op. cit., p. 215.

[2] Jaquet, op. cit., p. 89; Waldmeyer, E. Die schweizerische Salz- und Sodaindustrie, Ph.D. thesis (1929), pp. 218 ff.

[3] Ungewitter, op. cit., pp. 337–8, 346–8.

occupied, they should encroach on the business previously held by other firms which were also plagued by excess capacity.

Diversification and broadening the product range thus disturbed traditional marketing activities and relationships and so aggravated the problems caused by supply exceeding demand. This was not peculiar to Germany, but affected most countries where chemicals were made on any scale. Indeed it was made worse by the fact that with the war specialization in chemical production virtually ended. If we look at dyestuffs (though it applies equally, though less noticeably, to many other products) we can see that before the war the situation was at least stable, however unsatisfactory the near-monopoly of the Germans appeared to their rivals and their customers. But after 1918 there was no monopoly, no stability, and for many years no satisfactory business conditions. These long-term effects of the war will occupy our attentions in later chapters.

8

GOVERNMENT INTERVENTION

THE Great War marked the watershed in the relations between Government and industry. The most conspicuous and in the long run the most important forms of intervention were the measures taken to convert chemical manufacture from a civilian to a military economy and the encouragement and protection granted to some sectors of the industry during and after the war. They will be the main theme of this chapter, but we should note that the war and post-war acts and regulations were superimposed on earlier legislation that was occasionally of considerable significance for the industry.

The manipulation of patent laws, for the benefit of one company (and its licensees) and to the detriment of a whole branch of industry, was not confined to chemicals. But the connexion between chemicals and patents attracted attention, and the notoriety of some of the German practices in the dyestuffs business always kept this problem in the forefront. The leading dyestuffs manufacturers were able to defeat the spirit of the existing legislation, and skilfully used the numerous loopholes to maintain their dominating position. In the years before the war the main conflict was over the question of the compulsory working of patents granted to foreigners. We have seen that in Britain the usefulness of Clause 27 of the Patent Act of 1907 was largely nullified by the 'Eternit' judgement.[1] Elsewhere bilateral treaties between Germany and the United States, and Italy and Switzerland specifically exempted patents in these countries from the working clause on a reciprocal basis.[2]

In Germany the patent law of 1891 extended the protection by allowing the patenting not only of the process, but also of the product made by that process. The revised law thus resembled French and United States legislation. But the available statistics suggest that chemical process patents continued to grow, while the number of product patents remained unchanged: the former rose from 323 to 462 between 1900 and 1910, the latter was about 250 in both years.[3] Patents for chemical processes covered a wider field and could be used to block progress over a much larger number of end products. When such patents were taken out abroad, and this was the usual practice, their effectiveness was enhanced because it was more difficult to insist on compulsory working if several end products, instead of a single dye, were involved.

The modernization of the laws to prevent abuses took many years, and even now remains incomplete. In Germany a law was passed in 1911 that

[1] See p. 147. [2] Zimmermann, P. *Patentwesen in der Chemie* (1965), p. 124.
[3] *Z. angew. Chem.* **25** (1912), no. 48, 2474–5.

provided that a patent could be revoked if it was worked mainly or predominantly outside the Empire.[1] In France a major change was proposed in 1916 whereby the patenting of the product was to end, but the necessary legislation did not come into force until 1931.[2] The Russians took the extreme attitude confiscating all patents in 1919. Five years later, however, they introduced legislation modelled on German lines. The Soviet patent law was not widely used in the 1920s, nor were the 'certificates of authorship' introduced in 1931.[3]

After the war the French adopted what was then considered a novel policy towards German industrial property and secured the free use of German patents by Article 297 of the Treaty of Versailles. But in at least one instance, that of the Haber–Bosch process, possession of the patents was of little use without the accompanying know-how. So the French Government negotiated a special agreement with Badische in November 1919, whereby it obtained the technical details, and also help during the construction of the plant, in return for a lump sum payment of fr. 5 million (about £250 000 at the official rate of exchange), and a royalty, which fluctuated with the price of nitrates from 5 to 12 centimes a kilo N.[4] The French precedent of bilateral negotiations was followed by others. Licensing agreements, which included know-how, became more usual: they did not replace patents, but supplemented them and, as it were, increased their value. In the 1920s and 1930s the centre of interest moved from dyestuffs to high-pressure synthesis and plastics. Chemical engineering technology was the key factor in the development of the former, and the process details lent themselves readily to licensing and cross-licensing procedures. In this way Badische was able to retain its preponderance in high-pressure catalysis even after the expiration of Haber's ammonia patents. As regards plastics, the leading German firms were very active in this area, and in the years to 1930 IG Farben and its predecessors had taken out about 530 patents. This was nearly 13 per cent of all patents for plastics taken out in Germany, the United States, France, and Britain in that period, and represented a much higher proportion of those that later became commercially significant.[5]

The greater emphasis on agreements and licences modified, as it were, the rigours of the patent system. Generally speaking patents played a subordinate role after the war in determining the development of dyestuffs and intermediates production. In the United States, however, an attempt was made to administrate pre-war German patents for the benefit of American industry. This curious experiment in a country noted for its well-entrenched respect for

[1] Zimmermann, op. cit., p. 127. The German wording followed that of the British Act of 1907.

[2] Matagrin, A. *L'Industrie des produits chimiques et ses travailleurs* (1925), pp. 261–2; Fauque M. *L'Evolution économique de la grande industrie chimique en France* (1932), pp. 174 f.

[3] *The Listener*, 31 December 1959, pp. 1147–8.

[4] *Chem. Trade J.* **57** (1920), no. 1734, 235; Fauque, op. cit., pp. 229–30. The building of the factory was delayed for several years.

[5] Freeman, C. The plastics industry. *Rev. nat. Inst. econ. social Res.* (November 1963), 32, 37–8.

private property, merits brief mention. The initiative came from the Alien Property Custodian, an official created by the war, who did not think that tariffs could provide adequate protection against the post-war German industry. He therefore asked Congress for legislation enabling him to seize all German patents. Congress agreed, and in November 1918, about 5000 German patents were confiscated. The Custodian (hardly an appropriate title under the new circumstances) was authorized to sell these patents and remit the proceeds to the Germans. He sold them to the Chemical Foundation, which had been set up for the purpose of licensing German patents to American firms. The Foundation paid a mere $250 000, and licensed 735, mainly to Du Pont, National Aniline, and the Newport Co.; some patents (among them those for the Haber–Bosch process) were licensed to several firms.[1] The licences were non-exclusive, which greatly reduced the appeal of the scheme. The Foundation's income was to be used first to redeem its preference shares, and next to establish 'an institute of research which would be in a position to co-ordinate the efforts of the universities and of the dye works in this country as has been done in Germany'.[2] The intent was admirable, the proposal original, and the Chemical Foundation, run by Trustees consisting of bankers and lawyers, appeared to be an excellent combination of legality and capitalist enterprise created under official auspices to help companies. Unfortunately the patents were less valuable in practice than had been anticipated, and the scheme was soon attacked by representatives of German interests, and also by Congressmen. Its contribution to the development of the American dyestuffs industry was very small, and one of its few practical achievements was the financing of a research laboratory at Savannah (Ga.) where a process for making newsprint from the pulp of the Southern pine was evolved.[3]

Before the war several countries had Government laboratories in which some research was carried out that occasionally involved a good deal of chemical technology, as for example in the development of smokeless powder.[4] Among civilian laboratories one of the oldest, and possibly the most famous, was the Physikalisch-Technische Reichsanstalt in Berlin, established towards the end of the nineteenth century for the testing of materials, the determination of physical constants, and the standardization of measurements and instruments. In England a similar institution, the National Physical Laboratory, was founded in 1900 with Government assistance,

[1] Haynes, W. *American chemical industry* (1945), vol. 3, lists the patents on pp. 483–505.

[2] Hearings of the House of Representatives Committee on Ways and Means on H.R. 2706 and 6495, *Dyestuffs* (1920), Evidence of J. H. Choate (on behalf of the Chemical Foundation), pp. 106 ff.; Report of the Alien Property Custodian, 1919, quoted in *J. ind. Engng Chem.* (1919), p. 365.

[3] Bartlett, H. R. *The development of industrial research in the U.S.*, National Resources Planning Board (1941), pt. 2, p. 36.

[4] The Chemical Inspectorate of the War Office, created in 1854, was among the first Government scientific establishments in Britain. In 1914 it had a staff of about eleven. There was also the Research Department at Woolwich Arsenal, set up in the 1900s, which concerned itself with non-routine investigations. Norris, W. G. *Chemical Service in Defence of the Realm* (1957), pp. 5 ff.

though under the control of the Royal Society. A year later the United States Congress passed an Act that created the National Bureau of Standards. These three organizations were intended to help industry generally, but in the event the engineering industry was the main beneficiary. Their resources were at first minuscule: the National Physical Laboratory received a capital grant of £13 000, and initially an annual grant of £4000 from the Treasury. The National Bureau of Standards began with a staff of fourteen and an Appropriation (including buildings) of $392 000 (say £80 000). By 1913 the numbers had grown to 280 (including 50 qualified physicists) and $536 000 (£110 000) respectively. During and after the war the scope of the Bureau was considerably enlarged with consequent increases in staff and budget.[1]

There were also specialist departments, usually concerned with assisting in the enforcement of laws, which had some connexion with chemistry. Thus the Inland Revenue Laboratory in Britain dated back to 1842. In 1911 its duties were taken over by the Government Laboratory, which thereafter carried out investigations for all Civil Departments. Its head was the Government Chemist, who was responsible to the Chancellor of the Exchequer.[2] In the United States the Department of Agriculture established a Chemistry Division, later the Bureau of Chemistry, which had a staff of 425 in 1908, and was responsible for the enforcement of the Pure Food and Drug Act of 1906.[3] Similarly, the United States Bureau of Mines, created in 1910 as a branch of the Department of the Interior, carried out chemical investigations, and in 1914 F. G. Cottrell, an eminent scientific civil servant, became its chief chemist.[4]

The involvement of officialdom in the affairs of chemical manufacturers was at times extensive. It was most noticeable in Britain, where, from the 1860s onwards, the Alkali Acts checked the worst abuses of atmospheric and water pollution. By the eve of the war detailed regulations, enforced by the Alkali Inspectorate, covered many branches of the industry. There were also a few products subject to complicated excise duties, which required elaborate supervision. Thus the exemption of industrial alcohol from the British spirit duty in 1902 necessitated special safeguards to protect the Revenue, notably strict supervision by Customs and Excise staff and the addition of a denaturant. Four years later an allowance of 3*d.* a gallon was made to the users to compensate them for the extra costs resulting from these measures. Although the dyestuffs makers had demanded the concession, it is interesting to note that the exemption from duty had little effect on the consumption of industrial spirit.[5] In Germany the use of alcohol and of denaturants was examined by the Ministry of the Interior at about the same time. The Prussian excise had

[1] Melville, Sir Harry. *The Department of Scientific and Industrial Research* (1962), pp. 101 f.; Cochrane, R. C. *Measures for progress* (1966), pp. 231, 559.

[2] Melville, op. cit., pp. 113–14.

[3] Dupree, A. Hunter, *Science in the Federal Government* (1957), p. 176.

[4] Cameron, F. *Cottrell, Samaritan of science* (1952).

[5] Removal of Certain Excise Restrictions on Distillation and of Allowances on Industrial Alcohol and Exported Spirits (May Committee, 1945), Cmd 6622, p. 6.

delegated the administration to a producers' association, the Zentrale für Spiritusverwaltung, which handled about 80 per cent of the industrial spirit made, and so managed the business that the representatives of the chemical industry approved its policy of stable prices. Surpluses were sold abroad for what they would fetch![1] After the prohibition of saccharine manufacture in Germany, the Reich Ministry of Finance had to enforce the legislation and supervise the activities of the single licensed manufacturer, Fahlberg, List & Co., which was allowed to make saccharine for medicinal purposes. This German law had been passed in 1902 at the demand of the sugar-beet interests, and necessitated the payment of 4 million M (£0·2 million) compensation for the works that had been closed. An incidental and not insignificant consequence was the establishment of saccharine manufacture in Switzerland and the United States.[2]

Generally speaking, however, the influence of specific fiscal and protective legislation of the kind described and of Government scientific departments on the development of the chemical industry was small before the war. Nor is this really surprising when we consider the financial effort involved. It has been estimated that the United States Government's expenditure on research and development in 1900 was $4 million, or 1 per cent of Government expenditure as a whole. By 1910 it had risen to $16 million (say £3·3 million) or 2 per cent of total expenditure.[3] Corresponding figures have not been published for Britain and Germany, but it may be doubted whether the amounts were proportionately as high as the American.

The war transformed the situation. Government and industry increased their research efforts, and where previously there had been dearth, there was now superabundance. But this too caused problems, for official and quasi-official organizations multiplied, and soon their efforts overlapped. It was therefore necessary to co-ordinate the various research and development efforts. There was some talk in the United States of putting all these activities under one department, but, as in 1886, when a Department of Science had first been mooted, this solution was rejected.[4] Such a Department might well have been of great benefit, but the measures that were taken served their purpose well, and above all fitted in with the empirical and haphazard approach to official science.

In France and Germany, Government co-ordination of the scientific efforts was exercised indirectly during the war. In France the Académie des Sciences created a section for the applied sciences, and the Minister of Public Instruction set up a panel of scientists to co-ordinate military research work. Both organizations seem to have been of limited use, and it was not until

[1] *Kontradiktorische Verhandlungen über Kartelle, Heft 12: Verbände in der Spiritusindustrie*, (1906), pp. 335 ff.

[2] Vershofen, W. *Wirtschaftsgeschichte der chemisch-pharmazeutischen Industrie* (1958), vol. 3, pp. 107–8.

[3] National Science Foundation, *Scientific personnel resources* (1955), p. 7.

[4] Dupree, op. cit., pp. 230–1.

after the war that the Government intervened decisively by establishing the Office National des Recherches Scientifiques, Industrielles, et Agricoles et des Inventions. The long title describes the omnibus functions of the organization, which was to stimulate and coordinate applied research. Its staff were to serve as the Government's chief scientific advisers.[1]

Germany followed a different route. In that country there was less need to promote scientific research, since the universities and industry had extensive facilities. Moreover, the formation of the Kaiser-Wilhelm-Gesellschaft just before the war[2] with the support of industry and Government, provided the link between science and its application that other countries were trying to achieve. Thus the society could help in the scientific war effort, and its institute of Physical Chemistry and Electrochemistry, directed by Haber, became the centre of chemical warfare research and development. The institutional framework of the Kaiser-Wilhelm-Gesellschaft thus assisted the co-ordination of science during the war. Haber sought to establish even stronger links between the military and scientists by his proposal, in 1916, for the setting up of the Kaiser Wilhelm Stiftung für Kriegstechnische Wissenschaft, for short KWKW. His concept of a special body devoting itself to what came later to be called 'defence research' was original and acceptable. Among the foundation's leading members was the Prussian Minister of War and the then Secretary of State at the Ministry of Education, F. Schmidt-Ott. But the KWKW came too late to make an effective contribution, although its recommendation to establish a research institute for the iron and steel industry was adopted in 1917.[3] With the collapse of the Reich, the KWKW closed down, but the Kaiser-Wilhelm-Gesellschaft maintained the connexion. In December 1918, v. Harnack, its president, pointed out to the Ministry of Economics that continued support of research was essential, and that despite adverse conditions his society was actively considering setting up research institutes for non-ferrous metals, leather, cellulose, and textiles.[4] After some delay, justified by the exceptional circumstances, the Government agreed to help. The combination of State and private enterprise worked well in Germany, partly because of the personalities involved, and partly because the Kaiser-Wilhelm-Gesellschaft provided research facilities valued not only by pure scientists, but also by the Government and the different branches of industry.

In the United States the National Academy of Sciences took the initiative in setting up a coordinating organization, which was formally constituted as the National Research Council (N.R.C.) in September 1916. The Council included representatives of the Army, Navy, Civilian Departments, universities, research foundations, and firms. In March 1917 the President asked

[1] Moureu, C. *La chimie et la guerre* (1920), p. 300.
[2] See Chapter 3, pp. 49–50.
[3] Schmidt-Ott, F. *Erlebtes und Erstrebtes* (1952), pp. 144–5.
[4] Letter from Harnack to the Secretary of State in the Reichswirtschaftsamt dated 28 December 1918, quoted in *50 Jahre Kaiser-Wilhelm-Gesellschaft und Max-Planck-Gesellschaft* (1961), pp. 163–5.

the Council to act during the war as the Department of Science and Research of the Council of National Defense. The N.R.C., though recognized by the Government, was not in receipt of official funds, and its considerable administrative expenses were financed by private benefactions, especially by endowments from the Rockefeller and Carnegie foundations. The N.R.C.'s first task was the compilation of a national census of research and scientific manpower. Its Division of Scientific Work, divided into seven sub-divisions for each of the major disciplines, acted as a clearing house for the wartime scientific activities. Thus through the chemistry sub-division, the N.R.C. supported the production of pharmaceuticals, fine chemicals, poison gases, the exploitation of indigenous potash resources, and also the improvement of British and French military equipment. The N.R.C. maintained liaison with the War Industries Board, which, for good measure, had its own technical advisory staff and used the research facilities of the Mellon Institute. The N.R.C.'s post-war aims included the promotion of cooperative industrial research, but indifference, fear of the Anti-Trust Act, and the lack of cooperation from the big firms frustrated these projects.[1]

The British Department of Scientific and Industrial Research (DSIR) had its origins in an Order-in-Council of July 1915 that created the Committee of the Privy Council for Scientific and Industrial Research. The Committee initially relied on staff seconded from the Board of Education, but in December 1916 it set up a separate Department, the DSIR, which through the Lord President of the Council was responsible to Parliament. From the outset the DSIR was recognized as 'the principal body by means of which the Government seeks to promote and organize scientific research with a view to its application to trade and industry'.[2] In Britain then, official machinery was formally created to establish a scientific link between Government and industry. In this respect the DSIR was entirely novel and differed from the semi-official or non-governmental organizations for promoting scientific research in other countries. Just as significant was the substantial measure of official financial aid: an initial capital grant of £1 million was provided, and the new Department was allowed an annual budget out of public funds. Care was taken to form a strong Advisory Council, which assisted the Lord President in deciding on the soundness of the DSIR projects: this Council had considerable influence, and from the start made it possible for the DSIR to operate independently of other Government departments. It was made responsible for the National Physical Laboratory, and the Geological Survey in 1918, and later created new research establishments; for example, the Chemical Research Laboratory started life in the early 1920s as the Chemistry Coordinating Research Board. The Board relied on university laboratories at first, but in 1925 moved into its own building alongside the

[1] Yerkes, R. M. (Ed.), *The new world of science* (1920). Hunter Dupree, op. cit., pp. 314–17.

[2] Report of the Machinery of Government Committee (Haldane Report) (1918), Cd 9230 p. 29, para. 45.

National Physical Laboratory at Teddington. Since 1928 it has been called the National Chemical Laboratory.[1]

In general the research co-ordinating bodies of World War I were interesting but largely ineffectual forerunners of the elaborate defence research organizations of World War II. Their influence on the chemical industries of the belligerents was, with few exceptions, small. Indeed, during the war chemical research and development work was usually carried out by firms acting on their own initiative or at the behest of a Government department, and instructions from other organizations were less important. In Britain, for instance, dyestuffs research was carried out by British Dyes Ltd., and the work on ammonium nitrate was in the hands of Brunner, Mond. The considerable effort devoted to gas warfare in France was, at the demand of the Direction du Matériel Chimique, handled by university laboratories.[2] This caused administrative problems and also, we may be sure, some waste. But in France, as elsewhere, there was really no alternative, owing to the lack of senior officials with some understanding of applied chemistry. Among the very few exceptions was John Anderson, who had read science in Scotland and Germany before joining the Civil Service. There he was instrumental in directing the research and development facilities of the pharmaceutical manufacturers towards the production of aspirin and Salvarsan. Later he became the Government representative on the Royal Society committee on chemical research for war-time requirements.[3]

The fate of two war-time Government research projects that had an intimate association with chemical manufacture is of interest in this connexion. The British Nitrogen Products Committee, which was mentioned in the previous chapter, set up a laboratory at University College, London, where H. C. Greenwood (who had worked under Haber at Karlsruhe) was in charge of a small group. The Committee recommended the building of a pilot plant, but when the war ended, Greenwood's team had not got beyond a study of gas reactions, and little work had been done on the all-important problems of catalyst fouling and chemical engineering. In 1919, Government support ended and the remaining staff at University College joined Brunner, Mond.[4] But in the United States, the Government continued to finance research on nitrogen production for some time after the war. In March 1919, the Secretary of War renamed a laboratory of the Chemical Warfare Service the Fixed Nitrogen Research Laboratory, and instructed the scientists to help with the development of the Government nitrate works.[5] Two years later the Department of Agriculture took over the laboratory, because it was thought

[1] Melville, op. cit., pp. 26 ff., 116.

[2] Moureu, op. cit., pp. 35–7.

[3] Wheeler-Bennett, Sir John. *John Anderson, Viscount Waverley, 1882–1958* (1962). Anderson's later and most distinguished services to the country are outside the scope of this book. Suffice it to note here his connexion with Civil Defence in 1938 and with the atomic bomb project from 1941 onwards as head of the Scientific Advisory Committee.

[4] *Chem. Trade J.* **61** (1917), 459–60, 483–4; Parke, V. E. *Billingham—The first 10 years* (1957), pp. 4–5.

[5] Haynes, op. cit., vol. 2, p. 116.

that agriculture would be the chief beneficiary. In 1922 Cottrell became director of the laboratory, which employed about 100 people, roughly half being chemists. A considerable amount of applied research was undertaken, notably on catalysts and on the uses of nitrogenous fertilizers. But the success of the laboratory created difficulties with industrialists, who feared that the Government would become a potentially dangerous competitor of the ammonia works then being built by private enterprise. The climate of opinion changed, Government-operated chemical works became politically unfashionable, and Cottrell's best men, attracted by higher salaries, drained away to industry. At the same time there developed a difference of opinion among the staff between those who, like Cottrell, thought that the laboratory should confine itself to fundamental work, and those who wanted to extend the technological side and go into chemical engineering and plant design. In any case the laboratory's appropriations ran out in 1926, work was curtailed, and it finally became part of the Bureau of Chemistry and Soils of the Department of Agriculture. At this point, its work on nitrogen fixation came to an end.[1]

Although this aspect of Government intervention was not very significant during the war and its aftermath, it was recognized on all sides that research coordination could be beneficial. The effects of such intervention and co-ordination on the development of the industry are, in retrospect, hard to discern. Government and industrial laboratories went about their work after the war, but there was little sign of cross-fertilization. Attempts to form industrial research associations, to which the State contributed, were made in several countries, but the arrangements were not extended to chemical manufacture. The larger firms, which played such a dominant role in the industry, had many reasons for not participating in research associations: their own research and know-how were valuable and the great financial success of the various synthetic processes in the 1920s confirmed their attitude. With the big fish so uncooperative, the small fry lost interest.

During the war Governments in all belligerent countries were obliged to intervene directly in the affairs of industry. Intervention had many facets, from production and distribution controls to price regulation, and from financial measures such as grants or loans for building factories, to outright ownership of a few large works. The chemical industry, owing to its key importance to the war effort, was naturally affected.

Production controls were in the long run more significant than other Government measures. Such controls favoured the larger firms, not only because it was more convenient to deal with a few big companies than with many small businesses, but mainly because the larger firms generally had the modern equipment and so could put scarce materials and manpower to better use, and also deliver the goods on time. The operation of these official

[1] Cameron, op. cit., pp. 242 ff.

controls created conditions favourable to mergers and take-overs. Examples abound during the war: in Britain Brunner, Mond acquired a holding in Castner–Kellner and took over Chance & Hunt, in France Kuhlmann and the forerunners of Pechiney bought several firms, in the United States the dye-works combined to form the National Aniline & Chemical Co. In Germany the erstwhile rivals in the dyestuffs industry joined in an *Interessengemeinschaft* in August 1916; that same month also marked the turning point in the German war direction: Hindenburg and Ludendorff proclaimed the necessity for an all-out effort, which inevitably would rely heavily on the large industrial groups.

A by-product of Government intervention was the formation of new trade associations, or the strengthening of those in existence. The associations were useful for purposes of coordination or sometimes even for administering the various control measures. They also enabled chemical manufacturers to meet and discuss common problems, such as the interpretation of regulations, priorities, and quota systems. By extension they concerned themselves with tariffs, labour relations, and taxation. Towards the end of the war firms and State found trade associations of mutual advantage, and there were some who, like Walther Rathenau, considered them the key element in the industrial self-government of the future.

We may begin by looking at the development of the control apparatus in Germany. That country was not notably better prepared for war than the Allies, but the adaptation of the chemical industry to military requirements was accomplished earlier than elsewhere. The Reich Ministry of the Interior had set up a standing committee on industrial mobilization in December 1912, but no attempt had been made to survey the industrial potential or compile a record of stocks. The military anticipated a short war, and intended to replace losses from supplies in state arsenals, supplemented, as necessary, by calls on private enterprise, in particular Krupps. While the significance of nitrates was recognized, no official steps were taken to establish a stockpile.[1] A few days after the outbreak of war Rathenau called on von Falkenhayn, then Prussian Minister of War, and submitted a project for the creation of an office of industrial mobilization. It was a practical and timely proposal, and he was given full powers: the new department, the Kriegsrohstoffabteilung (KRA), started in mid-August 1914 with a staff of three. More were hurriedly recruited to organize a stocktaking of twelve scarce commodities; food and petrol were specifically omitted from the remit. The job was completed in mid-September and was followed by the requisitioning of the commodities. There was no interference with the manufacture of or trade in these products as long as they were destined for military purposes. The number of scarce materials subject to requisitioning orders rose during the war from the original 12 to 100.

[1] Reichsarchiv. *Der Weltkrieg: Kriegsrüstung und Kriegswirtschaft* (1930), vol. 1, pp. 339 395–6.

The staff of the KRA grew rapidly and by the end of 1915 there were 500 at headquarters and several thousand at the regional branch offices. They would not have been enough for the task if it had not been for the direct links between the KRA and firms in different sectors of industry. These links were forged by Rathenau when he set up the Kriegswirtschaftsgesellschaften (KWG). Each was under the chairmanship of a Commissar, appointed by the Reich, who had absolute veto powers. Under him the KWGs had the usual German twin board structure of supervisory (*Aufsichtsrat*) and executive (*Vorstand*) directors; in addition each KWG had an independent commission for preparing estimates and handling allocations. The commissions were headed by men drawn from the Chambers of Commerce or from the Civil Service who, it was expected, would maintain complete impartiality in their handling of these delicate and crucial problems. When Rathenau resigned on 1 April 1915, several KWGs were functioning and the KRA was working effectively as a supervisory body and less well as price-fixing authority.[1] It was also responsible for exploiting industries in occupied countries; the KRA was particularly thorough in Belgium, where chemical manufacturers were stripped of their equipment to such an extent that by October 1918 only five lead-chamber plants were still functioning.[2] The KRA operated throughout the war, but, though successful as a coordinating body and reasonably so as an agency for allocating supplies, it failed to control prices and from 1916 onwards was unable to stem inflation. Agriculture and food remained outside its scope, and food supply and distribution proved to be the weakest points in the entire German war economy.

About the middle of September 1914 the Chemie-Verein, the chief trade association, was put in charge of export licensing of dyes, intermediates, and materials for explosives. A year later all chemical exports became subject to licensing. Some attempts were made to barter chemicals for scarce materials such as oils, fats, and pyrites; in general export prices were well above domestic prices and the margin between them was used to compensate importers for the higher prices they had to pay.[3]

While the KRA had acted quickly in the matter of export licensing, the establishment of a KWG for the chemical industry encountered some hostility. The industry's objections were eventually overcome and the Kriegschemikaliengesellschaft (KCG) was founded on 26 September 1914. It was a non-profit making company, its capital being subscribed by the leading firms. The supervisory board consisted of representatives of the big dyestuffs

[1] Rathenau, W. *Deutschlands Rohstoffversorgung* (1917). Rathenau succeeded his father as chairman of the AEG during 1915. In the opening months of 1922 he was for a short time German foreign minister and negotiated the Rapallo agreement with the Russians. In June of that year he was murdered by German right-wing extremists. See also Count Harry Kessler. *Walther Rathenau* (1929).

[2] Denterghem, C. de Kerchoven de. *L'Industrie belge pendant l'occupation allemande* (1927), pp. 138–9.

[3] Ungewitter, C. (Ed.) *Ausgewählte Kapitel aus der chemisch-industriellen Wirtschaftspolitik* (1927), pp. 323–6.

makers and of the explosives manufacturers; Emil Fischer sat on the board as scientific adviser.[1] The Reich commissar on the KCG was H. Schmitz, who was also head of the chemical division of the KRA.[2]

Throughout the war the KCG handled, on behalf of the chemical division of the KRA, the supply of raw materials for chemical production, and supervised the manufacture and distribution of all chemicals required by the government. It was also made responsible for import licensing. The KCG did not build plants or finance extensions, and left this to the KRA or the Reich Ministry of Finance; the exception to the rule appears to have been Protol GmbH, which was established to make glycerine by a fermentation process. The KCG remained quite a small organization and early in 1916 employed only about 230 people, over half of them women who attended to the clerical work.[3] The executive staff at headquarters was on secondment from industry and divided the work along product lines (acids, nitrogen, etc.) or general supervision (administration, imports, etc.). The product sections concerned themselves with the implementation of the KRA's requisitioning orders. By June 1915 these covered explosives, nitrates (artificial and other), calcium cyanamide, ammonium nitrate, nitric acid, toluol, toluene, nitro-toluene, camphor, glycerine, sulphur and pyrites, sulphuric acid, oleum, and chlorine, as well as most compounds and intermediates derived therefrom.[4] A few months later the products were combined into seven main groups each under a Controller (Vertrauensmann). There were no additions to this list until July 1917 when a group was set up for hydrochloric acid.[5]

The KRA and the KCG never mastered the problem of price control.[6] For example, the price of sulphuric acid remained uncontrolled until the end of 1915, when the cost of the most expensive process was used to set the maximum price. Companies using cheaper processes were not allowed to profiteer and had to pay the difference between their costs and the maximum price to the KCG, which transferred it to the Reich. The Ministry of Finance then used the money to amortize factories built with state help to recover acid

[1] Ungewitter, op. cit., p. 334; *Z. angew. Chem.* **27** (1914), pt. 3, 651. Solvay-Werke, being Belgian-owned, did not subscribe to the KCG, nor did the potash mines, the pharmaceutical firms, Kunheim and Rhenania.

[2] Schmitz had been working in the audit department of the Metallgesellschaft before the war. In 1917 he left the KRA to become a director of Badische. Later he became chairman of the *Vorstand* of IG Farben. Schmitz was a defendant at the IG Nuremberg trial in 1947/8 and was sentenced to four years' imprisonment. Some of the other officials of the KRA and KCG also attained senior positions in IG Farben after its formation.

[3] The bulk of the enormous paper work was handled by the chemicals division of the KRA; towards the end of 1915 it issued 15 000 licences and demanded 5000 statistical returns every month. Ilgner, M. Rohstoffversorgung der deutschen Schwefelsäureindustrie, Ph.D. thesis (1923), pp. 78, 81. Ilgner was Schmitz' nephew and therefore well informed. Later he became director of intelligence of IG Farben.

[4] Kriegsministerium. KRA, *Bewirtschaftung von kriegswichtigen Stoffen 1914–18*, vol. 5 in a series of 12 volumes at the London School of Economics, no page.

[5] ibid.

[6] As far as I have been able to discover sodium nitrate was the first product subject to price control. A circular dated 3 March 1915 fixed the price at 240 M ton *ex port*, which may be compared with 180–190 M ton c.i.f. in 1914.

or make it from anhydrite. To add to the complication those manufacturers of acid that used low-cost processes to supply explosives and fertilizer factories were exempt from the regulations.[1] One can imagine the paperwork and the fantastic administrative complexities! It must have been difficult to operate efficiently the system, which had the added drawback of creating two prices for the same product: a low price for deliveries to munitions and fertilizer factories, and a high price for sales to non-essential users. The latter were not rationed or controlled, so that the price of the limited supplies available for these purposes soared. Manufacturers of non-essential products who required sulphuric acid would naturally look for cheaper substitutes. This happened in the metal treatment trade, where firms replaced sulphuric by hydrochloric acid to such an extent that supplies of the latter had to be controlled in 1917.

As the war dragged on, so the pressure mounted. On 11 January 1916 price control for fertilizers was introduced to placate the farmers. But no official mechanism for distribution was established. The KRA became involved because it was responsible for nitrogen compounds, and the Commissar in charge of this product group, Bueb,[2] had the job of dividing supplies between the military and the farmers. There was not enough to go round for both, and soon there was trouble: in the early months of 1917 some municipal gas works began to barter sulphate of ammonia for food. Officialdom reacted by introducing controls for sulphate of ammonia on 18 May 1917. The system of distribution remained in the hands of merchants and agricultural cooperatives, but the Commissar decreed that farmers who had used sulphate before the war would continue to obtain it, while those who had manured with nitrates would henceforth have to rely on cyanamide.[3] This was a pretty rough solution to an important problem, and, combined with the acute shortage of all nitrogenous fertilizers during 1917–18, led inexorably to grey markets in fertilizers, failures of distribution, dissatisfied farmers, and lower crop yields.

The Germans were first with a plan for operating the chemical industry in war-time. In principle they relied on central direction and coordination, with close industrial involvement at lower levels. Although the manufacturers were predisposed to collaboration and accepted the autocratic direction of the Commissars and their staffs, the elaborate organization was only partly successful. The weaknesses were in the first place irremediable shortages and in the second inefficient price controls. Similar problems arose in France and Russia. In Britain and the United States, however, supplies were less of a difficulty than in Germany, and for the first eighteen months of the war the main problem in Britain was the lack of appropriate production capacity. The division of responsibility between different Government departments was also hampering the war effort. The Board of Trade had been the Department

[1] Ilgner, op. cit., pp. 81–2.

[2] He later became a director of Badische.

[3] Kriegsernährungsamt. *Düngemittel im Kriege* (1917), pp. 46 ff.

in closest touch with industry before the war, and continued in that position in the opening months of hostilities. In addition various special committees were set up to deal with the supply of munitions and other war stores, but by the spring of 1915 these *ad hoc* procedures no longer sufficed and were replaced by the Ministry of Munitions, created by the Act of 9 June 1915. Munitions were defined as 'anything required to be provided for war purposes, and include arms, ammunition, warlike stores or material, and anything required for equipment or transport purposes or for, or in connexion with, the production of munitions'.[1] The scope and the inherent powers were greater than the German KRA, and, under a succession of extremely vigorous ministers,[2] the new Department played a decisive and successful role in the industrial war effort. The administrative structure was complex: by the second half of 1916, the Ministry had fifteen departments, and in 1917 aeroplanes, agricultural machinery, sulphuric acid, and fuel oil were added to its responsibilities. Later in that year the Ministry of Munitions was divided into ten Groups (or Councils), each with several Departments. The Headquarters staff grew from less than 200 in mid-1915 to over 12 000 two years later, and there were a further 40 000 in the rest of the United Kingdom and 8 000 in the United States.[3]

Some sections of the chemical industry came under the control of the Ministry of Munitions, which had taken over the Explosives Supply Department, which was descended from the High Explosives Committee established soon after the outbreak of war. J. F. Moulton[4] headed that Committee, and later became Director General of Group X in the Ministry, where he was in overall control of the manufacture and distribution of acids and coal-tar derivatives used in the manufacture of munitions. The Ministry of Munitions' machinery served admirably for supplying the ordnance factories, but created major difficulties for the parallel organization of the dyestuffs industry. These problems did not arise in Germany, where the production of dyes was cut back, and the personnel moved to the manufacture of explosives.[5] But in Britain the output of dyestuffs and explosives had to be raised simultaneously, and the enormous effort devoted to the latter was detrimental to the activities of British Dyes Ltd. and Levinstein Ltd. They and other makers of dyes remained nominally under the control of the Board of Trade, which had formed a Chemical Products Committee under Lord Haldane in August 1914 to advise on the best means of obtaining supplies of dyes and chemicals.

[1] 5 and 6 Geo. 5, Ch. 54.

[2] Lloyd George was Minister from June 1915 to June 1916. He was followed in turn by E. S. Montagu, Dr. C. Addison, and then by Winston Churchill in July 1917.

[3] Fairlie, J. A. *British war administration* (1919), pp. 106–7.

[4] Later Lord Moulton of Banks, 1844–1921. He had been a senior wrangler at Cambridge (1868) and became the leading patent attorney of his time.

[5] For instance Weiler–ter Meer's factory at Uerdingen was adapted to make TNT and phosgene. The Bayer plant at Leverkusen did not lend itself to such a conversion, and so a new works, solely for the manufacture of explosives, was built diagonally across the Rhine at Dormagen. It was operated by Bayer, and is still one of its main factories.

Later analogous committees were appointed to deal with specific products such as sulphate of ammonia, sulphuric acid and fertilizers, acetylene, etc.[1] These committees overlapped at certain points with the departments of the Ministry of Munitions, and certainly did not have the power or influence of that Ministry or of the German KCGs or their sections. Unlike Germany, the absence of an effective trade association link between Government departments and industry proved a serious handicap.[2] In addition to these committees, there was a special organization for dyestuffs headed by the Commissioner of Dyes, who was responsible to the Board of Trade. This official's authority (as distinct from his influence) was limited, and so he was unable to control effectively the activities of the leading manufacturers.

From August 1914 to the middle of 1916 the leading American explosives manufacturers, above all Du Pont, enlarged their works chiefly to meet Allied orders, and there was a free-for-all to obtain manpower and raw materials, with the result that prices soared. The creation of the Council of National Defense late in August 1916 was the first official step towards industrial mobilization. The Council had an Advisory Commission on which Bernard Baruch[3] sat as expert for raw materials. At his instigation, industry committees to advise on projects and coordinate their execution were generally adopted. The Committee on Chemicals was established in March 1917. Its chairman was W. H. Nichols of the General Chemical Co., and he was assisted by representatives from several important chemical manufacturers,[4] two Civil Servants, and a scientist from the National Research Council. The Committee spawned sub-committees for acids, alkalis, coal-tar by-products, synthetic phenol, electrochemicals, fertilizers, pyrites, and miscellaneous products. The subcommittees were primarily coordinating bodies, but they were entitled to allocate whenever the materials in question were subject to requisitioning orders, and also to impose priorities. They did not have price-fixing powers.[5]

In December 1917 the Committee on Chemicals ceased to be a branch of the Advisory Commission, but enlarged the scope of its activities by transferring to the Chemical Alliance Inc., a non-profit making company formed in July 1917 to handle the distribution of imported pyrites.[6] From these modest beginnings the Chemical Alliance developed into an industrial organization, divided into product groups, which for the remainder of the war linked the Chemical Divisions of the War Industries Board with the

[1] Fairlie, op. cit., pp. 179–80.

[2] In 1917 the Ministry of Reconstruction set up a standing committee on which, for the first time, the Association of British Chemical Manufacturers (now the Chemical Industries Association) founded in 1916, was represented.

[3] He was a financier, his special interest being mining ventures. In March 1918 Baruch became chairman of the War Industries Board. See Haynes, op. cit., vol. 2, pp. 42 ff.

[4] Excluding, however, Du Pont, Hercules and Atlas Powder, the largest firms in the explosives trade.

[5] The Chemical Alliance. *Historical Review . . . of the Chemical Alliance* (1920), p. 13.

[6] Pyrites and brimstone were transferred to the War Industries Board in 1918.

manufacturers. Thus the Chemical Alliance resembled the KCG in Germany, and it was broader in scope as well as more powerful than the various committees formed by the Board of Trade. The Alliance included the fertilizer industry, and so was able to help the farmers, which the KRA and the KCG in Germany signally failed to do. The War Industries Board was dissolved in January 1919, but the Chemical Alliance continued a little longer helping Government departments with the disposal of surplus stores and negotiating with the Tariff Commission. It was wound up in December 1919.[1]

The other belligerents, so far as I know, adopted with variations the methods used by the Germans, British, and Americans. In France and Russia the administrators had the additional problem that a large portion of chemical manufacturing capacity was located in enemy occupied territory.[2] In both countries major, and on the whole successful, efforts were made to re-establish production elsewhere. In France the Office des Produits Chimiques et Pharmaceutiques, a government agency established in October 1914, played a major role in this context. It was also responsible for allocating intermediates to the Soc. des Matières Colorantes de St. Denis and the CIBA factory at Lyon, attempted to restart the sequestrated German dye factories, and actively supported the formation of the Cie Nationale de Matières Colorantes & de Produits Chimiques.[3]

The Russian general staff began by commandeering all supplies essential to their needs. The resulting chaos took some time to sort out, but eventually the Chemical Committee of the Chief Artillery Administration emerged as the principal body responsible for the supply of explosives. Later more committees were formed to deal with other chemical activities. There was little coordination between them and it was only in 1916 that the Minister of War created the War Chemical Committee, which, under the chairmanship of Ipatieff, was put in charge of the supply of explosives, poison gas, gas masks, incendiaries, and inorganic acids. For a few months the Committee was able to work effectively, but growing transport difficulties during the winter of 1916/17 impaired its usefulness. It was not abolished at the Revolution and continued to operate until 1 June 1918.[4]

The creation of committees or other official bodies, directed by Civil Servants and representatives of industry to coordinate and control the affairs of the chemical industry, represented state intervention in its broadest sense. It was supplemented by other forms of intervention designed to achieve specific objectives, e.g. the development of dyestuffs manufacture or the

[1] Chemical Alliance. op. cit., pp. 17, 74 ff.

[2] The French estimated that 21 per cent of the employees in the pharmaceutical industry, 20 per cent in acid production, and almost 18 per cent in dyestuffs had been working in factories in *Départements* occupied by the Germans from September 1914 onwards. Fontaine, A. *L'Industrie française pendant la guerre* (1925), p. 23.

[3] Fauque, op. cit., p. 125; Matagrin, op. cit., pp. 56–7.

[4] Ipatieff, V. N. *The life of a chemist* (1946), pp. 216 ff. Ipatieff, 1867–1952, an outstanding worker in the field of high-pressure synthesis, played an important role in the Russian scientific establishment until the late 1920s when he emigrated to the United States.

building of an ammonia plant. These special measures generally had lasting effects; one of the most significant was taken in Britain and merits closer attention.

We saw in the last chapter that soon after the outbreak of war the supply of dyestuffs quickly became critical. The Government was well advised and the President of the Board of Trade was aware that the situation called for something more than a loan or a grant-in-aid. HMG intended to create a British dyestuffs industry and support its development after the war. Unfortunately personal antagonisms and the overriding demands of the Ministry of Munitions created unforeseen difficulties for British Dyes Ltd., the Government's chosen instrument. Formed in Spring 1915, it was capitalized two years later at £2 084 138 of which £1·7 million was in 4 per cent Debentures all held by the Treasury and £384 138 in ordinary shares held by the public. Controversy surrounded the company from the outset, and the two Government-appointed directors were unable to persuade their colleagues to manage the business so as to bring about the desired objectives. Towards the latter half of 1917 Whitehall became convinced (rightly, albeit belatedly) that a viable dyestuffs industry would not emerge until British Dyes was combined with their bitter rivals Levinstein Ltd. On 2 February 1918, Sir Albert Stanley,[1] then President of the Board of Trade, called a meeting of the two companies and told them that 'complete co-operation' between them was of 'vital importance'. It is doubtful whether the import of his remarks was fully appreciated. Nevertheless G. P. Norton and J. Turner on behalf of British Dyes and Mr. (later Lord) Harry McGowan and H. Levinstein on behalf of Levinstein Ltd. formed a committee to work out the details.[2] But little progress was made, and in May 1918 Stanley announced in the House of Commons proposals designed to help with research and extensions to existing plant or new buildings. As a further safeguard, imports would be subject to licensing for at least ten years after the war. He also recommended the merger of British Dyes and Levinstein as a desirable step in keeping with the Government's wishes. The following month at a meeting of British Dyes some directors warned that Levinsteins were forcing the pace after months of negotiations, but others hoped to prevent the merger by the formation of a community of interests on the model of the German dyestuffs industry. Their plan for various joint committees and profit pooling was defeated in August 1918, though a sizeable minority of the shareholders (over 25 per cent of the votes cast) supported the scheme. Sir James Falconer and some of his diehard colleagues thereupon resigned, and the remaining directors could now go ahead with the merger.[3] In November 1918 the British Dyestuffs Corporation was formally constituted and by the following June its share capital had been raised to £10 million. HMG's £1·7 million Debentures in British Dyes Ltd.

[1] Later Lord Ashfield, the founder of London Transport, and also a director of I.C.I. in the late 1920s.

[2] *Chem. Trade J.* **62** (1918), no. 1621, 475.

[3] *Chem. Trade J.* **62** (1918), no. 1622, 493; **63** (1918), no. 1630, 124–5; (1918), no. 1631, 137

were repaid, and the Treasury then took up 850 000 £1 7 per cent preference shares and 850 000 £1 8 per cent preferred ordinary shares. The Government thus held 17 per cent of the capital of the new company, compared with about 81 per cent of the combined loan and share capital of British Dyes. It could (and did) nominate two directors who served part time. They had special voting powers enabling them to prevent the Corporation from charging unreasonable prices, making chemicals for sale other than dyestuffs or signing contracts with foreign manufacturers.[1] Lord Moulton, who knew his way about the Corridors of Power and was familiar with the chemical industry, became the first chairman.

In November 1918 a Memorandum by the Board of Trade enlarged on the proposals announced by Stanley the previous May.[2] The Memorandum marks an important stage in the Government's relations with industry: in return for helping the British dyestuffs makers, the Government insisted on a large measure of intervention in the business of the recipients. The help offered included a scheme for twenty-year loans of up to 50 per cent of the net assets of the applicant company, to make dyes not hitherto produced in the United Kingdom. In addition there were repayable grants of up to 40 per cent of the cost of new buildings and plant; such grants also covered up to 40 per cent of the annual cost of research. Until the loans and grants had been repaid, dividends were limited to 8 per cent. In return the beneficiaries undertook to supply dyes at 'reasonable prices' (the Board of Trade being empowered to fix prices if necessary), the products were to be distributed in an 'equitable manner' (here again the Board of Trade could intervene), foreign shareholdings were limited to one-quarter of the voting power, and agreements with foreign manufacturers had to be approved by the Board of Trade. The Dyes Commissioner, whom we have already encountered, administered the scheme and was responsible to the Board of Trade for it. He was assisted by a Loans and Grants Allocation Committee and by a Trades and Licensing Committee. The latter consisted of four manufacturers and four users, under an independent chairman; it had a licensing sub-committee, which determined what types of colours and intermediates should be licensed for import, and in what amounts. It had the subsidiary function of advising the Commissioner which dyes he should recommend manufacturers to produce indigenously.[3] The main Committee was also to arrange for the collection of statistics on

[1] PRO, CAB/27/232. Memorandum by Government Directors of British Dyestuffs Corporation, circulated to Cabinet on 9 April 1924.

[2] Scheme for the Allocation and Administration of the Funds Provided by Parliament for the Development of the Dye Industry by Means of Financial Assistance to Companies and Firms in Aid of Developments, Extensions and Research, HMSO, (1918) Cd 9194.

[3] The Committee at its first meeting comprised H. Allen, M. S. Sharp, and Lennox B. Lee from the Colour Users' Committee, T. Taylor for the Paint and Varnish Manufacturers' Association, H. Levinstein and J. Turner for the British Dyestuffs Corporation, W. J. U. Woolcock for the Association of British Chemical Manufacturers (ABCM), W. H. Dawson of the Board of Trade, and Sir Evans Jones M.P., the Commissioner. At first the Board of Trade had omitted the ABCM, and Woolcock's inclusion was a belated acknowledgement of the role that chemical manufacturers generally could play. *Nature, Lond.* **102** (1918), no. 2562, 272–3; **103** (1919), no. 2568, 388.

production, scientific and technical manpower, and laboratories. The Committee was empowered to recommend restricting the output of those dyes that were being manufactured on a sufficient scale, in order to release resources for the production of new dyes. The reports of the Trades and Licensing Committee were to be the basis on which the Loans and Grants Allocation Committee would review the applications for assistance. Parliament accepted the Memorandum and voted £2 million for the development of the industry. Only £435 000 had been spent by 1922, when the scheme came to an end.[1]

The period of active financial participation lasted from 1915 to 1922, and Government money and pressure were responsible for a creditable achievement: the creation of a British dyestuffs industry. Admittedly it was not yet viable in 1922, and would certainly need continuing protection. The British Dyestuffs Corporation (with which HMG remained connected until 1925), though unprofitable, had the capacity and could form the nucleus of a larger group. The Government's intervention may have been clumsy at times, and the machinery was certainly cumbersome and often unpopular. But without official backing, dyestuffs manufacture would have withered away in the early post-war years.[2]

Outright state ownership of chemical works was rare, because the prospect of the Government operating alongside private enterprise in the same markets was unacceptable during the war, and even more so after it. But the novelty, scale, and risk element in nitrogen fixation were considered so exceptional that in this particular sector of chemical manufacture Government ownership often continued well beyond 1918. In Britain, for example, the synthetic ammonia project was conceived from the outset as a 'national effort', and was to be operated by the Ministry of Munitions. The Government plant at Billingham was to have a capacity of 60 000 tons a year of ammonium nitrate, but in January 1919 the project was drastically altered to 10 000 tons a year of sulphate of ammonia. Building of roads, drains, and some sheds began in January and February, but a few months later the climate of opinion moved decisively against Government operation of the mines and railways, and this resulted in a change of mind over the ammonia works. Accordingly Billingham was sold to Brunner, Mond & Co. early in 1920.[3] Abroad Government ownership was more enduring. Two large cyanamide plants had been built during the war by Bayerische Stickstoffwerke on behalf of the Reich: the one

[1] Committee on Industry and Trade. *Factors in industrial and commercial efficiency* (1927), pt. 1, pp. 415 ff.

[2] I have found only one other similar instance of extensive Government involvement in a British chemical enterprise. This was the large shareholding in British Cellulose & Chemical Mfg (later British Celanese, now part of Courtaulds). HMG's investment was criticized on the ground that it was helping a company that could become a rival to Courtaulds. This proved indeed to be the case some years later. HMG liquidated its holding and withdrew its directors in the early 1920s.

[3] *Chem. Trade J.* **66** (1920), no. 1720, 590–1; Parke, op. cit., pp. 5 ff. It may be noted that under the Trade Facilities Act, Synthetic Ammonia & Nitrates, the Brunner, Mond subsidiary in charge of the Billingham project, obtained a Government guarantee for a £2 million loan.

at Chorzow became Polish after the war, while the other at Piesteritz was operated by the Government until 1926, when it was leased to Bayerische Stickstoffwerke.[1] Later the company acquired a controlling interest in the works. In the United States the nitrate plants at Muscle Shoals were Federal property and after the war the Government proposed to complete and operate them for the manufacture of fertilizer nitrogen. The project did not come to fruition until many years later, when the Tennessee Valley Authority was created. The French Government, mindful of the essential role of nitrogen, intended to build a Haber–Bosch plant after the war. But the agreement with Badische was not implemented until Parliament passed the act setting up the Office National de l'Azote (later renamed ONIA and more recently changed again to Entreprise Chimique & Minière). Building began at Toulouse the following year and production started in 1927.[2] This works, now one of the country's largest chemical undertakings, remains state-owned and is jointly administered with the potash mines by the Entreprise Chimique & Minière.[3]

Official loans or special subsidies to chemical manufacturers were the most usual and least criticized manifestations of Government intervention. There were numerous instances. One of the most important was the credit of 260 million M by the Reich to Badische to assist with the building of synthetic ammonia plants. In return Badische undertook to supply nitrogen during the war at a fixed price.[4] The inflation reduced the company's liability to insignificant proportions. Another example is provided by Kuhlmann in France, which built, with public funds, a sulphuric acid plant at Port de Bouc and then leased it from the Government for thirty years.[5] Here again the depreciation of the franc was to make this a financially attractive project. Another method was used by the Japanese, where the Government guaranteed the dye makers a subsidy adequate to enable them to pay a dividend of 8 per cent a year. This scheme was introduced in 1915, but was replaced in 1925 by a straight subsidy of 1 million yen (say £100 000) a year, for four years.[6]

Whatever the form of state assistance, there is no doubt that without it, private capital would not have been forthcoming to an adequate extent. As a result the scale of the projects would have been smaller and the effect less significant. In the last resort then, the taxpayers' money was used at certain critical points to provide enough leverage to modify the structure of the

[1] *Enquete*, p. 205.

[2] Bricard, S. L'Industrie de l'azote dans l'économie française. *Chim. Ind.* **68** (1952), no. 6, 938–9.

[3] The German potash mines in France were sequestrated by the French in 1919. A proposal to sell them was rejected by the Senate in 1920, and four years later they were nationalized and until recently traded as Mines Domaniales. Gürber, H. Die elsässische Kaliindustrie und das Kaliweltkartell, Ph.D. thesis (1942).

[4] RM 260 million (£13 million at the pre-war rate) was the amount shown in the balance sheet on 31 December 1918 as advances from the Government. Waller (*Probleme der deutschen Chemischen Industrie* (1928), pp. 43 and 172) mentions 180 million without, however, giving the date. This may have been the amount lent at the end of 1917.

[5] Anon. *Les Etablissements Kuhlmann* (1925), pp. 70–1.

[6] USTC, *Census of Dyes* . . . 1924 (TIS 33) (1925), p. 158.

countries' chemical industries. If this seems fanciful, the reader is invited to imagine in the following chapters the position in Germany without Badische's lead in nitrogen fixation, or in Britain without the British Dyestuffs Corporation.

Between 1914 and 1918 the chemical industries of all the main countries had been greatly enlarged, and in particular the manufacture of dyestuffs and other organic compounds had been developed outside Germany and Switzerland. Now, in order to protect their 'infant industries' most countries adopted swingeing tariffs or introduced restrictions on imports. These measures profoundly influenced the flow of foreign trade and also the structure and development of the indigenous industry. A rough measure of the pattern and changes in some of the tariffs is given in the following table,

TABLE 8.1

Tariffs on selected chemicals in various countries, 1913 and 1926[1]
Maximum and minimum tariff rates in per cent ad valorem

		Germany	*Belgium*		*France*		*Italy*		*U.S.A.*	*Japan*
Indigo 20%	1913	nil	nil		64	36	nil		80	20
	1926	nil	15	0	161	40	100	53	99	112
Sulphur black	1913	nil	nil		123	62	nil		30	15
	1926	nil	15	0	271	68	59		95	201
Average for	1913	nil	nil		85	44	nil		20	14
aniline dyes	1926	nil	15	5	165	41	75	59	86	137
Sulphate of	1913	nil	nil		18	0	nil		nil	nil
ammonia	1926	nil	nil		15	0	15	4	11	0
Calcium cyanamide	1913	nil	nil		0	0	nil		nil	nil
	1926	nil	10	0	0	0	14		nil	nil
Bleaching powder	1913	13	nil		53	35	40		12	20
	1926	15	6	2	121	30	24		21	20
Soda ash	1913	17	nil		55	36	4		13	14
	1926	18	2	0·6	43	11	26		21	11

Note: the conversion of specific to *ad valorem* rates is approximate.

which was laboriously compiled by the German delegation to the International Economic Conference. Some of their tariff calculations were challenged, and it is a pity that their survey was restricted to twenty-one products that omitted the then rapidly developing group of solvents and plastics. But the broad conclusions were not questioned at the time and were undoubtedly correct: Germany and Belgium remained low tariff countries. The others had increased their tariff barriers yet further, and the figures in the above table illustrated the determination with which the French, the Italians, the Americans, and the Japanese were going to protect their dyestuffs

[1] League of Nations (Econ. and Financial Section), *The Chemical Industry* (1927), pp. 54 ff.

industries against all comers. The French for example revised their tariffs on dyes in 1919 and at the same time introduced import licensing for German dyes, initially for five years, but later extended to 1928. In that year the duties were raised again, and the rates on dyes that were also made in France were pitched much higher than the rates on those that were not made in the country.[1]

The United States and Britain had elaborate measures of protection, which, in view of the size of their industry, require more detailed examination. The United States Emergency Tariff Act of 1916 added to the rates of 15 per cent *ad valorem* on intermediates and 30 per cent on dyes, specific duties of $2\frac{1}{2}$ cents/lb and 5 cents/lb respectively. (Alizarin, indigo, and dyestuffs derived from anthracene remained duty free.) The Emergency Tariff was to remain in force for five years, and was then to be cut by 20 per cent annually until eliminated. A Tariff Commission was established to collect statistics and commercial intelligence, and also to administer a rather complicated provision of the Act, that after five years firms would forfeit protection on all those products of which they failed to make at least 60 per cent of domestic requirements.[2]

The Tariff Commission gradually organized the flow of statistics, which continues to the present day and is an unequalled source of information on the organic chemicals industry. When the war ended it had accumulated much data of value to the users as well as to the manufacturers of dyes in support of their conflicting demands. In June and July 1919 the House of Representatives held meetings to discuss the future of protection. The hearings take up some 700 pages and contain some fine specimens of political and industrial special pleading.[3] The representatives of the cotton industry favoured import licensing for dyes. The Chemical Foundation and the American Dyes Institute, which represented most of the makers, agreed with the spokesmen for cotton. The merchants and the wool industry opposed licensing, but supported tariff action. Republicans on the Committee felt that licensing would entail Government intervention in business, since the Tariff Commission would presumably have to adjudicate on United States and foreign prices before a licence was granted. And so the discussion went on. The bill, HR 6495, went to the Senate, where it was shelved. Time passed, and in May 1920 another Emergency Tariff Act had to be passed to tide over the situation until the whole question of tariffs had been reviewed. The outcome of the review was embodied in the Fordney–McCumber tariff of 1922. Its rates were much higher than those of the 1913 tariff. Protection was extended from dyes to the entire organic chemicals industry by specific as well as *ad valorem* duties. Indigo, alizarin, and the anthracene derivatives (the famous fast vat dyes), which had previously escaped the net, were now made dutiable, and the tariff classification was modernized.[4] Licensing was rejected, because

[1] Fauque, op. cit., pp. 235 ff.

[2] Haynes, op. cit., vol. 2, pp. 250 ff.

[3] Committee on Ways and Means on H.R. 2706 and 6495: *Dyestuffs* (66th Congress, 1st session, no. 7592), 1920.

[4] Haynes, op. cit., vol. 3, pp. 266–8.

it was argued that it could not operate without a considerable amount of intervention by officials in the business of makers and of users, and this the *laissez-faire* outlook of the United States in the 1920s rendered totally unacceptable.

The new duties on dyes and intermediates consisted of a specific rate, 7 cents/lb, and *ad valorem* rates ranging from 55 to 60 per cent. The Fordney–McCumber Act stated that after two years the *ad valorem* rates were to be reduced to 40–45 per cent. Such rates were stiff enough, but additional provisions made them virtually prohibitive. In the first place the *ad valorem* rate was based on the American selling price, and if there was no such price, on the estimated United States value. (Customs staff had to ascertain or determine the value.) Secondly, the President was empowered, on the advice of the Tariff Commission, to raise or lower the rates by 50 per cent. The provisions, when applied, in effect doubled or even trebled the basic rate of duty.[1] The tariff had the desired effect: imports of synthetic indigo and of alizarin fell to negligible levels within a few years, and the quantity of dyestuffs of all sorts that were still imported was maintained at about 2000 tons a year. The effect of the Fordney–McCumber Act may well have been to make dyes more expensive for the American than for the British user, but as the American textile industry was less export-oriented than the British, the effect of higher prices was confined to production for the home market. The scope of the Act was very wide, and certainly benefited the entire organic chemicals section of the industry, and so contributed to its rapid development in the inter-war years. One further point needs to be made: the American selling price method emerged forty years later as one of the principal non-tariff barriers during the Kennedy round of the Gatt negotiations in 1964–7, and proved to be a big obstacle to progress. It is a striking testimony to the longevity of protectionist attitudes that even in 1968 industry spokesmen appealed to the Ways and Means Committee of the House of Representatives to maintain this form of protection, though the need for it was not apparent, and the products in question had for long represented a negligible proportion of United States output.[2]

Britain adopted a system combining import licensing with tariffs, which marked a profound break with the policy of complete Free Trade maintained until 1915.[3] When the formation of British Dyes was discussed in 1915, M.P.s asked about protection after the war. Their attitudes reflected, in the main, the views of the parties: Conservatives thought that protection was essential, Liberals were of the opposite opinion. But the latter were in some quandary because they could not explain satisfactorily how British industry

[1] Hansen, K. Die Kennedy-Runde vor ihrer Bewährungsprobe. *Chem. Ind., Düsseld.* **16** (1964), no. 12, p. 837. By mid-1970 this obstacle had not yet been removed.

[2] *European Chemical News*, 12 July 1968, p. 6.

[3] There were some revenue duties on a few imported articles of food, drink, and tobacco. In all cases where these goods were also produced in the United Kingdom they were balanced by an excise duty. The element of protection in these duties was insignificant.

could withstand that of Germany without some protection. Accordingly some Liberals, among them Sir Alfred Mond (who was familiar with the chemical industry), while opposed to tariffs, recommended the adoption of a subsidy. But Government spokesmen, from the President of the Board of Trade down, were careful not to give any pledges about duties or subsidies after the war.[1] Later in 1915 the Chancellor of the Exchequer, Reginald McKenna, introduced a duty of $33\frac{1}{3}$ per cent *ad valorem* on motor cars, motor bicycles, and tricycles and their accessories, musical instruments, clocks and watches, and a specific duty on cinematographic film. The duties, it was said, were not for protective purposes, but to raise revenue and check consumption. Be that as it may, they remained in force after the war (with a brief suspension in 1924–5) and were later extended to commercial vehicles.[2]

Once Free Trade had been breached, some form of protection for the so-called 'essential industries' became acceptable. For example at the beginning of 1916, M. S. Sharp, the chairman of the Bradford Dyers' Association, though declaring himself to be a Free Trader and opposed to tariffs, strongly supported a group of measures that included a Government subsidy to dye makers, steps to stop dumping by Germans after the war, and some form of support for the Swiss manufacturers of colouring materials. He was accused of being impractical and told that any move to stop dumping would maintain high prices in Britain. The *Chemical Trade Journal*, which throughout the long debate supported the views of the chemical industry, in commenting on Sharp's remarks, suggested that tariffs as well as import restrictions were necessary to safeguard British dyemakers.[3] In April 1916 a committee representing many sectors of industry was appointed to consider the position of the textile trades after the war, with special reference to international competition. The Committee completed its work in May 1917, and though the report was not published until 1918, we may infer that officials were aware of it at the time of the British Dyes–Levinstein merger discussions. The Committee reviewed the progress made by the dyestuffs industry and advised HMG to 'coordinate the operations of the principal manufacturers', who should be helped by a subsidy or by loans. But even so, the country would still have to import some dyestuffs, and the Committee members were given to understand that after the war the Government would prohibit the import, except under licence, of the products of 'Key Industries'. The Committee recommended under this head that

> the importation of foreign dyes should be prohibited except under licences to be issued by an independent authority, whose duty it would be, while fostering the growth and maintenance of the colour producing industry, to safeguard the interests

[1] Hansard (House of Commons), **70**, 22 February 1915, cols. 95, 102–9, 110–14; and 11 March 1915, cols. 1682–99.

[2] Committee on Industry and Trade, op. cit., pp. 399–400.

[3] *Chem. Trade J.* **58** (1916), no. 1502, 209–10, 215. As late as 1920 Sharp opposed duties. By then he had also come to reject a subsidy, but instead supported import licensing. *Chem. Trade J.* **66** (1920), no. 1711, 300.

of colour users by means of the control it would exercise over the admission or exclusion of foreign dyestuffs.[1]

The connexion between 'Key Industries' and protection was spelt out in the Final Report of the Committee on Commercial and Industrial Policy after the war. Among its terms of reference was to consider 'what industries are essential to the future safety of the nation, and what steps should be taken to maintain or establish them'.[2] In an interim report, dated 16 March 1917, the Committee listed the products which it considered to be of 'key' or 'pivotal' importance. They were dyestuffs, spelter, tungsten, magnetos, optical and chemical glass, hosiery needles, thorium nitrate, limit and screw gauges, and certain pharmaceutical preparations. It proposed the establishment of a special board, which would watch over these products, assist their manufacture, or protect them from foreign competition. The committee in their conclusion stated bluntly that the Key Industries

> should be maintained in this country at all hazards and at any expense. No economic rules apply to the situation of these minor but important industries. They must be kept alive, either by loans, by subsidy, by tariff, by Government contracts or in the last event by Government manufacture. They will necessarily be under Government supervision.[3]

The two reports firmly backed Government intervention in the dyestuffs business: we have noted one aspect, the formation of the British Dyestuffs Corporation, and must now turn to the other, the post-war system of import licensing. The machinery has been described already (see pp. 235–6); and the various committees began their work in January 1919. A difficulty arose at once: the Government decided to end import controls generally on 1 September of that year, so some way had to be found of making the import licensing of Key Products lawful. The solution was found in a provision of the Customs Consolidation Act of 1876 whereby imports could be licensed by Order issued under Proclamation. It was perhaps old-fashioned, but legal, and the King signed the first Proclamation on 24 February 1919. It enabled the Trade and Licensing Committee to issue 'general open licences' for imports of dyes from France, Switzerland, and the United States; the open licences were revoked on 14 April and all imports had to go through the machinery of the Dyes Commissioner.[4] Later in the year the Board of Trade listed a broad range of organic compounds used in the preparation of dyes, pharmaceuticals, and photochemicals for which import licences would be required, and these products represented, as it were, the 'key' chemicals.[5]

[1] Report of the Departmental Committee . . . to Consider the Position of the Textiles Trades . . . HMSO, 1918, Cd 9070, p. 109.

[2] HMSO (1918), Cd 9034, p. 4. The Committee's report was in the hands of the Government in December 1917.

[3] ibid., p. 52, para. 254, section 2.

[4] *Chem. Trade J.* **64** (1919), no. 1658, 186; no. 1664, 325.

[5] Memorandum by the Board of Trade on the Articles of which the Importation in the U.K. is prohibited except under Licence; HMSO (1919), Cmd 454.

The licensing system was disliked.[1] The Licensing Sub-Committee's demand for statistics aroused fears that confidential information might be disclosed. The importers and merchants were also upset when they learnt that the Committee might decide to by-pass them. The complaints from the users were more serious however: with the end of the war the textile trades were booming, and 1919 would have been a golden year for the dyers if they had been able to buy all the supplies they needed. But those from the British Dyestuffs Corporation were inadequate and the range limited, not to speak of quality, and Reparation dyes were a mere trickle.[2] Fast wool dyes were particularly hard to come by, and the dyers alleged that their foreign rivals, unfettered by regulations, were buying cheaply in Germany and so prospering at the expense of the British. The dyers exaggerated, and the Government ignored all appeals to scrap licensing.

At this point a major event took place: in August 1919 Customs officers seized 5 cwt of pyrogallic acid consigned from Canada to Brown & Forth of Manchester, on the ground that under the Proclamation of February, a licence was needed for the import of this material. Brown & Forth took the matter to court, lost, and appealed. In December 1919 Mr. Justice Sankey upheld the appeal and ruled that the extension of the 1876 Act to include chemicals and other products was invalid and illegal.[3] The Sankey judgement effectively nullified the object of the Proclamation procedure, with the result that very soon, imports, unhindered by licensing, poured into the country from Germany and Switzerland. The British Dyestuffs Corporation was adversely affected by this totally unanticipated competion, the users were well-pleased, and the Government was in a difficulty: on the one hand it was committed to protection for dyes, and on the other it was under increasingly heavy pressure from users, who had forgotten what they had endorsed a few years earlier and were now clamouring for cheap imports. Something had to be done and the Board of Trade came up with a proposal for special legislation. The bill was vigorously contested in the Commons by the Lancashire M.P.s and elsewhere by the influential Manchester Chamber of Commerce. The fight went on right up to the Third Reading, and the Lords rushed the Bill through just before Christmas 1920. The Dyestuffs (Import Regulation) Act became law on 1 January 1921.[4] It prohibited imports of dyes and intermediates for ten years except under licence. Licences were granted if the dye was not made in the United Kingdom; they could also be granted if the price of the British dye was higher than that of the equivalent foreign product. (In 1922 the price differential was set at three times the foreign level, by 1930 it had been reduced to 1·75 times.) A Licensing Committee administered the

[1] Throughout 1919 there are frequent references in the *Chemical Trade Journal* to complaints from the users.

[2] 550 tons of Reparation dyes were received in 1919, and about double the amount in 1920. *Chem. Trade J.* **67** (1920), no. 1749, 717.

[3] 1920, 1 KB 773; *Chem. Trade J.* **65** (1919), no. 1688, 336; no. 1700, 720.

[4] *Chem. Trade J.* **67** (1920), no. 1751, 767–8, 776; (1920), no. 1752, 817; (1920), no. 1753, 840, 849.

Act. Licensing applied to all countries, and licensed imports were duty free. This last was an important concession to the users, because a high tariff on the American model would have raised their costs, and so diminished the competitiveness of British textile exports.

British imports of dyes in the 1920s are shown in the following table. While they were a fraction of the pre-war level, it is nevertheless surprising at first sight that the quantity increased steadily throughout the 1920s. Bearing in mind that licences were generally issued for dyes not made in the United Kingdom, one must conclude that throughout the period the industry was unable to supply an adequate range of speciality dyes. We will revert to this in a later chapter, suffice it here to note that some comfort could be found in the fall in unit value, which would benefit the users. The fall was partly due to changes in the rate of exchange and partly to excess dye-making capacity, which caused a decline in prices generally.

TABLE 8.2

Licences granted by the Dyestuffs Advisory Committee, 1921–30[1]

Year	*German dyes*			*Swiss dyes*			*Other dyes*		*Total*		
	million lb	£000	£/ton	million lb	£000	£/ton	million lb	£000	million lb	£000	£/ton
1921	0·7	198	636	1·8	763	949	0·2	82	2·7	1043	865
1922	1·3	376	648	1·6	695	973	0·3	33	3·2	1104	773
1923	1·8	493	614	1·4	460	736	0·5	36	3·7	989	599
1924	1·8	398	494	1·2	364	679	negl.	9	3·0	771	570
1925	2·2	335	341	1·2	308	575	negl.	9	3·4	652	420
1926	2·9	599	463	1·2	334	623	0·1	11	4·2	944	503
1927	3·7	711	430	1·2	307	573	0·1	16	5·0	1034	463
1928	3·5	729	466	1·4	335	536	0·1	9	5·0	1073	481
1929	3·9	744	427	1·8	403	502	0·1	11	5·8	1158	447
1930	3·8	713	420	1·5	360	538	0·2	14	5·5	1087	443

Negl. means less than 100 000 lb (about 45 tons).
Reparation dyestuffs are excluded.

In November 1930 the Government, supported by the Colour Users' Association, announced that it would allow the Dyestuffs Act to lapse at the end of the year. Their decision was challenged in Parliament and the Government eventually accepted a proposal of the Lords to continue the Act for a year.[2] Soon afterwards the price criterion was altered and the licensing system became more liberal, which met the demands of the users. In 1932 a duty of 10 per cent was imposed, but was abandoned (except for intermediates) on the recommendation of the Import Duties Advisory Committee and by Act of

[1] Report of the 12th Annual General Meeting of the Colour Users' Association, Manchester, 17 July 1931. (The figures were provided by the Board of Trade.)

[2] *Chemy Ind.* **50** (1931), 642–3.

1934 the legislation of 1920 was made permanent.[1] With the outbreak of Hitler's war 'open general licences' were introduced and remained in force until the European Free Trade Area Act formally ended import licensing for dyestuffs and replaced it by a duty of 33⅓ per cent *ad valorem*.[2]

A different method was adopted to protect 'key' chemicals other than dyestuffs and intermediates. When the Sankey judgement had put an end to licensing by Proclamation, the Government decided on tariffs. Accordingly Part I of the Safeguarding of Industries Act provided that duties, commonly referred to as 'Key Industries Duties', be levied at the rate of 33⅓ per cent *ad valorem* for a period of five years. The tariff covered a wide range of organic chemicals as well as a number of other 'key' products. The Act was renewed in 1925 for ten years. Part II of the Act provided protection, if requested, against imports from countries whose currency was depreciating. The measure was directed against Germany, and remained in force until 1924, but no application for such protection was made on behalf of chemical products. The Act also established machinery for investigating allegations of dumping, and if proved, the Board of Trade could impose a duty of 33⅓ per cent on the goods.[3] Only the superphosphate producers among chemical manufacturers applied for anti-dumping protection. Their demand was rejected in 1925 on the ground that they were inefficient and must first put their house in order.[4] They tried again a few years later, and on that occasion the Import Duties Advisory Committee agreed that they be granted a duty of 10*s*. a ton.[5]

Import licensing, Key Industries Duties, and anti-dumping legislation did not, of their own, create an organic chemicals industry in Britain, but they did provide conditions favourable to growth. The Chadwick Committee looking at the situation in the mid-1920s concluded that 'great advances' had been made, but that the industry had not yet reached the stage 'at which it can stand by itself'.[6] Only the ignorant and the impatient would expect to achieve in ten years what had taken the Germans two generations. Whether these different forms of protection were equally justified in the 1930s, when effective international cartels regulated world trade in dyestuffs and nitrogenous fertilizers, is more doubtful.[7] Another question that may be raised at this point is whether a government bent on fostering dyestuffs and intermediates

[1] Dyestuffs Industry, HMSO (1933), Cmd 4411, pp. 7 ff.; Dyestuffs (Import Regulation) Act 1934; 24 Geo. 5, Ch. 6.

[2] *Nature, Lond.* **186** (1960), no. 4728, 829–30. The duty is being reduced from 1968 onwards under the Kennedy Round agreement of Gatt.

[3] Committee on Industry and Trade, op. cit., pp. 406–10.

[4] Board of Trade (Safeguarding of Industries), Report of the Superphosphate Committee, HMSO (1925), Cmd 2475.

[5] Packard, W. G. T. The history of the fertiliser industry in Britain, n.d. *Proc. Fertil. Soc.* no. 19, pp. 12 f.

[6] Safeguarding of Industries Act, Report of a Committee appointed by the Board of Trade, HMSO (1926), Cmd 2631, p. 36, para. 136.

[7] Richardson, H. W. The development of the British dyestuffs industry before 1939. *Scott. J. polit. Econ.* **9** (1962), no. 2, 127–9 mentions that prices were raised after I.C.I., the successor to the British Dyestuffs Corporation, joined the cartel. But the company's manufacturing activity was maintained at a time of serious unemployment.

should also intervene to develop some of the newer branches such as the products of high-pressure synthesis or plastics. It is an interesting subject for debate, but we should ask ourselves whether the Board of Trade possessed sufficient commercial competence to anticipate technological change. And even if it had, was it part of the functions of a Government department to intervene continuously in the business of the chemical manufacturers? The answer must surely be that such a policy would not have been in keeping with the times and would have invited hostility. Under the circumstances the form and extent of Government intervention were just about right and timely.

9

FROM INSECURITY TO CARTELS

The aftermath of war

THE early post-war years were extremely difficult for the entire chemical industry. Contrary to expectations, repair and reconstruction were among the lesser problems. As Allied troops marched to the Rhine in the last weeks of 1918 they passed through battle-devastated areas in northern France and Western Belgium, and further east the Germans had dismantled or dynamited many factories. But chemical manufacture recovered quickly: in December 1919 employment in the Belgian chemical industry, which had suffered most severely, was within three-quarters of the pre-war level.[1] Across the border in the region of Lille, Kuhlmann's factories had been extensively damaged. But by the end of 1920 virtually all the war damage had been made good at a cost of around 55 million fr. (say £1·1 million). The French Government refunded the expenditure by cash payments, interest-bearing notes, and also, to a large extent, by transferring to the company the title deeds of works built during the war and managed by Kuhlmann.[2] Elsewhere the physical damage was negligible, but everywhere the upkeep had been neglected. The shortage of coal was more serious. The mines of northern France were out of commission for months. In Germany strikes and military occupation reduced supplies, and so prevented a return to full production. Lack of coal particularly affected the output of by-product sulphate of ammonia, which remained for several years below the level attained in 1913.[3] For the same reason the Bayer company was only able to achieve 60 per cent of the pre-war output in 1919, and the management found it very difficult to organize work for the demobilized soldiers and the people from the nearby explosives factory at Dormagen.[4] The sudden cancellation of military contracts caused chaos in the labour market. In the United States these things were done on a huge scale, and Du

[1] Denterghem, C. de Kerchoven de. *L'Industrie belge pendant l'occupation allemande* (1927 p. 273.

[2] Annual report of the Kuhlmann company for 1920.

Production of by-product sulphate of ammonia (000 *tons* N)

	1913	1919	1920	1921	1922
Germany	108	55	74	77	78
United Kingdom	90	83	85	54	75
France	15	9	10	11	13

[3] *Source*: Grossmann, H. and Weicksel, P. *Die Stickstoffindustrie der Welt* (1930), p. 36.

[4] Flechtner, H. J. Von der. 'Kleinen IG' zur IG Farbenindustrie, in *Beiträge zur hundert ährigen Firmengeschichte 1863–1963* (1963), vol. 3, p. 45.

Pont, the largest of all suppliers, reported that orders worth $260 million (about £53 million) were unexpectedly cancelled. The company cut its payroll with equal ruthlessness from 86 000 on Armistice Day to 28 000 on 27 December 1918.[1]

In the Allied countries the elaborate wartime machinery for priorities, allocations, and production controls was quickly dismantled. It endured, however, in Germany. The Office of Economic Demobilization, and its chemical department under Fritz Haber, played an important role for some time. The chemical industry was divided into eleven sections, each headed by a committee of delegates from the Chemie-Verein and the trades unions. The committees concerned themselves usefully with fixing priorities for coal deliveries, with the allocation of raw materials, and the re-employment of ex-soldiers.[2] Some products long remained scarce and appropriate controls were retained for many years: on sodium sulphate until July 1920, on chlorine and bleaching powder until September of that year, on caustic soda, soda ash and caustic potash until October. For sulphuric acid a different system was devised. The war-time organization was replaced by the Schwefelsäure-Ausschuss in Spring 1920: this association of producers and users was given statutory powers to fix maximum prices and control distribution. The Ausschuss was dissolved at the end of June 1923 and thereafter the production of and trade in sulphuric acid was entirely free.[3] An even more powerful body was the Aussenhandelstelle Chemie, founded in June 1919 and given legal powers in 1920: it was responsible for all foreign trade licences and extended its activities to the control of export prices and foreign exchange earnings. At its height in 1921/2 the organization employed about 250 people who dealt with about half a million export licences a year.[4]

Reparation deliveries created another difficulty and extended beyond Germany by contributing to the problem of world trade in chemicals. Annex VI to Part 8 of the Treaty of Versailles dealt with dyestuffs, intermediates, and pharmaceutical preparations. Its main provisions were: 1. the Reparation Commission would seize one-half of the stocks of dyes and pharmaceuticals held in Germany on 15 August 1919; 2. up to one-quarter of the production of a specified dyestuff or drug had to be offered for sale to the Commission until 1 January 1925; 3. prices for products were to be fixed by the Commission, having regard to the level of pre-war prices.[5] The intention was to help dye users in the short run, and weaken the German producers in the long run.

[1] Annual Report of the company for 1918.

[2] *Z. angew. Chem.* **31** (1918), pt. 3, 637–8, 649.

[3] Ebert, W. *Die chemische Industrie Deutschlands* (1926), p. 65; Ilgner, M. Die Rohstoffversorgung der deutschen Schwefelsäure-Industrie, Ph.D. thesis, Frankfurt (1923), p. 92.

[4] Ungewitter, C. (Ed.) *Ausgewählte Kapitel aus der Chemisch-Industriellen Wirtschaftspolitik* (1927), pp. 329 ff. The Schwefelsäure Ausschuss and the Aussenhandelstelle were autonomous corporations, *Selbstverwaltungskörper*. These, and similar organizations in other industries were the principal manifestations of industrial self-government, backed by statutory powers, which had an influential following in some European countries after the war.

[5] *Fifth report on the work of the Reparation Commission from 1920 to 1922*, HMSO (1923), p. 230. (Hereinafter referred to as *Rep. Com. 5*.)

Neither objective was attained. The users hoped that large stocks of dyestuffs (above all the fashionable lighter colours) would be made quickly available. Instead there was much confusion and wrangling over the stocks. They were said to amount to 30 000–35 000 tons, but the Commission later explained that there were only 20 700 tons. The distribution of these stocks in 1919 and 1920 was accompanied by complaints. While French, Belgian, and Italian dyers and merchants did relatively well, the British were handicapped at first because of their system of import licensing; but they caught up later.[1] The demand for high quality dyes continued until 1921, and was indeed so pressing that the Germans were able to secure valuable concessions, which greatly mitigated the effects of the second and third Treaty provisions mentioned above. From August 1919 to the end of 1922 the Germans supplied via the Commission altogether 23 000 tons out of stock or production. The value was about £3 million, that is a little over one quarter of their dyestuffs export trade in 1913.

The Treaty of Versailles also laid down that Germany was to deliver 30 000 tons a year of sulphate of ammonia to France. Between 1920 and 1922, 80 000 tons were dispatched, corresponding to 20–30 per cent of the country's annual consumption. In 1923, despite the Ruhr occupation (or possibly because of it), the deliveries rose to 49 000 tons, and in the first eight months of 1924 reached 55 000 tons.[2] The London Agreement of 1924 (Dawes Plan) put Germany's payment obligations on a different basis, and dyestuffs, pharmaceuticals, and fertilizers became part of a wider settlement. Foreign buyer and German seller henceforth negotiated at arm's length: the former paid the Agent General for Reparation Payments, the latter was reimbursed by the Reich Treasury. In this way considerable sales were made to France, Belgium, and some other countries.[3]

[1] *Chem. Trade J.* **66** (1920), no. 1705, 105; *Rep. Com. 5*, pp. 231, 237. Ultimately Britain and France obtained 2300 tons each, the United States 1900 tons, Italy 1600 tons, and Belgium 1100 tons. The total value was £0·6 million. The average at around £130 a ton was a bargain under the circumstances.

[2] Rabaté, E. Les Engrais, in Gérard, J. (Ed.) *Dix ans d'efforts scientifiques et industriels* (1926) vol. 1, pp. 1253–4.

[3] *Reparation deliveries in the first 3 years of the London agreement, in million Gold Mark*

	Years ending 31 August 1925		1926		1927	
	Dyes and pharma-ceuticals	*Fertilizers*	*Dyes and pharma-ceuticals*	*Fertilizers*	*Dyes and pharma-ceuticals*	*Fertilizers*
France	5·2	19·6	2·4	47·9	3·3	47·3
Italy	8·9	—	3·9	—	3·0	—
Belgium	3·7	0·3	4·8	2·9	6·1	10·8
Other	8·4	—	0·1	—	0·1	2·7
Total	26·2	19·9	11·2	50·8	12·5	60·8

Source: Report of the Agent General for Reparation Payments, HMSO, vol. 11, p. 114, vol. 15, p. 123.

On their own and on an annual basis, the Reparation deliveries and the Dawes Plan transactions were quite small (they reached about £3 million in 1926/7), but they need to be considered in the broader context of the dyestuffs and fertilizer trade of the 1920s. There was, as we shall see, considerable excess capacity, especially in dyestuffs manufacture. The resale of German stocks and of a portion of output through quasi-official channels, was detrimental to the efforts by British, French, and American manufacturers to broaden their base, and compete on quality as well as price with the Germans. Looked at from the German producers' point of view, the dyestuffs and fertilizer sales gave the firms a basis of operation at the very moment when the home market was deeply disturbed by inflation, labour unrest, and political strife. The business was not profitable, at least not until 1924, but it kept the works going, the men employed, and abroad maintained (or perhaps restored) the image of German quality products.

None of these consequences were foreseen by the experts at Versailles, nor for that matter by the chemical manufacturers. In 1919 shortages were still acute, and in the United States an unexpected boom absorbed the products of the chemical industry. With few exceptions that year and the next yielded good profits. American Cyanamid reported excellent results from fertilizer sales, Davison Chemical raised prices for sulphuric acid and made a profit of $2 million in 1920, and Du Pont's sales of leathercloth and similar materials by the Fabrikoid group jumped from $6 million in 1918, to $10 million a year later, and the return on investment improved even more spectacularly from 6·6 to 15·8 per cent.[1] In Europe conditions were less favourable, and the large firms in the heavy chemical sector were not doing particularly well. The dyestuffs business, however, was exceedingly prosperous. The British Dyestuffs Corporation, whatever else was wrong with it, at least earned and declared dividends. Swiss exports bounded ahead, the value rising from 98 million fr. (£4 million) in 1918, to 136 million fr. a year later, and reached a peak of 196 million fr. in 1920.[2] In general prices of dyestuffs remained high throughout 1919 and for most of 1920; in the United States, where the Tariff Commission collected such information, the average prices of indigenously produced dyestuffs changed little between 1919 and 1920.[3]

In 1920 the bubble burst. Trade slowed down, prices fell. The chairman of Kuhlmann referred to the slump as '. . . un orage économique, on peut même dire un cataclysme . . .'. The works of his company were operating on average at only half their capacity during 1921, and the net profit being negligible, no dividend was paid.[4] The deterioration was unexpected and widespread.

[1] Annual reports of American Cyanamid; Davison Chemical Co. *A century of progress* (1926), p. 14; GTX 106. Short History of Fabrikoid; memorandum by R. R. Carpenter to Finance Committee of Du Pont 4 November 1921.

[2] Jaquet, N. *Die Entwicklung und volkswirtschaftliche Bedeutung der Schweizerischen Teerarbenindustrie* (1922), Appendix, Table 4.

[3] USTC, *Census 1921* (TIS 26) (1922), p. 75.

[4] Reports of the company for 1920 and 1921.

Exports of Chilean nitrates fell from a peak of 2·8 million tons in 1920 to 1·1 million tons a year later, and in 1922 recovered only to 1·4 million.[1] The turn-round was abrupt for American Cyanamid: the fertilizer business slumped in the autumn of 1920, and the following year the cyanamide plant at Niagara Falls was working at a mere fifth of capacity.[2] Similarly du Pont's Fabrikoid, which a short time earlier had been so prosperous, reported that sales were two-thirds of the 1920 level, and losses were made in 1920 and 1921.[3] The collapse of the dyestuffs boom reflected, at one remove, the slump in textiles. The results were disastrous. The British Dyestuffs Corporation was caught unprepared, and though import licensing was reintroduced, the company made a loss. In the United States, the production of dyes halved from 39 000 tons in 1920 to 17 500 tons a year later. Average prices fell from \$1·08/lb in 1920 to \$0·83/lb a year later and slid further to \$0·60 in 1922.[4] Even the Swiss, whose business included many speciality dyes and was therefore thought to be depression-proof, lost over two-thirds of their dyestuffs exports in a single year, when their sales plummeted from almost £10 million to just under £3 million.[5]

The chemical depression lasted only about eighteen months, roughly from the middle of 1920 to the end of 1921. Though short, it was extremely severe and drew attention, more emphatically than any pre-war crisis, to the problem of excess capacity. Even before 1914 it had never been easy for chemical manufacturers to dovetail the capacity of their works and the seasonal and cyclical fluctuations. During the war the factories had been enlarged (or new ones built) and the extent to which they would and could be occupied aroused concern even then. It was obvious that the equipment ought to be operated at a high and steady level. Once output falls, profitability is likely to fall more than proportionately, and this diminishes the cash flow and so imperils new investment. Eventually there comes a point at which the company fails to

[1] Hernandez, R. *El Salitre* (1930), p. 174.

[2] Annual Reports of the company.

[3] GTX 106, op. cit.

[4] USTC, *Census for 1921*, pp. 75–6, and *1923*, pp. 34–5. The weight of dyestuffs made was calculated according to a different formula from European practice, and this accounts for the large tonnage.

[5] The following figures speak for themselves, but it is worth remembering that the Swiss firms were developing energetically manufacture in other countries, and this must have affected the business of the parent companies in Basle:

	1920		1921		1922	
	(000 t)	(million fr.)	(000 t)	(million fr.)	(000 t)	(million fr.)
Total exports	8·5	195·6	3·7	57·7	3·9	55·3
of which to						
U.K.	3·4	62·8	0·6	10·6	0·8	13·1
France	1·4	49·2	0·7	12·9	1·0	13·9
U.S.A.	0·6	16·1	0·7	11·3	0·7	12·7
Belgium	0·5	15·5	0·2	2·6	0·3	2·6
Italy	0·6	10·7	0·3	3·3	0·2	2·0

Source: Swiss foreign trade statistics.

renew its equipment, makes do with the old and mended process plant, and so becomes technologically backward. In the 1910s the problem was more serious in the inorganic sector than in dyestuffs, where the equipment was less costly, but the research expenditure was at risk. There was general awareness of the difficulties that the new, war-time capacity in heavy chemicals would cause once defence orders ceased. In Britain a Departmental Committee was appointed in February 1917 and instructed to examine the situation of the fertilizer and sulphuric acid trades after the war. The committee published its findings two years later: the excess of production over demand was placed at about 650 000 tons of sulphuric acid (as 100 per cent H_2SO_4). The committee recommended that the Government-owned works be closed, that some of the older privately-owned works be dismantled and their owners compensated out of public funds, and lastly that the consumption of sulphate of ammonia be encouraged. Though writing in 1919, the committee did not think that the newly formed dyestuffs industry, which was then using about 30 000 tons a year of acid (100 per cent H_2SO_4), would significantly increase its demand for sulphuric acid and oleum.[1] Elsewhere in Europe the large works put up after 1914 were changing the scale and pattern of production. The chemical industry in south-eastern France, exemplified by the new factories at St. Auban and Pont de Claix, owed its expansion primarily to the demands of the military.[2] The war-time extensions to the German capacity for making inorganic chemicals, chiefly chlorine and nitric acid, greatly exceeded the needs of the dyestuffs factories. The mission sent by the Association of British Chemical Manufacturers saw the potential danger and warned that competition with Britain's traditional heavy chemical exports (notably bleaching powder and alkalis) would become even more severe, because the Germans, deprived of markets for their dyestuffs, would seek with their customary vigour additional business for their inorganics.[3] Events were to show a few years later that competition was severe in both.

The problems of excess capacity disappeared for a short time after the war, only to reappear in 1920, and nowhere more so than in dyestuffs. This branch of the industry was deemed to be of 'key' importance, yet the scale of surplus capacity was such as to make even the most enthusiastic supporter of self-sufficiency pause and think. The consumption of dyestuffs in the world was estimated at around 110 000 tons in 1922 and 160 000–170 000 tons three years later.[4] The estimates are not particularly reliable, but I am quoting them to show the order of magnitude. On the supply side the capacity was around 280 000 tons. In some countries demand barely sufficed to keep half the

[1] Board of Trade (on behalf of the Minister of Munitions), Departmental Committee on Sulphuric Acid and Fertiliser Trades, HMSO (1919), Cmd 23, pp. 4, 12–13.

[2] Simonin, G. *et al. L'Industrie chimique française*: *Lyon et Sud-Est* (1959), p. 56.

[3] ABCM. *Report of the British Chemical Mission on the Chemical Factories in the Occupied Area of Germany* (1919), pp. 39, 41.

[4] The estimate excludes the Soviet Union and the lesser European countries, among them Switzerland. USTC, *Census 1922* (TIS 31), (1923), p. 125, and British Dyestuffs Corporation, Review of the dyestuffs industry, 1927.

capacity going. The experience of the British Dyestuffs Corporation was, with differences of degree, representative of the dyestuffs industry as a whole: in 1922 the company obtained, on average, 36*d*./lb for its colouring materials, a year later 28*d*., and in 1925 only 25*d*.[1]

TABLE 9.1

Dyestuffs capacity and production in the principal countries[2]
(000 tons)

Country		*Capacity*	*Output in 1924*
Germany	(1923)	156	70
U.S.A.	(1923)	52	30
U.K.	(1924)	24	15
Switzerland	(1920)	13	13
France	(1924)	25	17
Japan	(1919)	9	2
Italy	(1922)	5	5
Other		n.a.	2
Total		284	154

The boom-and-bust trading conditions of 1919–22, the German monetary chaos, and, not least, the aftermath of the Russian Revolution engendered a feeling of uncertainty and insecurity among businessmen. In retrospect 1913 seemed a golden year, and if only the stability of the pre-war era could be regained, all might yet be well. The fluctuations in the economic environment thus added their weight to the industry's internal problems, conspicuous among them excess capacity. The early post-war years were a time when chemical manufacturers, like their colleagues in other industries, thought in terms of maintaining the status quo, of strong trade associations, of caution rather than of daring enterprise. In such an atmosphere cartels and agreements flourished, and mergers to eliminate competition were common. But before we come to these examples of restrictive practices and their proliferation, it is necessary to illustrate, if only briefly, the post-war deterioration that had taken place in the chemical companies' finances.

Financial problems

The directors of chemical companies would see the fluctuating condition of their industry, or more precisely of their sector, reflected in the monthly movements of their sales returns and at the end of the year in the profit and

[1] Chairman's speech at the Extraordinary General Meeting on 13 November 1925 quoted in Papers Relating to the Liquidation of the Government's Interests in the British Dyestuffs Corporation Ltd. HMSO (1925), Cmd 2545, p. 18.

[2] USTC, *Census 1923* (TIS 32), (1924), pp. 9–10, 124; League of Nations, *The chemical industry* (1927), pp. 29, 85.

loss account and balance sheet of their enterprise. When we try to follow them we encounter many difficulties. In the first place financial records are often incomplete, so that a long run of years is available for only a few companies. Secondly the presentation of the accounts and, more significantly, the extent of their disclosures, were different forty or fifty years ago when legislation on these matters was more lenient than it is now. It is only recently that inter-firm comparisons have been systematically standardized, and international comparisons continue to be hamstrung to the present day because accounting practices and auditing requirements vary considerably from one country to another. In short, when we look at the finance of chemical companies just before and just after the war, we have a woefully small sample, the information is inadequate and the conclusions must necessarily be tentative. But despite these major weaknesses the effort is still worth making, because the financial experience of the companies in 1920–2 was a major—perhaps the major—influence on the decisions taken by management for the rest of the decade and even beyond.

TABLE 9.2

Net assets and profitability of six companies in the chemical industry, selected years 1912–22

Company	*A Assets*[a] *P Profitability*[b]	1912 *or* 1912/13	1913 *or* 1913/14	1917 *or* 1917/18	1918 *or* 1918/19	1920 *or* 1920/1	1921 *or* 1921/2
Brunner,	A £ million	4·9	5·9	10·3	10·8	14·6	17·3
Mond[c]	P as % of A	16	13·5	11	9·5	6	6
Badische	A million M	98	100	180	166	382	752
Anilin[d]	P as % of A	15·5	15	16·5	6·5	17	22
CIBA[e]	A million fr.	21	21	31	36	35	37
	P as % of A	3	3	15	15	6	1
Dow[f]	A million $	2·1	2·1	5·2	4·5	5·6	5·2
	P as % of A	5	5	21	loss	loss	loss
American	A million $	5·7	5·9	16·9	16·7	15·7	15·5
Cyanamid[g]	P as % of A	4	5	9·5	6	2	1·5
Monsanto[d]	A million $	0·4	0·4	4·3	6·2	10·8	8·6
	P as % of A	21	19	18	5	7	loss

Notes:

[a] Assets are net assets (excepting Monsanto), i.e. fixed assets after depreciation, investment in subsidiaries, goodwill and patents, plus net current assets. For Monsanto the item includes gross current assets and reserves.

[b] Profitability is net profit after depreciation, taxes, interest on loan capital, and (where shown) directors' fees. All figures have been rounded to the nearest half per cent.

[c] The financial years end on 31 March 1912, 1914, 1918, 1919, 1921, 1922.

[d] The financial year corresponds to the calendar year.

[e] The financial years end on 30 June 1912, 1913, 1917; thereafter they end on 31 December 1918, 1920, 1921.

[f] The financial years end on 31 May 1913, 1914, 1918, 1919, 1921, 1922.

[g] The financial years end on 30 June 1913, 1914, 1918, 1919, 1921, 1922.

Company reports have been used throughout.

The common financial denominator in chemical manufacture was the rapid growth of fixed assets during the war. The objective was, of course, to earn a good return on these assets. The preceding table shows for six companies net assets and net profits as a proportion of net assets for the last two pre-war years, for 1917–18, and for two post-war years. Brunner, Mond, it will be remembered, was a large manufacturer of alkalis; Badische Anilin made dyestuffs and intermediates before the war, but after 1914 nitrogen compounds became increasingly important; CIBA specialized in dyestuffs; Dow made a growing range of inorganic and organic chlorine compounds, while American Cyanamid and Monsanto concentrated on fertilizers and organic intermediates respectively.

The assets have been shown in national currencies, since exchange rates fluctuated considerably after the war, and have little meaning for Germany. Although the six companies differed greatly in their development, size, and product range, all show considerable increases in their net assets. Their pre-war profitability varied, rose considerably at the height of the war, but fell back later; the American companies did especially badly. Badische had a disastrous year in 1918, but then made a notable recovery, and its remarkable profitability can be attributed in large part to the sulphate of ammonia business.

The above figures prompt the question to what extent net assets represented the real worth of the company, and what sort of return the directors might be looking for. Both are relevant to the situation in the 1920s: the former requires a factual answer, the latter a knowledge of people's intentions. Very little information has been published on the replacement value of fixed assets. It would appear that in Germany it was the practice to write down plant and machinery as quickly as possible. Information from Badische shows that the original cost of fixed assets at the end of 1912 and 1914 was £7 million and £9·14 million respectively. But the published accounts show that the book value was only £1·7 million and £3 million. In the case of Kuhlmann the balance sheet at the end of 1921 showed net assets of 187 million fr. (£3·6 million), but the board estimated that the replacement value of these assets was roughly 306 million fr. (£5·8 million).[1] If we calculate the net profit on the higher valuation of assets and ignore book values, the profitability of Badische before the war is halved, and that for Kuhlmann falls from 6 to 4 per cent in 1921. All we can safely deduce from this is the obvious statement that the rate of return depends on the definition of assets. The directors, however, would be less interested in definitions, and more in a return to pre-war trading conditions and steady profits.

These considerations would lead them to review their marketing policy. The sales were indeed well above pre-1914 levels, but the profit on sales was down, owing to higher costs, excess capacity, or more severe competition, or indeed a combination of these factors. Particulars of sales and of profits are

[1] Company reports and Badische archives.

not readily available, so that only two of the companies mentioned below were also shown in Table 9.2; Badische Anilin has been omitted, CIBA, Brunner, Mond, and Monsanto did not publish turnover figures. Although the basis for comparison is clearly unsatisfactory, and I would not claim that it is representative of the industry, the figures illustrate the deteriorating situation after the war.

TABLE 9.3

Sales and profitability of four companies in the chemical industry, selected years 1912–22

Company	*S Sales*[a] *P Profitability*[b]	1912 *or* 1912/13	1913 *or* 1913/14	1917 *or* 1917/18	1918 *or* 1918/19	1920 *or* 1920/1	1921 *or* 1921/2
Norsk	S million N Kr.	n.a.	28	138	112	99	86
Hydro[c]	P as % of S	n.a.	19·5	16	15	loss	loss
Du Pont[d]	S million $	36·5	26·7	270	329	94	55
	P as % of S	19	20	18	13	16	12·5
Dow[e]	S million $	1·0	1·1	11·8	12·2	6·1	4·2
	P as % of S	10	9	9·5	loss	loss	loss
American	S million $	n.a.	n.a.	5·5	5·5	5·5	3·8
Cyanamid[f]	P as % of S	n.a.	n.a.	29	18	5·5	5·5

Notes:

[a] Sales of the U.S. companies are net sales and exclude freight and rebates; Norsk Hydro's include sales of electricity.

[b] Profitability is net profit after depreciation, taxes, interest on loan capital, and (where shown) directors' fees. All figures have been rounded to the nearest half per cent.

[c] The financial years end on 30 June 1914, 1918, 1919, 1921, 1922.

[d] Fabrikoid Group included in 1920, but excluded in 1921. Net receipts or net earnings have been used for profitability. Du Pont's figures are for the calendar years.

[e] The financial years end on 31 May 1913, 1914, 1918, 1919, 1921, 1922.

[f] The financial years end on 30 June 1918, 1919, 1921, 1922.

Company reports have been used, except for Norsk Hydro.

Manufacturers faced with large sales, but small profits (or even losses), began to look again at market-sharing and other protective devices and hoped that the Government would raise tariffs to protect them from foreign competition. Some decided to diversify their activities: they hoped that by adapting themselves to technical change they could restore their profits; a few did, many were over-hasty and regretted it. Among the important and eventually successful instances of diversification was Brunner, Mond in Britain, which added ammonia and nitrogenous fertilizers to its alkali business, and Monsanto in the United States, which greatly widened its range of organic chemicals. The extreme in diversification is exemplified by Dynamit AG in Germany: the company lost its military explosives business after the war, and converted its factories to make miscellaneous manufactures ranging from lenses to gas meters and from cellulose lacquers to bobbins. In the end only wire drawing and steel cables proved profitable, and the other experiments had to be abandoned.[1]

When diversification failed or was not practicable, other routes to success were tried. Merck & Co., then the leading manufacturer of fine chemicals, cut costs by reducing the range from almost 11 000 different products to about 8000 and was able to make further savings by introducing standard packages.[2] Few companies appear to have possessed the combination of resources and skill of Du Pont and Badische in adapting themselves to changing circumstances. Du Pont's Finance Committee anticipated that the immensely profitable military explosives business would end. In 1917 the return on capital had been exceptionally high at 16 per cent, compared with $11\frac{1}{2}$ per cent before the war. The lower figure was henceforth used as the minimum acceptable return on new investment, and the Treasurer recommended late in 1917 that the company buy shares in General Motors and Chevrolet.[3] Badische went a different way and decided to stake its future on the products obtained from ammonia. We have already seen how quickly its factories were enlarged during the war, and building continued at Leuna after the Armistice. The work, though paid for in depreciating Marks, was expensive, and the enormous investment in nitrogen fixation capacity was an example of large-scale risk-taking.

Diversification and new business ventures were not only risky, but they involved directors and managers in unfamiliar problems and unusual situations. Among the most important were stocks. Before the war manufacturers were often obliged to make for stock, but now some of them had additionally to cope with the seasonality of the fertilizer trade or the occasional interruptions in the dyestuffs business. It was, in short, not so much a change in kind, but in degree, and the figures in Table 9.4 on page 258, taken from the accounts of companies that we have looked at in a different context, show the extent of the change.

For all the companies shown in the table the proportion of stocks to net assets was higher after the war, and in the case of Badische, CIBA, and Dow the ratio was much higher. Stable trading conditions would have eased this particular difficulty. But under post-war conditions the pressure of stock levels on the requirements for working capital was another factor that led manufacturers of dyestuffs and of soda to regulate their trade. The pressure was even greater in fertilizers. The seasonal cycle of the fertilizer trade was not, of course, a new phenomenon. Before the war the makers of superphosphate had solved it by engaging extra labour and working overtime in the busy season, and closing the works in summer and early autumn. Chilean nitrate shipments were adjusted to the rhythm of European farming. But synthetic ammonia could not be made in that way: production was continuous and had

[1] *Enquete*, p. 372. [2] ibid, p. 457.

[3] GTX 124, memorandum from the company's treasurer (J. J. Raskob) to the Finance Committee, 19 December 1917; Annual Report for 1917.

TABLE 9.4

Value of stocks and proportion of stocks to net assets of five companies in the chemical industry, selected years 1912–22

Company	*St Stocks* *A Assets*	1912 *or* 1912/13	1913 *or* 1913/14	1917 *or* 1917/18	1918 *or* 1918/19	1920 *or* 1920/1	1921 *or* 1921/2
Brunner,	St £ million	0·5	0·8	1·3	1·6	2·7	2·1
Mond	St as % of A	10	13·5	12·5	15	18·5	12
Badische	St million M	21	25	64	81	319	332
Anilin	St as % of A	21·5	25·5	35·5	49	83·5	41
CIBA	St million fr.	6	6	5	17	26	22
	St as % of A	29	29	16	47	74·5	59·5
Dow	St \$ million	0·2	0·3	1·7	1·3	2·1	1·5
	St as % of A	10·5	12	33	29	37·5	29
American	St \$ million	0·3	0·5	1·3	2·1	1·9	1·0
Cyanamid	St as % of A	5	8·5	7·5	12·5	12	6·5

Notes:

For definition of net assets and particulars of the companies' financial years see Table 9.2.
The proportion of stocks to net assets has been rounded off to the nearest half per cent.
Company reports have been used throughout.

to be stocked at the works in specially built silos, and to a lesser extent in railway warehouses or under cover at merchants' depots. Although manufacturers varied their prices to encourage farmers to buy in the off-season, the bulk of the demand for nitrogenous fertilizers was concentrated in the late winter and early spring. As a result Badische and later IG Farben, and on a lesser scale the other firms, had to hold and finance very large stocks. In 1929, for instance, the value of IG Farben's stocks of fertilizers ranged from a low point of 150 million RM (£7·5 million) to a peak of 300 million RM.[1] The financing charges for such stocks were considerable, and the movements in stocks could have a disproportionate effect on the company's entire finances.

It may be rash to generalize from such incomplete and imprecise information, but it seems to me that a few conclusions emerge and can be sustained. Thus the assets of the companies we have been looking at increased markedly during the war, but post-war trading difficulties were reflected in disappointing returns in monetary terms and also relative to the pre-war performance. Secondly, the value of sales (at current prices) was much higher than before the war, but the profit on sales was poor. And finally the value of the stocks carried was burdensome and in the case of nitrogenous fertilizers was, at times, exceptionally high. In brief, the aftermath of war, far from removing financial difficulties, intensified them and caused new ones to emerge. Caution seemed advisable, prudence was at a premium. Business appeared to be more risky, a wrong decision more serious in its consequences than ten years earlier. Under such conditions arrangements to organize sales so as to give

[1] Evidence of Bueb to the Enquete, pp. 140, 169.

each member of the group or cartel some assurance of profitable trade flourished and multiplied. We must now look at trade associations and combines and note their increasing effectiveness in the 1920s.

Trade associations and restraint of trade

People of the same trade seldom meet together even for merriment and diversion, but the conversation ends in a conspiracy against the public or in some contrivance to raise prices.

Wealth of Nations, Book 1, Chapter 9

A century and a half after Adam Smith's classic the methods had become a good deal more sophisticated, and their effects ranged far beyond national boundaries. We will look first at the most conspicuous, though by no means the most significant of these 'contrivances'. Trade associations had scores of members, which, between them, encompassed virtually all branches of chemical manufacture. But arrangements to control prices, production, or sales were generally unsuitable subjects for the associations, which in any case rarely had sufficient staff to initiate and administrate such agreements. Of necessity the associations had objectives that were sufficiently broad to unite the majority of the members behind specific programmes for action.

One of the oldest, and assuredly the most influential, of the associations was the German Verein zur Wahrung der Interessen der chemischen Industrie Deutschlands, for short the Chemie-Verein, which dated back to 1877, and from the outset played an active and successful role in tariff negotiations and patent legislation. The Chemie-Verein was the mouthpiece of the large firms. Early in its history it formed six associations for the main product groups, and later, during the war, organized and managed the export control system. For a few years after the war the Chemie-Verein was overshadowed by the Reichsarbeitsgemeinschaft Chemie, essentially a combination of employers and unions for promoting the economic and social interests of the industry. The new body, with its broad aims, was given statutory powers, but its heyday was short, and in 1922 responsibility for wages and conditions of work was transferred to a special negotiating body, the Tarifgemeinschaft Chemie. The power and influence of the Reichsarbeitsgemeinschaft were gradually eroded, so enabling the Chemie-Verein to resume its former activities. German associations extended their scope considerably between the wars, and the central association, the Reichsverband der deutschen Industrie (founded in 1919), thus wielded very large, quasi-political power. The Chemie-Verein, after some initial hesitation, joined the Reichsverband.[1]

The German chemical industry's association was a powerful lobbyist and pressure group. It was copied elsewhere, but if the imitators were less successful that was in part due to the lesser economic and political role of the chemical industry in other countries, and perhaps also to the parochialism of

[1] Ebert, W. *Die chemische Industrie Deutschlands* (1926), pp. 16–17, 64–5, 86–7; Greiling, W. *75 Jahre Chemieverband* (1952); *Chem. Ind., Düsseld.* **42** (1919), no. 3/4, 28; no. 7/8, 67–8.

some large manufacturers. In Britain, for example, the Association of British Chemical Manufacturers (ABCM) was founded in 1916 to further the common interests of the industry.[1] We can only marvel now that it took so long for British firms to form such a body, but at least the ABCM was an immediate success, for a year later the Ministry of Reconstruction's Committee on the Chemical Trade concluded that 'it is the most representative' association 'at present in existence in this country' and recommended that the Ministry and the ABCM set up a standing committee to deal with post-war problems in general, and specifically with the statistical needs of the trade.[2] No action followed on the recommendation. There was a setback, and critics pointed out that as the ABCM did not represent the fertilizer trades[3] it carried little weight. Sure enough, its recommendation to integrate dyestuffs with chemical manufacture generally, as in Germany, was ignored by the Board of Trade, and its request to be consulted on the distribution of Government funds to the dyestuffs industry was turned down. But the ABCM was able to secure the appointment of W. J. U. Woolcock, M.P., as its representative on the Dyestuffs Trades and Licensing Committee, and in the 1920s it sent a delegation to give evidence before the Balfour Committee.[4]

In France the Syndicat Général des Industries Chimiques was converted into the Union des Industries Chimiques (U.I.C.) in 1920.[5] The Union confined itself to non-controversial matters such as the publication of trade directories, the organization of exhibitions, and the compilation of statistics. Something more was needed, and so the large firms created the Comité des Industries Chimiques in 1927. Initially it had only about twenty members, but they represented the leading undertakings. The Comité's main function was to maintain contact between the members collectively and large concerns and associations in other industries, and also with the Government over such matters as tariff negotiations and help for exports. There were reports at the time that the Comité had participated in negotiations with IG Farben on dyestuffs and the Swiss firms on pharmaceuticals.[6] If the reports were correct, they would indicate a degree of collective action in matters of commercial detail that neither the German nor the British associations were then authorized to handle on behalf of member firms.

[1] Miall, S. *A history of the British chemical industry* (1931), p. 263. The name has recently been changed to Chemical Industries Association.

[2] Ministry of Reconstruction, Committee on the Chemical Trade, HMSO (1917), Cd 8882, pp. 1–2.

[3] Packard, W. G. T. *The history of the fertiliser industry in Britain*, n.d. (1952?), pp. 35–6. This sector was championed by the Fertiliser Manufacturers Association which dated back to 1877 and was incorporated in 1919.

[4] *Nature, Lond.* **102** (1918), no. 2562, 272–3; *Chem. Trade J.* **63** (1918), no. 1647, 38. Appendix 1 of the Final Report of the Committee on Industry and Trade (HMSO 1929, Cmd 3282, pp. 315 f.) shows that the team consisted of Woolcock, Armstrong (formerly of the British Dyestuffs Corporation) and Dr. Frankland F.R.S.

[5] Matagrin, A. *L'industrie des produits chimiques et ses travailleurs* (1925), pp. 297–8.

[6] Reagan, D. J. *The French chemical industry and trade in 1928* (Dept. of Commerce Tariff Information Bulletin No. 652) (1929), pp. 3–4; Ogburn, W. F. and Jaffé, W. *The economic development of post-war France* (1929), p. 468.

Similar bodies were established in other European countries during or after the Great War, when joint action for political or broad economic purposes became necessary. In Holland, a meeting attended by representatives from twenty-three companies to discuss the possibility of buying a permanent stand at the Utrecht fair, which was then being planned, led to the formation of the Verenigung van de Nederlandse Chemische Industrie (VNCI). An *ad hoc* committee first met in spring 1918 and was soon becoming involved in discussions with the Government, being asked to advise on export licensing of raw materials to Germany. A do-it-yourself atmosphere prevailed in the early years when the member firms shared all the secretariat work; a permanent staff was not appointed until 1934.[1] The destruction of most of their industry and the problems of reconstruction caused the Belgian manufacturers to form an association, and in 1919 they set up the Fédération des Industries Chimiques Belges.

The Chemical Alliance, during its short existence from 1917 to 1919, was the American version of the Chemie-Verein's wartime activities. The Alliance, unlike the Chemie-Verein, was wound up because its members had no use for it in peace, and also because they were afraid that further collaboration might bring down upon them the lawyers from the Department of Justice and prosecutions.[2] There were, however, some remarkably effective organizations in the United States, the oldest being the Manufacturing Chemists' Association founded in 1872, partly for social functions and partly to deal with tariff questions. From these convivial and also practical beginnings it extended its activities and by 1914 concerned itself with safety regulations and standard forms of contract. (Under Prohibition it represented the industry in matters connected with industrial alcohol.) Another lobby was the American Dyes Institute, which had been set up in the war. By 1919 its 46 members represented virtually all the dyestuffs manufacturers. Self-interest and anxiety about tariffs united them briefly and the Institute sent a spokesman to the hearings before the House of Representatives on the proposed new tariff. The delegate recommended import licensing as a 'temporary expedient', and further proposed that intermediates, as well as dyestuffs, should be given full protection.[3] The Institute was dissolved in 1921, and its place was taken by the Synthetic Organic Chemical Manufacturers' Association (SOCMA). This new organization took under its wing a heterogeneous collection of firms making all manner of organic chemicals, including compounds for the pharmaceutical industries. It was not so easy to achieve a common purpose, but SOCMA negotiated with the Tariff Commission the detailed implementation of the customs code and was successful 'as an instrument of cooperative

[1] *Chem. Ind., Düsseld.* **20** (1968), no. 7, 451.

[2] The Alliance was briefly resuscitated in the early 1930s under the National Recovery Act. Hempel, E. H. *The economics of chemical industries* (1939), p. 81.

[3] Written statement by J. M. Matthews on behalf of the American Dyes Institute in Hearings before the Committee on Ways and Means on H.R. 2706 and 6495: *Dyestuffs* (1920), pp. 182–3, 186 7.

action among manufacturers of coal-tar products'.[1] (In the 1960s it played a lively and not insignificant role in the discussions on the Kennedy Round at Geneva.)

All these trade associations had broadly similar objectives, among which tariffs, customs procedures, factory legislation, and fiscal regulations predominated. Below the central or national organizations there were many specialist groupings, which dealt with problems such as safety regulations, transport, and standardization. In the United States there were over fifty of them towards the end of the 1930s, ranging from the Carbon Dioxide Institute to the Society of Plastics Industry and the National Fertilizer Association.[2] Such fragmentation was the rule in Europe where Adam Smith's remarks on the habits of businessmen were particularly apposite. For lack of evidence (which under the circumstances is not readily available) it is hard to tell where the overt aims ended and the covert intentions to regulate the trade began. One thing, however, is certain. Price and production restrictions, whatever their form, were not new phenomena that followed the Great War. So it is necessary in the following pages to turn back occasionally to the 1900s, or conversely to look to the 1930s and 1940s, particularly as in more recent times special investigations have shed much light on the work of associations.

The first official investigation of cartels seems to have been the detailed German enquiry of 1903–5. There was another one in the late 1920s, the famous Enquete, which, though meticulous in many respects, was remarkably unenlightening on this subject. The Austrians also had a cartel enquiry in 1912. In Anglo-Saxon countries the first comprehensive enquiry was set on foot by the Ministry of Reconstruction in 1918, and was followed by the numerous reports of the sub-committees appointed by the Standing Committee on Trusts under the Profiteering Act.[3] Finally the Balfour Committee addressed itself to the problem.[4] Across the Atlantic the Temporary National Economic Committee issued dozens of reports on restraint of trade and related subjects. To round off the account we must not omit the League of Nations, which, on the occasion of the International Economic Conference of 1927, commissioned academic economists to write short studies.[5] All these documents were concerned with manufacturing industry as a whole, and only incidentally with chemicals. The first German enquiry was based on evidence collected between 1903 and 1905: its list of cartels in the chemical industry is certainly incomplete.[6] Twelve years later the Enquete returned to the subject and its third sub-committee published a 500-page report on the chemical

[1] Haynes, W. *American chemical industry* (1945), vol. 3, p. 272; (1948), vol. 4, pp. 238–9.

[2] Pearce, C. A. *et al. Trade association survey*, T.N.E.C. no. 18 (1941), pp. 447–8.

[3] Report of the Committee on Trusts, HMSO, 1919, Cd 9236. In 1920 and 1921 about thirty separate reports were published.

[4] Committee on Industry and Trade, *Factors in industrial and commercial efficiency*, pt. 1, HMSO, (1927): the first chapter is devoted to Combination in Industry and Trade.

[5] For example: MacGregor, D. H. *International cartels*; Wiedenfeld, K. *Cartels and combines*; these and other monographs were published by the Economic and Financial Section in 1927.

[6] Ministerium des Inneren, *Denkschrift über das Kartellwesen* (1906) vol. 1; see below, p. 267.

industry.[1] The diligent reader will look in vain for a list of cartels or combinations (the single exception being the nitrogen syndicate), or for an account of the associations in the industry and their aims. But he can console himself with thirty-nine memoranda submitted by the principal companies in the industry: they are a mine of useful information, but contain no word of criticism. Elsewhere the reports of the TNEC, though diffuse, at least discuss restraint of trade and criticize it. In the United Kingdom two reports of the sub-committee of the Standing Committee on Trusts dealt with important branches of chemical manufacture.[2]

These enquiries were not launched because it was felt that the organization of trade was, of itself, a worthwhile subject for investigation. Political pressures lent weight to what many (probably most) businessmen knew perfectly well: competition was becoming less vigorous and prices were showing remarkable rigidity. There may also have been the feeling that the creation, often with state support, of new industries during the war had not been followed by the anticipated results, and it was suspected that international combinations were defeating the original purposes. Finally, there were sensitive areas of national interest, notably explosives (including the adjuncts) and nitrogenous fertilizers, where price fixing could have politically undesirable effects.

What were the results of the investigations, specifically those concerned with chemical products? In Europe, they were negligible. With the lapsing of the British Profiteering Act in 1921, no further action was taken. In Germany the Decree of November 1923 against the Misuse of Economic Power provided for a Cartel Court. So far as I know neither the law nor the court were invoked to deal with a chemical manufacturer.[3] As for France, combinations were prohibited by article 419 of the Penal Code of 1810. Its applicability to industry was restricted, for the courts only took action if fraudulent means had been used to manipulate prices. The law was amended in 1926, without however altering the situation to any significant extent. While the law was no deterrent to businessmen, it is a fact that agreements on prices and markets were not publicized, so that the evidence, with few exceptions, is extremely scanty.[4]

In the United States the situation was different, both in law and in practice, and warrants more detailed attention. The Sherman and Clayton Acts were intended to prevent restraint of trade.[5] The intent, however, was weakened by

[1] 3. *Unterausschuss, die deutsche chemische Industrie* (1930).

[2] Report on Dyes and Dyestuffs, HMSO (1921), Cmd 1370; Report on the Explosives Industry, HMSO (1921), Cmd 1347. The reports run to 27 and 12 pages respectively.

[3] Decugis, H., Old, R. E., and Tschierschky, S. *Review of the legal aspects of industrial agreements* (League of Nations Economic and Financial Section) (1930), pp. 50–1.

[4] The French involvement in the potash and dyestuffs cartels will be mentioned later in the chapter. See also Fauque, M. *L'Evolution economique de la grande industrie chimique en France* (1932), pp. 198–9.

[5] The Sherman Act (1890) declared illegal every contract, combination or conspiracy in restraint of trade. The Clayton Act (1914) prohibited price discrimination, sales or leases involving the exclusion of a competitor's products, interlocking directorates, and the acquisition of shares in a competing business.

the Webb–Pomerene Act of 1918, which exempted from the Sherman Act Associations solely engaged in the export trade; corresponding exemption from the Clayton Act was granted to companies. It must have been difficult for businesses operating under the Webb–Pomerene exemption to keep home and export trade in absolutely water-tight compartments, and there must have been many occasions when the right hand knew very well what the left hand was doing.

The Federal Trade Commission was set up in 1914 to police the legislation. In the Harding–Coolidge era of the 1920s business-baiting was out of favour and the *laissez-faire* attitude of the period is illustrated by the following statement issued by the Commission in 1924:

> There seems to be no reason why a Webb–Pomerene association . . . might not adopt a trade arrangement with non-nationals reaching the same market providing this market was not the domestic market of the United States and the action of this organization did not reflect unlawfully upon domestic conditions.[1]

Official condonation as well as lack of enforcement staff provided an environment wholly favourable to the growth of agreements.[2] The Alkali Export Association (ALKASSO) was formed under the Act in 1919; it comprised all the leading manufacturers in the United States and controlled the export of alkali through quotas based on members' shares in production. ALKASSO acted as sales agent on behalf of its members and owned warehouses. In the same year export associations were also formed by two groups of companies in the phosphate-rock trade, but neither was comprehensive, and they were only partly effective. (With the formation of a world cartel for phosphate rock, the associations immediately became more useful.[3]) The Sulphur Export Corporation was established in 1922 and continues to the present day: its purpose was to fix prices, allot export orders, and split the profit on overseas sales among the producers of brimstone in proportion to their shareholdings. In 1923 the Export Corporation reached agreement with the Sicilian producers of brimstone, and secured three-quarters of the world market outside the United States and Italy.[4] The arrangement was successful and enduring, for the two groups controlled the production of sulphur for many years. There were probably other, more or less formal, understandings between American chemical manufacturers and their erstwhile rivals, but the Depression and Roosevelt changed all this. Congressional investigations began to ferret out all sorts of information that pointed to agreements and violations of the law. The Nye Committee's enquiry (1934–6) showed that Du Pont, against whom it was directed, had not only links with foreign producers of

[1] Quoted by Wilcox, C. *Competition and monopoly in American industry*, TNEC no. 21, (1941), pp. 221–2.

[2] The Anti-Trust Division of the Department of Justice had a staff of 23 in the 1920s. Twenty years later about 280 were employed. Wendell Berge (formerly head of the Division) in *The Times* (London), 6 December 1949.

[3] Federal Trade Commission, *The fertilizer industry* (1950), pp. 67–8.

[4] Federal Trade Commission, *The sulphur industry and international cartels* (1947), pp. 49 ff.

explosives, but was a party to agreements covering many other branches of chemical manufacture. Further revelations followed in the TNEC reports and, as if this was not enough, more incriminating material was dug up in evidence before the Committee on Patents (sometimes called the Bone Committee).[1]

Inevitably there were prosecutions. Sometimes the defendants pleaded *nolo contendere*, which saved them from trial but leaves one with the impression that the Government had established its point. For instance the civil suit of the United States *v.* Allied Chemical & Dye Corp. was terminated by a consent decree in May 1941 whereby the company agreed to sell not more then 35 per cent of the United States production of sulphate of ammonia after 1945. An even more important case concerned IG Farben, I.C.I., and Du Pont, which the Department of Justice won and thereby broke up and terminated agreements on patents and markets in force since the 1920s.[2] I.C.I., IG Farben, and Solvay were cited as co-conspirators of ALKASSO by the Department of Justice in 1944. It was the first time that a Webb–Pomerene association had been challenged by the Government, which alleged that ALKASSO, apart from its export functions, had also restricted competition between the manufacturers within the United States and had illegally fixed the price of caustic soda. The Department of Justice won the case and in 1951 ALKASSO was dissolved.[3]

The effect of such vigorous assertion of the anti-trust laws in the 1940s on the post-war chemical industry is well outside my scope. I have run ahead of the story to show that restraint of trade could co-exist for many years with legislation specifically directed against it if the Government was sufficiently tolerant. Moreover restraint of trade was a fact in the United States chemical industry, which generally and outwardly gave the impression of operating in a highly competitive commercial environment.

After this digression on the situation in the United States, we must now turn to Europe, where the legal constraints were negligible and long tradition condoned the operation of restrictive agreements. Apart from associations representing the chemical industry as a whole, there were many specialist groupings, which after the Kaiser's war began to play an increasingly important role. For example the French *syndicats* united firms in the same trade for purposes of organizing exhibitions, maintaining standards of quality, and introducing uniform packages and containers. Beyond these

[1] The Bone Committee's hearings run to many thousands of pages. An excellent summary and interpretation will be found in Stocking, G. W. and Watkins, M. W. *Cartels in action* (1946), particularly chapters 9–11.

[2] The indictment was filed in 1944 (U.S. *v.* I.C.I. *et al.*, Civil Action No. 24-13; U.S. District Court for the Southern District of New York). The results were summarized in the *Wall Street Journal* of 29 September 1951 and *Chem. Age* **65** (1951), no. 1687, 617.

[3] Federal Trade Commission, *International cartels in the alkali industry* (1950). Three sentences from Judge Kaufmann's summing up in the matter of caustic soda prices are appropriate in the context of price agreements. 'A suggestion is made that the stabilisation of prices was not "artificially enhancing or depressing" prices. The claim is baseless. Interpretation of the Anti-Trust laws does not turn upon such semantic niceties.' *Chem. Age* **61** (1949), no. 1572, 281.

admirable objectives there were others, less publicized, which were concerned with prices. The German pharmaceutical manufacturers started in 1905 the Interessenverband der chemisch-pharmazeutischen Fabriken and the Zentralauskunftstelle für Markenschutz; the names were abbreviated a little later to Verband der chemisch-pharmazeutischen Grossindustrie and Zentralstelle für Markenschutz. The two organizations overlapped to some extent: their main objectives were to reach and maintain agreements on wholesale and resale price margins, to negotiate with the cooperative buying associations of chemists and druggists, and to protect the trade names and patents taken out by members in foreign countries.[1] In Britain the first specialist trade association (if we exclude fertilizer makers) was the National Sulphuric Acid Association. It was the threatening excess capacity that brought the manufacturers together to talk about some form of joint action in March 1919. But it took almost another year before the Association was formally constituted with the encouragement of the Board of Trade.[2] It represented most of the acid makers, and apart from the publication of very useful statistics, handled the purchase of brimstone on behalf of the members. The Association was thus one of the Sulphur Export Corporation's important negotiating partners, and provides an early example of the replacement of arm's length bargaining between individual buyer and individual seller, by contract negotiations between associations. We have already mentioned the Fertiliser Manufacturers Association, which spoke for the British manufacturers of superphosphate and concerned itself with the supply of raw materials, terms of sales, and freight rates; its local sections dealt with prices. At home the FMA fought competition from imported superphosphate, abroad it secured control over a Belgian producer the more effectively to retaliate on the Continent. The FMA also played a leading part in the formation of the International Superphosphate Manufacturers Association in 1926. Despite all the vigorous activity, the FMA was unable to bring prosperity to this branch of the fertilizer industry in the 1920s, which had to wait until tariff protection and amalgamations brought an improvement in the next decade.[3] The British Sulphate of Ammonia Federation (BSAF) dates back to 1920, though there was an association in existence just before the war. Initially it was the sales agent for by-product sulphate made by gas works and coke ovens, but when Brunner, Mond began making the fertilizer from synthetic ammonia, the scale of their operations enabled them, as it were, to take over the Federation from 1926 onwards. Brunner, Mond, and their successors I.C.I., thus became at an early date the principal factor in the sulphate of ammonia business. The BSAF's principal objective was to obtain the best possible price for its

[1] Vershofen, W. *Wirtschaftsgeschichte der chemisch-pharmazeutischen Industrie* (1958), pt. 3, pp. 132 ff.

[2] *Chem. Trade J.* **62** (1918), no. 1610, 263, 265–6; Board of Trade, . . . Committee on Sulphuric Acid . . ., op. cit., p. 7.

[3] Monopolies Commission, *Report on the supply of chemical fertilisers* HMSO (1960), pp. 20–1.

members, and this it achieved, particularly after 1932, when an import duty on nitrogenous fertilizers was introduced.[1]

The proliferation of trade associations connected with marketing was a conspicuous example of that search for stability among chemical manufacturers to which we referred earlier in the chapter. Another was the growing number of agreements linking firms. Cartels and combinations were not post-war discoveries, but their role was enhanced in the 1920s. We may conveniently start with Germany, for the size and diversity of its industry, and also the sympathetic attitude of its leading men to this sort of arrangement, provided an environment in which methods for regulating trade could flourish. A start had been made with the principal heavy chemicals in the 1900s (and in a few cases rather earlier), and after the war conventions and production cartels came to include a broad range of chemicals. Under pressure from the larger firms, the smaller manufacturers had to give up their independence and conform to agreements if they wanted to remain in business. The rules and conditions of cartels were, generally speaking, reviewed annually, and only in very rare cases did they remain unchanged for five years. The primary purpose of cartels was to fix prices, export prices being usually lower than prices in the home market. Another object was to arrange a system of production or sales quotas. When several manufacturers formed a single sales agency, it was necessary to establish an administrative organization, sometimes reinforced by a pooling system from which members were compensated if they sold less than their full quota and into which other members paid a penalty if they exceeded theirs. The cost of running such fairly elaborate schemes never came to more than 3 per cent of the sales price in Germany.[2]

The cartel enquiry of the German Ministry of the Interior unearthed forty-seven cartels in the chemical industry. Thirteen were concerned with acids, alkalis, and other inorganic chemicals, two with dyestuffs, and the remainder with other products of the industry, notably explosives (4), wood distillation (6), and paint and glue (5). The two cartels in dyestuffs was certainly an understatement, and the Committee did not enquire or maybe were not told of the understandings and agreements between the two principal German groups of manufacturers, the independents, and the Swiss firms.[3] With rare exceptions the cartels were of the price-fixing or joint selling type; less than a third had provisions for restrictions on output. A few years later Erban prepared another list of cartels, which differed from that of the official enquiry: possibly a few new cartels may have been created after 1905, but in the main the difference stems from Erban's great familiarity with the

[1] ibid., pp. 49–52.

[2] Link, A. E. Economic Study of IG, Section 4—Sales Data (U.S. Military Government of Germany, Document TP/200/P/6, August 1945). K. von Heider, a senior member of the IG Farben staff, wrote the section on cartels: his stout defence of the system was based on a wide knowledge of the German practice.

[3] *Denkschrift über das Kartellwesen*, op. cit., vol. 1, pp. 114–32.

dyestuffs business.[1] He mentioned alizarin, indigo, congo red (until 1898), methylene blue, the naphthols, hydrosulphite, and many bulk chemicals, among them soda ash, caustic soda, bleaching powder, sulphate of alumina, sodium bichromate, sulphuric acid, oleum, acetic acid, ammonia, and fertilizers. Not all these products were commercially significant, and in some instances the cartels may have been short lived.

Among inorganic chemicals, the Syndikat Deutscher Sodafabriken, founded in 1887 and reorganized in 1900, was certainly one of the most important. The cartel enquiry mentioned that it had ten members who sold their soda ash, caustic soda, and bicarbonate through the syndicate. There were friendly connexions with the German salt mines, so that the Syndikat could exercise pressure (if necessary) on supplies to non-members. Deutsche Solvay-Werke dominated the group, and took care that exports were managed in such a way as not to harm Solvay companies in other countries. After the war Chemische Fabrik Kalk joined the Syndikat, which thereupon came to include all German manufacturers. It would appear that there were no production quotas until after the war.[2] Electrolytic alkalis and bleaching powder formed part of another cartel that extended to production quotas; Chemische Fabrik Griesheim–Elektron was the sales agent for the members. Consolidirte Alkaliwerke Westeregeln, one of the larger firms in the trade, resigned from the cartel in 1911, but returned in 1924.[3] Sales of by-product ammonia had been organized as early as 1895, when the principal coke-oven operators combined to set up the Deutsche Ammoniak Verkaufs–Vereinigung (DAVV). As the production of sulphate of ammonia grew, the DAVV gained in importance, and by the eve of the Great War it was handling most nitrogen compounds except nitrates. An interesting situation arose when Badische introduced synthetic ammonia and sulphate of ammonia in 1913. At first Badische had little difficulty in marketing its anhydrous ammonia through Kunheim & Co. and Chemische Fabrik Kalk, but the sulphate presented some problems. Badische launched it in eastern Germany. The local syndicate, at the behest of DAVV, undersold Badische, which thereupon bought a fertilizer mixer and so secured a captive market. The other producers were greatly impressed and soon came to an agreement with the new rival.[4] The outbreak of war transformed the situation and at its end Badische was assuredly the largest producer of fixed nitrogen in Germany. The Government, in order to shore up the position of the cyanamide makers that it had financed and to curb the growing power of large concerns exemplified by Badische, established a statutory cartel, the Stickstoffsyndikat, in May 1919. It was a

[1] Erban, F., *Kartelle und Konventionen in der chemischen Industrie* (1910). Erban had been a colourist with Hoechst and later became a lecturer at the university of Vienna.

[2] Grossmann, H. *Die Bedeutung der chemischen Technik für das deutsche Wirtschaftsleben* (1907), pp. 32–3.

[3] Anon. *Fünfzig Jahre Consolidirte Alkaliwerke Westeregeln* (1931), pp. 217–18, 249; Waller, P. *Probleme der deutschen chemischen Industrie* (1927), pp. 89–90.

[4] Voigtländer-Tetzner, W. and Pinnow, W. Geschichte der BASF—Kaufmännische Entwicklung, typescript, n.d. (c. 1939), p. 345a.

non-profit making sales organization for the entire fertilizer nitrogen business at home and abroad, and about eighteen months later it also began marketing technical nitrogen products. Within Germany the Stickstoffsyndikat dealt principally with the various buying organizations of the German farmers' unions and with the manufacturers of superphosphate, who required sulphate for mixing purposes. The Syndikat was controlled by a board of 10 members: 3 appointed by the Reich, 3 by Badische and 4 by the other producers; later the number was raised to 17, but the votes were so apportioned that Badische had a clear majority, and from 1924 onwards controlled 20 out of 33 votes.[1] Thus the Stickstoffsyndikat was in effect the instrument of Badische (and later of IG Farben), while the DAVV, which before the war had been the dominant element, now held only 6 votes. The cartel gradually lowered its prices, which represented a compromise between the low-cost synthetic ammonia works and the high cost cyanamide works. If prices had been lowered too much the cyanamide makers would have been out of business and even the makers of by-product sulphate would have suffered. Badische certainly benefited from the arrangement, but there is no published information on the profits it made from the fertilizer business in the 1920s.

The potash cartel, reorganized in 1919, was another statutory combination, but it differed from the 1910 cartel by providing for the compulsory restructuring of the German potash industry. It is possible that the 1919 cartel resulted from the combined pressure of socialists and technocrats (such as Rathenau) whose temporary association wielded considerable political influence. At any rate the new law established the Reichskalirat, the potash commission, to take charge of the industry. The Reichskalirat reported directly to the Ministry of Economics and supervised the implementation of the law: the mine owners were granted a monopoly of mining and marketing of potash, but they were obliged to rationalize production, and under a decree of 1921 they could be directed to close uneconomic pits, the remaining firms paying them appropriate compensation. During the 1920s scores of pits were shut or placed in reserve, and by 1928 the number of mining groups had been cut from 27 to 8 large concerns, which between them worked 61 pits.[2]

When we turn to the German organic chemical industry, the first characteristic to note is the large number of agreements and conventions. By contrast there were relatively few cartels with many member firms: the number of dyestuffs manufacturers was small and they tended to specialize, so that bilateral or trilateral negotiations commonly sufficed to deal with the matter. For instance the Auramine-G convention between Badische, CIBA, and Geigy fixed the minimum prices for that colouring material until 1907. Chrome dyes—black, later also blue-black—were the subject of a convention signed about 1907 between Geigy, Bayer, and Badische; in this case there was

[1] *Enquete*, pp. 165–7 (evidence of Bueb and Warmbold of the Stickstoffsyndikat); Waller, op. cit., pp. 172–3.

[2] Paxman, H. *Die Deutsche Kaliwirtschaft* (1929), pp. 46 ff.

no minimum price, but a profit pool in which Geigy's share was 9 per cent. The agreement was renewed in 1924. Other post-war agreements to which Geigy was a party covered Auramine, Rhodamine, and Victoria blue, the Swiss company's quota being 12–13 per cent.[1] An important bilateral agreement was the synthetic indigo convention between Badische and Hoechst, which had provisions for output as well as prices.[2]

Another characteristic was the difficulty of making binding agreements for aniline dyes, for in this branch of the industry there was a relatively large number of manufacturers. The larger among them tried repeatedly to agree on prices for intermediates. This was done less for their own sake, for they were practically self-sufficient in intermediates, or even altruistically to support the independent specialists, but above all to prevent exports at low prices to foreign rivals. In the end conventions for aniline oil and aniline salt were arranged, but the outcome was to raise prices to such an extent that foreign manufacturers were able to make the materials at a profit and sell them to dyemakers, with the result that exports from Germany declined. Here was an unexpected flaw in the marketing strategy of the big German companies; unfortunately we do not know what steps were taken to change the situation.[3]

The manufacturers of dyes, and later also those of pharmaceuticals, discovered that agreements limited to prices were not always sufficient, especially when the patents expired, and outsiders were able to move in and sell at slightly less than the convention's price, thereby disrupting the status quo. The remedy was the adoption of the 'pool' system combined with sales quotas: the members paid a fixed amount for every kilo sold into a fund, which was shared out at the end of the year among the members in proportion to their quotas.[4] In some cases there were arrangements for compensating members for abandoning production, a notable instance being the Alizarine Convention. It dated back to the early 1880s, had collapsed after a few years, been reformed in 1900, and survived to the outbreak of war. It was, incidentally, international in scope and besides Bayer, Badische, and Hoechst, included British Alizarine and the Verein für chemische & metallurgische Production at Aussig (Bohemia). In addition to the five manufacturing members, four companies received compensation for having stopped production.

Conventions and agreements are often mentioned in the context of German industrial history, but there is evidence of similar arrangements in Britain. For instance the English Caustic Soda Makers' Association included all the producers and met periodically before the war. The meetings provided an opportunity for discussing prices and for reviewing the progress of negotia-

[1] Bürgin, A. *Geschichte des Geigy-Unternehmens von 1758 bis 1939* (1958), p. 254, note 24; Heitz, W. *Erinnerungen. . . .*, n.d. (1937?), pp. 44–5, 127.

[2] *Denkschrift über das Kartellwesen*, op. cit., vol. 1, pp. 114 ff.

[3] Voigtländer-Tetzner and Pinnow, op. cit., pp. 180–1. Higgins, S. H. *Dyeing in Germany and America* (2nd edn. 1916), p. 100, mentions that the small independents sometimes undersold the convention members.

[4] Doermer, O. Die Konventionen und ihr Einfluss auf die Farbenfabriken, in *Geschichte . . . der . . . Fr. Bayer & Co. . . .* (1918), pp. 514–17.

tions with foreign manufacturers. The Association was unable to reach agreement with the Americans and the proposal to form a single export sales organization, mooted in 1910 and so the precursor of ALKASSO by nine years, was not adopted in the end because Brunner, Mond and United Alkali would not handle each other's exports.[1] The two firms collaborated smoothly in the home trade, where the largest group of customers were the soap makers. It was the usual practice to sign long-term contracts with them and—

> The evidence of several soap makers indicated that neither of the two alkali companies was desirous of supplying a soap maker who had previously been supplied by the other alkali company. We are nevertheless assured by the alkali manufacturers that *although there is an agreement between them, by which a percentage of their aggregate deliveries is allotted to each company, that agreement does not limit output*, and does not prevent either company from supplying any particular soap maker.[2]

Caustic soda and soda ash were major products of the pre-war British chemical industry and accounted for a substantial share of its total output. Besides these two, there were agreements on production, volume of sales, or prices (or on all three) for several other inorganic chemicals, specifically soda crystals, hydrochloric acid, ammonium carbonate, sal ammoniac (ammonium chloride), and sodium sulphate.[3] There was also an agreement between Oldbury Alkali Co. and the Anglo-Sicilian Sulphur Co. whereby sulphur recovered by the Chance & Hunt process was sold to Anglo-Sicilian at a fixed price.[4]

Britain and Germany were obviously not the only countries with a well-developed system of output or price restrictions, but we have no reliable and detailed information for France, Switzerland, and Russia before the war. According to Glaser, a soda cartel established in Austria in 1888 was still in existence on the eve of the war. There were at least nineteen cartels in the Austrian chemical industry at the time of the official enquiry in 1912.[5]

The concentration movement in the inter-war chemical industry, to which we will devote the next chapter, did not make cartels and trade associations superfluous, but had the effect of broadening their range from a single product or a group of related products to entire sectors of the industry. By extending agreements across national boundaries and thereby nullifying, or at the least

[1] Notes of meeting of English Caustic Soda Makers Association at Liverpool on 15 December 1910 taken by R. G. Perry of Chance & Hunt Ltd., in Chance & Hunt archives at Oldbury. See also chapter 6, pp. 138–9.

[2] Report on the Soap Industry, HMSO (1921), Cmd 1126, p. 15 (my italics).

[3] Chance & Hunt records at Oldbury. Owing to flood damage in 1960 the records are unfortunatelyi ncomplete.

[4] Board Minute Book of Oldbury Alkali Co. for 1896/7. Anglo-Sicilian was the sales company or Sicilian brimstone, and in 1906 was replaced by the government-supported export cartel for which see Squarzina, F. *Produzione e commercio dello zolfo in Sicilia nel secolo* (1963), pp. 95 ff.

[5] Glaser, J. *Die Chemische Industrie Oesterreichs und ihre Entwicklung* (1918), pp. 190, 192.

modifying, the effect of tariffs, the chemical concerns gave the post-war cartels a new dimension and political significance.

We may conveniently begin by referring to some significant trends. In the first place, the rapid development of the new sectors of the industry just before and during the war sharpened competition after it. Nitrogenous fertilizers were one instance (and I shall refer to them more fully later), pharmaceuticals were another. After 1905 the leading German manufacturers were associated in a community of interests, but challenged by the independents with their novelties they set up numerous cartels: Merck was a member of twenty-six, Schering of nineteen, Riedel de Haen of ten.[1] Here then we have an example of diversity of production leading to a multiplication of agreements. On the other hand the creation of giant concerns made it possible to dispense with agreements, simply because all the manufacturers were henceforth under the same management: this was the situation in the British alkali trade after the formation of I.C.I. or the German dyestuffs businesses after the formation of IG Farben. And it is interesting to observe that the latter claimed to be a member of only seventeen cartels in 1937, while one of its predecessors, Farbenfabriken Bayer, had been a member of twenty-five thirty years previously.[2]

Equally significant, but potentially more menacing, was the increasing number of international understandings. Before the war there were few, usually confined to a single product, for example alizarin dyes. But even in 1914 there already existed some associations with great and extensive powers. Nobel's licensees in Europe dominated the explosives business, and in turn were governed by the Nobel Dynamite Trust Co. Ltd. and the Soc. Centrale de Dynamite. And in their field of operations Solvay's licensees were just as powerful and closely-knit, for the Belgian partnership held important minority interests when it did not own them outright, and in addition exchanged technical information and process improvements with all the firms.[3] After the war ALKASSO was able to reach an agreement with the European ammonia-soda makers, the majority of whom were members of the Solvay connexion. Towards the end of the 1920s the American exporters raised their quota of world trade to 25 per cent and with the Europeans arranged to have '... complete cooperation to avoid competition in any part of the world'.[4] The phosphorus trade was just as tightly controlled, because agreements between Albright & Wilson, Griesheim, and Coignet before the war, and again after it, 'guaranteed a steady and very satisfactory commercial development'.[5] We will round off this account by observing that with the loss of Alsace, the German potash monopoly came to an end. But in 1924 the Kalisyndikat and

[1] *Enquete*, pp. 466, 440, 455; Waller, op. cit., pp. 82–3.

[2] Doermer, op. cit., p. 511; publicity brochure of IG Farben, undated, c. 1938.

[3] Trump, E. N. 50 years in Ammonia Soda Alkali. *Chem. metall. Engng* **40** (1933), no. 3, 126. Trump had been on the staff of the Solvay Process Corporation since 1882.

[4] *International cartels in the alkali industry*, op. cit., pp. 37 ff.

[5] Pistor, G. *Hundert Jahre Griesheim* (1958), p. 116 tr.

the sales company for the Alsatian mines agreed on prices and export quotas for sales to the United States. Two years later the agreement was extended to all countries, and in 1929 the Germans and the French joined forces to buy control of the American Potash & Chemical Co., which, having taken over an unsuccessful producer and revived it, now threatened to become a troublesome rival to the European monopolists.[1]

In the dyestuffs and nitrogenous fertilizer trades, international alliances were formed after the war, and have been well documented, so that their history and significance can be examined in some detail. They represented areas of chemical manufacture that by reason of their past growth and their importance in the late war were considered to be of strategic importance by industrialists as well as politicians. The demand for dyestuffs and fertilizers was presently to grow at a slower rate and the wartime extensions had, as their corollary, much spare capacity after 1918. In both sectors the Germans took the initiative.

As early as November 1921 representatives of the British Dyestuffs Corporation (B.D.C.) and of the German IG met in Paris. The IG people proposed an exchange of processes in return for an undertaking that the British firm restrict its activities to the United Kingdom. Not surprisingly the B.D.C. people refused, and further discussions the following May were equally inconclusive. Nevertheless the British were still hopeful, for the chairman said at the Annual General Meeting that the informal contacts with the Germans might lead to a 'working arrangement'.[2] With the deterioration of B.D.C.'s trading position an agreement seemed unavoidable, and indeed the only solution short of complete reconstruction. So both sides got together again to work out a draft agreement; it was completed in November 1923 and was a comprehensive document. The main provisions were as follows: 1. IG would not manufacture in the United Kingdom and conversely B.D.C. would not manufacture outside the British Empire; 2. IG undertook not to trade in the United Kingdom without the permission of B.D.C.; 3. British users of dyes would buy directly from the IG, which would compensate B.D.C., by purchasing from it an equivalent quantity of dyes, class for class, for sale overseas; 4. the two companies would exchange patents and know-how; 5. B.D.C. undertook to buy all those intermediates it did not make itself exclusively from IG; 6. IG promised not to make an agreement covering organic chemicals with any other British firm (inorganic chemicals, notably ammonia, were specifically excluded); 7. for the first five years B.D.C. would pay 50 per cent of the net profit above £200 000 to IG, after five years it would pay 50 per cent of £150 000 or over; losses, however, would not be shared; 8. committees would be formed to deal with policy, and technical and commercial matters; they would consist of three men from each side;

[1] *The fertilizer industry*, op. cit., pp. 96–8; Ogburn and Jaffé, op. cit.; Gürber, H. Die elsässische Kaliindustrie und das Kaliweltkartell, thesis Zurich (1942), pp. 117–20, 123.

[2] *Manchester Guardian Commercial*, 26 September 1922, pp. 813–14.

chairmen would rotate, except the chairman of the policy committee who would always be British and have veto powers; 9. the agreement would last for fifty years.[1]

The objects and scope of the draft agreement were totally different from the pre-war cartels. If carried out, IG would have dominated the manufacture of organic chemicals in the United Kingdom. But B.D.C.'s Articles of Association provided that the Government had to give its approval before signing an agreement of this sort. The British anticipated some delay and as an interim measure the Germans promised not to lower prices. B.D.C., in return, undertook not to raise capacity; as it was working well below capacity, the likelihood of extensions was remote! This temporary arrangement lasted until the end of 1924 and then lapsed.[2] In due course the draft agreement came before Baldwin's outgoing Cabinet, which passed it on, without a decision having been reached, to the incoming Labour Government. It was an awkward business for the new Cabinet. Lloyd-Greame, Baldwin's President of the Board of Trade, had not objected to the terms, nor, so it would appear, did the Licensing Committee, on which B.D.C. as well as the Board of Trade were represented. The Baldwin Cabinet were in no doubt that B.D.C. would become 'mere selling agents' if the proposal went through unchanged. Sydney Webb, MacDonald's President of the Board of Trade, assumed office in February 1924, and not surprisingly was unsure as to which course to take. So a Cabinet Committee under Lord Haldane was appointed to consider the agreement.[3] Haldane may well have felt that he had been through this once before, but at any rate he was familiar with the problem. The Colour Users Association were now asked for their views, and it emerged that they were very much opposed to the deal with IG. This clinched matters for Webb, until B.D.C. pointed out to him that unless there was an agreement with the Germans the company would have to be reconstructed. The Cabinet now had two problems before it: the proposed agreement, and the reorganization of B.D.C., with which was associated a writing down of the capital (partly held by the Treasury) and a complete withdrawal of the connexion with H.M.G. There matters stood during the summer of 1924, and a few months later MacDonald's Cabinet fell before a decision had been reached. The Labour Ministers were united in opposing the German agreement, but were divided in their attitude to the B.D.C. reorganization.[4] Baldwin's men returned towards the end of the year, and appointed another committee. It eventually reached a decision and recommended that the Government withdraw from B.D.C., and the Cabinet agreed at last on the matter in the latter half of 1925.

[1] Memorandum by Government Directors of British Dyestuffs Corporation to Cabinet, circulated 9 April 1924. PRO, CAB/27/232.

[2] British Dyestuffs Corporation, Review of the dyestuffs industry (1927).

[3] PRO, CAB/23/46 (Cabinet Conclusion 5 (24) item 3, pp. 492–3) and PRO, CAB/24/165.

[4] The Cabinet Committee reported on 26 July 1924, the members disagreeing on the terms of the Government withdrawal from BDC. PRO, CAB/24/165, Report 54–5.

[5] The sequel to the story will be found in the next chapter.

Thus four years of negotiations and marking time had led to no result except the limited and temporary agreement for 1924. While B.D.C. had no prepared positions on which to fall back, the bargaining strength of the IG had grown from 1921 onwards. In that year the Germans negotiated the first treaty with the Cie Nationale de Matières Colorantes (CNMC). The agreement with the French resembled the B.D.C. proposals: the Germans assigned their patents and know-how to CNMC in return for $16\frac{3}{4}$ million gold fr. (nearly £0·7 million) spread over ten years. They undertook not to compete with CNMC in metropolitan France, and in return the French firm agreed not to compete with the IG outside the French dominions. The IG became entitled to one-half of CNMC's profits until 1965.[1] When the French army occupied the Ruhr in 1923, the Germans considered that they were no longer bound by the agreement and also ceased making Reparation deliveries of dyestuffs. The following year Kuhlmann took over CNMC and on their side cancelled the agreement. As the change of ownership did not sufficiently improve the outlook for French dyes it was necessary to resort to new negotiations. The outcome was another agreement with the Germans, this time involving the newly formed IG Farben. The terms, amended in 1927, emphasized the dominant position of the Germans, for they were awarded four-fifths of the combined exports. The unequal partners agreed on a common marketing policy, and the French even undertook to stop the production of vat dyes (except synthetic indigo) in return for a larger quota of the cheaper sulphur colours.[2]

We have already referred to the relations between German and Swiss firms, and the existence of several well-established links may explain why formal negotiations for a general agreement on dyestuffs only started several years after the war. At the beginning of 1925 the German representatives suggested to the delegates of the three Swiss companies that they should combine and take over the fourth company, Durand & Huguenin, which was then owned by the IG. Such 'rationalization' measures were fashionable at the time, and the Germans thought of it as the first step in a wider scheme whereby the combined Swiss producers would exchange the majority of their stock for shares in IG Farben. It was an ambitious proposal, of the sort that IG Farben was putting persuasively (if not forcefully) to German firms. If carried out it would have concentrated the bulk of world dyestuffs capacity and virtually, the entire output of quality dyes, under single ownership. But CIBA, Geigy, and Sandoz valued their independence, and their hostility to the German proposal strengthened the bonds between them. They turned it down.[3] IG Farben dropped the matter until 1928, when negotiations were resumed: on this occasion the Germans suggested an export cartel, and after investigating

[1] Memorandum by Sir Percy Lloyd-Greame to the Cabinet, dated 15 January 1924; PRO, CAB/24/164.

[2] USTC, *Census 1927* (TIS 37), (1928), pp. 160–1; Fauque, op. cit., pp. 211–12; ter Meer, F. *Die IG Farben* (1953), p. 53. I do not know what caused Kuhlmann's changes of policy in 1924 and again two years later.

[3] Bürgin, op. cit., pp. 254–5.

the proposition the Swiss firms accepted. They were given 8 per cent of the combined exports, the CNMC got 17 per cent and IG Farben held 75 per cent. But the Swiss firms, unlike CNMC, retained their independent sales organization, and were granted a privileged position in southern Europe. The French were given special advantages in Latin America and the Germans reserved to themselves Russia and the large markets of the Far East. Thus the world was carved up and regional privileges allotted to the three parties to the cartel. The United States, however, were excluded from the scope of the agreement, partly because it was no longer a significant importer and partly because the Swiss and the Germans were manufacturing there on a substantial scale; for similar reasons the United Kingdom and the British Empire were excluded.[1]

Although I.C.I., the successors to B.D.C., had not reopened negotiations with the Germans, there is evidence of at least one agreement, for chrome dyes, between the British and the German concern.[2] But even I.C.I. was not able to hold out indefinitely, and in 1931 it joined the cartel to sell dyes 'in an orderly manner'.[3] The German–British–French–Swiss group virtually controlled world trade in dyestuffs, because neither the United States, nor Japan, nor Italy exported dyestuffs to any significant extent. The cartel, which comprehensively regulated the dyestuffs business until 1939, was more elaborate than the earlier and more limited versions. In particular the agreement made allowances for different types of dyes and also their application. Thus within the overall framework of quantity controls, there was scope for the further development of specialities or new uses, so that there was a good deal of incentive for the member firms to pursue research.[4] The British and the Swiss in particular benefited from this aspect of the 1931 agreement.

The first international nitrogen convention came at the very end of our period, being founded in 1929 by IG Farben, I.C.I., Norsk Hydro, and the Chilean producers' syndicate. Unlike the dyestuffs cartel there was no long gestation period, because until 1928 IG Farben and the Chileans between them dominated world trade in nitrogenous fertilizers and nitrates respectively. But in 1928/9 the completion of several large nitrogen fixation plants threatened to throw the export trade into disarray, while at the same time there were signs that demand was lagging so that pockets of excess capacity were threatening. (Soon both were to become the conspicuous manifestations of the Great Depression.) An additional factor was that Norsk Hydro, traditionally a large exporter, was converting its plant from the arc to the Haber–Bosch process and had become, rather against its will, an associate of IG Farben.

The first cartel, known as the DEN group (i.e. Deutsch–Englisch–Norwegische Gruppe) began operations in the fertilizer year 1929/30. It was a

[1] USTC, *Census 1928* (TIS 38), (1930), pp. 146–8.

[2] Report of the Dyestuffs Industry Development Committee. *The present position and development of the dyestuffs manufacturing industry*. HMSO (1930), Cmd 3658, pp. 7, 44.

[3] Chairman's speech at the Annual General Meeting; *The Times*, 14 April 1932.

[4] Bürgin, op. cit., p. 257.

fairly simple arrangement—essentially an agreement on export quotas. Thus the Norwegians got 27·2 per cent of combined exports up to 250 000 tons and 25·3 per cent above that figure.[1] The group then bargained with the Chileans. The parties did not have production or sales quotas, and all they could agree on was an across-the-board price cut of 5 per cent, and on rebates. They also undertook not to compete with each other.[2] This cartel was impractical, because it did not impose quantity restrictions on the Chileans and took no account of the smaller European producers of nitrogenous fertilizers. In 1930 a new agreement was made and under the name of Convention Internationale de l'Azote (CIA) it operated in the fertilizer year 1930/1; a special agreement linked the CIA with the Chileans. CIA had a system of sales quotas and compensation for unfilled quotas. It was better than the DEN group, but then an improvement was essential, since the fall in agricultural purchasing power was accompanied by a cut in fertilizer consumption. The Chileans nevertheless thought they could do better on their own and in 1931 refused to renew the agreement. A price war followed and led to retaliatory tariffs or import licensing by several countries. A third agreement was signed in 1932 and with some modifications lasted until the outbreak of the Second World War. The CIA's primary purpose was to stabilize the export prices of fertilizer nitrogen (home prices were taken care of by the national cartels) by maintaining restrictions on output. The plan, though simple in concept, did not work smoothly at first, and in 1933/4 the Chileans withdrew once again. But from the mid-1930s onwards the reviving agricultural demand provided more favourable conditions for operating the international agreement.[3]

Conclusion

The difference between the pre- and the post-war cartel was less one of kind than of degree, or rather scope. The change was due to excess capacity and low return on capital, which led businessmen to seek closer ties and more effective alliances. The international cartels and the big concerns that emerged in the 1920s importantly influenced the structure of the industry up to and even beyond Hitler's war. We have not sufficient information to determine whether restraint of trade, as practised in the 1920s, was economically justified. Some contemporaries had no doubts and asserted that the measures were necessary and desirable. Waller thought of them as a form of production planning, a sort of capitalist riposte to the Socialist planners. In Germany, he wrote, conventions led to 'a single direction of output as orders are allotted according to the quotas to the different factories'.[4] Sir Alfred Mond, though once a Liberal, had no reservations:

> . . . the combinations . . . which are taking place are literally forced upon us . . . In fact you want to eliminate unnecessary capital expenditure. You want to eliminate

[1] *Norsk Hydro history*, pp. 298–9. [2] Grossmann and Weicksel, op. cit., pp. 59–60.
[3] Stocking, G. W. and Watkins, M. W., *Cartels in action* (1946), pp. 143 f.
[4] Op. cit., pp. 107 8, tr.

unnecessary competition. You want to eliminate unnecessary fluctuations in production, and you want to be able to foresee the probable needs of consumption. All this is to the interests, not merely of the producer and capitalist, but it is to the interests of everybody in a reasonably organized community to-day.[1]

That the benefits of the agreements extended beyond the members' profit and loss accounts was obvious to all. Manufacturers who sought stable conditions of production and sales thereby safeguarded the jobs of their men and so the standard of living of their dependants. Work for some, at a time when unemployment was high, was a valuable result of cartels. Restraint of trade need have no detrimental effect on research. During the 1930s cartels and agreements were particularly active in the chemical industry, but the decade was equally fruitful in major chemical advances.[2]

The agreements and the other measures suggest that in the 1920s a period of rapid advance was ending, that there was a pause before the next stage in growth took place. The cartels solidified, as it were, the *status quo*, for they did not alter the character of the member firms, and indeed their 'product mix' hardly changed. The big firms were able to remain big and almost everywhere replaced trade associations as the *de facto* spokesmen for the chemical industry. The small firms were able to survive but were prevented from growing and becoming a threat to the big ones. And so the aims listed by Sir Alfred in 1927 were indeed achieved.

[1] Mond, Sir Alfred, *Industry and Politics* (1927), p. 250.

[2] One student of the plastics industry has come to the conclusion that IG Farben probably 'spent more on plastics research and development than any other firm in the world'. Freeman, C. The plastics industry: A comparative study of research and innovation. *Nat. Inst. econ. social Res. Rev.*, November 1963, no. 26, pp. 33–4.

10

FROM CARTELS TO CONCERNS

ONE way of looking at the chemical industry in the inter-war years is in terms of changes in its organizational structure. Rationalization and concentration were vogue words at the time, but they describe quite inadequately the amalgamation of substantial enterprises into large and powerful concerns that came to dominate entire sectors of the industry. This chapter will summarize the evolution and main characteristics of the chemical concerns. Nowadays they are familiar to wider circles than readers of *Fortune* and stockbrokers' circulars, but some forty years ago the emergence of giant firms was one of the most significant results of developments in applied chemistry after the turn of the century. We have already referred to the technological advances, which, intensified by war, were the major element of change and form the general background to the story here. I am mindful that I have largely omitted the growth of demand, which influenced unevenly, though powerfully, the growth of the different sectors and so of individual firms. It would require a separate study to do justice to the detailed relationship between demand and growth in the principal areas. Suffice it to say here that the chemical concerns were usually production- rather than marketing-oriented.

IG Farbenindustrie AG

We have seen in Chapter 5 that the German dyestuffs industry, which was the pacemaker of that country's chemical industry as a whole, consisted of eight firms: BASF, Bayer, and AGFA were associated in the Dreibund; Hoechst was the leader of another group that included Cassella and Kalle; Griesheim–Elektron (with important interests outside organic chemistry) and Weiler–ter Meer were not connected, either financially or administratively, with the others. After the outbreak of war the dyestuffs business fell abruptly and was only partly compensated by orders from the military for inorganic acids, pharmaceuticals, auxiliaries for high explosives, and photographic chemicals and accessories.

Duisberg, the architect of the Dreibund, somewhat disenchanted by this form of association, advised his colleagues in the industry to seize the opportunity presented by war-time co-operation between former rivals, to consider the future organization of the dyestuffs trade. In particular he recommended, in a circular dated July 1915, the amalgamation of the Dreibund with the Hoechst group to form an *Interessengemeinschaft* (IG). But no action was taken on the circular, because its recipients discounted the threat of post-war competition from foreign producers, and in any case were still thinking of an early German victory. A year later, however, victory was receding, while developments in Britain and the United States pointed to excess capacity and competition, and so caused alarm to the German manufacturers.[1] There were also good domestic reasons for some formal association, notably the intensification of the war effort and the spread of the control and licensing machinery, which inevitably strengthened collaboration between firms. Lastly, financial problems of some magnitude were looming up: cash flow declined while the investment in new chemical ventures proved extremely costly. Notable instances were the ammonia plants of Badische at Oppau and the even larger one under construction at Leuna; Hoechst was developing the manufacture of acetic acid from acetylene and in this connexion was buying the Knapsack carbide works; Griesheim, in partnership with the Metallgesellschaft and the Reich was building aluminium factories and enlarging the magnesium works. Hoechst was probably least affected, but Griesheim and Badische were in a quandary, and the latter in particular urgently needed very large sums.

The moment was thus opportune for re-examining Duisberg's proposals, and his adaptation of the principles of the Dreibund to a larger grouping now proved acceptable. The agreement to form the Interessengemeinschaft der deutschen Teerfarbenfabriken (sometimes called Erweiterte IG and here shortened to IG) was signed by the eight firms on 18 August 1916. Several smaller companies also joined the IG at that time.[2] The duration of the agreement was initially fifty years, but in 1920 was extended to 1999. The corporate identity of the member firms as well as their internal organization were left untouched and the key provisions of the accord concerned the powers of the general council (*Gemeinschaftsrat*) and the arrangements for the pooling of profits. The general council consisted of directors delegated by the member companies and its first chairman was C. von Weinberg, then the chairman of Cassella. The council met periodically to approve new projects and the raising of additional capital, to consider the acquisition of other companies or of licences, the exchange of know-how, and the renewal of cartels, conventions, etc. All resolutions required 70 per cent of the votes to be carried, but decisions affecting the closure of works, restrictions on output, or changes in the organization of sales required unanimity.[3] Each company calculated profits according to standard rules and the profits so ascertained were then pooled and shared out as follows: Badische, Bayer, and Hoechst got 24·82 per cent each, Cassella 9·81 per cent,[4] AGFA 8·08 per cent, Griesheim 6 per cent, and

[1] There is no detailed account of the negotiations leading to the IG; there are some references in Flechtner, H. J. *Carl Duisberg* (1960), p. 278 and in his Vom Dreibund bis zur 'kleinen IG' in Jacobi, F. (Ed.) *Beiträge zur hundertjährigen Firmengeschichte* (1963/4), p. 43.

[2] A. Leonhardt & Co. of Mühlheim, Carl Jaeger GmbH of Düsseldorf, Wülfing, Dahl & Co. AG of Barmen, and Chemikalien Fabrik Griesheim.

[3] ter Meer, F. *Die IG Farben* (1953), pp. 18 f. ter Meer was a member of the original General Council.

[4] Cassella was owned by the other firms and in turn held their preference shares.

Weiler–ter Meer 1·65 per cent. The shares in the profit pool determined the voting rights in the general council and also the entitlements to the proceeds from new capital issues.[1] The chairmanship of the council rotated every two years, and the chairman's company provided during that time the considerable secretariat services.

The agreement excluded from the scope of the IG the activities of Badische in the area of nitrogen fixation, those of Hoechst in acetylene and its derivatives, and Griesheim's non-ferrous metals. These three represented very important operations indeed and it is a pity that the details are not available, because they would shed interesting light on the financial aspects of these growth sectors and the policies of the firms engaged in them. Some writers have stated that the profits (or losses) from these activities were expressly excluded for a number of years from the IG profit pool. Yet there is some evidence that the IG supplied the finance for them, and there is obviously an area of uncertainty until definite information becomes available.[2] As far as nitrogen fixation was concerned a new arrangement was made in 1920 when Ammoniakwerke Merseburg GmbH was formed by the member firms, each subscribing to the 500 million M capital in proportion to its stake in the IG profit pool. (It is not clear whether the large works at Oppau were included in the new company.)

The foreign reaction to the IG was, as might be expected, hostile:

It cannot be doubted that this enormous engine of commercial warfare has been created expressly for the expected war after the war, and that it is intended to undertake still more efficiently and on a larger scale the various methods by which German attacks upon all competition were carried on.[3]

The British mission visiting Germany in 1919 took intentions for facts and reported that production was concentrated in the most efficient firms, which supplied the others at cost plus 10 per cent.[4] The report was unnecessarily alarmist, for the IG, because it was an 'enormous engine', moved slowly and achieved rather less than it had set out to do.

Most of IG's activities concerned the traditional dyestuffs business. It had been intended at the time of the negotiations in 1916 to concentrate the manufacture of dyes, hitherto made in several works, in the two most efficient factories of the group. In this way it was hoped to effect economies and yet provide a ready means for cost comparisons. The implementation of the policy inevitably entailed closures for which unanimous approval had to be obtained. But the general council dithered, and eventually appointed a

[1] Delahanty, T. W. *The German dyestuffs industry* (1924); the percentages shown were to be altered slightly after ten years.

[2] Flechtner, op. cit., p. 280; ter Meer, op. cit., p. 18; Beer, J. J. *The emergence of the German dye industry* (1959), pp. 138–9. Dr. E. Fischer of the Hoechst archives and Miss M. Frosch of the Badische archives kindly investigated the matter at my request, but unfortunately were not able to locate the precise details asked for.

[3] Report of the Alien Property Custodian; *J. Ind. Engng Chem.* April 1919, p. 355.

[4] Association of British Chemical Manufacturers, *Report of the British . . . Mission . . . in . . . Germany* (1919), p. 36.

special sub-committee for the purpose in 1920. A year later it decided to shut down the alizarin red department at Hoechst. The company was given Kalle's azo-section to compensate for the loss.[1] Another committee was rather more successful or effective in cutting down the number of dyes made from the 35 000 available before the war. But it was impossible to achieve the coordination of export sales because the member firms, the 1916 agreement notwithstanding, retained their foreign agents and apparently condoned competition between them.[2] At home however some measure of uniform marketing was introduced in 1922 when the vat dyes for the cotton trade made by Badische, Bayer, and Hoechst were sold under the single trade name, originally registered by Badische, of 'Indanthren'.[3] It will not have escaped notice that these immensely successful colours were made in three works, one of which ought to have been closed. But the proposal to implement the 1916 principles might have jeopardized the entire agreement, and so nothing was done.

In its relations with foreign manufacturers, the IG was not notably effective. There were negotiations with the British, the French, and the Swiss, but they led to no lasting agreements. Special committees were founded to handle trade with Russia and Japan, possibly because the two countries had central buying organizations. From 1919 onwards the IG became responsible for handling all matters concerning Reparation dyes and for this purpose created the Zentrale der Farbstoff Fabriken in Frankfurt. Later it took over the responsibility from the German Government for implementing the revised Reparations Agreement on dyestuffs.[4]

The IG seems to have been of some use in the newer areas where the old attitudes of the dyestuffs trade were less entrenched. For example the financial help for the extensions at Leuna and the rebuilding of Oppau after an explosion in 1921 must have been welcome at a critical time. Another example of fruitful collaboration was the joint development by Badische and Griesheim of methods to cheapen the manufacture of phosphoric acid after 1922 which led a few years later to the construction of a large plant at Piesteritz and the expansion of compound fertilizer production.[5] Two other achievements of the IG require mention: the bickering and quarrelling over trade marks and patents ceased, and in 1920 an insurance company was launched, and Pallas GmbH, as this effort at diversification was called, brought about considerable savings.[6]

[1] Bäumler, E. *Ein Jahrhundert Chemie* (1963), p. 96. The detailed statistics reproduced by Delahanty (op. cit., pp. 42 f.) show that alizarin red production at Hoechst ceased in the first half of 1920. But there is no evidence up to August 1923 of Kalle's loss. This part of the deal may have been postponed until 1924 or 1925 for which years detailed figures are not available.

[2] Voigtländer-Tetzner, W. and Pinnow, H. *Geschichte der BASF: Kaufmännische Entwicklung*, n.d. (c. 1938–9), pp. 333, 336.

[3] Bayer, H. BASF, das Werden eines Grossunternehmens. *Die BASF* **14** (1964), no. 4, 181.

[4] This was in June 1922. See the *5th report on the work of the Reparation Commission from 1920 to 1922*, HMSO (1923), p. 235.

[5] Pistor, G. *100 Jahre Griesheim* (1958), pp. 116–17.

[6] ter Meer, op. cit., pp. 19, 49–50.

In summing up the effectiveness of the IG in the eight years of its existence, 1916–24, one must remember that circumstances prevented the attainment of the objectives formulated in 1916. It had been designed to maintain the companies' pre-war domination over the dyestuffs business, or more loosely, their leadership in organic chemicals manufacture generally. But the post-war environment differed from the one anticipated, and the IG was ill-suited to adapt itself. Within Germany dyestuffs were becoming a relatively less important sector, because technological progress affected nitrogenous fertilizers, plastics, photographic products, and man-made fibres, rather than colouring materials.[1] Outside Germany the post-war competition in the world dyestuffs market did not materialize, while French, British, and American manufacturers, whatever the quality of their goods, could not be eliminated in their respective home markets because their Governments would not have tolerated this. So the IG had to accept the existence of new rivals and seek a *modus vivendi* with them. In this it was only partly successful. A further unforeseen problem was civil war in China, which caused a decline in German exports to the area.

These difficulties might have been solved had there been organizational changes that took account of the alterations in the structure of the members' business and their profitability, and altered the objectives accordingly. But the IG's constitution was inflexible. In law it was a terminable association so that any member could give notice and leave, thereby injuring the others.[2] During the inflation very little could be done, although Duisberg, for the third time in his life, prepared a memorandum in which he drew attention to the need for a drastic overhaul of the IG's organization (December 1923). His opportunity came in July 1924 when he became chairman of the General Council and his son, Curt, head of the IG secretariat services at Leverkusen. He presented his proposals again at the council meeting and elaborated them in a memorandum circulated in October. Duisberg realized that a more effective organization was required. But he rejected an outright amalgamation of all the companies because of the high cost involved (stamp duties and fees might come to £2 million) and the consequential administrative changes. Instead he recommended a management company that would act on behalf of the eight member firms. The advantages of the scheme, even with hindsight, are hard to discern, except that the firms would retain their corporate identity and so their famous names as well as their own administration. It is not clear how Duisberg expected the middle management of the firms to work with the supervisory staff of the management company, particularly on matters of purchases and sales, which would devolve from the former to the latter. The expense of setting up the new company was almost £3 million.[3]

[1] One example of potential commercial conflict was the growth of Bayer's photo-chemical, sensitive paper, and camera business, which overlapped Agfa's. Stenger, E. *100 Jahre Photographie und die Agfa* (1939), pp. 37 f.

[2] ter Meer, op. cit., p. 23.

[3] Flechtner, op. cit., pp. 335 ff., summarizes the project. It was the forerunner of the elaborate regional management companies that some large American firms established in the 1960s to

Duisberg's plan met with qualified approval from the Hoechst board, but on the whole was not well received by the others. Bosch, who had become chairman of the executive board of Badische in 1919, attacked it on the ground of impracticability. He had just returned from a trip to the United States and was convinced that central management and rationalization of production (concepts which were then being vigorously promoted by Anglo-Saxon in-industrialists) was the only way out for the German firms.[1] There was also another element in the attitude of Bosch: he was about thirteen years younger than Duisberg, and temperamentally the opposite of his senior, a pragmatist where the older man believed in well-designed systems. Above all Bosch was determined not to give way on the question of a complete fusion. He prevailed, and in November 1924 the general council voted by a majority for amalgamation. Duisberg thereupon resigned as chairman of the council and Bosch, who succeeded him, for it was the turn of Badische after Bayer to take the chair, now became responsible for the preparation of the merger arrangements. The details took months, one of the most difficult to settle being the composition of the management of the new concern. The matter was solved in the end when all the members of the general council, except Bosch, agreed to become members of the supervisory board of the new company. By kicking themselves upstairs they left Bosch in charge of the management and he became automatically the chief executive.[2]

The amalgamation of the eight companies to form IG Farbenindustrie AG was announced in September 1925, and the new company started business on 1 January 1926. It had a capital of 454 million RM (£22·7 million), which may be compared with the combined capital of 214 million M (£10·7 million) of the predecessor companies on 31 December 1913.[3] The preliminaries to the merger were laborious and the financial work immense, but the results of the first year showed its huge size: the turnover in 1926 was 1027 million RM (about £51 million). Business improved in the next few years to 1266 million RM in 1927 and to around 1420 million RM (£71 million) a year in 1928 and 1929. Sales of dyestuffs and textile auxiliaries amounted to 436 million RM in 1928, three-quarters being for export.[4] The traditional dyestuffs business thus accounted for very roughly one-third of total turnover, and though the lack of data prevents comparison with pre-war years, we can safely assume that

---

coordinate the activities of their European enterprises. Unlike Duisberg's company, the new companies do not trade, for this would result in their being liable to taxation in the country where they are established.

[1] Management, like bringing up children, has its fashions: some twenty years previously Duisberg had returned from the United States, certain that an American-style trust was needed for the German chemical industry. See Chapter 5.

[2] Flechtner, op. cit., pp. 340–50. His account is more informative than K. Holdermann's, *Carl Bosch* (1953), pp. 195–207.

[3] Waller, P. *Probleme der deutschen chemischen Industrie* (1928), pp. 27–8. Ebert, W. *Die chemische Industrie Deutschlands* (1926), p. 84; USTC, *Census 1924* (TIS 33), (1925), pp. 148–9. Cassella and Kalle were not included in the amalgamation, as the other companies held their shares.

[4] ter Meer, op. cit., p. 73.

the relative importance of the business had greatly declined. The measures that the IG had failed to take were now swiftly carried out. A single dyestuffs sales office was established at Frankfurt and an extensive rearrangement of production programmes was put in hand. The principle of manufacturing a dye or group of related dyes in two works, which had not been vigorously applied before 1926, was not pursued by the new management either, and was gradually abandoned after 1930. Nevertheless there was a measure of rationalization: production of dyestuffs ceased at Elberfeld, Uerdingen, and Biebrich, and changes were made at Hoechst and at the Offenbach works of Griesheim. As a result of the rearrangements the Leverkusen works of Bayer increased their make of dyes by over a quarter, while Badische gained a little and Hoechst lost a little.[1]

The manufacture of synthetic ammonia and of nitrogenous fertilizers grew quickly, and in 1928 sales of these products reached 540 million RM (£27 million), of which nearly one-half was exported. The factories at Oppau and Leuna employed about 35 000 people in 1927/8, that is roughly a third of the concern's entire labour force. Output, measured in terms of nitrogen, rose from 117 000 tons in 1920 to 350 000 tons in 1925; two years later it reached 450 000 tons and in 1928 was 640 000 tons.[2]

The other manufacturing activities were reorganized, except for heavy inorganic chemicals, where the high cost of transport made it uneconomic to concentrate production. (IG Farben retained the IG practice, itself inherited from the old régime, of buying considerable quantities from independent producers.) But the structure of the heavy organics business was thoroughly modified. Thus the manufacture of aniline oil was stopped in four works, but enlarged in the other two belonging to IG Farben. The factory at Uerdingen was reconstituted and became a supplier of organic intermediates in bulk. Hoechst was assigned responsibility for all chemicals from acetylene, and in this connexion took over the Griesheim patents and technique for polyvinyl acetate manufacture; the production of this important new compound began at Hoechst on the commercial scale in 1928.[3] Elsewhere the competition between Bayer and AGFA was stopped by the simple expedient of adopting the trade name AGFA for the photoproducts of IG Farben wherever they were made. The sales of this branch were rising fast, and production came to be concentrated at Leverkusen (photographic paper), Munich (cameras), and Wolfen (film and chemicals). Griesheim shed many of its chemical activities

[1] Bäumler, op. cit., p. 103; ter Meer, op. cit., pp. 34–6. It is not clear to what extent Cassella at Mainkur and Agfa at Berlin were affected by the changes.

[2] These figures may be compared with about 100 000 tons N obtained from by-product recovery of coke-oven operations and 95 000 tons N from cyanamide plants. It should be noted that Leuna also produced synthetic methanol and in the late 1920s was beginning the manufacture of petrol by the hydrogenation of coal. As a result a substantial portion of the manpower at the works was not concerned with nitrogen compounds. ter Meer, op. cit., p. 39; *Enquete*, pp. 137, 150, 154, 156; Marcus, A. *Die grossen Chemiekonzerne* (1929), p. 54.

[3] Hauser, H. *Geschichte des Werkes Uerdingen der Farbenfabriken Bayer* (1956), pp. 85, 89; Pistor, op. cit., pp. 83–4.

and concentrated on light metals. IG Farben made fibres at Dormagen (cuprammonium) and Berlin (acetate), and through Köln-Rottweil AG operated three viscose factories. The turnover of photoproducts, metals, and fibres benefited from the vigorous policies pursued by the concern, and in 1928 amounted to roughly 130–140 million (£6½–7 million).[1]

IG Farben also displayed considerable activity abroad, and in particular adopted and carried out systematically a unified export marketing policy. Whereas the IG had encountered much hostility carried over from the war, its successor benefited from the friendlier atmosphere characterized by the Locarno spirit. The establishment of subsidiaries abroad became possible once again. A start was made on neutral ground by buying the small Swiss firm of Durand & Huguenin, which had pioneered a new type of soluble indigo dye in 1924. In the United States the dyestuffs agencies of the predecessor companies were brought together in the General Dyestuffs Corporation at the end of 1925. The factories previously owned by the representatives of German interests were enlarged and in 1929 became part of the American IG Farben Chemical Corporation.[2] Another major investment was in Norsk Hydro, which thereby came for the second time under German influence: in 1927 IG Farben granted the Norwegian company the use of the Haber–Bosch process and helped with the know-how, and in return obtained a 25 per cent holding.[3] The shares in the various foreign enterprises were held by IG Chemie, a holding company registered in Switzerland. Much litigation was caused by this arrangement after Hitler's war, and was only concluded in 1965 by a compromise settlement between the American government and the successors to the Swiss company.

This brief account has necessarily been confined to the more notable developments that followed the merger of the companies. The administrative methods adopted to run this huge enterprise will be reviewed in the next chapter. Here it is appropriate to look at the effect of the rationalization measures on the activities of the predecessor companies, and, at greater length, at the role of IG Farben in the German chemical industry.

The formation of the concern and the reorganization that affected most of its branches, entailed the multiplication of committees and the proliferation of new administrative and supervisory functions. Most of the reshuffling was completed by about 1930, its chief feature being the pronounced tendency towards product specialization (except for heavy inorganic chemicals) in the main works. Two examples must suffice: Bayer became the concern's principal manufacturer of the azo-group of dyes and took the lead in that business; Badische became the specialist in high-pressure synthesis in all its aspects. This specialization stemmed from the technical competence displayed by the production teams at plant level over a period of many years, and the directors of

[1] Brand, P. Photographische Erzeugnisse, in Jacobi, F. (Ed.), op. cit., p. 177; Greisenheyner, M. *Schaffende Gedanken—Schaffende Hände* (1952), p. 28; ter Meer, op. cit.

[2] Haynes, W. *American chemical industry* (1948), vol. 4, pp. 233 ff.

[3] *Norsk Hydro history*, pp. 241–6.

IG Farben thus recognized and officially confirmed arrangements that had proved their worth over a long time. In some instances, fairness and the need to keep the labour force together may have led the directors to assign to a works that had lost its traditional business, a new group of products: thus Kalle lost dyestuffs, but gained cellophane film.[1] It is not possible to be precise about the detailed effects, but in general and qualitative terms the following conclusion appears valid: the reorganization of the 1920s gave the main works a characteristic product mix, and allowed them to enlarge their product range within the broad lines assigned to them. For the next thirty years or so the factories developed along the course charted for them in 1926–30. Recently a number of changes have blurred the demarcation lines, but it is still possible to discern product groups in which a works has a monopoly or dominating position by descent, as it were, from those early days.

Neither the predecessor companies nor the IG had been financially connected with other chemical manufacturers. But owing to its size and its undoubted commercial power IG Farben pursued a different policy. Although Bosch and his colleagues were not aiming to eliminate all competition, they made arrangements with other companies that, under the pretext of orderly marketing, in practice minimized it. A profit-pooling agreement was signed with Dynamit AG and an agency contract made for a range of Kali-Chemie's products, and so sales of explosives and of a number of inorganic chemicals were 'rationalized'.[2] There were many other arrangements whose primary effect was to set limits beyond which either party would not go. In this way IG Farben kept off the manufacture of sodium carbonate and derivatives, metallic sodium, cyanides, peroxides, barium-, borax-, and bromine compounds, straight superphosphate and potash fertilizers, lacquers and paints, ceramic colours, and some pharmaceuticals.[3] As a matter of general policy, increases in the capacity for making inorganic chemicals (except, of course, synthetic ammonia) were only sanctioned if other manufacturers could not supply more cheaply, and then the additional plant was for captive use only. In this manner the board of IG Farben hoped to sustain firms making heavy chemicals.

The policies carried out between 1926 and 1930 gave IG Farben a secure hold, if not a monopoly, over the growth sectors of the modern chemical industry, and left the others in possession of the nineteenth-century business. The generalization simplifies the picture and the conclusion may be harsh. Nevertheless in a growth industry such as chemicals, the ownership of new technologies, and the ability to commercialize them, is the basis of financial power, which in turn provides the cash flow to sustain further research and development. The later history of IG Farben, and from the 1950s onwards of

[1] 'Ozalid' dye-line paper for blueprints was another growth product assigned to Kalle. The equipment and lay-out of the factory at Biebrich had to be altered considerably.

[2] Waller, op. cit., pp. 50–2, 88.

[3] The list, though incomplete, represents the main products.

the successor companies, testifies to the benefits secured by the delimitation agreements of the 1920s.

The concentration and rationalization movement among the other chemical manufacturers requires only brief comment. We may start by noting that Deutsche Solvay-Werke, next to IG Farben the largest firm, remained untouched by these changes. DEGUSSA (which specialized in cyanides, sodium compounds, and products for the ceramics industry), Metallurgische Gesellschaft, and Metallbank organized an *Interessengemeinschaft* in 1922 that appears to have been modelled on the IG. The association ended in 1926 and DEGUSSA drew closer to IG Farben and the soap firm of Henkel & Co. These two held one-third each of DEGUSSA's preference shares and entered into long-term contracts with it for the sale of metallic sodium and sodium perborate respectively. In this way DEGUSSA's commercial viability was assured and its take-over by foreign interests prevented. The identity of the foreign bidder was not revealed, and perhaps it was just another pretext for establishing an orderly marketing scheme for the company's group of products.[1] Another large firm was Kali-Chemie, which emerged in 1928 when Kaliwerk Friedrichshall took over Rhenania–Kunheim. The former was one of the smaller groups in the German potash industry, but had done well since the reorganization of potash mining after the war. The latter was itself the result of a merger of companies in the heavy chemicals trade. Kunheim had been in business for almost a century, and made a large range of products; its management, which had long been dominated by the family, had not been particularly successful. In 1924 it had exchanged shares with Rhenania-Verein Chemischer Fabriken. Rhenania had been a famous name in the great days of the Leblanc industry, but its conversion to new technologies had been unsuccessful. Rhenania had been shored up by the Verein Chemischer Fabriken Mannheim, another old-established firm, which profitably made soda ash by the ammonia-soda process on a scale equivalent to an eighth of the country's output. Kali-Chemie comprised altogether sixteen works, but their combined sales were only about £1½ million in 1928. The amalgamation came too late to revive the business and ensure its independence: during the 1930s Solvay-Werke obtained a controlling interest and thereby completed their hold over the German alkali trade.[2] The superphosphate business did badly in the 1920s: Rhenania's lime–phosphate fertilizer, made from indigenous materials, would not sell and so proved a weakness to the group as a whole. But even those firms that used imported phosphate rock were in trouble. As in Britain, foreign competition was blamed, but it is likely that IG Farben's aggressive marketing of compound fertilizers hurt the sales of straights. There were some mergers in the sector, but they did not improve the situation of the manufacturers.[3]

[1] *Enquete*, pp. 351–2.

[2] Rüsberg, F. *50 Jahre Kali-Chemie AG* (1949); Spiegel, L. *100 Jahre Kunheim* (1925), p. 43; *Enquete*, pp. 250 ff.

[3] The most important were those of Chemische Produkten Fabriken Pommerensdorf-Milch,

IG Farben, through Bayer and Hoechst, occupied an important position in the pharmaceutical industry. Despite the great fragmentation that characterized it even after the war (according to the occupational census of 1925 there were no less than 1219 establishments with altogether 24 000 employees), a handful of firms dominated the field: these big companies between them owned six works, which taken together employed 7200 people. They also owned valuable trade names and above all maintained a priceless goodwill with the medical profession. Businesses such as Merck could not be eliminated by IG Farben, and as if to make doubly sure it was linked with C. F. Boehringer & Söhne and Knoll & Co. in an *Interessengemeinschaft*.[1] J. D. Riedel, another large pharmaceutical manufacturer, took over E. D. de Haen AG (which specialized in fine chemicals) in 1923 and five years later employed altogether 1800 people at Berlin and Seelze (near Hanover). The amalgamation of Schering with Kahlbaum, which took place at about that time, was a similar union of manufacturers of drugs with fine chemicals. The company formed part of a much larger organization, the Kokswerke & Chemische Fabriken, which would now be called a conglomerate. Kokswerke brought Voigtländer & Sohn, Germany's oldest camera makers, into the group, reorganized the country's roofing-paper trade, and in 1929 formed the Duco AG with Du Pont, for making the American company's finishes in Germany. The less profitable enterprises, notably the superphosphate plants, were sold.[2]

It is difficult to assess IG Farben's role and position objectively: throughout its short life and even beyond, it aroused emotions that one does not usually associate with a limited liability company. Publicists passionately took sides, and as early as 1927 it was the centrepiece of a Ph.D. dissertation.[3] IG Farben formed the subject of two novels.[4] Since the war, Dr. Fritz ter Meer, a former director, has written an apologia, which is at least informative.[5] Trade unionists were initially friendly, but around 1930, when many employees were made redundant, their attitude became extremely critical.[6] They were shortly to lose their voice, but American congressional committees took over and dug up some damaging information. After the war, most of the

which was active in the area of Stettin, and of Guano-Werke, which had factories in Hamburg and Lübeck. Both took place in 1927. *Enquete*, pp. 250 ff.

[1] Schmitt, R. Die pharmazeutische Industrie und ihre Stellung in der Weltwirtschaft, Ph.D. dissertation, Frankfurt (1932), pp. 139–41; Waller, op. cit., pp. 61, 82 ff.

[2] Kokswerke retained substantial holdings in Silesian coke-oven enterprises. Holländer, H. *Geschichte der Schering AG* (1955), pp. 47 ff.; *Enquete*, p. 429; Waller, op. cit., pp. 57–9.

[3] Waller's dissertation Entwicklungsziele der deutschen chemischen Industrie was submitted to the Faculty of Law and Political Science at Hamburg in 1927 and published a year later, the title being changed to *Probleme der deutschen chemischen Industrie*.

[4] K. A. Schenzinger's *Anilin* was published in 1936 and is said to have sold nearly 1·8 million copies. The print order for his *Bei IG Farben* (Munich 1953) was only 20 000. Thus do tastes in fiction change.

[5] ter Meer's little book has been reprinted at least twice since 1953.

[6] cf. *Enquete*, pp. 492 ff.

IG Farben records fell into Allied hands and were thoroughly aired by conscientious investigators. Some of the directors were prosecuted at Nuremberg, and the concern was eventually broken up.[1] Its assets in West Germany were handed over to the successor companies, Badische, Bayer, Hoechst, and Hüls, but its shares, though the company is in liquidation, are still being traded on the German stock exchanges; some people are evidently hoping that the former assets in East Germany will be returned or compensation paid for them!

Soon after the formation of IG Farben there was a tendency in some quarters to belittle the significance of the concern. Professor Wiedenfeld wrote in 1927, '. . . to identify the German chemical industry with the dye trust, as is generally done, is to disregard obvious facts'.[2] And Dr. C. Ungewitter, who certainly knew better, declared, 'In Germany, the IG Farbenindustrie AG only represents a fraction of the whole chemical industry, although it includes the entire German aniline dye production'.[3] Perhaps the intention was to lull, rather than to mislead, but the statements give a totally false impression of the concern's role. The Enquete addressed itself to the matter, and the employment statistics for the leading German firms show the gulf that separated them from IG Farben.

TABLE 10.1

Employment in large German chemical companies, 1913, 1925, 1928[4]
(000s)

	1913	1925	1928
IG Farben	46*	84	114
Deutsche Solvay-Werke	5·5	6·7	6·3
Bayerische Stickstoffwerke	0·5	4·7	5·3
DEGUSSA	0·5	1·9	2·3
Kali-Chemie (excl. potash mines)	3·3*	3·5*	2·1
Schering–Kahlbaum	1·2*	2·7	3·0
E. Merck & Co.	2·1	2·7	3·0
Dynamit AG	3·9	3·5	1·8
German superphosphate manufacturers	n.a.	4·7	2·4

*Employed by predecessor companies

In the latter half of the 1920s, the German chemical industry employed nearly 0·3 million people, and on this reckoning IG Farben accounted for

[1] A. Krischan has listed about 400 articles in the German language on the break-up of IG Farben from 1947 to August 1952. For details see *Chem. Ind., Düsseld.* N.S. **4** (1952), no. 10, 899 ff.

[2] Wiedenfeld was then professor of economics at Leipzig. *Cartels and combines*, League of Nations (1926 II 70), (1927), p. 15.

[3] He was head of the Chemical Group in the Reichsverband der deutschen Industrie. *The chemical industry*, League of Nations (1927 II 4), (1927), p. 53.

[4] ter Meer, op. cit., p. 80, for IG Farben; *Enquete*, for the others.

over a third.[1] We have less material on the financial situation, and inter-firm comparisons were unreliable in those days. But the tabulation of assets shown below emphasizes once again the size of the concern.

TABLE 10.2

Assets of large German chemical companies, 1925, 1928, 1935[2]
(million RM)

	1925	1928	1935
IG Farben	1120*	1753	1574
Deutsche Solvay-Werke	124	134	102
DEGUSSA	45	55	78
Kali-Chemie	43	56	56
Schering-Kahlbaum	42	63	56

*1st January 1926.
Note: The figures for 1935 are on a slightly different basis from those for 1925 and 1928.

Attempts have also been made to relate IG Farben's turnover to the value of the German chemical industry's production. In 1927 this gives a figure of about 35 per cent, which includes, however, a significant element of double-counting, and is further swollen by the inclusion of products not properly classified as chemicals, such as matches.[3] A detailed estimate, prepared after the war, gives IG Farben a share of 28–30 per cent of output, but 47–48 per cent of invested capital.[4] The basis of all these calculations is arguable, but the proportions show the order of magnitude, and, more important, give the clue to IG Farben's role and significance: it was a capital-intensive business. In this industry the plant and machinery and the additions made thereto are a sure indicator of growth. While IG Farben grew, the rest of the industry, with a few exceptions, marked time. The underlying structure changed less than might have been expected in the period considered by this book: the dyestuffs makers of the Kaiserreich were now under new management, and their businesses expanded and diversified into technologically advanced and commercially promising areas. The rest, but for the handful of exceptions, continued very much as before, and because they were weak, IG Farben was able to confine them to sectors that developed rather more slowly.

Imperial Chemical Industries Ltd

I.C.I. was formed in December 1926, and from that time on has been the largest chemical business in the United Kingdom. The old statistical records

[1] *Enquete* (p. 6) shows that total employment in the industry, excluding soap, rubber manufacture, and chemists' shops, was 0·26 million in 1925 and 0·29 million in 1929.
[2] *Enquete*, and *Chem. Ind., Düsseld.*, Gemeinschaftsausgabe, August 1937, p. 189.
[3] ter Meer gives the turnover of IG Farben, while *Enquete* (p. 8) estimated the value of production at 3600 million RM.
[4] Gross, H. *Material zur Aufteilung der IG Farbenindustrie* (1950), pp. 27 f.

appear to be incomplete, but the available data suggest that the firm's turnover was of the order of £25–30 million in the late 1920s; the number of employees was about 40 000.[1] In both respects, I.C.I. was considerably smaller than IG Farben, but it is nevertheless worth noting that if we take the lower figure of £25 million, sales a head for I.C.I. were £625, while IG Farben's were barely higher at £640. The two concerns, although they have often been compared, differed importantly in their product range, their structure, and not least in their attitude to the development of technologically advanced products. I.C.I.'s four predecessor companies determined to an even greater extent than IG Farben's the evolution during the early years.

The first of these, Brunner, Mond, was the most important in the chemical industry, its net assets at the time of the merger being about £18 million. Sodium compounds remained its principal business, and the company's hold over the trade had been strengthened during and after the Great War: the Castner–Kellner Alkali Co., with large electrolytic works at Runcorn and Wallsend-on-Tyne, became wholly owned in 1920, and incidentally provided a link with the Cassel Cyanide Co.; Chance & Hunt, which made acids and alkalis at Oldbury, was also acquired. By the mid-twenties Brunner, Mond and its satellites probably accounted for over three-quarters of the alkalis made in the United Kingdom.

The company's development of the ammonia synthesis, first at Runcorn and later at Billingham, was without doubt its principal achievement in the 1920s. The first commercial-sized plant, the so-called No. 2 Unit, started at Billingham in December 1923. A bigger works, the No. 3 Unit, was built between 1925 and 1927 and went into service in February 1928. Billingham was, however, considerably smaller than IG Farben's nitrogen fixation works. The No. 2 Unit was producing roughly 80 000–90 000 tons a year of sulphate of ammonia (16 000–18 000 tons N) in the period 1927 to 1930, and the No. 3 Unit was working up towards 230 000 tons (46 000 tons N) a year in 1929/30.[2] The works included a power station, fertilizer silos, and an anhydrite mine (a large deposit having been found there providentially soon after the war). All this was very expensive, and in 1925 Brunner, Mond converted Synthetic Ammonia & Nitrates Ltd., the operating subsidiary, into a public company for the purpose of raising £2 million by a debenture. Although the interest on the loan was guaranteed by the Treasury under the Trades Facilities Act, one-third of the issue was left with the underwriters. Evidently the public thought little of the venture, perhaps because investors were unimpressed with the parent company's profitability since the war. Sir Alfred Mond, the first Lord Melchett, who became chairman in 1926, certainly

[1] I am very much indebted to Mr. P. L. Morgan for guidance on the sales figure; the number of payroll workers was 33 100 on 1 January 1927 and there were about 10 000 in staff grades. About 18 months later the number of workers was 40 000 according to a statement made by Sir Alfred Mond at the Annual General Meeting on 31 May 1928.

[2] Parke, V. E. *Billingham—The first ten Years* (1957). The No. 1 Unit was at Runcorn and had a capacity of 21 tons a day of ammonia.

faced considerable financial problems, not all of them due to the activities at Billingham. He was no chemist, but a lawyer and politician, with great experience of mergers, his most recent success having been the creation of the Amalgamated Anthracite Collieries. There were additional difficulties caused by the resignation of the previous chairman, Roscoe Brunner, and by a major row with Lever Brothers, which was eventually settled out of court at great expense to Brunner, Mond.[1]

Sir Alfred and most of his colleagues were therefore responsive to the proposal for amalgamation, which came from Sir Harry (later Lord) McGowan, the chairman of Nobel Industries Ltd. McGowan was not a chemical manufacturer either, and his knowledge of the industry, explosives excepted, was, to say the least, scrappy. In 1920 he had been briefly chairman of British Cellulose & Chemical Manufacturing Co., a war baby formed to make cellulose acetate dope, which was not adjusting itself smoothly to the production of film and fibre, and later, after a complete overhaul, was renamed British Celanese. At about the same time he had also been made a director of the British Dyestuffs Corporation. McGowan was essentially a specialist in mergers: his first great success was the reorganization of the Canadian explosives trade in 1910, and in 1918 he combined the United Kingdom explosives and ammunition manufacturers into Explosives Trades Ltd., renamed Nobel Industries Ltd. in 1920. Nobels was essentially a holding company: it had taken over the shares of seventeen firms, which, with their subsidiaries, made up a group of several score of companies. From 1919 onwards many works were closed, a few were combined, and this process of rationalization was accompanied by vigorous diversification through the acquisition of shares in other companies. The most notable holding was a large block of General Motors stock, which had been bought on the advice of Du Pont's directors, and proved to be a very lucrative investment indeed. Nobels also owned companies making leather-cloth, cellulose lacquers, automobile radiators and carburettors, zip fasteners, and, rather oddly, the Scala cinema chain. The group was an early example of the now fashionable conglomerate. The investments were shrewdly chosen: Nobels predecessor firms were familiar with cellulose chemistry and with solvents, so that leather-cloth and nitrocellulose lacquers were sensible and practical extensions of the traditional business. They also held commercial promise. Similarly the ammunition companies in the group (among them Eley Brothers Ltd. and Kynoch Ltd., two famous names in the Midlands) had gone into non-ferrous metal rolling and fabricating, and so had become connected with companies making components for cars and motor bicycles.[2] By 1926 Nobel Industries had net assets of around £20 million and unlike Brunner, Mond and the other

[1] Relations with Lever Brothers were rarely friendly, and this latest trouble stemmed from the accidental discovery that Brunners had failed to implement an agreement to give Levers 'most favoured nation' treatment on prices. Levers insisted on a refund. They were paid £1 million. Private information.

[2] Trotter, R. (Ed.) *The history of Nobel's Explosives Co. Ltd.* (1938).

firms in the I.C.I. merger, its profits were on a rising trend during the early 1920s.

The other two companies that became part of I.C.I. were very different from the two senior partners. The United Alkali Co., the outcome of the merger in 1890 of the principal Leblanc soda makers, had soon fallen on hard times. The war boom restored its fortunes temporarily and it managed to earn profits of about £0·7 million a year in 1923–5. The miners' strike of 1926 cut them by half. United Alkali's main works were at Widnes, where it manufactured sulphuric acid, chlorine, saltcake, and several organic chlorine derivatives. The Leblanc process was finally abandoned after the war, but the company owned a big ammonia-soda works at Fleetwood, with a capacity of 200 000 tons tons a year, and an electrolytic alkali plant was opened at the Allhusen works at Gateshead-on-Tyne in the early 1920s.[1] United Alkali had tried for many years to attain success through product diversification, but did not succeed, partly because much of its plant was old and partly because demand at home failed to rise, while competition was severe abroad.

The short history of the British Dyestuffs Corporation (B.D.C.) had been very troubled. Sir William Alexander was appointed chairman in 1921, and, though no chemist, he knew all about the chemical trade, through Charles Tennant & Co. He quickly diagnosed the firm's troubles: personal antagonism between the managing directors, too rapid expansion in 1919–20, failure to develop high quality dyes, large stocks, factories too big for the requirements of the business, excessive overheads, German competition. He did something about the first and obtained the resignations of Herbert Levinstein, the technical managing director, and of Sir Joseph Turner, the commercial managing director.[2] The firebrands departed to air their quarrels in public, but left behind warring factions. Sir William was less successful with the other problem areas. And as to checking German competition, the only solution was to make a deal with the IG. But, as we saw in the last chapter, the Government stepped in and blocked it.[3] In the home market B.D.C. encountered hostility from the Colour Users' Association, a powerful pressure group, which feared the effects on the supply of dyestuffs resulting from an Anglo-German combination.[4] There was also some rivalry, if not enmity, from other dye makers in the United Kingdom. The Swiss-owned Clayton Aniline Co. made about 3500 tons of dyes in 1925, equivalent to 20–25 per cent of United Kingdom output; the works of L. B. Holliday,[5] and the British Alizarine Co. between them accounted for a further 15 per cent or so of production. Although the three companies bought some intermediates from B.D.C., they

[1] Hardie, D. W. F. *A history of the chemical industry in Widnes* (1950).

[2] *Manchester Guardian Commercial* **4**, 29 June 1922, 813/14.

[3] See pp. 273–4.

[4] In addition the manufacturers of fine chemicals were alarmed by the prospect of Germany re-entering the market through the agency of the British Dyestuffs Corporation. Memorandum by Sidney Webb, dated 6 May 1924, PRO, CAB/24/165.

[5] He was the grandson of Read Holliday whose business had become the Government-sponsored British Dyes in 1915. He made explosives during the war.

competed with it in finished colouring materials.[1] There were also several smaller firms, among them Scottish Dyes Ltd., which had established a great reputation for fast colours.[2]

The difficulties at home and abroad led the board of B.D.C. to consider a reconstruction and a complete break with the Government, which still held a considerable investment and in particular had special voting powers. The negotiations began in the first half of 1925 and proceeded slowly. Baldwin set up a Cabinet Committee and also sought advice from the Committee on Civil Research. The Conservative Ministers (like their Socialist predecessors) remained adamant that an agreement with IG Farben was inadmissible, but in November 1925 they authorized the Treasury to settle with B.D.C. Accordingly the Government holding of £1·7 million in B.D.C. was sold for £0·6 million. In return the Government demanded and obtained a guarantee that not more than 25 per cent of the voting power would be allowed to pass into foreign ownership, and that the company would 'keep in touch' with the Board of Trade on technical and research matters.[3] The decision at last opened the way to a complete reorganization under yet another chairman, Lord Ashfield, who as Sir Albert Stanley had brought about in 1918 the merger of British Dyes with Levinstein and so created B.D.C. Sir Alfred Mond also joined the board, and Dr. E. F. Armstrong F.R.S., late of Joseph Crosfield & Sons, became managing director. Although some of the shareholders and even the High Court Judge who sanctioned the capital reconstruction remained critical,[4] the new team were determined to use the available resources effectively and make a fresh start. At the beginning of 1926, B.D.C. bought control of Scottish Dyes and so raised its total production to about 7300 tons a year. And what was more important than mere volume, the company secured at last a dominant position in the manufacture of British vat dyes.

Doubts still remained in some quarters, and Reginald McKenna of the Midland Bank, which had been involved in the capital reconstruction, approached McGowan about that time with the suggestion that Nobels take over B.D.C. That indeed would have enabled the creditors to sleep soundly. McGowan saw Ashfield on the matter and after consideration formed the opinion that the takeover should form part of a wider regrouping in the

[1] British Dyestuffs Corporation, Review of the dyestuffs industry (1927).

[2] Its Caledon Jade Green (patented in 1920) was a famous anthraquinone dyestuff. In 1924 the company developed a process whereby the anthraquinone dyes could be rendered soluble in water and the following year the 'Soledon' soluble vat dyes were first marketed. Abroad only Durand & Huguenin had a similar product; IG Farben spotted the winner and bought it. Morton, J. *A history of the development of fast dyeing* (1929), p. 48.

[3] Report of the Chairman of the Committee on Civil Research, 16 October 1925, reproduced in PRO, CAB/24/175; Papers Relating to the Liquidation of the Government Interests in the British Dyestuffs Corporation Ltd., HMSO (1925), Cmd 2545.

[4] On 23 March 1926 Mr. Justice Eve, in the Chancery Division of the High Court, in granting the petition to reduce the capital, concluded the proceedings by saying '. . . I rather take the view of those who lost their money, I am rather glad the country is out of it, because I do not know how much more we should have lost', and he made B.D.C. pay the Government's costs!

chemical industry. He had talks with Sir Alfred Mond and his colleagues from Brunner, Mond about this project, but by July 1926 no decision had been reached.[1] The delay may have been caused by the political and economic situation after the General Strike, but there was also some hesitation among the Brunner, Mond people. Mond and some of his associates were deeply interested in oil-from-coal processes and so were favouring an alliance with IG Farben, which was then about to start constructing a plant for hydrogenating tar and lignite. He discussed the proposal with Bosch and other IG Farben men as well as with Solvays in Brussels in August. A measure of agreement was reached: Brunner, Mond would take over B.D.C. and this British group, Solvays, and Allied Chemical would then be linked with IG Farben by an exchange of shares. The German company would thereby secure the stability in the chemical, and particularly the dyestuffs trades that was not yet in its grasp, while the other partners would obtain access to IG Farben's hydrogenation patents and its know-how. Shortly afterwards Mond and Bosch sailed severally to the United States, the former to see Allied Chemical and obtain its consent, the latter to witness the event and clinch the deal, and also take up an invitation extended earlier in the year by W. C. Teagle of Standard Oil to accompany him on the autumn inspection trip through the Esso empire.

Mond's activities had not gone unnoticed, and McGowan followed the parties across the Atlantic. Thus he was at hand at the crucial moment when Allied Chemical refused to consider an exchange of shares with IG Farben. So the project for an international company collapsed and relations between Allied Chemical on the one hand and Brunner, Mond and Solvays on the other deteriorated. Bosch went on his inspection trip, which was followed some months later by the first of several IG Farben–Standard Oil agreements. These circumstances made McGowan's task easier. Mond was readily persuaded to join him in the formation of a purely British group of which the core would be Nobels and Brunner, Mond. The discussions between the two men and their advisers began in the last September days and were continued on the voyage home, so that the main principles and objectives of the proposed merger had been established before the 'Aquitania' reached England.[2]

On 22 October Sir Alfred issued a long statement in which he set forth the reasons for combining Brunner, Mond, Nobel Industries, United Alkali, and B.D.C.: he spoke of a 'united front' (presumably against troublesome competitors at home and abroad), and the avoidance of duplication. The identity of the individual units was to be maintained, but the new board would co-

[1] Obituary of Lord McGowan in *The Times*, 14 July 1961; 25 years ago. *I.C.I. Magazine* **29** (1951), no. 182, 356.

[2] W. J. Reader's *History of I.C.I.*, vol. 1, Chap. 19, describes the foundation of I.C.I. Mr. Reader kindly allowed me to draw on his typescript for information on Mond's interest in hydrogenation and the plan for a link with the other firms. The German version of these events and the details of their parallel discussions with Standard Oil must await the publication of the Bosch papers. Meanwhile K. Holdermann, (*Carl Bosch* (1953, pp. 225–9)) and F. A. Howard, (*Buna Rubber* (1947, pp. 10–19)) refer to these German–American contacts over coal hydrogenation.

ordinate and provide overall policy as well as financial resources. He concluded by stating that 'instead of leaving it to individual units to make arrangements for the world's competitive conditions as best as they can, [we] will give them all the authority and prestige and advantages of a great combination'.[1] There were few facts, but the statement was nevertheless heady stuff for the investors.[2] *The Times* approved the proposed merger, but its business leader managed the remarkable feat of avoiding all reference to chemical manufacture in its comment on the amalgamation! *Chemistry & Industry*, in a leading article, addressed to professional chemists, was pathetically jejune.[3] So much then for the publicity accompanying one of the major industrial events of the 1920s.

Negotiations between the directors of the four companies took place during October and November. There were some last minute doubts about the wisdom of including B.D.C., but these were resolved partly, it would appear, by an appeal to emotion, for as Mond later explained it was 'our duty' not to leave dyestuffs 'in an undefended position'.[4] The death of Roscoe Brunner supplied a touch of personal drama: he had been unwell for some time and suffered a breakdown. His name was not put forward as a director of the new company, and unknown to him his wife had been canvassing his claims with the press. The discovery precipitated the final crisis, he shot his wife and then killed himself.[5]

In December the last details were settled and I.C.I. was incorporated: its issued capital was £56·8 million, only a little less than that of Lever Brothers and about the same as IG Farben's after the new issue in the autumn of 1926. I.C.I. was a holding company, exchanging its preference, ordinary, and deferred shares for those of the predecessor companies, whose share capital aggregated about £38 million. The new concern certainly required a substantial turnover and buoyant profits, because the owners of Brunner, Mond and Nobel Industries had been promised that the share exchange would give

[1] *The Times*, 23 October 1926.

[2] Movement in the price of the ordinary shares of the predecessor companies:

	24 September 1926[a]	*22 October 1926*[b]	*1 November 1926*[c]	*1 December 1926*[d]
Brunner, Mond & Co.	35/6	40/9	40/–	36/6
Nobel Industries	29/–	34/–	37/3	37/–
United Alkali	30/–	37/6	37/6	32/9
British Dyestuffs Corporation	11/–	17/–	16/10½	16/9

(*a*) Activity first developed; (*b*) Mond's statement; (*c*) eve of announcement of share exchange terms; (*d*) a few days before the formation of I.C.I.

[3] Amalgamations. *Chemy Ind.* N.S. **45** (1926), no. 44, 789–90.

[4] Speech at the 3rd Annual General Meeting of I.C.I., 29 April 1930.

[5] Sir Max Muspratt of the United Alkali Co. wrote an obituary note to *The Times*, but his Brunner, Mond colleagues remained silent. *The Times* 5 and 9 November 1926. Roscoe's brother, Sir John Brunner, was appointed to the board of I.C.I. in December. He resigned less than a year later.

them the same income on their holding in I.C.I. Indeed, without this promise the merger might not have been achieved. I.C.I. issued £31·1 million ordinary £1 shares, £16·2 million 7 per cent cumulative preference shares, and £9·5 million 10 shillings deferred shares. The directors undertook to pay 7 per cent on the ordinary shares so that altogether £3·3 million had to be found for the dividends on the first two classes of capital. Events were to show that the promoters had been overconfident.

In 1926 and later the leaders of I.C.I. justified the merger on the ground that 'rationalization' (then a vogue word) in the chemical industry was essential. Old works were to be closed and profitability improved by eliminating inefficiency and superfluous plant. Among other novel features was the creation of a central staff department, and even more important, the establishment of central financial control, which would make possible the direction of large funds to the growth points in the concern. But the structure of the predecessor companies limited the extent to which the objectives of the 'rationalizers' could be attained, and in this respect I.C.I. had less scope for reorganization than IG Farben. The greatest potential savings lay in the alkali business: United Alkali's big factory at Fleetwood was taken over by I.C.I. (Alkali) Ltd., the successor to Brunner, Mond. In exchange the Alkali company gave up Castner–Kellner and Chance & Hunt to I.C.I. (General Chemicals) Ltd., the successor to United Alkali. This was a reshuffling of assets, and not a systematic rationalization that related capacities to future demand. The moves left the character of the two firms unchanged and their weaknesses untouched. It is worth recalling that Brunner, Mond and United Alkali had reduced competition in the soda business to a minimum years before the I.C.I. merger. As if to underline their traditional distinctiveness both continued to make caustic soda, the one by the lime–soda process, the other by electrolysis. Nevertheless some important decisions were taken: Cassel Cyanide Ltd. of Glasgow was bought in 1927, and the Tyneside works of United Alkali, the 'cradle' of the Leblanc soda industry, and of Castner–Kellner were closed. All three were transferred to Billingham and combined there in a new factory which from 1929 onwards operated as the Cassel works. Elsewhere duplication and some inefficient enterprises survived, notably in dyestuffs and among the numerous branches of Nobel Industries.[1]

Rationalization proved to be an enormous undertaking, and except for the measures just described, had made little progress by 1932 when McGowan admitted that '. . . it was not yet complete . . . and much remains to be done . . . in the way of eliminating disadvantages which arise from the separate existence of different companies'.[2] The process continued for the next thirty years, testifying to the endurance of traditional attitudes and organizations.

[1] In addition to the four works of B.D.C. (two each at Manchester and Huddersfield), there were dye works at Grangemouth in Scotland and Trafford Park (Manchester), the last-named being bought from the British Alizarine Co. (1931) which ceased to make colours in London.

[2] At the 5th Annual General Meeting of I.C.I. on 14 April 1932.

At the beginning of 1964 the amalgamation of the Alkali and General Chemicals Divisions to form the Mond Division and the segregation of the metals business marked the culmination of this particular development.

The introduction of central financial control was easier to accomplish, and its considerable benefits outweighed the drawbacks of bureaucratic administration. For instance the board authorized the building of the No. 4 and No. 5 ammonia units at Billingham towards the end of 1927, and thereby greatly increased the scale of I.C.I.'s fertilizer business. The plants were completed at the height of the slump, but such unfortunate timing at least gave employment to a large (and growing) number of workers, many of them ex-miners, who would otherwise have been on the dole, as the north-eastern pits and shipyards suffered terribly in the 1930s. The example is given to show not the company's employment policy, but its command of resources and its determination to carry through a major project. Both were major factors in the concern's later growth.[1]

There was another, intangible aspect to the I.C.I. merger: power. The senior men on the board were not primarily concerned with chemical technology, but with the establishment of a company sufficiently influential to play a big role in the chemical world. This may seem to some a curious reason for amalgamating, but McGowan for one had no doubts about the matter. Years later he wrote:

> . . . the formation of I.C.I. in 1926 was accelerated by the need for some strong British chemical unit to stand up to the foreign combines which were threatening to divide the world. . . . It is my pride that I.C.I. has done this. I say definitely that had I.C.I. not been formed Great Britain would before this war have been a second class power to Germany and the U.S.A. in the production of heavy chemicals.[2]

But he failed to mention that the signing of the ALKASSO–I.C.I. agreement on exports of alkali provided for a gradual reduction in I.C.I.'s quota for soda ash.[3] IG Farben was an even more dangerous rival than the American soda makers. I.C.I.'s directors may have hoped to overawe the Germans, but failed, because the merger added nothing to the strength and viability of the dyestuffs business or that of organic chemicals manufacture generally. The IG Farben managers knew the facts, and saw no need for changing their policy. In the spring of 1930, Melchett made much play about the progress achieved in the dyestuffs business since the merger and declared that after litigation with IG Farben, I.C.I. now had access to wider markets without joining the cartel.[4] Within a year I.C.I. was a cartel member with a quota of 7 per cent, as against 66 per cent for the Germans, 16½ per cent for the Swiss,

[1] Financial control from London was accompanied by unnecessarily strict attention to small items of expenditure. According to Parke (op. cit., p. 104) expenditure at Billingham above £2000 had to be approved by Head Office from 1927 onwards. The additional paper-work can be imagined.

[2] *Sunday Times*, 25 July 1943.

[3] Federal Trade Commission, *International cartels in the alkali industry* (1950), pp. 37 ff. See pp. 264–5 for ALKASSO

[4] Speech at third Annual General Meeting, 29 April 1930.

and $10\frac{1}{2}$ per cent for the other European producers.[1] Apart from dyes the other major points of contact were explosives and nitrogenous fertilizers. As regards the latter Billingham's size commanded sufficient respect to secure I.C.I. a big role in the first nitrogen cartel (1929). In this important area, the claim of increased bargaining strength was certainly justified.

The objectives of the merger took years to come to fruition and 'rationalization' remained a target rather than an attainment. McGowan described I.C.I. as a 'managing company' that controlled the subsidiary operating companies.[2] Alkalis and other inorganic chemicals, explosives, and dyestuffs remained the principal product groups, as indeed they had been in 1927 when they represented nine-tenths of the turnover.[3] During the 1930s the enlargement of Billingham increased the relative importance of nitrogenous fertilizers, though they retained a subordinate position by comparison with IG Farben. Similarly the German firm's interests in plastics were on a much larger scale than I.C.I.'s, whose first venture in that sector was a part holding in Croydon Mouldrite Ltd.—makers of phenolic resins—bought in 1933.[4] A year later I.C.I. built a small plant for the production of polymethyl methacrylate ('Perspex') at Billingham. There were many other differences between the two firms, which Herbert Levinstein summed up a few years after the merger. I.C.I., he said,

> depend for success on making comparatively few products extremely well and selling them at prices usually fixed by international agreement.

IG Farben, by contrast

> relies on making a large variety of substances constantly changing in range as new products appear in the laboratory and are taken up for large scale manufacture because they seem to have profit-making capacity.[5]

The judgement was not friendly, but remained true to Hitler's war and well into the forties.

While I.C.I. and its predecessors dominated the British chemical scene in the 1920s, some important developments took place among other firms, such as the closure of old works, improvements in marketing, and general modernization of the trade. This was particularly necessary in the superphosphate business. After the war the many small manufacturers that made up this business suffered severely from foreign competition. British production in 1924 was about 0·5 million tons, compared with 0·8 million tons in 1913. Consumption was lower than before the war. Employment in this

[1] ter Meer, op. cit., pp. 75–6. The Swiss share includes foreign branches of the Basle firms.

[2] Statement at the fourth Annual General Meeting on 31 March 1931.

[3] In 1927 alkalis and heavy chemicals accounted for 62 per cent of sales, explosives for 20 per cent and dyestuffs for 8 per cent. Chambers, Sir Paul. My twenty years with I.C.I. *I.C.I. Magazine* **46** (1968), no. 341, 41.

[4] Anon. *This is your concern* (1947), p. 39.

[5] From within the dyestuffs industry. *J. Soc. chem. Ind., Lond.* **50** (1931), 251T.

sector ranged from 4300 to 7500 according to the time of year.[1] The manufacturers applied for protection under the Safeguarding of Industries Act. It was refused and so they turned to amalgamation. In 1928 I.C.I. took the lead by combining half a dozen Scottish superphosphate makers into Scottish Agricultural Industries.[2] The next year the firms making superphosphate in the Eastern Counties combined to form Fison, Packard & Prentice Ltd., now trading as Fisons Ltd. A further nineteen companies were formed or taken over before the war, among them the large Anglo-Continental Guano Works Ltd. By 1939 Fisons interests extended beyond the Eastern Counties, and it accounted for a third of the superphosphate sold in the United Kingdom.[3] There were also combinations between coal-tar distillers, then a much more important group than now, operating in the same region. Examples are the South Western Tar Distillers, formed in 1918, the Midland Tar Distillers (1923), and the Yorkshire Tar Distillers (1926).[4]

For the rest chemical manufacture remained fragmented and many small firms survived. In dyestuffs besides I.C.I., the largest business, several independents continued to trade. The Swiss-owned Clayton Aniline Co. was of considerable importance and by 1933 had almost 1200 employees.[5] British Alizarine, later bought by I.C.I., L. B. Holliday, and Brotherton & Co. were the principal other manufacturers. The last-named had only begun the production of dyestuffs when it took over the works of Mersey Chemical Co. at Liverpool, which had passed from Badische to the Custodian of Enemy Property. Brothertons specialized in ammonium salts, tar-distillation products, sulphuric acid, and hydrosulphite used in vat dyeing.[6]

The Distillers Co. and Charles Tennant & Co. between them accounted for most of the fermenting capacity for making general purpose industrial solvents, while the much smaller firm of A. Boake Roberts made a speciality of cellulose lacquer solvents. In 1928 Tennants bought Synthite, which was then perhaps the largest maker of phenolic resins in the British Isles.[7] B. Laporte & Co. and Albright & Wilson Ltd. dominated respectively the manufacture of hydrogen peroxide and of phosphorus compounds, the former from Luton, the latter from works at Oldbury. Before the Great War Albright & Wilson's main product was amorphous phosphorus for matches, but in the 1920s sodium orthophosphate and calcium phosphate, both used in the foodstuffs industry, rapidly gained in importance and became the largest selling items. From about 1925 onwards carbon tetrachloride, a solvent for dry

[1] Board of Trade (Safeguarding of Industries), Report of the Superphosphate Committee, HMSO (1925), Cmd 2475, pp. 7–8, 11.

[2] It is now a large public company controlled by I.C.I.

[3] Monopolies Commission, Report on the supply of chemical fertilizer, HMSO (1960), pp. 55 ff.

[4] Miall, S. *A history of the British chemical industry* (1931), p. 198.

[5] *CIBA Blätter* **15** (1957), no. 152, 5.

[6] Edward, first Lord Brotherton, had founded the business in 1878; he was also a shareholder in the Cassel Cyanide Co. and in the British Alizarine Co. In 1951 the company merged with others to become Associated Chemicals Ltd., which was taken over by Albright & Wilson in 1965.

[7] Anon. *Enterprise, an account of the activities of the Tennant Group . . .* (1945), pp. 69, 93–4; Miall, op. cit., p. 110.

cleaning, was being made on a growing scale. The company's payroll was about 550 in 1920, which gives an indication of its relatively small size.[1]

Concentration tendencies in other European countries

A period of near stagnation in the French chemical industry was abruptly terminated by the Great War, which importantly altered its location as well as structure. Much was damaged in the north, some of it destroyed, and when the factories were rebuilt after 1918 they were up to date and incorporated many labour-saving devices. At the other end of the country works were established to make chlorine and sulphuric acid. The fertilizer business suffered between 1914 and 1918, but then improved and remained satisfactory because the post-war farmers were more systematic in their manuring and the applications generally more intensive. At the same time the country began to benefit from the development of the North African phosphate deposits, and from the expansion of the Alsatian mines that had been sunk on the eve of the war by German interests. About the middle of the 1920s France made 2·3–2·4 million tons of superphosphate a year and so occupied second place after the United States. She also produced 0·4 million tons of potash salts (as K_2O) in 1927; it was a fraction of the German output, but met the country's needs. The substantial imports of nitrates, sulphate of ammonia, sulphur, and phosphate rock emphasize the bias towards fertilizers, and it is a fair generalization that the chemical industry catered for and was geared to the needs of French agriculture.[2]

Small firms predominated. According to the occupational census of 1926, 30 000 people were employed making industrial chemicals, a third of them in the Paris region, a sixth in the north. Only 14 per cent of all the factories employed more than fifty people, and although the proportion was low, it represented considerable progress, for twenty years earlier it had been a mere 8 per cent.[3] One might have expected that concentration would be particularly conspicuous in the newer branches of the industry, for instance nitrogen fixation. But one would have been wrong, and this post-war development exhibited the characteristic fragmentation of most other sectors. In 1929 there were no less than twenty-two works, of which fifteen (including four being built) worked the synthesis of ammonia and between them made 147 000 tons of sulphate of ammonia. Nine factories used the Casale process, five Claude's, and the state-owned Office National Industriel de l'Azote (ONIA), the Haber–Bosch. The last-named, at Toulouse, was by far the largest installation, and its initial capacity was reported to be 4000 tons N a year, but two years later, in 1929, after extensions had been commissioned,

[1] Threlfall, R. E. *The story of 100 years of phosphorus making* (1951), pp. 178 ff.

[2] Ogburn, W. F., and Jaffé, W. *The economic development of post-war France* (1929), pp. 458 ff.; Reagan, D. J. *The French chemical industry and trade in 1928* (Dept. of Commerce Trade Information Bulletin No. 652), (1929), pp. 5 ff.

[3] Fauque, M. *L'Evolution économique de la grande industrie chimique en France* (1932), p. 188.

it jumped to about 18 000 tons N.[1] Besides the ammonia plants, there were five cyanamide factories, with a sixth under construction, and one firm still employed the arc process. Production averaged at the most a few thousand tons per works, and was obviously on a scale that prevented the efficient utilization of the technological potential of the processes.

When speaking of concentration in France we must bear in mind the size of the firms and avoid all comparisons with IG Farben or I.C.I. Large factories employing several hundred workers were still comparatively rare even in the 1920s, and belonged usually to Kuhlmann, St. Gobain, or Pechiney.[2] The combined capital of these three at the end of 1929 amounted to 800 million fr. (£6·5 million), while that of the thirty-seven firms in the chemical and allied trades that were quoted on the *bourse* was about 1400 million fr. (say £11·3 million). But the stock exchange valuation was of a different order of magnitude altogether, being 6800 million fr. (£55 million) for the Big Three and 13 600 million fr. for the quoted companies.[3]

Only Kuhlmann with a capital of 312 million fr. (£2·5 million) at the end of 1929, was entirely in the chemical business. The company accounted for a significant proportion of French fertilizer production, its output rising from 370 000 tons in 1923 to almost 480 000 tons five years later; virtually all of it was superphosphate. Kuhlmann broadened the product range in the 1920s by buying shares in the Soc. d'Electrochimie de Brignoud, which made cyanamide (1922), and formed the Soc. de Produits Chimiques Anzin-Kuhlmann with a coke-oven company to work the Casale process (1924).[4] One might have thought that these new ventures were part of a wider project to make compound fertilizers, but the location of the new works prevented their integration with the modern superphosphate factories around Lille or Port de Bouc. Similarly the take-over of the Cie Nationale de Matières Colorantes (CNMC) at the end of 1923 does not appear to have been part of a deliberate policy of amalgamation, but merely a move dictated by financial and political considerations. CNMC had been started during the war to make dyes, and had been assisted by the Government and the banks. Its early years had been difficult, but the occupation of the Ruhr in 1923 put an end to the deliveries of German dyes under the Reparation Agreement and correspondingly boosted CNMC's production from 4100 tons the year before to 6000 tons. With Kuhlmann in control, there was another increase to 9000 tons in 1924; then the growth of sales slowed down and four years later 9800 tons were made. In addition CNMC turned out about 10 000 tons a year of intermediates in the late 1920s.[5] French dyestuffs production was 17 000 tons in

[1] Bricard, S. L'Industrie de l'azote dans l'economie française. *Chim. Ind.* **68** (1952), no. 6, 938 f.; Grossmann, H., and Weicksel, P. *Die Stickstoffindustrie der Welt* (1930), pp. 133–4.

[2] I have used the present name of the company. In those days it was referred to as Cie de Produits Chimiques & Electrométallurgiques Alais, Froges & Camargue.

[3] Fauque, op. cit., pp. 183–5; Reagan's figures (op. cit., p. 2) for 1927 and 1928 are a little higher than Fauque's, but refer to forty companies.

[4] Anon. *Les Etablissements Kuhlmann 1825–1925* (1925), p. 60, and annual reports of the company. [5] ibid.

1924 and almost 15 000 tons in 1928, so that Kuhlmann represented 53 and 65 per cent respectively. During those years it bought Cie Française de Produits Chimiques & Matières Colorantes de St. Clair du Rhône and the old-established Soc. de Produits Chimiques & Colorants Français of St. Denis, and so obtained complete control over the French-owned dyestuffs makers, without, however, appreciably raising the output of colouring materials.[1] The only rival was the CIBA company, which operated a large works at St. Fons near Lyon.[2] Kuhlmann's dominant position in dyestuffs was useful in the negotiations with IG Farben, and in the cartel agreement of 1927 the French company secured one-fifth of the combined export trade.

Kuhlmann's dyestuffs business was under a separate managing director, and in general no attempt appears to have been made by the other branches of the concern to develop the manufacture of chlorine, caustic soda, metallic sodium, or coal-tar distillation products in order to supply the dye and intermediate works. In brief, dyes illustrate the tendency of the company at that time not to rationalize its operations, but on the contrary to enlarge them by diversification through purchase or investment in other companies. As a result published net profits more than trebled from 11·8 million fr. in 1920 to 39·4 million fr. (say £0·3 million) in 1928.[3] Employment figures have not been published, but we can safely assume that there was a considerable increase in the number of workers and by the end of the period they may have numbered 6000 to 8000 at the peak of the fertilizer season.

St. Gobain and Pechiney were certainly much larger than Kuhlmann but we have no information enabling us to separate their considerable interests in glass and non-ferrous metals respectively. St. Gobain owned about twenty-five chemical factories and remained the country's largest producer of superphosphate, accounting for about two-fifths of French output.[4] There were a few instances of diversification during the 1920s: St. Gobain opened a cyanamide works at Modane in the Alps in 1923, and together with Air Liquide owned the Grande Paroisse company, which made synthetic ammonia by the Claude process. Among its investments were holdings in the St Denis dyestuffs works and a firm making acids and pigments at Thann in Alsace.[5]

The Pechiney company was the outcome of a merger of the two leading French producers of aluminium in 1921. Although one of the predecessor companies had large chemical works at St. Auban and Salindres in south-

[1] Reagan, op. cit., pp. 32 f.; USTC, *Census 1927* (TIS 37), (1928), p. 159; *Census 1928* (TIS 38), (1929), p. 161; British Dyestuffs Corporation, op. cit.

[2] About 500 people were employed in 1933. CIBA brochure, n.d. (1934?).

[3] The growing role of subsidiary and associated companies appears from the balance sheets. At the end of 1920 investments represented about 11 per cent of the company's net assets and eight years later almost 30 per cent. The next year there was a measure of consolidation and the proportion fell to 23 per cent.

[4] Reagan, op. cit., p. 26; his figures tend to be on the high side.

[5] Choffel, J. *St. Gobain* (1960), pp. 86 ff.; Gérard, J. (Ed.) *Dix ans d'efforts scientifiques et industriels* (1926), vol. 2, pp. 1231 ff.

eastern France, Pechiney's investments between the wars were in power stations and reduction plants, and aluminium continued to be the mainstay of the business. In chemicals Pechiney specialized at first in aluminium compounds and electrolytic alkalis, later it began making monochloracetic acid and trichloroethylene. Like the other large firms it went into nitrogen fixation: in 1922 Pechiney and Mines de Lens formed Société Ammonia to work the Casale process and about two years later it joined Mines du Nord in setting up Huiles, Goudrons & Dérivés to distil coal tar and later to develop the hydrogenation of coal.[1]

There were also mergers among companies of the second rank. The most important instance occurred in 1928 when Poulenc Frères merged with Société Chimiques des Usines du Rhône. It was a union of firms in the same branch of the industry, but with a complementary range of products. Poulenc, with works in the Paris area, was well known for its fine chemicals, enamels, and colours for the ceramics industry, and above all for its pharmaceuticals, notably arsenical compounds. Usines du Rhône, located in the area of Lyon, had made dyestuffs until 1905, and then specialized in pharmaceuticals (mainly febrifuges and sedatives), synthetic essences and flavours, organic acids and intermediates, photographic developers, and cellulose acetate film and sheet. Rhône–Poulenc, as the new company was called, employed about 3000 people, and was the largest manufacturer of organic chemicals, dyes excepted, in France.[2] Société Bozel–Malétra was the outcome of a merger in 1925 of the old-established Malétra company, which made fertilizers, hydrochloric acid, and bleaching compounds, with the Bozel–Lamotte group, the successor to several of the pioneering firms in electrolytic alkali and chlorates. Two years later the amalgamation of the Société Industrielle des Matières Plastiques (which, in turn, was a union of celluloid makers in 1924) with the explosives companies belonging to the Société Centrale de Dynamite led to the formation of Société Nobel Française.[3]

The examples suffice to show that mergers were a frequent occurrence in the post-war French industry, and that they took place in all the principal sectors. But the purpose as well as the scale differed from that of Britain and Germany. No information is available to me on closures of works, the reorganization of marketing, and other characteristics of rationalization that were then being partly or wholly implemented across the Channel and the Rhine. By the late 1920s the mergers, whilst scarcely reducing the number of firms, did lead to the emergence of several important companies, notably Kuhlmann, St. Gobain, Pechiney, and Rhône-Poulenc. Alongside these

[1] Gignoux, C. J. *Histoire d'une entreprise française* (1955), pp. 159 ff. In 1960 St. Gobain and Pechiney combined their chemical activities, but nine years later Pechiney sold most of its shares in the Pechiney–St. Gobain company to Rhône–Poulenc.

[2] Gérard, J. (Ed.), op. cit., vol. 2, pp. 1289–91. Since the war the plastics and fibres divisions have grown very fast, and the company is now the leading French chemical business.

[3] In 1957 Bozel–Malétra combined with Nobel Française and now trades as Nobel–Bozel. Brochure published by company in 1960.

substantial enterprises, and often linked to them by agreements, banking, or family ties, there were scores of lesser companies, and their survival testifies to the endurance of the small entrepreneur. These small chemical producers were probably not very efficient by comparison with the bigger businesses, but efficiency and profitability were not always their main objectives. It would be interesting, but beyond our scope, to examine the motivations and attitudes of French chemical manufacturers and compare them with their American and German counterparts. As for Kuhlmann, St. Gobain, and Rhône–Poulenc, they have remained the leaders of the industry.

In Germany and in Britain, the entire chemical industry was greatly influenced by the creation of IG Farben and of I.C.I. In France, however, there was no dominant firm, so that in comparing the structure of the industry in the different countries one must take into account the fragmentation so characteristic of the French scene until the 1960s. One further important feature remains to be noted. After the Great War the French Government and the coal-mining companies began to play a growing role in the chemical trades, the former in potash, ammonia, and nitric acid, the latter in high-pressure synthesis generally and in tar distilling. It has been estimated that in 1938 the mines with their chemical partners accounted for 46 per cent of the nitrogen and 90 per cent of the methanol made.[1] When the mines were nationalized after the war the French Coal Board became a large factor in the country's chemical industry, a situation not equalled in Britain. The existence of the state-owned synthetic ammonia works together with the numerous nitrogen fixation plants connected with coke-oven enterprises meant that this important sector was only partly controlled by chemical companies. The divided ownership combined with the small unit size of plants in such a key sector, may well have been among the contributory factors in slowing down the development of the leading firms. Nitrogen fixation and the derivative technologies of methanol synthesis and coal hydrogenation were perhaps the most important growth points for IG Farben and I.C.I. in the 1930s. Their technical and commercial dispersal in France limited the benefits to be gained by concentration and mergers in other areas and so perpetuated the fragmented structure of the country's entire industry.

In Italy, Montecatini[2] emerged during the 1920s as by far the largest chemical enterprise. Its rise to prominence was remarkably fast: until the war the firm had been small and known chiefly for mining and trading in pyrites. From there it went into the manufacture of sulphuric acid, but until about 1917 the business remained extremely modest. Then a period of great activity began: as a first step many superphosphate factories were bought, and

[1] Bedhome, R. L'Evolution de la carbochimie dans le bassin du Nord et du Pas-de-Calais. *Chim. Ind.* **87** (1962), no. 4, 439.

[2] Its full name was Montecatini Stà Generale per l'Industria Mineraria e Chimica. In the 1960s it merged with the Edison group.

Montecatini soon became the largest Italian manufacturer of the fertilizer. The preparation of copper sulphate (of which Italy was then the world's first producer) was also vigorously promoted. From the enrichment of the soil and the protection of the vine, the company branched out, and between 1922 and 1924 took over the country's explosives industry; later it built the first Italian synthetic ammonia plant, which used Fauser's process. Marble works and interests in the manufacture of aluminium rounded off the activities of the group on the eve of the slump. Later it took over the Aziende Chimiche Nazionale Associate (ACNA), the most important of some half dozen firms in the dyestuffs trade, which had been formed in 1928 by a combination of several small producers. For a few years during and immediately after the war, dyestuffs, particularly sulphur black, had been profitable, but the firms were too small to afford expensive research and sales support services and so totally unable to withstand the Germans. ACNA proved to be a temporary expedient, and despite high tariffs was soon in difficulties and had to be rescued in 1931 by Montecatini.[1] Here we have another instance of dyestuffs manufacturers requiring the protection of a more prosperous and powerful concern to survive the rigours of German marketing methods.

By British and German standards, Montecatini was minuscule. Even as late as 1928 its capital was only about £1·1 million. As in France the issued capital masks the true state of affairs. Sales figures are not available, but the group's turnover probably accounted for a larger share of the country's chemical business than I.C.I.'s or IG Farben's in theirs.

In Belgium, Solvay & Cie and the large banks brought about in 1928 the amalgamation of some of the principal superphosphate and sulphuric acid works into the Union Chimique Belge.[2] This still left substantial areas of the chemical industry in other hands. For example there were in 1929 at least three enterprises working the synthetic ammonia process in association with coke-oven plants; all these works were quite small.[3] Solvay's own enterprises were relatively small, but their interests and connexions within Belgium as well as outside were extensive.[4]

Concentration stopped short of amalgamation among the Swiss firms, but in all matters concerning dyestuffs they acted as one: the Swiss IG was

[1] Marcus, A. *Die grossen Chemiekonzerne* (1929); brochure published by the company after the Second World War; Parravano, N. (Ed.) *La Chimica in Italia* (1938); USTC, *Census 1928* (TIS 38), (1930), p. 163.

[2] Lemoine, R. J. *La concentration des entreprises en Belgique* (1929), pp. 24, 27.

[3] The Société Belge de l'Azote was associated with Grande-Paroisse and used the Claude process; Societé des Fours à Coke de Selzaete, a subsidiary of Kuhlmann, used the Casale process; Société Ammoniaque & Dérivés was controlled by Evence Coppée & Cie (a large firm of coke-oven operators) and Montecatini held a minority interest. It used the Fauser process.

[4] Official and private sources shed little light on the activities of this international enterprise which between the wars played a key role in the European alkali trade. It held, *inter alia*, substantial minority interests in I.C.I. and Allied Chemical, which had taken over Brunner, Mond and Solvay Process, and its subsidiary, Deutsche Solvay-Werke, was the second largest German chemical enterprise. A comprehensive history of Solvay is badly needed and would fill a major gap in our knowledge of the development of the European chemical industry.

important, lasted longer than the German and requires detailed treatment. It was formed by CIBA, Geigy, and Sandoz, after nine months of negotiations, in September 1918.[1] The three companies, which accounted for virtually the entire Swiss output of dyestuffs had been busy and prosperous during the war, and two years previously had collaborated in setting up the Säurefabrik Schweizerhall for supplying oleum. The Swiss were, of course, fully aware of the broader implications of the formation of the German IG in 1916 and of the British Government's intervention in the United Kingdom dyestuffs business, and their IG was the response of the Basle firms to the anticipated changes in the post-war dyestuffs industry. Its provisions were at least partly modelled on the German precedent. The agreement provided for cooperation in technical and commercial areas, the elimination of competition, mutual support, and the exchange of know-how. In addition there were three specific measures: 1. raw materials were to be bought on joint account; 2. inter-company sales were to be invoiced at cost; 3. products were to be made in the works reporting the lowest costs. The IG was to last for fifty years. The key feature was the apportioning of profits, and some surprising facts emerged during the discussions on the subject. In the seven years to 1914, the combined annual sales of the three companies averaged 21·5 million fr. (say £0·85 million), which shows how small they were even by the standards of those days. During the war the turnover aggregated 132 million fr. (£5·3 million) a year. CIBA contributed roughly one-half of the total throughout the period, but Geigy had a setback and its share of the turnover fell from about 35 per cent before the war to 13 per cent during the war, correspondingly Sandoz improved its position from 15 to 35 per cent. A firm of chartered accountants was then called in, and after examining the books of the three companies decided that the gross profit should be divided as follows: CIBA 52 per cent, Geigy 20 per cent, and Sandoz 28 per cent. In 1920 the proportions were changed to 52–24–24, where they remained for the next thirty years.[2]

While all three companies made dyestuffs, they tended to specialize, so that the amount of overlap was not excessive. CIBA and Geigy also manufactured pharmaceuticals; Geigy had a line of vegetable colours and tanning extracts; which were dependent on imports and so contributed to the unsatisfactory performance after 1914. These subsidiary activities barely represented 10 per cent of the total turnover of the three firms in the 1920s, and during the early years of the IG the differences in the product range were not particularly significant.[3]

Whereas the German IG did not really achieve its objectives, the Swiss counterpart was a good deal more successful, at any rate in the early years when the three firms worked together quite smoothly. Thus Sandoz and

[1] Durand & Huguenin was small and in the mid-1920s came under German ownership. F. Hoffmann–La Roche & Cie did not make dyestuffs and therefore did not join the IG.

[2] Baumgartner, R. *Die wirtschaftliche Bedeutung der chemischen Industrie in Basel* (1947), pp. 36–7; Bürgin, A. *Geschichte des Geigy- Unternehmens von 1758 bis 1939* (1958), pp. 244 ff.

[3] Ibid., p. 251.

Geigy took shares in the Clayton Aniline Co. in 1919, and this erstwhile CIBA subsidiary was now enlarged to make dyestuffs for the Swiss firms' customers in Britain and the Empire. The following year they bought Ault & Wiborg's dyestuffs factory in the United States, and formed the Cincinnati Chemical Works to run it.[1] The third joint venture was Stà Bergamasca per l'Industria Chimica, established in Italy in 1925 to make sulphur dyes on a small scale.[2] The foreign works served a useful purpose by providing manufacturing facilities in highly protected markets, and were also a most valuable bargaining counter in later years at the time of the IG Farben negotiations. Collectively the Swiss manufacturers were an influential and powerful group, and although internal dissensions were increasingly making themselves felt, they retained their autonomy *vis-à-vis* IG Farben. In the first dyestuffs cartel (1928) they secured a 19 per cent share of the world export markets.[3]

The agreement on dyestuffs provided conditions for stable and profitable trading, and so the means to develop the marketing effort, as well as the business in other than dyestuffs. CIBA and Sandoz in particular benefited or perhaps seized the opportunity. At any rate by 1930 their non-dyestuffs sales represented roughly 27 per cent of the total turnover of the Swiss IG.[4] They derived an additional advantage from an arrangement concerning dyestuffs, whereby CIBA was allotted vat dyes and Sandoz the alizarin group, presumably because the two had the most efficient manufacturing facilities for these colours. Meanwhile Geigy was hanging back and showing little sign of inventiveness, and its traditional natural tanning extract and dyewood business fell off sharply owing to competition from new synthetic materials. The result was that CIBA and Sandoz put more into the profit pool than they took out and, as it were, 'carried' Geigy. Sandoz appeares to have been particularly aggrieved by the situation, and there was more ill-feeling over the project, sponsored by IG Farben, but opposed by Sandoz, to merge the three Swiss companies into a single combine. There matters rested, but, after Hitler's war, Sandoz demanded that the 1918 agreement be altered. Negotiations took place in 1947–8, but were broken off by Geigy. With its enormously successful D.D.T. it could look forward to a renewal of prosperity, and so demanded that the IG dispute be referred to arbitration. The court, after hearing the case, ordered that the Basle IG be terminated with effect from 1 January 1951.[5]

[1] In 1960 the Cincinnati Works were moved to Toms River and the name changed to Toms River Chemical Corp.

[2] Bürgin, op. cit., pp. 296 f.; CIBA, *From the annals of the Society of Chemical Industry* (1939), p. 16.

[3] ter Meer, op. cit., p. 75. In 1931 the share was cut to 16½ per cent when the cartel was enlarged.

[4] The total turnover has not been published, but from partial evidence it appears likely that it did not exceed £7 million in the late 1920s.

[5] Anon. *Sandoz 1886–1961: 75 years of research and enterprise* (1961), p. 125. Although the agreement was ended, some of the factories continued to work for the joint account of the three firms. They also retained their combined majority interest in Durand & Huguenin (acquired from IG Farben) until 1969 when Sandoz bought out CIBA, Geigy, and the remaining shareholders. In 1970 CIBA and Geigy decided to merge their businesses.

The chemical corporation in the United States

The American chemical industry, though developing rapidly in the 1920s, remained small relative to motor manufacturing and electrical engineering. Employment in 1925 ranged from around 60 000 to 120 000 depending on the definitions used, by comparison with almost 200 000 in motor vehicles and nearly a quarter million in electrical engineering.[1] There were numerous mergers in the period, and the concentration movement was at least as vigorous in the United States as in Europe. In the early post-war years the amalgamations had as their prime objective the enlargement of manufacturing capacity under a single direction, although this might be contrary to the spirit and certainly to the letter of the Clayton Act. Later, however, the objectives became less clearly defined, and were blurred by the general business euphoria in which even unsound schemes were found to have attractions. We should observe that the mergers were of a local or regional character, and that there was no union of firms on a country-wide scale as exemplified by I.C.I. or IG Farben, or by Du Pont at the turn of the century.

In Britain and in Germany, and to a lesser extent elsewhere, the development of high-pressure synthesis was one of the important factors in the concentration movement and had an important bearing on the early years of the new concerns. In the United States, by contrast, the fixation of nitrogen played a subordinate role and throughout the 1920s Chilean nitrates continued to dominate the fertilizer nitrogen business. In 1924 the output of synthetic nitrogen was only 10 000 tons N, and even as late as 1928 it was only 23 000 tons N; the following year some large units made their first contribution, and the figure jumped to 75 000 tons.[2] The rationalization of dyestuffs manufacture, which was a significant element in the European restructuring, had little bearing on the emergence of the big American chemical corporation.

The very word corporation nowadays implies size and power, but in the 1920s only Du Pont and Allied Chemical could be compared with the European giants.[3] The extremely rapid growth during and after the Second World War of Dow Chemical, American Cyanamid, and Monsanto has obscured the fact that in the decade following the Kaiser's war they were small. One way in which the American chemical concerns can be related to their industrial environment is by ranking of total assets: in 1920 Du Pont occupied seven-

[1] Brissenden, P. F. *Earnings of Factory Workers 1897 to 1927* (Dept. of Commerce, Bureau of the Census, Monograph No. 10), (1929), pp. 377–83. Brissenden's figure of 56 000 wage earners in the chemical industry excludes fertilizers, explosives, and some other sectors.

[2] Crane, J. E. Development of the synthetic ammonia industry in the U.S. *J. ind. Engng Chem.* **22** (1930), no. 7, 796.

[3] I am not including Union Carbide & Carbon, an important amalgamation in 1917 of several firms in the acetylene, industrial gases, and electrode carbon business. In the 1920s, its chemical activities were extremely original and comprised the manufacture of aliphatic compounds from acetylene and petroleum fractions. But in 1929 chemical sales were estimated by *Fortune* (June 1941) at only $7 million out of a total turnover of $142 million (say £29 million). Nowadays Union Carbide's chemical business has become so large that it is often classified as part of the industry. This was certainly not the case up to 1930.

teenth place, in 1930 it had moved up to thirteenth, while Allied Chemical was not counted in the former and rated seventeenth in the latter year. The other chemical companies did not appear among the twenty-five largest American manufacturing corporations, which were then headed by steel and oil.[1] In Europe, on the other hand, IG Farben, I.C.I., Kuhlmann, and Montecatini would have been among the first half dozen in their respective countries, whatever the basis of comparison with other large enterprises.

Du Pont was then and remains to the present day the largest firm in the American chemical industry. We shall concern ourselves here with its diversification and acquisitions, and leave for the following chapters its management and research policies. The piecemeal approach can be justified on the ground that in all these areas Du Pont's activities were significant, frequently controversial, and often pioneering.

The acquisitions can be quickly summarized. During the war Du Pont bought several companies, notably Harrison Bros & Co. (1917), a large manufacturer of paints and sulphuric acid. After the war the principal take-over was the Grasselli Chemical Co., for which about \$64½ million (roughly £13·2 million) was paid in 1928. Grasselli owned twenty-four works and its sales of industrial chemicals amounted to \$43 million (almost £9 million).[2] Two later purchases illustrate the trend in diversification: Roessler & Hasslacher Chemical Co. (1930) gave Du Pont access to the electrochemical business and the Remington Arms Co. (1933) to light weapons.[3] The company was associated at first with French and later with other foreign firms in various American enterprises. The transatlantic links in order of appearance were as follows. Agreements with the Comptoir des Textiles Artificiels to make viscose and cellophane were dated 1920 and 1923 respectively and during 1924 Du Pont established Lazote Inc. with Air Liquide and Grande Paroisse and started building a plant to work the Claude process at Belle (W.Va.) in 1925. At about the same time Du Pont together with Pathé Cinema S.A. and its American licensee formed Du Pont Pathé Film Manufacturing Corp. to make cine-film, and with the National Distillers Co. went into the industrial alcohol business. In 1928 it obtained a licence from Rhodiaceta in France to make cellulose acetate yarn and joined IG Farben in the Bayer–Semesan Co. to make seed dressings. To round off this short account, Du Pont began making directly, or through newly acquired companies, dyestuffs and coated fabrics during the war, cellulose lacquers for use as automobile finishes in the early 1920s, and synthetic resin coatings, the 'Dulux' brand of paints, from 1927 onwards.[4] There were also several trade investments, most notably the

[1] National City Bank of New York, Monthly Newsletter, June 1949, p. 70.

[2] Winkler, J. K. *The Du Pont dynasty* (1935), p. 278; annual reports of the company for the years to 1929. Du Pont did not buy the dyestuffs interests of Graselli, which had been hived off in 1924 to become the Grasselli Dyestuffs Corporation. That company was the agent for Bayer, and later became the manufacturing subsidiary of IG Farben's American branch.

[3] Dutton, W. S. *Du Pont* (3rd edn. 1951), pp. 275–7.

[4] Dutton, op. cit., p. 278; Haynes, W. *The American chemical industry* (1948), vol. 4, p. 343,

holding in General Motors. The first purchases were made in 1919, and later the investment was increased, and finally it amounted to 23 per cent of General Motors' share capital. (The du Pont family holding companies also had shares in General Motors.) The dividends received from the investment were not retained by Du Pont, but passed on to its shareholders.

The short description is enough to show that Du Pont followed a different course and that its product structure differed greatly from that of IG Farben and I.C.I., with which it has been often, though erroneously, compared. Until the take-over of Grasselli, the company's turnover was modest (the war years when the explosives contracts boosted sales excepted): much smaller than IG Farben's and a little below I.C.I.'s. In 1925 the turnover was $84·6 million (£17½ million), and three years later $119·0 million (£24 million). Employment, excluding the workers in the cine-film factory and the distillery, rose from 15 000 to 21 000 between 1925 and 1928. The picture changed when Grasselli came to be included, and in 1929 the company's sales rose to $203 million (about £41 million) and the average number of employees increased to 37 600.[1]

Du Pont's activities and successes have often been described, and with added embellishments have become part of the folklore of American free enterprise. But the firm's development also had its less glamorous side and what is remarkable, perhaps even exceptional, in the routine of its policy, is the consistency with which the long-term objectives were pursued. These objectives were in the first place to lessen the dependence on the explosives trade, secondly to improve the profitability of the newly acquired subsidiaries by placing Du Pont staff in key positions, and thirdly, though not least, to re-invest profits rather than distribute them to shareholders.[2] Initially the policies adopted to achieve these ends yielded disappointing returns. It was indeed logical for Du Pont to seek opportunities in nitro-cellulose derivatives of which, as a manufacturer of explosives, it had great experience. Hence attempts were made even before the war to make artificial silk by buying the necessary know-how, and some attention was also paid to the manufacture of paints. But contrary to expectations the company made a loss on the paint business from 1917 to 1921. And in those four years no chemical businesses were bought 'because there was no opportunity to purchase any concern at a reasonable price';[3] instead the available funds were invested in General

and anuual reports of the company. Du Pont had been making its own types of seed dressings since 1924.

[1] Annual reports of the company.

[2] See also Dale, E. Du Pont: pioneer in systematic management. *Admin. Science Quarterly* **2** (1957), no. 1, 45 f. The attitude towards the shareholders was not so high handed as it would appear, because the dividends from General Motors were rising and so compensated the shareholders for their patience. Another factor was the du Pont family. Its members collectively were the largest group of shareholders—as late as 1961 they held about 30 per cent of the equity—and they could afford to wait.

[3] GTX 285, memo from General Manager Paint Department to Irénée du Pont dated 4 November 1921, and GTX 1244, report from the Development Dept. to the Executive Committee dated 31 March 1917.

Motors. Similarly the dyestuffs and ammonia works were disappointing for many years: as regards the first, $43 million was spent between 1916 and 1934 before a satisfactory return was obtained, and as to the second, the nitrogen fixation plant cost $24 million and took about ten years to attain profitable operations.[1]

Although the early years of diversification were thus extremely costly and marked by unexpected difficulties, they were nevertheless of the utmost significance for the future, for they witnessed the slow but nevertheless far-reaching transformation in Du Pont's product structure. By the middle of 1921 some considerable progress had been made along this direction. At that time assets were $275 million (say £72 million); if the goodwill and the book value of the General Motors holding are deducted, we are left with $176 million (£46 million). Of this total, the traditional explosives business accounted for only a quarter. By contrast the plant for making dyestuffs represented just over 16 per cent, and celluloid, coated fabrics and cellulose lacquers between them for another 15 per cent. The remaining assets, a little over two-fifths of the total, were represented by the Harrison company, some smaller firms making paints and various subsidiary and associated companies. The profit made was $14·6 million, that is 8·3 per cent on assets (excluding goodwill and General Motors, on which a dividend was received).[2] The overall result is a creditable achievement in the middle of a slump and also illustrates the lessened dependence on the explosives trade.

In the transitional phase from explosives to diverse chemical manufacture, the dividends from General Motors played an important role in the shareholders' returns. Presently this investment was to yield further valuable benefits by providing if not a captive then at least a friendly market for Du Pont's cellulose finishes, celluloid, and coated fabrics. Details of Du Pont's sales to General Motors, including Fisher Body, in the 1920s have not, so far as I know, been published in full, but figures are available for the years to 30 June 1926 and 1928. In the earlier year the value of sales was about $8 million, in the latter $15·2 million, which was equivalent to about 9 and 14 per cent respectively of Du Pont's turnover in those years.[3] The sales continued in the 1930s and 1940s and these trading links with General Motors caused the Department of Justice to claim that Du Pont had violated Section 7 of the Clayton Act by obtaining 'non-competitive preferences' from its associate.[4] The matter was taken to trial, and after litigation ending in 1959, Du Pont was ordered to get rid of the General Motors shares; the final

[1] Dutton, op. cit., pp. 288–9; Haynes, op. cit., p. 87.

[2] GTX 166, Du Pont memo, source not given, dated 17 August 1921. A rate of exchange of $3·8 to the £ has been used for the 1921 figures.

[3] Du Pont Trial Brief (Civil Action No. 49 C 1071, U.S. District Court, Northern District of Illinois), filed 9 October 1953, pp. 231, 237. The proportions have been derived from the average net sales in 1925/6 and 1927/8.

[4] In its pre-1950 version the clause provided that a corporation was not to acquire stock in another where the effect 'of such acquisition may be to substantially lessen competition' between the two '. . . or to restrain such commerce' between them.

distribution of these shares to Du Pont's shareholders was made in January 1965.

For our purposes the immediate practical effects of the investments and take-overs are more important. They point clearly in one direction: Du Pont was making, and making on a growing scale, precisely those products that were among the principal beneficiaries of the consumer-goods boom of the 1920s. In 1924 the sale of new products, notably dyestuffs, rayon, automobile lacquers, leather-cloth, and celluloid, amounted to almost $28 million, and so had nearly caught up with the $33 million of the explosives business.[1] A few years later the former had decisively overtaken the latter. In two cases, nitro-cellulose lacquers and rayon, we have statistics that show for the United States how quickly they passed from the stage of interesting novelties to the scale of a major branch of industry. Thus between 1914 and 1925 the production of nitro-cellulose lacquers and varnishes rose from 0·9 to 12·3 million U.S. gal, while the value of the rayon output grew from $40 million to $88 million between 1919 and 1925.[2] There is no doubt that Du Pont was carried along by this technical and, even more so, commercial advance. It had not created these changes, luck certainly played a role in causing the firm to acquire and to promote in collaboration with others, but not to invent, a few extremely profitable products. The management's engineering skill and undoubted financial shrewdness helped in exploiting the opportunities as they occurred, but the main-spring for the company's extremely rapid development at the time was not, as was usual in Germany or in Switzerland, the utilization of its own inventiveness.[3]

The Allied Chemical & Dye Corporation (now called Allied Chemical) was a totally different concern. It was the largest single amalgamation in the American chemical industry in the inter-war years. The union took place in the second half of 1920 when five firms (some of them the result of a merger) agreed to combine with effect from the end of the year. They were the Solvay Process Co., the leading producer of ammonia soda, the Semet–Solvay Co., which operated coke ovens, the Barrett Co., a large firm of coal-tar distillers, the General Chemical Co., an important manufacturer of heavy chemicals, and the National Aniline & Chemical Co., which had been established during the war to make dyestuffs.[4] The new company was primarily a manufacturer of heavy inorganic chemicals, with subsidiary interests in tar products and dyestuffs. Unlike Du Pont it had few connexions with the consumer-goods

[1] Company's report for 1924.

[2] Department of Commerce. *The American chemical industry* (Trade Promotion Series No. 78), (1929), pp. 10, 22.

[3] The matter has been investigated at some length by W. F. Mueller in The Origins of the Basic Inventions underlying Du Pont's major Product and Process Innovations 1920 to 1950. His study appears in a report of the National Bureau of Economic Research entitled *The rate and direction of inventive activity* (1962), pp. 323–58.

[4] Cosby, J. History of allied chemical (paper presented to the American Chemical Society in Chicago on 11 September 1958); Federal Trade Commission *International cartels in the alkali industry* (1950), p. 33; company's annual report for 1920. Solvays held just under 17 per cent of the share capital of Allied Chemical.

industries. At the time of the merger the consolidated fixed assets were \$140·4 million (about £38 million), and increased slowly to \$202 million (£41 million) nine years later. In that time Allied did not buy other companies, nor did it diversify except in one important respect: it was the first United States company to work the ammonia synthesis. A small plant, with a capacity of just over 7000 tons a year of ammonia, was started at Syracuse (N.Y.) in 1921. A few years later the building of a much larger works began at Hopewell (Va.) and was completed towards the end of 1928; this factory had a capacity for making 45 000 tons annually of ammonia, which was later raised to a considerably higher figure.[1] Towards the end of the 1920s nitrogenous fertilizers and technical nitrogen compounds joined alkalis and sulphuric acid as the principal heavy chemicals made.

It is not possible to relate the company's activities sector by sector to the growth of the main branches of the United States chemical industry. But the alkali business ought to have been very active, for the United States production of caustic soda more than doubled in the 1920s, while that of soda ash rose by nearly 80 per cent.[2] But Allied's profits showed little change throughout the period. The combined gross income before tax of the constituent companies was \$30 million in 1920, but fell to \$13·1 million in 1921. From this inauspicious beginning, Allied Chemical gradually recovered to around \$28 million in 1927 and \$33 million two years later. The company was unusually secretive and published no sales figures, but the turnover is unlikely to have exceeded Du Pont's.[3] Allied's product range resembled I.C.I.'s, with which it had other points in common, and both companies used the Solvay process and kept in touch over the ammonia synthesis.

We have seen in an earlier chapter that the manufacture of dyestuffs was revived during the Great War. Three undertakings launched at that time were ultimately to become large producers. Du Pont's interests were considerable from the start, but, as has already been mentioned, took many years to become profitable. The National Aniline & Chemical Co.'s business was even larger and supplied 55–60 per cent of the United States demand for dyestuffs in the year of the great shortage, 1919.[4] The third firm was the Calco Chemical Co. It fell on hard times in the slump of the early 1920s, but recovered and a few years later took over some makers of dyestuffs which did not manufacture intermediates and in this manner secured a captive market for its considerable production of organic chemicals. Calco was bought by American Cyanamid in 1929.[5] There were, finally, several small American-owned manufacturers

[1] Grossmann and Weicksel, op. cit., p. 193. General Chemical had developed its own variant of the Haber–Bosch process for Syracuse.

[2] Between 1919 and 1929 the make of caustic soda grew from 280 000 to 680 000 long tons and soda ash (excluding that for causticizing) from 1·3 to 2·4 million long tons. Federal Trade Commission, op. cit., p. 7.

[3] Sales were probably of the order of \$150 million (say £30 million) towards the end of the 1920s.

[4] Anon. *The General Chemical Co. after 20 years* (1919), p. 51.

[5] McMurray, J. H. History of Calco, unpublished typescript at American Cyanamid (1943).

which, considered separately, were of little importance; they managed to survive behind the high tariff wall and the elaborate protective measures devised by the Tariff Commission. In addition there were two foreign-owned companies which played a considerable role: the Cincinnati Chemical Works belonged to the Swiss IG and the General Aniline Works (later the General Aniline & Film Corp.) to IG Farben. General Aniline had a short, but complicated history. It represented the spearhead of the renewed German commercial effort in North America from 1925 onwards, and the business was very considerably enlarged in 1928 when it finally took over the Grasselli Dyestuffs Corporation's factories at Albany and Grasselli.[1]

Although the manufacture of dyestuffs was carefully sheltered from foreign competition, the United States market in the 1920s was not sufficiently large to sustain profitably the five medium-sized manufacturers as well as the handful of smaller ones. The volume made was about 40 000 tons in 1920 and 43 000 tons worth $46 million (almost £9·5 million) in 1928. In that year the industry was able to supply 92 per cent by weight and 80 per cent by value of United States consumption.[2] Some measure of rationalization as in Germany or Britain would have been of advantage by creating fewer but more economic units. But with the exception of Calco and General Aniline no such move was made in the 1920s when the political climate was not hostile to amalgamations. In the following decade conditions were less favourable, and at the outbreak of Hitler's war, this branch of the American chemical industry was still fragmented. General Aniline and Du Pont each accounted for roughly one-quarter of the value of the dyestuffs trade, National Aniline and the Swiss works contributed 15 per cent each, and the rest was divided between Calco and the other United States firms.[3] It is possible, though the evidence is fragmentary and the proposition needs to be sustained by further research, that the inventiveness displayed and progress made by the British, Swiss, and German producers in the period 1930–60 is attributable to the powerful backing of the chemical combines, while conversely the fragmentation in the United States was a factor inhibiting research and development in dyestuffs.

The situation in the United States dyestuffs industry resembled superficially that obtaining in the French chemical industry. It is worth stressing that in the 1920s the predominance of the small chemical works was the characteristic feature of the United States industry. Haynes has stated that there were about 500 mergers in that decade,[4] but overwhelmingly they combined the small or the minuscule, and, with very few exceptions, the big firms did not seek the

[1] Haynes, op. cit., pp. 233–6. In 1965, after years of litigation, the U.S. Government abandoned its claim to control over the firm, which it had exercised ever since the outbreak of war. The stock was sold and General Aniline is now a public company.

[2] USTC, *Census 1928* (TIS 38), (1930), p. 5; Haynes, op. cit., p. 24.

[3] Stocking, G. W. and Watkins, M. W. *Cartels in action* (1946), p. 508. The figures were obtained from Du Pont by the Bone Committee (Hearings on Patents, pt. 5, p. 2481). General Aniline's production in the U.S. was relatively small, but it imported a substantial tonnage from Germany, and this explains its quarter share in the value of sales.

[4] Op. cit., p. 46.

companionship of other large enterprises. The legal restrictions were then not the primary deterrent, and the causes must be sought in other, less obvious explanations. But whatever the reasons, there has been a sea change since the 1930s and especially since the war, when the size and position of Du Pont and Allied has been challenged by other firms, including some whose primary interest lies in the oil industry. But in the 1920s it must have been difficult for contemporaries to discern the growth potential of the small rivals. Dow Chemical, however, did grow at a fast rate. It had about 660 employees in 1922 and almost 1400 six years later, while its sales rose from \$4·6 million to \$15·8 million (£3·2 million) in the same period.[1] Monsanto Chemical, now another giant of the industry, had only 700 employees in 1925 and up to 1928 its net income was less than \$1 million. A major accretion in size came in 1929 with the take-over of the Merrimac Chemical Co., a large chemical business operating mainly in New England.[2] American Cyanamid, which is now among the half dozen biggest chemical enterprises, had a turnover of \$8·5 million in the year 30 June 1920. The slump cut down trade, and five years later the sales were still relatively small at \$13 million. The main business throughout the period was the manufacture of cyanamide on the Canadian side of Niagara Falls. The capacity of the works, substantial by any standard, was further raised from 120 000 tons a year of cyanamide to about 360 000 tons between 1927 and 1930. Cyanides and sodium phosphates were also produced, but the company's growth through acquisitions dates from the 1930s.[3] Among other medium-sized firms, the Hercules Powder Co. (the off-spring of the Du Pont break-up just before the war) extended its stake in the production of pine oil, turpentine, and wood resin in the 1920s. From this there developed later the company's large interests in chemicals for the paper trade.[4] Westvaco Chlorine Products Co. was a combination, put together in the mid-1920s, of several businesses making electrolytic chlorine, and became a large manufacturer of alkali phosphates.[5]

The list could be lengthened. But our purpose here is not to catalogue the scores of chemical enterprises, but to show that the pattern of development followed rather different lines from that in Europe. Du Pont and Allied

[1] Company's annual reports and public relations department.

[2] I am indebted to Mr. B. Thomas and others of the Monsanto Chemical Co. for this information.

[3] Haynes, W. On the chemical frontier—the Cyanamid story, unpublished typescript at American Cyanamid; Grossmann and Weicksel, op. cit., p. 154. Apart from Calco which has been mentioned previously, the principal acquisitions in 1929/30 were the Kalbfleisch Co. and Lederle Laboratories, which made pharmaceuticals. Another subsidiary was the Chemical Construction Co., bought by American Cyanamid in 1929. Chemical Construction had previously taken over the Nitrogen Engineering Corp., a process plant contractor, which made a name for itself building synthetic ammonia plants in Japan, Norway, France, and Russia. It is piquant that one branch of American Cyanamid was promoting the newer nitrogen fixation process, while the firm as a whole still relied for the bulk of its turnover on the older.

[4] Hercules Powder Co. *A History* (1947), p. 4.

[5] Haynes, W. *Chemical Pioneers* (1939), pp. 194–5. In the late 1940s Westvaco was taken over by an engineering company and is now the Chloralkali Products Division of the F.M.C. Corporation.

Chemical emerged during the 1920s as easily the largest firms in the industry, and both could stand comparison with the leading European groups. Neither, however, occupied in the country the dominant heights that IG Farben, I.C.I., Montecatini, and Kuhlmann and St. Gobain did in theirs. The different situation encouraged competition, especially in the 1930s, when the markets for industrial chemicals began to develop quickly and maintained a high rate of growth. This provided opportunities that the smaller firms seized and exploited, and so, in a surprisingly short time, enabled them to approach in size the ranks of the senior members.

11

COMMERCIAL AND ADMINISTRATIVE PROBLEMS

New patterns of trade

WITH broad brush strokes Rostow sketched the circumstances that cause a country's industries to 'take off'. He was referring to industry generally and not to chemicals specifically, but it is legitimate to follow at least part of the way and list the conditions that had to be fulfilled in our period to start the evolution of chemical manufacture and to sustain it so that 'growth becomes its normal condition'.[1] In the first place there needs to be a diversified range of products, which ought to include fertilizers as well as organic chemicals, among them dyestuffs or at least some of the more important intermediates. Secondly there must be a fair amount of trade between the main sectors of indigenously made bulk products such as acids and alkalis. Thirdly there needs to be a substantial home market for chemicals. Fourthly, the quality of the chemicals made and their price should be such as to provide them with an export market. All four conditions need to be met if growth it to be maintained and become a normal condition. This was the case in pre-war Germany, and, with some qualifications, in Britain and France; the circumstances were favourable in Russia and the United States, but not all the conditions had been met there. The war changed the direction and the rate of growth, so that technically as well as commercially the industry was markedly different from what it had been at the turn of the century. One new element was the general tendency for supply to exceed demand, especially as post-war trade barriers restricted imports. Another was that demand occasionally lagged and the slack in one sector could not always be compensated by greater activity in another. The post-war slump was in this respect merely the advance warning of the Great Depression. It showed that the industry, contrary to widely held belief, was not depression-proof, and its leaders reacted negatively by seeking safety in restrictive practices, and positively by seeking diversity through research. In fostering research as an antidote to stagnation chemical manufacturers introduced a special element into the development of their industry.

The following figures illustrate the scale of the changes in the chemical industry, notably the rapid growth in many countries and the extent to which it was being dispersed. The statistics are probably not too reliable (because of the definitions adopted, not to mention incomplete data) but they are, as far as I know, the only ones available in such detail.

[1] Rostow, W. W. *The stages of economic growth* (1960), pp. 7 and 36 ff.

TABLE 11.1

Estimated value of chemical production in the principal countries, 1913, 1927[1]

	1913		1927	
	£ million	%	£ million	%
U.S.A	170	34	472	43
Germany	120	24	180	16
U.K.	55	11	115	11
France	43	9	75	7
Italy	14	3	36	3
Japan	8	1	28	3
Canada	6*	1	27	2
Belgium	13	3	23	2
Holland	8	1	18	2
Switzerland	9	2	16	2
Sweden	6	1	10	1
Norway	5	1	5	negl.
Others (incl. Russia)	43	9	85	8
Total	500	100	1090	100

*1911

Note: The figures have been rounded off. The United States figures include oil refining and seed crushing.

The emergence of the United States as easily the leading producer of chemicals needs no emphasis, but some of the other changes require comment. The redrawing of European frontiers at Versailles and the ensuing upheavals still left their mark in the late 1920s. The share held by Germany fell by a third, in part owing to the loss of Alsace-Lorraine and of Upper Silesia. Interestingly, France and Belgium, though raising the value of their output quite considerably, did not manage to improve their relative position. Before the war the Habsburg Empire had possessed a scattered, but in the aggregate substantial, chemical industry, which was split up by the treaties of St. Germain and Trianon. Czechoslovakia emerged from the wreckage as the leading producer and so continued the pre-war pattern, when Bohemia was reported to account for three-quarters of the Empire's chemical production.[2] Solvay's ammonia-soda works were now dispersed among the successor states (excepting Hungary, which lost the one it had in Transylvania), and in general the capacities were no longer related to the markets they had been initially designed to serve.

The value of Italian chemical production increased one and a half times. She gained the Tyrol with its water-power resources (which were shortly to be harnessed to cyanamide production) and also Istria where chemical works had been established by the Austrians at Trieste and Monfalcone. These

[1] *Enquete*, p. 8.

[2] In glass and ceramics the predominance was even more pronounced. See also Schicht, H. The development and present situation of the chemical industry in Czechoslovakia. *J. Soc. chem. Ind., Lond.* **50** (1931), 276–7 T.

newly-won provinces offered additional incentives to the few manufacturers (notable among them Montecatini) who had greatly increased their activities during the war and made good profits. During the 1920s the Italian chemical industry (with the exception of some dyestuffs firms) did well: the growth of superphosphate production was particularly striking, and there was a remarkable proliferation of factories making cyanamide and ammonia, the latter by the processes of the Italian inventors Casale and Fauser. The manufacture of dyestuffs grew, and gradually was extended beyond sulphur colours; in 1926 for example the production of synthetic indigo was begun. The development of the dyestuffs sector helped the growth of other branches of organic chemistry, notably the preparation of chlorine compounds, acetic acid, formaldehyde, and industrial alcohol.[1]

The above table does not give figures separately for Russia, presumably because no reliable estimates could be obtained for the years shown. In 1912 production was estimated at £11 million, a considerable proportion of the output being derived from imported raw materials or intermediate compounds, and on the eve of the war imports of chemicals were valued at £8 million. The loss of Poland and of the Baltic states in 1917/18 removed several active centres of production, and the recovery of the industry under the Soviet Government was long delayed; imports of chemicals in 1925 were valued at £5 million.[2] This state of affairs was not due to lack of skill in replacing damaged or worn-out equipment, for many of the most able chemists and chemical engineers of the Czarist period continued their work under the new regime. The main factor was policy decisions at the highest level that assigned to chemical manufacture lower priority than railways, electricity, and steel. Under the New Economic Policy (NEP) several trusts were formed either to operate factories making similar products, such as the Anilo Trust for dyestuffs, or to manage a group of works in the same area such as Chimugol in the Lissichansk district of the Ukraine. The arrangements were not so very different from the war-time measures, or indeed the pre-war free-enterprise days. Additionally there were regional and central organizations for finance, purchases, imports, and sales to other trusts and combines.[3] They soon degenerated along Parkinsonian lines into bureaucratic bodies in no way suited to the needs of the industry. Foreign, particularly German, know-how and components were badly needed, and one would have thought that after the meeting at Rapallo there would be numerous fruitful opportunities for Russo-German collaboration in chemical manufacture. But no concessions were granted to foreign chemical manufacturers, so marking a clear policy distinction from the treatment of foreign

[1] Parravano, N. (Ed.) *La chimica in Italia* (1938); Ginori-Conti, P. L'Industrie chimique en Italie—son developpement. *Chim. Ind.* (Special Number, Feb. 1929), 729, 731.

[2] See Chapter 6, pp. 169–73. League of Nations (Economic and Financial Section). *The chemical Industry* (1927), pp. 41–2.

[3] *Chem. Ind., Düsseld.* 47 (1924), no. 17, 196–7; no. 33, 427; Ipatieff, V. N. *The life of a chemist* (1946), pp. 320–1.

entrepreneurs in other areas. Even after the Russo-German commercial treaty of 1925 trade remained small because the Russians demanded long-term credits above the limit the Germans would grant.[1] The IG set up a special Russian section, but it seems to have concerned itself exclusively with the sale of dyestuffs and pharmaceuticals.[2] There were, however, good relations between the scientists of the two countries and the Osteuropa Gesellschaft, originally founded in 1913 as the Deutsche Gesellschaft zum Studium Russlands, was active in Weimar Germany.[3]

The Russian chemical industry lacked external assistance, and domestic conditions prevented or at least delayed its recovery until the middle and end of the 1920s. In particular the production of textiles, paper, soap, and glass was late in regaining and surpassing pre-war levels, so that the post-war demand from these industries for chemicals developed slowly.[4] For many years the farmers were unable to use more fertilizers, and the pre-war trials on sugar-beet and cotton, which had yielded promising results, were not resumed. The following production figures for some important chemicals,

TABLE 11.2

Russian production of chemicals, selected years 1913 to 1928/9[5]
(000 tons)

	1913	1924/5	1927/8	1928/9
Caustic soda	55	36	59	65
Soda ash	160	98	217	239
Sulphuric acid (as SO_3)	226	187	384	481
Superphosphate	61	39	111	145
Sulphate of ammonia	14	n.a.	11	17
Dyestuffs	4	2*	10	13

*1923/4

[1] Hilger, G. and Meyer, A. G. *The incompatible allies, a memoir history of German–Soviet relations* (1953), pp. 165–86. Hilger was head of the economic department of the German embassy in Moscow from 1923 to 1941.

[2] Towards the end of 1924 it was reported that a group of IG chemists had gone to Russia. No more was heard of this mission, and it is possible that it never went. USTC, *Census 1924* (TIS 33), (1925), p. 147.

[3] Its president was Schmidt-Ott, an important personage in the German academic establishment. The Cologne branch was directed at the time by Dr. Adenauer. Schmidt-Ott, F. *Erlebtes und Erstrebtes* (1952), pp. 169–70, 225–7.

[4] The output of these goods illustrates their slow recovery:

	1913	1924/5	1927/8	1928/9
Cotton fabrics (000 million metres)	2·6	1·7	2·7	3·0
Paper (000 tons)	197	211	284	385
Window glass (million square metres)	24	16	34	40
Soap (000 tons)	254	n.a.	360	390

Nutter, G. W. *et al. Growth of industrial production in the Soviet Union* (1962), pp. 426, 430, 455, 541.

[5] ibid., pp. 423 ff.

though they must be treated with care, indicate that the turning point in recovery occurred between 1924/5 and 1927/8. Towards the end of the 1920s, the country's chemical production rose above the pre-war level (adjusted for changes in the frontier), but given the size and the general industrial development of the Soviet Union, its chemical manufactures were relatively small.[1]

Just before and during the first Five Year Plan, the Government took steps to improve productivity in the industry, develop indigenous raw materials, and adopt new processes. One notable achievement was the doubling of output per employee in sulphuric acid works, by replacing the old lead chamber installations by modern contact acid units. But for the industry as a whole, productivity was scarcely higher in 1928 than in 1913.[2] In other areas, however, the planners had considerably greater successes. In about 1930 the indigenous potash and phosphate-rock deposits were worked for the first time, the former in the Urals, the latter in the Kola peninsula of northern Russia. As for up to date technology, Ipatieff, one of Russia's most distinguished chemist–administrators, failed to get a licence from Badische in 1925 for the Haber–Bosch process. Soon afterwards he obtained a licence from Casale. There was, however, a little difficulty in that the Soviet Government was at first unwilling to pay royalties to a citizen of Fascist Italy; but the matter was soon settled. A plant using Casale's process was opened not far from Gorki in 1928. The factory, the first in Russia to make synthetic ammonia, had an initial capacity of 20 000 tons a year.[3] A few years later the Nitrogen Engineering Corporation built a rather larger works, based on their design, in the Solikamsk area.[4] The manufacture of dyestuffs, which, as we have seen in an earlier chapter, had got under way in the last pre-war years, restarted soon after the Civil War, but little progress was made until the preparation of coal-tar distillates and intermediates was resumed. Dyestuffs as well as pharmaceuticals were made principally in the Moscow region in the factories previously owned by German companies.[5]

The war years boosted the chemical manufactures of Japan and Canada, and by 1927 further growth placed them respectively in the sixth and seventh places among the producing countries. Superphosphate manufacture developed very rapidly in Japan to compensate for diminished foreign supplies after 1914 and also as a substitute for fish meal, which had been widely used.

[1] The Soviet Union, alone among industrially advanced countries, neglected the development of chemical manufacture until about the 1960s.

[2] Kats, A. M. *Voprosy ekonomiki osnovnoi khimicheskoi promyshlennosti SSR* (1939), pp. 125–6. According to Kats the output (in 000 tons SO_3) per worker went up from 3·8 a year in 1913, to 7·3 in 1927/8. Nutter *et al.*, op. cit., p. 172.

[3] Ipatieff, op. cit., pp. 410 ff.

[4] Kats, op. cit., p. 162; Hirsch, A. *Industrialized Russia* (1934), pp. 77–8. Hirsch was an American and was in Russia from 1931 to 1933 as a consultant to the Chemical Construction Trust.

[5] Ioffe, Ya. A. *et al. Fkonomika i planirouaniya khimicheskoi promyshlennosti* (publication No. 11 of the Molotov Institute of Economics and Engineering), (1940), p. 31.

Nitrogen fixation plants multiplied, and by 1930 six synthetic ammonia works were in operation or under construction, the total capacity being about 250 tons a day of ammonia, equivalent to roughly 70 000 tons N a year. There were in addition several factories making cyanamide. But many of the new enterprises were languishing, because their production costs, owing to their small size, were too high.[1] Considerable diversity was added to the range of chemicals made by the construction of electrolytic alkali works during the war and of an ammonia-soda factory a few years later. The manufacture of dyestuffs, stimulated by subsidies and protected by import licensing, was considerable. Initially only sulphur dyes were made, but towards the late 1920s the first vat dyes appeared, though the quantity—about 1000 tons out of a total of around 9000 tons—was still relatively small.[2] Relations with Germany became closer and culminated in a commercial treaty in 1928. The links between academic chemists created by Fritz Haber in 1924 had their industrial counterparts. For example IG Farben agreed with Japanese firms not to compete in their home market, in return for which the Japanese agreed to keep out of India and Indonesia.[3]

The value of Canadian production, according to the figures in Table 11.1, more than quadrupled between 1911 and 1927. Whatever the reliability of the estimates, and in the case of Canada it cannot be put very high, it is a fact that particular local conditions resulted in the development of a narrow range of products on an extremely large scale. Thus the mining boom created a big demand for explosives and so for inorganic acids, and the huge hydro-electric resources were utilized for some of the largest electrochemical projects in the world. The works at Niagara Falls of American Cyanamid, which had a capacity of 25 000 tons N in 1927, was trebled within the next three years. There were also large factories for the production of chlorates, phosphorus, and, at Shawinigan Falls, for the manufacture of acetic acid and acetone from carbide acetylene. By contrast superphosphate and sulphate of ammonia were not made to any significant extent in the 1920s, nor was there an indigenous production of dyestuffs and of intermediates. The first ammonia-soda factory, at Amherstburg, opposite Detroit, was completed in 1919 by Brunner, Mond.[4]

Even before the war, a few chemical enterprises had been established in Africa. The largest were the explosives works near Johannesburg, Durban, and in the Western Cape, which between them employed 3000 people and

[1] Schumpeter, E. B. (Ed.) *The Industrialization of Japan and Manchukuo 1930–1940* (1940), pp. 329 ff.; Grossmann, H. and Weicksel, P. *Die Stickstoffindustrie der Welt* (1930), p. 148; Waeser, B. *Die Luftstickstoffindustrie* (2nd edn. 1932), pp. 172–3.

[2] USTC, *Census 1923* (TIS 32), (1924), p. 145; *1925* (TIS 34), (1926), pp. 185–7; *1928* (TIS 38), (1930), p. 165; Schumpeter, op. cit., p. 359.

[3] Wickel, H. *IG Deutschland* (1932), pp. 163, 166. Although Wickel wrote in a sensational style, his remarks about Japan are based on fact. The IG Farben agreement was probably limited to dyes and to nitrogenous fertilizers.

[4] Warrington, C. J. S. and Nicholls, R. V. V. *A history of chemistry in Canada* (1949); Waeser, op. cit., pp. 157 f. Brunner, Mond sold their factory to Allied Chemical in the mid-1920s.

represented an investment of £2 million. Those owned by Nobels and Kynoch were, like their British parents, combined in 1918. The new firm combined with De Beers in 1924 to form African Explosives & Chemical Industries Ltd. Raw materials had to be imported, but the new firm, which was soon to derive considerable help from I.C.I., began making fertilizers, cattle and sheep dips, and paints. In 1930 it began the construction of a synthetic ammonia plant, so as to make nitric acid from ammonia instead of from imported nitrates.[1] I.C.I. also materially assisted the development of the Magadi Soda Co. This firm had tried to work the deposits of *trona* (sesquicarbonate of soda) in Lake Magadi in Kenya, but had collapsed. Brunner, Mond & Co. took over, revived the business, and by the end of the 1920s nearly 100 000 tons of soda were produced annually. The colony suddenly emerged as an important factor in the world alkali trade, the exports from East Africa being mainly directed to Japan and to Australia.[2]

The purpose of these short descriptions has been to show that after the war the manufacture of chemicals grew quickly in what, for want of a better word, we may call the lesser producing countries. So the industry came to be dispersed widely in Europe and elsewhere. A few instances related to individual products such as alkalis, cyanamide, superphosphate, and dyestuffs will illustrate the process.

As regards alkalis, Solvays remained the dominant enterprise in Europe, but their works became involved in the redrawing of frontiers. In 1914 they had factories in six countries, the Austrian Empire counting as one; in the 1920s they had factories in twelve, excluding Russia where they lost their properties in the Revolution.[3] Outside Europe modest beginnings had important consequences. We have already noted the start of soda-making in Japan and Kenya. The Nippon Soda Co., after only a few years in business, was making 60 000 tons of alkalis in 1928, and the next year it completed extensions that raised the volume to 100 000 tons. Even so the country was not yet self-sufficient, and though imports were declining, 122 000 tons of sodium compounds were still imported in 1929.[4]

Nitrogen fixation will serve as another example of the diffusion of chemical technology after the war. We noted in Chapter 4 that the first factories were built in the 1900s and that some relied on hydro-power. After 1918 the needs of defence and agriculture led to a spate of new projects, which in turn promoted the development of other chemical activities. The growth of nitrogen fixation was extraordinary, production rising from a mere 34 000 tons N in 1913 to 567 000 in 1925 and almost 1·1 million tons four years later. Throughout the period Germany was the largest producer, and even in 1929 accounted for three-fifths of the total, being followed at some considerable distance by the United Kingdom, the United States, and France.[5] But my

[1] Cartwright, A. P. *The Dynamite Co.* (1964). [2] Hill, M. F. *Magadi* (1964).
[3] Bolle, J. *Solvay 1863–1963* (1963), pp. 120–1.
[4] Report of the Mission on Japanese Combines (U.S. Department of State Far Eastern Series No. 2628), (1946), p. 230, Schumpeter, op. cit., pp. 359, 361. [5] *Enquete*, pp. 38–9, 157.

purpose here is not to review once more the progress of the technologically advanced countries, but to note the advance made in this sector by the lesser producers and in particular countries where chemical manufacture in general was a novel industrial phenomenon. In this context the development of the cyanamide process is particularly relevant, its technology was fairly simple, and the know-how readily available. By contrast the Haber–Bosch and similar processes required expensive licences, and the technology was complex.

TABLE 11.3

Production of cyanamide in the principal countries 1922–8[1]
(000 metric tons calcium cyanamide)

	1922	1925	1927	1928
Germany	320	350	440	450
Japan*	102	125	120	160
Poland	68	86	143	155
Canada	32	82	90	125
Norway	—	—	34	57
Italy	31	38	39	55
France	29	55	53	50
Yugoslavia	—	34	30	32
Others	52	78	90	78
Total	634	848	1039	1162

*Including cyanamide converted to ammonia. There is probably some element of double-counting in the figures.

Germany was easily the leading producer. There were only four factories: Knapsack (IG Farben), Trostberg and Piesteritz (Bayerische Stickstoffwerke), and Waldshut (Lonza-Werke, a Swiss company). IG Farben and Bayerische Stickstoffwerke, though operating on a very large scale, were working at only 70–80 per cent of capacity.[2] The works at Waldshut were small, and this was the norm everywhere, excepting only Poland and Canada. It has not been possible to obtain comprehensive information about Japan. Production grew significantly, and by the end of the 1920s there were apparently eight or nine factories in operation or under construction with a total capacity of 200 000 tons of cyanamide. A substantial share of the cyanamide was converted to ammonia, but this introduces statistical complications that cannot be cleared up. There is a likelihood that the Japanese figures are considerably inflated, and a conservative estimate of output in 1929 is 180 000 tons.[3]

It is surprising to see Poland occupy the third place. The reason was as follows: the Germans built a factory at Chorzow near Bytom during the war. After the Silesian plebiscite of 1921/2 the town became Polish, and the works were taken over by the Government and enlarged to an ultimate capacity of

[1] *League of Nations Statistical Yearbook 1931/32* (1932), p. 166.
[2] Waeser, op. cit., pp. 20–1. [3] ibid., p. 173.

44 000 tons N a year. By the late 1920s they were working at about three-quarters of their capacity. In addition there were three synthetic ammonia plants in the coal–steel district of upper Silesia. Thus, on the eve of the slump, Poland had emerged as one of the large producers of nitrogenous fertilizers, and exported a substantial tonnage of sulphate of ammonia.[1] The Canadian cyanamide production reflects the output of a single works at Niagara Falls belonging to the American Cyanamid Co. Like the others it was not working near full capacity; the product was sold in the United States.[2]

The situation in Norway was unusual in the 1920s. There were two manufacturers: A/S Bjølvefossen in the south-west of the country had been started during the war, but economic conditions forced it to abandon production and it became an electricity supply undertaking. Odda Smelteverk A/S was founded in 1924 with the help of A/S Tyssefaldene, a large electricity supply company, with the object of restarting the cyanamide factory of the bankrupt North Western Cyanamide Co. located to the east of Bergen. The new company's works had a capacity of around 80 000 tons a year of cyanamide, but was working well below capacity.[3] It could not compete successfully with the other manufacturers of nitrogenous compounds, among which Norsk Hydro was, geographically, its nearest rival. That company gradually increased the tonnage of nitrogenous fertilizers produced by the arc process to 215 000 tons in 1924 and around 240 000 tons a few years later. But the arc process was no longer really economic, and in 1925 the directors sanctioned the construction of a small ammonia plant to the design of the Nitrogen Engineering Corporation. Two years later a more important policy change was made: the directors reached an agreement with IG Farben, and were authorized to adopt the Haber–Bosch process. A large works was built at Rjukan, and the first stage of the plant, with a total capacity of 50 000 tons N, was commissioned in 1929.[4]

There were no less than six cyanamide factories in post-war Italy, with a combined capacity of around 20 000 tons N. They were operating well below capacity. There were also nine synthetic ammonia plants that used Fauser's or Casale's process, and as a rule obtained their hydrogen very expensively by electrolysis. The combined capacity of the ammonia plants was about 83 000 tons N, but they were only making half that tonnage, and their operations were certainly uneconomic not only on account of the cost of hydrogen, but also because of the small size and low rate of utilization.[5] The

[1] Grossmann, op. cit., pp. 169 ff. The fact that Poland's first president, Moscicki, had developed an arc process before the war may have had something to do with the vigorous support of nitrogen fixation in the new state.

[2] Warrington, op. cit., pp. 186–7. The first synthetic ammonia plant, a unit working the Casale process, was opened at Sandwich in 1930.

[3] Norges Industri (pub.), *Norwegian hydro-electric industry* (1932), pp. 74–5. Odda was later bought by the British Oxygen Co.

[4] *Norsk Hydro history*, pp. 218 ff., 260, 459.

[5] Ginori Conti, art. cit., p. 728; Parravano, op. cit., p. 8; Grossmann, op. cit., p. 144; Waeser, op. cit., pp. 141 ff.

situation was not greatly different in Yugoslavia, which inherited two cyanamide factories from the Austrian Empire. Local demand was insufficient to sustain the works, which were therefore dependent on export sales. Thus it came about that for some years Yugoslavia was the third largest exporter of cyanamide in the world, her foreign sales amounting to 15 000 tons in 1924 and 27 000 tons in 1928.[1] Elsewhere, in Czechoslovakia, Switzerland, and Sweden, there were small cyanamide works, but it is surprising that the Netherlands with their intensive agriculture and large demand for fertilizers were one of the last countries to have a nitrogen fixation works. A subsidiary of the Royal Dutch-Shell group, Mij. tot Exploitatie van Kooksovengassen, opened a synthetic ammonia factory, which used coke-oven gas as a source of hydrogen, in 1929.

The technology of superphosphate manufacture is very much simpler than that of nitrogen fixation and the investment in plant and machinery much lower. It is not surprising therefore that manufacture developed quickly during the 1920s in some countries that up to 1913 or even later were making the fertilizer on a very small scale and did not possess in effect an indigenous chemical industry.

TABLE 11.4

Production of superphosphate in the principal countries, 1922, 1925, 1928[2]
(million long tons)

	1922	1925	1928
United States of America	2·5	3·5*	4·1
France	2·1	2·4	2·4
Italy	0·9	1·5	1·2
Japan	0·5	0·7	0·9
Spain	0·5	0·7	0·9
Australia	n.a.	0·8	0·9
Germany	0·6	0·7	0·8
Netherlands	0·3	0·6	0·6
United Kingdom and Ireland	0·7	0·5	0·5
Belgium	0·4	0·6	0·4
Poland	0·1	0·3	0·3
Sweden	0·1	0·2	0·2
Czechoslovakia	<0·1	0·2	0·2
Total	8·8	12·7	13·4

*The Census figure is 3·7 million tons.

The figures illustrate two aspects of the changes in the fertilizer trade. In France, Italy, Germany, Britain, and Belgium superphosphate production remained stationary or fell in the later 1920s. On the other hand Japan, Spain, Australia, the Netherlands, and Poland emerged as large producers. Their progress can be attributed to the success of measures taken to attain self-

[1] Grossmann, op. cit., pp. 158–9.

[2] League of Nations, op. cit., p. 167.

sufficiency in this product. One important consequence of the expanding output of the newcomers was that intra-European trade of superphosphate fell. Those producers who made the fertilizer in small and old-fashioned plants were particularly hard-hit, and this probably accounts for the distress in Britain and Belgium and their lower exports. In Belgium, the capacity for making superphosphate was said to be in excess of 1 million tons, so that the factories were operating at less than half their potential.[1]

Dyestuffs will serve as the final example of the changing pattern in the chemical industry. It will be recalled that before the war production had been concentrated in Germany and Switzerland; in a few other countries the factories belonging to German or Swiss firms dominated the local scene. During the war manufacture was started on a large scale in the United States, France, and Britain, and on a lesser scale in Russia, Italy, and Japan. They were joined by Poland and Czechoslovakia after the war. The statistics in the table below summarize the transformation that took place in the 1920s: Germany and Switzerland increased their production by 11 000 tons only, the other countries by 26 000 tons; Britain, France, and Japan accounted for the bulk of the higher volume.

TABLE 11.5

Estimated production of dyes in 1923, 1925 and 1928[2]
(000 tons)

	1923	1925	1928
Germany	65	74	74
United States of America	42	38	43
Great Britain	15	15	23
Switzerland	9	8	11
France	11	16	15
Italy	5	6	7
Japan	6	8	9*
Russia†	2	8	10
Total‡	155	173	192

* 1929.
† The years are 1923/4, 1925/6, and 1927/8.
‡ There was also some production in Poland and Czechoslovakia, probably less than 1000 tons in each country.

Despite elaborate protectionist devices, the manufacture of dyestuffs was unprofitable in many countries. We have already looked at the situation in Britain, France, and the United States;[3] here we may note that among the

[1] Lepersonne, H. Quelques aspects de l'industrie chimique en Belgique. *Chim. Ind.* (Special Issue, March 1931), pp. 87–8.

[2] USTC, *Census 1925* (TIS 34), (1926), p. 154; *1928* (TIS 38), (1930), p. 142; Third Report of the Dyestuffs Industry Development Committee, HMSO (1932), Cmd 4191, pp. 17 ff.; Schumpeter, op. cit., p. 359, and Table 11.2 for Russia.

[3] In Chapters 9 and 10.

smaller producers the problems were, if anything, even greater. In Japan and in Italy, for example, sulphur dyes, which represented a large proportion of the output, were primarily sold to local cotton dyers, and so dependent on a restricted and fluctuating market. Sulphur dyes are not suitable for silk or wool, so that the two countries (and probably also Poland) were obliged to import a substantial volume of dyestuffs suitable for those fibres.

Dyestuffs, like fertilizers and alkalis, had a different trade pattern after the war, compared with the period up to 1914. The composition as well as the direction of the trade changed. The change is inadequately documented, not so much owing to the lack of data, but rather because of the absence of a common definition of chemicals. The following table, based on a German compilation, suffers from this defect and is therefore not very accurate. But subject to this qualification the export statistics indicate in broad terms the changes between 1913 and the late 1920s.

TABLE 11.6

Chemical exports of the principal countries, 1913, 1925, 1929[1]
(£ million)

	1913		1925		1929	
	Value	%	Value	%	Value	%
Germany	43	27	46	23	67	28
United States of America	15	10	32	16	38	16
United Kingdom	25	16	30	15	33	14
France	16	10	26	13	28	12
Chile	24	15	27	13	27	11
Belgium	9	6	9	4	8	3
Switzerland	3	2	5	3	7	3
Others	22	14	26	13	32	13
Total	157	100	201	100	240	100

Note: The figures have been rounded off.

The value of exports increased by about 60 per cent between 1913 and 1929. By contrast production, which is shown in Table 11.1, grew by 120 per cent between 1913 and 1927. (The comparison, however, is not entirely fair, because of the different definitions used.) Germany retained its lead, but the growth of United States exports will be noted; it reflects the large post-war business in naval stores, brimstone, and alkalis. Britain and Belgium lost ground, and the decline in the position of Chile reflects the lessened significance of nitrate shipments. Among other countries, Italy and Japan made particularly rapid progress. In 1913 the former exported chemical products

[1] Ausschuss zur Untersuchung . . . der deutschen Wirtschaft. *Der Deutsche Aussenhandel unter der Einwirkung weltwirtschaftlicher Strukturwandlungen* (1932), p. 200. Referred to hereafter as *Deutscher Aussenhandel*. Data for 1927 for comparison with Table 11.1 are not available, nor are the details of the adjustments made to the statistics of the various countries.

worth £3·2 million, the latter £2·75 million; twelve years later their trade had grown to £8·5 and £4·5 million respectively.[1]

In some countries the changes in foreign trade draw attention to the effects of new chemical technologies. For example, Germany imported chemicals valued at £22 million in 1913, but only £7·8 million in 1925 and £11·3 million three years later. The reduction was due principally to the lessened dependence on imported nitrates, pyrites, and sulphur, which were replaced by synthetic ammonia or indigenously produced materials such as anhydrite. By contrast its exports of nitrogenous fertilizers grew phenomenally. They amounted to a mere £1·7 million (that is equivalent to 4 per cent of the total) in 1913. In 1925 the trade was valued at £7·2 million (16 per cent of the total), and three years later reached £14 million, corresponding to more than a fifth of the country's chemical exports.[2] The diffusion of dyestuffs manufacture wrought many changes in the foreign trade of that group of products. Before the war only Germany and Switzerland exported dyestuffs and intermediates on a large scale, and their principal customers were the United States, China, and Britain in that order. The three countries between them took roughly half of Germany's dyestuffs exports, and about 35 per cent of Switzerland's by value. But after the war the United States and Britain were no longer among the major importers (except of speciality dyes with a high unit value), and India and China now headed the list. Indeed during the 1920s the United States, Britain, and France were becoming dyestuffs exporters of some significance. American exports, though usually consisting of the simpler dyes of low unit value, ranged between £1·2 million and £1·5 million a year in the late 1920s.[3] British shipments of dyes and intermediates were erratic in the early 1920s, then declined, and later recovered: in 1926 the volume was 3800 tons valued at £0·6 million; it rose to 5200 tons worth £0·8 million two years later and in 1929 increased further to 7800 tons and £1 million.[4] French exports reached a peak of 5000 tons in 1927, the largest single group being vat dyes, probably synthetic indigo.[5] The emergence of the new exporters brought some

[1] League of Nations, op. cit., p. 41. The League of Nations work is more detailed than that of the *Deutscher Aussenhandel*, but ends in 1925.

[2] *Deutscher Aussenhandel*, op. cit., pp. 199 and 207. There are some differences between this report and others (notably the *Enquete*), which are due to differences in the definitions used, but they do not affect the argument.

[3] The quantities and values were as follows:

Year ending 30 June	*000 tons*	*$ million*
1925	9·2	6·6
1927	11·8	5·8
1929	15·2	7·2

Haynes, W. *American chemical industry* (1948), vol. 4, p. 497.

[4] Third Report of the Dyestuffs . . . Committee, op. cit., p. 29.

[5] Ogburn, W. F., and Jaffé, W. *The economic development of post-war France* (1929), p. 465; Fauque, M. *L'Evolution économique de la grand industrie chimique en France* (1932), pp. 126 and 127.

competition into markets such as India, China, and the Dutch and French colonies, which had hitherto been the exclusive domain of the Germans. The IG and their successors were naturally alarmed by the trend and strenuously attempted to reintroduce orderly marketing methods. They successfully negotiated agreements with all the other producers and arranged for quotas or other arrangements to check competition in the important overseas markets.

Even an old-established trade, where technological change was no longer a determining factor, such as Britain's exports of alkalis, altered direction as some traditional customers were lost and others had to be found. Between 1913 and 1929 the tonnage sold overseas rose by two-fifths: sales to India, Sweden, China, and the other British territories doubled. Japan was, for a short time, Britain's largest single customer of sodium compounds, but as its own soda industry developed, its imports fell steeply. Exports to the group of other foreign countries fell by 10 per cent between 1913 and 1929, owing to

TABLE 11.7

British exports of sodium compounds, 1913, 1925 and 1929[1]
(000 tons)

	1913	1925	1929
India	30·0	48·7	73·1
Sweden	28·7	45·7	68·2
Japan	47·4	111·9	57·6
China	25·2	43·5	51·8
Argentina	28·3	21·8	33·5
Australia	20·6	37·5	33·0
Other British	32·5	49·3	63·1
Other foreign	150·6	104·5	133·2
Total	363·3	462·9	513·5

Note: Sodium compounds comprise soda ash, caustic soda, sodium bicarbonate, sodium cyanide, sodium silicate, sodium sulphate, and other unidentified sodium compounds.

the loss of some pre-war business. Exports to the Commonwealth were remarkably buoyant throughout the period. As the needs of the alkali-using industries of those countries grew, so their demand for imported sodium compounds increased and Brunner, Mond and I.C.I. were able to obtain the full benefit of this increase. With the exception of Canada and Kenya there were no alkali works in British overseas territories in the 1920s, and it is probable that there were tacit or perhaps even written agreements with European producers whereby the Empire was recognized as the preserve of the United Kingdom manufacturers.[2] In a similar manner agreements

[1] Board of Trade, Statistical Tables Relating to British and Foreign Trade and Industry, HMSO (1931), Cmd 3849, pp. 358–9; and Trade and Navigation Accounts.

[2] This may also have applied to Sweden, where the paper industry was becoming a big user of alkali.

reinforced tariff policy in the French colonies, where the companies in metropolitan France held a secure position. French exports of alkalis rose from 97 000 tons in 1913 to 210 000 tons in 1929, the increase being mainly due to higher sales in French overseas territories.[1]

After 1918, self-sufficiency, fostered and cosseted in every possible way, restricted the movement of chemicals between European countries, Britain being a partial exception, and between Europe and the United States. Manufacturing capacity, however, generally exceeded domestic requirements, so that the development of export markets was imperative if operations were to be at all economic. The capacity problem existed in most European countries, the differences being of degree not of kind, and the export opportunities appeared most promising in Latin America (although the markets were fragmented), India, and the Far East. But China, rent by civil war, became relatively unimportant after 1928. All chemical products were affected to a greater or lesser extent by the need to boost exports, even pharmaceuticals and fine chemicals, which until 1914 had been traded almost wholly in Europe and North America.[2] Although the available information is probably incomplete, it would appear that Japan and India, taken together, increased their imports of chemicals from £8½ million to £20·7 million between 1913 and 1925; China, in its days of peace, added a further £8·25 million.[3] These figures represent a substantial shift in the direction of trade, and sharpened further the rivalry between the leading producers, a rivalry to which the solution was either the rapid expansion of that market or more often various arrangements to share the market regardless of whether it was growing or not. But even where agreements existed, the experience of the 1920s pointed to a more intensive marketing effort abroad and an effort that had to be spread over a wider range of products and countries.

Changing business ways

The method of managing chemical businesses reflected, after a time lag, the need for more elaborate administrative structures to cope with and preferably to master successfully the consequences of the technical and commercial changes in the industry. There was no uniformity except that inexorably as the business grew and became more complex, so the founders, or their heirs, increasingly depended on outside professional assistance. The separation of ownership from control proceeded throughout the period with which we are dealing here. The process has been analysed in the American context elsewhere,[4] our object here is to draw attention not only to the survival

[1] Fauque, op. cit., pp. 44–5.

[2] *Enquete*, p. 461. The statement was made by Dr. Merck on behalf of his firm. It may reasonably be assumed that it applied also to others.

[3] *Deutscher Aussenhandel*, op. cit., p. 199.

[4] The classic on the subject is A. A. Berle and G. C. Means, *The modern corporation and private property* (1932). There are, incidentally, very few references to chemical companies in the book.

of the old ways, but to the emergence of modern systems of management and their culmination in the colossal elaborations of IG Farben and I.C.I.

Ownership combined with control was becoming rare among large businesses, but there were still some conspicuous examples, notably Solvay & Cie. The founders' male descendants continued to own the concern, but had delegated its management to a small number among them, the *gérants commanditaires*.[1] Solvays, though very large, specialized in a small number of products, so that the direction of production and of sales, though on a European scale, was not too complicated. There were other companies with a far wider product range where family connexions still predominated. Kunheim, an old-established medium-sized business active in northern Germany, was until 1921 directed by the descendants of the founder, though of late their activities had not been especially profitable.[2] Geigy of Basle dated back to 1758, and only became a limited liability company in 1901. Legal responsibility was thereby transferred from the partners to the board of management, but in practice there was little difference, because the owners or their relations constituted the board of directors (on which there was only one outsider), and also held all the shares. The board handed over actual management to four men, three of them descendants of the founders: two dealt with all technical matters and two with all commercial and other questions. Until 1921 the transfer of shares was subject to the approval of the board, and until 1924 there were no regular meetings of the board or the departmental heads.[3] This quasi-patriarchal organization underwent few other changes in the 1920s.

There were other, similar instances, even in the United States, where, to take the case of Dow, H. H. Dow, its founder, was also for more than thirty years its chief executive. The Monsanto Co., though growing quite rapidly after the Great War, was managed until about 1925 by four men: E. Queeny, the founder and chairman, was responsible for financial and commercial matters, and his Swiss associates, L. Veillon, G. Dubois, and J. Bébié, occupied respectively the posts of general superintendent, technical director, and head of research.[4]

Throughout our period more elaborate forms of organization were evolving and marking the gradual break between ownership and control. This became unavoidable when the founders or their heirs had to rely increasingly on men who were neither related to them nor even co-owners, but hired managers. There are many instances, among British and German chemical firms, of this

[1] There were six in 1956, according to a brochure published by Solvay & Cie in that year (pp. 14–15). A generation earlier Ernest Solvay had said: 'Notre société est une société fermée . . . Elle s'est en réalité constituée avec l'aide de notre famille et de personnes que mon frère et moi pouvions considérer comme des amis. Elle est une commandite simple, n'ayant ni actions, ni titres, rien que des parts écriturées . . . qui ne peuvent sortir du groupe social intime.' Solvay & Cie, *Cérémonie de commémoration du 50ème anniversaire de la fondation* (1913), p. 29.

[2] Spiegel, L. *100 Jahre Kunheim* (1925), p. 22.

[3] Bürgin, A. *Geschichte des Geigy Unternehmens von 1758 bis 1939* (1958), pp. 221–2, 268.

[4] Private information from Dow Chemical and Monsanto Chemical.

change in the last quarter of the nineteenth century. On the Continent the two-tiered structure of company administration was often adopted at the time: at first the owners would be represented on the management board (variously called *Direktionsrat*, *Vorstand* or *Direction Générale*) as well as on the supervisory board (*Verwaltungsrat*, *Aufsichtsrat*, *Conseil d'Administration*). Later they sat only on the supervisory board, whose main function was to approve the accounts and appoint the directors of the management board. At Hoechst the last descendants of the founders to play a key role in management died in 1913 and 1919 respectively, and that large firm became entirely management controlled.[1] The change at Du Pont is well-documented and shows the gradual evolution of the firm's management system. Until 1902 the business was run by a small group of du Ponts who took their instructions from the all-powerful president, who was, of course, also a du Pont. This organization of idyllic simplicity lasted for roughly a century, but by 1902 the machinery was no longer functioning properly. The four leading du Pont men, Coleman, Alfred I., Pierre S., and Francis I. set up the Executive Committee and invited H. Barksdale, A. Haskell, and A. J. Moxham, three dynamite specialists, to join them.[2] The event created no less than three precedents: the delegation of the company's top administration to a committee broke with family tradition, the inclusion of three men from outside the family was novel, and thirdly, in order to reward these outsiders, they were given bonuses in the form of shares, thus dispersing the ownership outside the family group. The bonus scheme provided for purchase of Du Pont stock by the company for resale to members of the Executive Committee (and later to others) at special concessionary prices well below the stock exchange value.[3] (The scheme caused controversy later and was one of the reasons for Alfred's litigation with his cousins.) The bonuses and the growth of the du Pont family gradually dispersed the ownership of the company's stock, and by 1915 the firm had 3840 shareholders, 1122 of them employees. In order to maintain some form of family control, Du Pont Securities (the name was changed later to Christiana Securities) was established in 1915. It held 27·6 per cent of the Du Pont stock and was thus the largest single shareholder. Pierre, Irénée, and Lammot, between them, owned over one-third of Christiana's shares. By the device of pyramiding, the three with only a minority interest were effectively

[1] They were G. von Brüning who was General Direktor (chief executive) from 1901 to 1913 and H. E. A. von Meister who held that post from 1913 to 1915 and then sat on the *Aufsichtsrat* until his death. Fischer, E. Die Gründer der Farbwerke Hoechst AG. *Nassauische Lebensbilder* **6** (1961), 252 f.

[2] The ramifications of the family are extensive. Coleman, Alfred, and Pierre were cousins. After the withdrawal of the first two, Pierre S. (1870–1954) and his two younger brothers, Irénée (1876–1963) and Lammot (1880–1952), became the principal men in the business. The presidents, that is chief executives, up to the Second World War were: Coleman (1902–15), Pierre (de facto 1914, de jure 1915–19), Irénée (1919–26), and Lammot (1926–46).

[3] The price of the stock varied. Early in 1915 Alfred, in a transaction outside the scope of the scheme proposed to sell 1000 shares at $120 to J. J. Raskob. The lowest market price in 1915 was $184 and by October it was $840. Dutton, W. S. *Du Pont* (3rd edn. 1951), pp. 185 ff.; James, M. *Alfred I. du Pont* (1941), pp. 258–9.

the controlling group of shareholders and in this way were able to maintain the role of the family.[1]

It was probably not anticipated in 1902 that the powers of the Executive Committee would grow so quickly and extend so widely. Perhaps the frequent absences of Coleman du Pont, the committee's first chairman, who had some absorbing outside interests, enabled the other members collectively to obtain and secure a dominating position in the firm's management. In 1914, Pierre, who had to all intents and purposes replaced Coleman as president of Du Pont, laid down that the Executive Committee was to be in complete charge of the company's affairs, subject only to the Board of Directors and to the Finance Committee. At the time the committee had eight members who were all directors, and excepting Irénée, its chairman, also heads of departments; they met weekly and convened monthly for longer discussions. The Finance Committee was created a few years before the Great War. It consisted of four members of the family, and was responsible for dealing with large budget items, stock issues, and the salaries of the members of the Executive Committee.[2] During the war the pressure of work increased and membership had to be enlarged, so that towards its end the Executive Committee consisted of nine men and the Finance Committee of five, among them three du Ponts.[3] In March 1919 the Executive Committee appointed a special sub-committee on organization; it reported in due course, but two and a half years were to pass before the administration was thoroughly overhauled. Five (later seven) operating departments or Divisions were formed.[4] Each was put under a general manager (who was also a director) with complete responsibility for manufacture and sales, and the profit and loss account of his department. At the same time eight (later fourteen) functional departments were established and directors put in charge of them. The departments were responsible for the different 'functions' such as purchases, research, accounts, legal matters, and so on. The directors with divisional or functional duties formed a separate group (perhaps layer would be a more picturesque description) from the Executive Committee, whose members were henceforth freed from all departmental duties and responsibilities. The Committee jointly or severally now dealt with broad policy questions and 'advised' the departments.[5] One could speculate about the meaning of 'advice' and the extent to which policy went into departmental detail, but one thing is certain, the directorate grew: in 1922 it had twenty-six members, three years later there were twenty-seven, and in 1928 they numbered thirty-two.

[1] Winkler, J. K. *The du Pont dynasty* (1935).

[2] GTX No. 27, Letter by P. S. du Pont to all directors, dated 11 September 1914.

[3] Dutton, op. cit., p. 273. The oldest man on the Finance Committee was Irénée, who was then forty-three. J. J. Raskob and R. R. M. Carpenter, who were on the Executive Committee as members for finance and development respectively, were the two outsiders on the Finance Committee.

[4] Explosives, cellulose products, 'Pyralin' (i.e. leather-cloth), paint, dyes. In 1922 explosives were divided into smokeless powder and explosives.

[5] Annual Report of the company for 1921 and 1922; Dutton, op. cit., pp. 281–3; Lessing, L. P. Du Pont. *Fortune*, October 1950.

During the 1920s the Executive Committee, on which du Ponts were now in a minority, gained even more power under the chairmanship first of Irénée and then of Lammot. Du Pont was certainly fortunate in that its senior directors were in the prime of life so that the membership of the key committees hardly changed between 1920 and 1930. Such continuity was a great advantage and so probably was the simplicity of the line of command, with the board of directors ratifying as it were the broad plans and the major decisions of the Executive and Finance committees. The organization of Du Pont's management was, for a company of its size, simple, which made for clear lines of communication and probably helped to speed up the taking of decisions. The company's profitability was rising in the 1920s and the directors and the methods they used can certainly take credit for that.[1]

The management system with which we are most familiar nowadays is one where ownership is separate from control, where decisions are taken by committees, and the numerous committees are the symptom of elaborate administrative structures. One of the first steps towards such a system was the increasing size of the directorate. In the 1920s the European companies generally adopted (where they had not already done so) the two-tiered board structure. For example Kuhlmann had a supervisory board of about twenty-five, mostly big industrialists and bankers. There were too many for effective supervision, let alone decision-taking, and a rather smaller Comité de Direction, consisting of the Chairman, the three vice-presidents, four directors, the two general managers, and the company secretary, met weekly to monitor affairs and submit proposals to the board as a whole. The proposals came from two directorates, the Direction Générale Produits Chimiques and the Direction Générale Produits Organiques. Each was under a general manager who was assisted by works managers, sales directors, and other senior executives. In all matters concerning dyestuffs the Comité de Direction was advised by a special committee of directors.[2] CIBA had a similar system comprising the Verwaltungsrat with supervisory functions, the management board which consisted of two managing directors, and a group of about fourteen deputy directors. The Verwaltungsrat appointed the directors, who took their orders from the duumvirate of J. Brodbeck-Sandreuter, the commercial chief executive, and his technical colleague, G. Engi. They were autocrats and they produced results, but at the cost of stifling initiative among the deputy directors and lower down the hierarchy. This management system was in force throughout the inter-war years and ended only when Brodbeck died and Engi retired during the 1939–45 war.[3]

[1] Dale, E. Du Pont: pioneer in systematic management. *Admin. Science Quarterly* **2** (June 1957), no. 1, 51 ff. According to Dale, the committees were 'got together in a democratic manner'. This is a surprising statement to make about the selection of a group that, of necessity, must be authoritarian rather than democratic in its constitution.

[2] Anon. *Les Etablissements Kuhlmann 1825–1925* (1925), pp. 51–2.

[3] Annual reports of the company and information from the late Dr. Schulthess, a director of the company in those days.

In the United States the General Chemical Co. evolved by the end of the Kaiser's War a fairly elaborate structure, at whose apex was a group of directors acting as the Executive Committee. It met once or twice a week to handle financial matters, and endorse the decisions of the Operating Committee, which consisted of the departmental heads. These men, in turn, delegated some of the responsibility to the weekly meeting of the works superintendents. General Chemical also had a Finance Committee, which included several bankers, to advise the board of directors, a secretariat to coordinate board matters, and a number of sub-committees to deal with specific problems.[1]

The complex nature of modern chemical operations, combined with the increasing size of the factories, did not so much raise many novel difficulties, but rather increased the dimensions of those with which alert directors had long grappled. Lord Melchett summarized the position clearly:

> The real problem of rationalization and merging of big enterprises consists in effective central control with sufficient elasticity lower down to allow action to be neither arrested nor delayed, and to prevent the chiefs at the head from being encumbered by too much detail, at the same time keeping a grip on all essential features of policy, such as finance, expenditure, sales, and relations with other great groups of industrialists.[2]

Could the circle be squared? Melchett was thinking of I.C.I., but his remarks applied equally well to other large concerns. The objective was stated clearly, the question remained to what extent it could be achieved. IG Farben's experience is appropriate in this context: its predecessors had coordinated some of their activities for about ten years previously, though not always with great success, and the German company served as a model to other chemical firms. One of its directors wrote after the last war that IG Farben's organization was

> a very skilful structure of technical, commercial and administrative committees, which worked smoothly together or on their own and which dealt competently with the countless problems of the concern or presented them for a final decision by the board.[3]

How was this paragon among organizations created and, more important, did it really work as ter Meer described it? Duisberg, who loved to organize things and people, had much to do with the concern's structure and its key feature: *regionale dezentralisierte Zentralisation.* Translated it sounds gibberish, but the concept was theoretically admirable: four large companies were amalgamating, but their works would be grouped by regions that would be responsible for the technical direction, labour, and (initially) sales. But financial matters, patents, purchases of raw materials, and commercial in-

[1] Anon. *The General Chemical Co. after 20 years* (1919), pp. 23 ff.
[2] Speech at Third Annual General Meeting of I.C.I., 29 April 1930.
[3] ter Meer, F. *Die IG Farben* (1953), p. 47, tr.

telligence were to be handled centrally.[1] This structure proved impracticable and was soon abandoned. Another proposal was designed to enlist the experience of the supervisory directors of the predecessor companies, and incidentally also to pacify those among them who were disgruntled by the merger. This interesting stratagem provided sinecures for scores of ageing gentlemen, and resulted in a huge and very expensive supervisory board.

Towards the late 1920s, IG Farben's administration set in its ultimate form and developed bureaucratically. Under Duisberg as chairman of the supervisory board there was a gradual pruning of the numbers, and what powers the board did possess were further whittled down, until it was left only with the ratification (a euphemism for rubber-stamping) of the appointments to the directorate. The more active members of the supervisory board, a group consisting of about eleven men, attended the meetings of the management board as advisers. In other companies the management board was fairly small and was collectively *the* executive of the firm. But in IG Farben it consisted initially of over eighty men drawn from the executive boards of the predecessor companies. Bosch chaired this assembly from the start, but as it was obviously quite unwieldy he organized a smaller executive committee, which was called at first the Geschäftsführender Ausschuss and later the Arbeitsausschuss. This group still numbered thirty men and so was unsuited for the principal management functions, and Bosch, who disliked meetings, created a small inner group (later called the Zentralausschuss), which was the real seat of decision taking.[2]

The board secretariat played a most important role. With so many directors operating at different levels, planning, control, and coordination were delegated to three Divisions, four regional groups, several policy-making committees and about a dozen departments. The Divisions (or Sparten) represented the main product groups of the concern, that is 1. nitrogenous compounds (including fertilizers) and oil, 2. chemicals, dyestuffs, and metals, and 3. photographic products and man-made fibres. The chairmen of the Divisions did not have responsibility for the profit and loss account of their product groups, and in this important respect they differed from Du Pont's General Managers. But the chairmen were responsible to the board for the research and development programmes, all questions relating to salaried staff, and for technical matters of their Divisions.[3] The regional groups or Betriebsgemeinschaften combined the works in the four main centres of chemical production: the upper Rhine (Ludwigshafen), the middle Rhine (Hoechst, etc.), the lower Rhine (Leverkusen, etc.), and central Germany

[1] Waller, P. *Probleme der deutschen chemischen Industrie* (1928), p. 18. Duisberg's ideas on restructuring the industry went back to 1904.

[2] ter Meer, op. cit. Towards the end of the war the inner group had only seven members, including the chairman of IG Farben.

[3] Technical matters within each division were handled by standing committees of experts, which concerned themselves with process improvements, application techniques (for the particular products of the division), chemical engineering questions, packaging, and storage.

(Leuna, Wolfen, Berlin, etc.).[1] Each group management was responsible for the administration of the works, raw materials, fuel, power, production coordination, labour relations, marketing plans, and application research. (Some of the arrangements seem odd, for instance the inclusion of application research, but it must be remembered that the groups took over laboratories specifically designed for this function in the large works.) Among the policy-making committees, those for technical matters (Technischer Ausschuss) and for sales were the most important. They were divided into several sub-committees and collectively were extremely powerful, supervising as they did in the case of sales the activities of the marketing combines, that is to say the control bodies for the marketing of dyestuffs, chemicals, etc. The works invoiced the marketing combines at cost, and the proceeds from the sales, after deducting expenses, were credited to the concern's central financial administration. Under such conditions it must have been difficult to monitor costs efficiently. Finally there were several head office departments, in Berlin and elsewhere, which advised on or were responsible for legal matters, taxation, public relations, accounts, insurance, finance, and commercial intelligence.[2]

This colossal organization employed thousands in the chief offices at Frankfurt (the seat of the board), Berlin, Heidelberg, and Ludwigshafen. The extremely wide product range called for a complex administrative structure, but the system was too elaborate and under the circumstances communications and effective control were often clogged by paper. No doubt action was frequently delayed, and there must have been times when parts of the machinery failed to perform satisfactorily. Bosch was well aware of the weaknesses and on occasion used to appoint plenipotentiaries who could by-pass committees and groups, investigate, and secure an early decision.[3] In the quarter century since the collapse of the Third Reich, there have been several reports that the IG Farben would be reconstituted. They were false alarms and the merger of the successor companies BASF, Bayer, and Hoechst seems unlikely: IG Farben was not an administrative success, except perhaps in one respect. The Betriebsgemeinschaften were in fact the old operating units writ large. They were the successors to the famous and very successful dyestuffs companies, and continued under a new management structure, but retained the same personnel, the same sites, and, with minor changes, their traditional business. IG Farben was superimposed on business organizations that had managed perfectly well on their own, and the post-war history of the successor companies is sufficient testimony that their prosperity has not been impaired.

I.C.I. did not follow the same pattern as IG Farben. We do well to remember that some of its predecessors, far from having a history of success-

[1] At one time there was a fifth group for the works in Berlin.

[2] Link, A. E. Economic study of IG Farben (Section 4: Sales Data) (U.S. Military Government of Germany TP 200/P6, August 1945); ter Meer, op. cit.; Combined Intelligence Objectives Sub-Committee Report, IG Farben AG (Item No. 30, File 23—15), HMSO (1946).

[3] ter Meer, op. cit., p. 56.

ful achievement and so corporate self-confidence in their technical and commercial ability, had fared badly for many years. I.C.I.'s management therefore had to provide not only central direction, but also had to stimulate enthusiasm and encourage a progressive outlook. This may have been among the reasons why the directors of the British concern at first delegated very little autonomy to the 'Group Boards', which corresponded to Divisions or Sparten elsewhere. After about 1930 these boards were given more power to take the initiative and decisions, though all matters concerning accounts, finance, and taxation continued to be handled in London. In the mid-1930s, I.C.I. had a compact board of seventeen men, ten of them full time. The latter, like their senior Du Pont colleagues, had no specific executive functions. Policy decisions were taken by the General Purposes and Finance committees, and merely ratified at the monthly meetings of the full board. The structure at the top resembled that of Du Pont, but the proliferation of committees at the next level followed IG Farben's example. The coordination of the Group or Divisional activities was handled by the Central Administration Committee, which was chaired by a director, and consisted of the chairmen of the Groups and of the head office departmental managers. The Central Sales Committee was the coordinating body for all marketing policy. The Special Sub-Committee on New Products was considered so important that its members included three main board directors. Committee meetings were usually held at fortnightly intervals.[1]

These policy-making committees were formed in or about 1931, and represented the first steps towards decentralization. The system was evidently found satisfactory, because a few years later six Executive Committees were established and empowered to deal with policy matters affecting the company as a whole in respect of finance, personnel, overseas, research, technical, and commercial matters.[2] These and other committees must have taken up much of the directors' time, and by involving them in departmental functions must have eroded the freedom from specific executive functions of a few years earlier. As with IG Farben, we may safely assume that delegation of authority was accompanied by the creation of new administrative tiers, long chains of command, and delays in reaching decisions. But it helped to bring about a corporate approach to problems and so helped to create a common purpose among the heterogeneous predecessor companies.

The management structure and organization of Du Pont, IG Farben, and I.C.I. thus differed in many respects, and changes were made in them in response to particular needs or the pressures of powerful groups. A firm's business performance reflects the appropriateness of its organization and the quality of its top management. Unfortunately these important determinants

[1] Mitchell, H. J. Methods of inculcating modern management principles and practices in large scale undertakings, paper presented to the International Management Congress, London (1935).

[2] Mitchell, H. J. Relationship between headquarters' departments and the cooperating groups of I.C.I. Ltd., paper presented to the 7th International Management Congress, Washington (1938), pp. 42 ff.

cannot be measured, and even qualitative comments must be in fairly general terms. Thus most people would agree that Du Pont was successful in the 1920s, and the managers are entitled to take credit for this. But conditions were certainly favourable to Du Pont in those years: it had ample cash resources, a tied market for some products, and a rapidly growing one for others. These factors, plus managerial ability and the careful measurement of its performance by assigning to each General Manager responsibility for the department's profit and loss account, seem to have been Du Pont's recipe for success. Neither IG Farben, nor I.C.I., nor, so far as I know, the big Swiss and French firms adopted Du Pont's system of accountability. But, apart from financial autonomy, a key factor was (and remains) the choice of the managers and directors. In this area, IG Farben was very skilful. The looked-for qualifications were chemical knowledge, an understanding of industrial economics, and the ability to take charge of plant operations. Men who met such specifications moved up from superintendent to manager and works director, and so, with luck, to a seat on the board. Recruitment was biased towards scientists and the pay was good.[1] IG Farben's staff policy followed broadly the practice of the predecessor companies, and it is possible that the quality of the senior personnel compensated to some extent for the bureaucratic organization of the concern.

In those days it was rare for senior staff to be recruited from outside the chemical industry. Similarly there is little evidence of a drift to other industries, if only because the salaries elsewhere would be lower.

Except in very rare cases we do not know when firms introduced modern business methods, but for Du Pont some of the relevant information is fortunately available. The company took the first steps during the 1900s, and the initial measures included the adoption of regular cost comparison between the firm's powder mills and the systematic approach to accident prevention in all the explosives factories. Lammot du Pont, H. M. Barksdale, and the Haskell brothers were men schooled in the technologically advanced dynamite trade, and on being promoted to senior positions at headquarters they began transforming the traditional business ways. Early instances of the new approach were the establishment of a section to collect and analyse statistical data, the unification of prices, and the introduction of a marketing programme that provided for the gradual replacement of agents by the company's own salesmen.[2]

There were also interesting developments among the German companies. The formation of the Dreibund in 1904 opened the way to an exchange of experience between the companies, from which the partners benefited. For instance Badische now learnt that the Bayer people had a separate costing section, and promptly established one at Ludwigshafen; stock control was introduced about 1904. In general Bayer and AGFA had the better, or, more

[1] Kahan, J. L'essor de l'IG Farben et le recruitement de ses cadres superieurs. *Chim. Ind.* **61** (1949), no. 5, 502–4.

[2] Dale, art. cit., pp. 35 ff.

precisely, more 'numerate', financial administration, but both could, and did, learn from Badische when it came to sales organization.[1] But in many places the old, counting-house way of trading survived into the 1920s: Geigy did not have departments with specific functions until 1924, a legal section until 1925, and a statistical group until 1932.[2]

The founders of the chemical industry in the nineteenth century stamped their personality on the early development of their firms. The impress weakened when companies grew larger and managers took over control from the owners. When we get to the modern chemical concern few individuals can be singled out for special mention: seen in historical perspective the large boards of directors appear to be a group of conformists, overwhelmed by paper and by committee sessions, slow to decide and usually very prudent. Levinstein, who knew them, was scathing: 'It is not so important never to fail, as occasionally to succeed in establishing something new.'[3] But it was becoming difficult for the individual to initiate major changes. Often the chemist turned businessman was being succeeded by the administrator whose special skills lay in matters financial rather than chemical. To this general rule there were some important exceptions. The younger du Ponts, Pierre, Irénée, and Lammot, had not only technical qualifications (they were chemists or engineers), but were also interested in finance and management, and this combination of skills contributed to the firm's success in the inter-war years.[4] Their rivals on the board of Allied Chemical & Dye were little known and it has been reported of Orlando Weber, the company's president from 1920 to 1935, that he never made a speech, gave a lecture, or granted an interview. From sporting beginnings as a bicycle-racing champion he had graduated to successful car salesman and was now ruling the second largest American chemical enterprise autocratically.[5]

IG Farben and I.C.I. offered even less scope for the display of brilliant initiatives. Carl Bosch (1874–1940) was an exceptionally gifted chemical engineer, but one feels that his best years were behind him when he became chairman of the new concern's management board in 1925. By temperament reserved and abrupt, he was not interested in people and now found himself immersed in administrative problems for which he had little enthusiasm. He suffered increasingly from depressions and towards the end of his life he may have felt that the machine had taken over and that he was merely its figurehead. He is now remembered as the man who developed the ammonia synthesis and high-pressure technology generally, rather than as the first

[1] Voigtländer-Tetzner, W. and Pinnow, H. Geschichte der BASF—Kaufmännische Entwicklung, typescript, n.d., pp. 218 ff.

[2] Bürgin, op. cit., pp. 268–9.

[3] Levinstein, H. From within the dyestuffs industry. *J. Soc. chem. Ind., Lond.* **50** (1931), 253 T.

[4] Dale, op. cit., pp. 41–2, 53.

[5] Cosby, J. History of Allied Chemical, typescript of a paper read to the American Chemical Society, 11 September 1958 at Chicago. For a chatty though unfriendly report see *Time* **17**, 2 March 1931, 43; **25**, 27 May 1931, 63–4.

head of IG Farben.[1] His I.C.I. colleague and contemporary, Alfred Mond (1868–1930), first Lord Melchett, was a lawyer and lifelong politician; he had been a junior Minister in Lloyd George's cabinets, but had not been made the Liberals' leader, and, having quarrelled with the party over land policy, joined the Tories in 1926, but obtained no political advantage from the move.[2] It was said of him 'that he cannot have had any real feeling for the industries with which he was connected', first the anthracite mines of South Wales and, late in life, I.C.I.[3] The technical and scientific potential of the new company seems to have left him unmoved, but he was captivated by Zionism and the Jewish home in Palestine. His deputy Harry (Lord) McGowan (1874–1961) was cast in a different mould: a Horatio Alger character, but no chemist, he 'was by temperament and conviction an autocrat, and a man of vision as well'. He was familiar with production problems, and as he was also patient, knew what he wanted and 'could see and seize an opportunity', he became the architect of several big mergers in the explosives industry and later was one of the founders of I.C.I.[4] The company owed to Melchett its first, highly centralized organization and to McGowan the proliferation of committees.

Among the leaders of the chemical industry, only Melchett was a politician of national stature. Neither Duisberg nor Bosch participated actively in German politics. The heads of Du Pont were usually Republicans, but their political role, though large in the State of Delaware, was unimportant in Washington.[5] The political activities of the chemical tycoons can safely be ignored, but their general business philosophy, though it may appear woolly to us, requires some comment. Few looked at business through the eyes of the intellectual, but Rathenau did and his views attracted some attention. In his opinion the modern manager looked for fulfilment in terms of business achievements and of power, rather than solely in terms of material rewards. Rathenau's manager was at the apex of a hierarchy of lesser officials, whose rules and autocratic methods resembled those of the contemporary state, or more specifically imperial Germany.[6] The analogy of industry with state came to include by extension summit meetings of captains of industry, a sort of businessmen's Locarno where all was reason and good fellowship. The meetings of industrialists, in imitation of the periodic foregatherings of statesmen, presupposed, however, some form of association or combination, for as Melchett (then still Sir Alfred Mond) said, 'You cannot discuss big prob-

[1] Holderman, K. *Carl Bosch* (1953).

[2] He was first Commissioner of Works, 1916–21, and Minister of Health, 1921–2. Bolitho, H. *Alfred Mond* (1933).

[3] Armstrong, H. E. The Monds and chemical industry, a study in heredity. *Nature, Lond.* **127** (14 February 1931), 238 ff.

[4] Obituaries in the *Guardian* and *The Times* of 14 July 1961.

[5] Winkler, op. cit., pp. 296 ff. Coleman du Pont was a senator in the 1920s. Pierre contributed to the improvement of schools in the Delaware legislature.

[6] Kessler, Count H. *Walther Rathenau* (1928), p. 151. Kessler summarizes Rathenau's *Von Kommenden Dingen* (Berlin, 1917), which contains a discussion of the separation of ownership and control.

lems of industry with other countries until your own industries are organized first . . .'[1] It is not surprising that a man who spoke in this way should look at the aims of large concerns through the eyes of a visionary, and we find him pontificating at the first Annual General Meeting of I.C.I. that large firms had duties first to their country, then to the shareholders, next to the worker, and finally to the consumer. McGowan, however, changed the order of priority, and declared that the managers must so direct the large concern that it serves 'the best interests of its employees, its customers, and its shareholders'.[2] In a long speech he said that the main purpose of industry '. . . must be to provide the highest possible levels of employment and output, to produce goods of the best quality, and to sell at the lowest economic prices . . . .' Government, for its part, 'should accept its responsibility by providing conditions which are favourable to capital expenditure at a time when activity in the production of consumable goods appears to be falling off'.[3] The journey was certainly a long one from Melchett's industrial jingoism to McGowan's interpretation of Keynes. Lammot du Pont also agreed with lower costs, the maintenance of high wages, and the improvement of quality. But when he placed '. . . the bulk of responsibility for social betterment squarely on the shoulders of free enterprise where it belongs . . .'[4] he showed a lack of understanding of the nature of these problems. Europeans, on the other hand, had long been aware of them and the 1930s heightened the awareness. According to Duisberg the aims of industry should be the pursuit of business, and the highest ideal to the striving businessman should be the improvement of relations between Capital and Labour.[5] Many agreed with him, the more so as after 1918 the traditional paternalism of the chemical employers did not prevent strikes and unrest in their factories. Bosch as chief executive of Badische and later IG Farben, devoted considerable effort to the improvement of relations with trade unions in 1919 and again in 1930. Unfortunately he could not abide the IG Farben works council, probably because he hated long discussions over points of details.[6] His good intentions accordingly resulted in only partly successful measures. Mond, though not the originator, lent his name to the meetings with Trade Union leaders, which became known as the Mond–Turner talks, from 1927 to 1929. They were interesting and novel, especially in the context of the aftermath of the General Strike of 1926. But they petered out eventually without result.[7]

[1] Mond, A. *Industry and politics* (1927), pp. 249–50 (from a speech to the Institute of Chartered Accountants on 23 March 1927).

[2] Article in the *Sunday Times* of 25 July 1943, entitled Combines and cartels.

[3] Speech at Cardiff on 21 March 1944.

[4] Speech made in 1940 and quoted by Dutton, op. cit., p. 375.

[5] The German wording is vague, perhaps deliberately so: '. . . das soziale Problem der Lösung näherzubringen'. Duisberg, C. *Meine Lebenserinnerungen* (1933), p. 139.

[6] Holdermann, op. cit., pp. 158 f., 250 ff.

[7] Bullock, A. *The life and times of Ernest Bevin* (1960), vol. 1, pp. 392, 402 f. McGowan and Lord Ashfield (who was an I.C.I. director at the time) joined Mond at these meetings.

We must now see to what extent the performance of the chemical corporation matched the intentions of its leaders. Their principles and *Weltanschauung* cannot be translated into figures, but the profitability of their companies, as reflected by the return on assets, does give us a yardstick.

Although little work has been done on the finances of chemical companies, a few generalizations can be made despite the inadequate material. In the first place limited liability had virtually replaced partnerships, and only Solvay and Merck among the large businesses retained the old legal form. German firms had at first relied almost completely on equity finance, with local or provincial banks as large shareholders. Later, additional funds were obtained in the form of loan capital.[1] This was the typical pattern in the dyestuffs industry where the firms were large, profitable, and had sufficient assets to allow some gearing. Badische, for instance, was able to raise altogether £1·25 million in loans at 4½ per cent between 1901 and 1910. In these ten years the shareholders benefited from a rights issue with a large bonus element, a doubling of dividend income, and an increase in the value of the shares (par 1200 M) from 4700 to 5700 M. But firms in the heavy chemical industry were considered more risky, and were not able to rely on debt finance to the same extent. The war, however, and the troubled years that followed, marked a complete break: the inflation destroyed debt and until 1928 loan capital played a negligible role. Nevertheless the financial resources of the average firm were considerably higher in the late 1920s than on the eve of the war: in 1913 paid up share capital and reserves of chemical companies averaged about £0·22 million, but in 1928, when the figure is distorted by IG Farben, the figure had grown to £1·05 million.

TABLE 11.8

Capital structure of firms in the German chemical industry, 1912/13, 1926–8[2]

	1912/13	1926	1927	1928
No. of companies making returns	158	108	105	72
	(£ million)	(£ million)	(£ million)	(£ million)
Paid up share capital plus reserves*	34·4	68·3	76·6	75·2
Loan capital	7·0	1·3	1·2	13·5
Loan capital as percentage of share capital plus reserves	20	2	1½	18

*Excluding pension funds.

The sudden jump in loan capital in 1928 was caused by the issue of IG Farben's 6 per cent convertible loan stock to the amount of 250 million RM (say £12½ million). It was not only the largest fund-raising operation in the

[1] Grabower, R. *Die finanzielle Entwicklung der Aktiengesellschaften der deutschen chemischen Industrie* (1910), pp. 140, 152.

[2] Summarized from returns published by the German Statistical Office and published in *Z. angew. Chem.* **27** (1914), pt. 3, 745–9; *Enquete*, pp. 96–8, 105–7. All the large companies are included.

European chemical industry, but also the first convertible loan issue, and not least a great financial success. It enabled IG Farben to borrow cheaply from the public and to by-pass the banks.[1] Only IG Farben, owing to its size and prestige, could adopt this novel approach to borrowing, the smaller firms had to rely, as before, on self-finance or help from the banks.

Similar statistics for companies in the American chemical industry are not available, but there are some indications that the situation was different. The cost of borrowing was relatively high, and chemical firms relied principally on equity finance, ordinary shares and a large proportion of preference shares. General Chemical, which ranked next to Du Pont in the industry before the war, had only preference and ordinary shares: the former paid 6 per cent a year and had been issued in 1899 to pay for the fixed and current assets of the predecessor companies, the latter were issued for the intangible assets. American Cyanamid obtained some of its early finance from the London private bankers Spurling & Co. Until 1927/8 the company's capital consisted of \$5·6 million in ordinary shares and \$6·6 million in 6 per cent cumulative preference shares. American Cyanamid's first loan stock was issued in 1927/8 when \$5·8 million was raised by means of a 5 per cent gold debenture. (A subsidiary company, Amalgamated Phosphate, had to pay 6 per cent on Mortgage Bonds to the value of \$0·8 million).[2] In the case of Du Pont, preference stock represented a third of the equity capital before the war, and Allied Chemical's 7 per cent preference stock was the principal liability in the 1920s. Loan stock was not particularly important, except for Du Pont, where it represented in the last four pre-war years about 27 per cent of the combined share and loan capital. The proportion was higher after the war, when despite its high standing Du Pont had to pay 7½ per cent in 1921. Three years later loans reached a peak of 48 per cent of the total capital, and then declined to 31 per cent in 1928. Du Pont's ordinary shares were first extensively traded during the Great War, and were the first chemical stock to be included in the Dow Jones Industrial Average, the famous index, in 1925. Union Carbide followed three years later.[3] In Europe only I.C.I. and IG Farben, and their precursors, were regularly quoted on the stock exchanges.

Most firms in the industry relied on depreciation and retained earnings to finance their growth. Rights issues were infrequent, because the market for the ordinary shares of the lesser companies was insufficiently developed at least until well into the 1920s. Even by that late date there was still a widespread impression that chemical shares were more risky than traditional stocks such as railways and public utilities. So it is not surprising that the General Chemical Co. paid for all extensions between 1899 and 1919,

[1] Conversion into ordinary shares was on a sliding scale starting at 200 per cent of par and improving to par in 1941. Waller, op. cit., pp. 31–2; Wickel, op. cit., pp. 130–1; ter Meer, op. cit., pp. 28–9.

[2] Annual reports of the company; Haynes, W. On the chemical frontier—the Cyanamid story, unpublished typescript (1957/8).

[3] The Morgan Guarantee Survey, July 1964, p. 6. The Index comprised altogether thirty different stocks.

\$31 million (say £6·2 million), out of earnings.[1] American Cyanamid in the five years to 30 June 1927 spent nearly \$4·9 million on plant and equipment, all of it out of its own resources. But in the financial year 1927/8 the capital budget of \$2·6 million exceeded the anticipated cash flow and it was decided to raise a loan in October 1927.[2] Badische provides us with a European example: its fixed assets grew by 48 million M (almost £2½ million) in the three years to 31 December 1913, but it was not necessary to seek either equity or loan capital during this period;[3] a few months later, however, a rights issue was made and brought in 19 million M (£1 million).

No broad generalization about the profits of chemical manufacture in the different countries can be made, and, even within each country, the average masks a wide range between the sectors. Technological change or a change in demand (or a combination of both) would improve the outlook for one sector (or for a few firms within it) and conversely lead to a deterioration in another. At the level of the individual company the particular range of products made was obviously of great importance at any given time, and in the longer run the ownership of a natural or a patent monopoly, such as potash deposits or the synthetic ammonia patents, was of great value; finally the existence or absence of efficient cartels and agreements had a direct bearing on profitability. Published prices were only meaningful for a minority of the products, and in the absence of a free market, prices were usually determined for months and sometimes years by arm's length bargaining between maker and user. Spot sales were not especially important. Export prices were usually set well below domestic levels, often in order to get rid of surplus production. No index exists that measures with any reliability changes in unit prices in the first thirty years of the century. Prices of some important products fell, and in a few areas, notably nitrogenous fertilizers, there was an appreciable drop.[4] Improvements in manufacture, the elaboration of know-how, and, not least, marketing ability were key factors in improving profitability. Where they were absent firms performed unsatisfactorily even if their trade was protected by tariffs and agreements.

The Germans collected statistics on the profitability of their industry. The average profit trebled between 1912/13 and 1928, but this reflects once again the effects of the concentration movement, which eliminated the dead wood and, as we have already seen, raised average assets. The return on equity

[1] The General Chemical Co., op. cit., p. 85.

[2] Annual reports of the company.

[3] The figures refer to the original value of plant and equipment, which rose from 120 to 168 million M (£8·4 million). The published accounts show only the book value, which grew modestly from 29 to 54 million M (£2·7 million). The accumulation of hidden assets thrown up by the large difference between the two valuations was an important characteristic of the German industry.

[4] Some exceptions must be recorded. Deutsche Solvay-Werke told the Enquete (p. 321) that German prices for soda ash and caustic soda were 42 and 35 per cent higher in 1927 than in 1913. The U.S. Tariff Commission, which was well informed on dyestuffs, reached the conclusion that American prices were higher in the 1920s than in 1913 and attributed this to the higher tariffs.

TABLE 11.9

Profitability of firms in the German chemical industry, 1912/13 and 1926–8[1]

No. of companies making returns	1912/13 158	1926 108	1927 105	1928 72
Net profit, £ million	5·7	4·2	7·0	7·2
as per cent of share capital plus reserves	17	6	9	10
as per cent of net assets	6	4	7	7

Notes:
Net profit = profit after published depreciation and tax.
Net assets = balance sheet total less current liabilities.
All percentages have been rounded off to the nearest whole number.

capital, although improving in the late 1920s, was significantly below the 1912/13 level, and that on net assets remained virtually unchanged. The 1926–8 figures are weighted by the inclusion of IG Farben, whose prosperity appreciably raised the average.

In a few cases it is possible to compare the pre-war and post-war profitability of American companies. The definition of net profits is different from the German, so that the above table cannot be compared with the data below. Du Pont stands out as the most profitable business. The management introduced novel techniques of financial analysis and investment appraisal before the war and expected a minimum return of 11½ per cent on their investment

TABLE 11.10

Profitability of selected American firms 1913, 1925–8[2]

		1913 %	1925 %	1926 %	1927 %	1928 %
Du Pont	Net income : net assets	7·5	8·0	13·6	13·4	15·2
	Earnings : net sales	19·7	13·6	17·7	17·0	20·7
Allied Chemical	Net income : net assets	—	6·6	7·3	7·2	7·6
Union Carbide	Net income : net assets	—	9·5	10·7	10·6	11·4
Dow Chemical	Net income : net assets	4·7	13·2	11·9	13·5	14·5
	Net income : net sales	9·1	8·7	9·6	10·8	11·8
American Cyanamid	Net income : net assets	5·0	8·6	11·9	9·5	7·2
	Net income : net sales	n.a.	13·2	14·4	13·0	n.a.

Notes:
Net income = profit after depreciation and taxation, and represents amount available for dividends and transfer to surplus.
Net assets = total assets less current liabilities.
Du Pont = earnings are after tax and depreciation on manufacturing operations; net income includes dividends received from General Motors and other investments.
Dow Chemical's year ends on 31 May 1914, 1925, 1926, 1927, 1928.
American Cyanamid's year ends on 30 June 1914, 1925, 1926, 1927, 1928.

[1] Source: as for Table 11.8. [2] Source: company reports.

in newly acquired companies.[1] The objective was not reached until 1925, because the policy of expansion was at first financially disappointing. Du Pont's Paint & Varnish Department will serve as illustration: it was responsible for Harrison Brothers and the Flint Varnish Co., both bought during the war. There were losses in 1917–19 and the new managers had to lower their earnings target to 10 per cent net on sales, which was well below the return for the company as a whole.[2] The improvement in Du Pont's profitability (the causes of which we have previously described) in the late 1920s brought the earnings-to-sales ratio back to pre-war levels. The other companies, although also doing better, were unable to match Du Pont's performance. Allied Chemical, despite its size, was particularly disappointing, and the conservative policies of the management, the slower growth of some of the heavy chemicals, and the difficulties with dyestuffs must share the blame. The fluctuations in American Cyanamid's profitability reflect the cyclical nature of the fertilizer trade; the contribution of the newly acquired dyestuffs and pharmaceuticals enterprises only became significant in the 1930s.

Technological change accelerated in the 1920s and entailed the provision of larger sums for obsolescence and investment in additional capacity. Correspondingly the allocations to reserves were larger than before the war. For example IG Farben's accounts show that during the first four years the published transfers to reserves and provisions for depreciation amounted to 362 million RM (£18·1 million), which may be compared with an increase of 425 million RM in net assets. In the United States the accumulation of internal resources represented by the surplus gathered speed in the 1920s, and in the case of Allied Chemical was pushed to the extreme of the surplus being equivalent to about one-half of the net assets. Allied's directors were exceedingly cautious, and I.C.I.'s followed their lead: they established large reserves and supplemented them by a Central Obsolescence Fund. Mond, following Marx, said of this Fund that 'each subsidiary will contribute according to its ability and receive according to its needs'.[3]

Investments remain to be mentioned as a significant aspect of company finance. Chemical firms were not investment trusts or even industrial holding companies. Generally speaking they acquired shares in other firms in order to secure a measure of control over them. Thus IG Farben had a holding of 37·4 per cent in Rheinische Stahlwerke in order to secure supplies of steel and coal on specially favourable terms.[4] Solvays held substantial blocks of shares in the licensees of the ammonia-soda process such as the Solvay Process

[1] Dale (art. cit., pp. 39–40) mentions a 15 per cent return, but a company memorandum from J. J. Raskob to the Finance Committee dated 19 December 1917 (GTX 124) stated that 11½ per cent was the minimum acceptable return on investments made out of war-time profits.

[2] GTX 280, Development Dept. memorandum to W. S. Carpenter, dated 15 March 1920, entitled Paint and Varnish Situation and Outlook; GTX 285 memorandum by H. Grubb (Paint Department) to Irénée du Pont, dated 4 November 1921, entitled The Paint and Chemical Industries.

[3] Speech at second Annual General Meeting of I.C.I. on 18 April 1929.

[4] *Enquete*, p. 119.

Co. and Brunner, Mond. Later they were exchanged into shares of Allied Chemical and I.C.I. respectively.[1] Du Pont's purchases of large parcels of General Motors stock in 1918 and 1919 were conceived as a promising investment for the profits made out of Government contracts during the war. The directors then became involved in the post-war problems of the company and as a result the Du Pont holding was increased until it reached 37 per cent at the end of 1921. Later sales cut it to 23 per cent. This investment provided Du Pont not only with a captive market for its products, but was also remarkably lucrative: at the end of 1926 it was carried in the books at $93 million, on which $23·6 million was received in dividends, and two years later the amounts had grown to $140 (£28·6 million) and $37·9 million (£7·7 million) respectively. The dividends were passed on to the Du Pont shareholders, and so enabled the directors to retain the bulk of the earnings from manufacturing operations in the business, whilst satisfying the shareholders.[2] At the suggestion of Du Pont, Nobel Industries also invested in General Motors, and when I.C.I. liquidated Nobels, the holding was sold in 1928/9. A very large profit over book values was thrown up and was used to increase the general reserve, establish the Central Obsolescence Fund and write off I.C.I.'s preliminary expenses.[3] I know of no other case in the chemical industry of that period of such a large and remunerative investment, and of its use to strengthen the finances and so the development of Du Pont and I.C.I.

[1] *The Financial Times* reported on 13 December 1965 that in 1962 Solvay held 12 million I.C.I. shares, that is about 4½ per cent of the shares then outstanding. In the latter part of 1965 the Belgian concern still held 6·2 million I.C.I. shares, equivalent to 1½ per cent.

[2] The General Motors shares were distributed in kind to the Du Pont shareholders in 1963–5. In the late 1920s, the dividend from General Motors accounted for over 70 per cent of Du Pont's distributions to shareholders, but about 30 years later the proportion had dropped to roughly a third.

[3] Sir Alfred Mond at the Annual General Meetings of I.C.I. on 31 May 1928, 18 April 1929, and 29 April 1930.

12

RESEARCH IN INDUSTRY AND UNIVERSITIES

AT a time when management by objectives and cost-benefit analysis is not only fashionable, but also profitable, it is not surprising that research within and without industry should be closely scrutinized. One instance is the study of research effectiveness, which is being extended as far as the psychology and sociology of invention, another the growing literature on the subject.[1] But all this interesting and valuable activity should not obscure the fundamental change in the attitude to chemical research and development by industry outside Germany from the Great War onwards. Previously neglected, research was henceforth encouraged, and while the benefits at least in the 1920s were still small, they were unmistakable in the 1930s and 1940s. Accordingly no account of the chemical industry would be complete without a description of the development of industrial research and parallel with it the development of academic chemistry.

Industrial research

Large scale, systematic research by chemical companies began in Germany in the 1880s. Despite criticism of neglect, and in the face of sustained exhortation, it made little headway in Britain and the United States until the Kaiser's War. In Britain, after 1914, money and resolution were suddenly and simultaneously combined with startling effects. For example British Dyes Ltd. created an Advisory Council in 1915 to distribute the Government grant of £100 000, spread over ten years, for the promotion of dyestuffs research, and also to maintain regular contact with universities. The chairman of the Advisory Council and head of the new company's research department was appropriately Professor W. H. Perkin jnr., the oldest son of the discoverer of mauve. Perkin was an eminent academic chemist, and he provided a valuable link between university and industrial work in organic chemistry. The Treasury's grant was used to assist the department of dyestuffs chemistry at Leeds University under Professor Green and later A. G. Perkin, W. H.'s younger brother, and the department of organic chemistry at Liverpool under

[1] Perhaps the most imaginative approach to the whole question is Arthur Koestler's *The act of creation* (1964). Other writers have made extensive use of the case study method, notably J. Jewkes *et al.* in the *Sources of inventions* (1958) and in some of the papers presented to a conference called by the National Bureau of Economic Research and published under the title of *The rate and direction of inventive activity* (1962).

Robert Robinson. Graduates in the two departments were thereby enabled to continue with their work; some later accepted three-year contracts from British Dyes and became research assistants with security of tenure, which was itself a novel arrangement.[1] Levinstein Ltd. in this as in other matters were not to be outdone by their rivals: they secured Green from Leeds as the director of their Research Department and allowed him to become head of a new department for postgraduate work in dyestuffs at the Manchester College of Technology.[2] The scale of these industry–university contacts was initially quite small, but they represented a start, the men in charge were of high calibre, and, most important of all, the work dealt with organic chemistry, which had long been neglected in Britain.

In the United States the research effort was on a massive scale. There too Government assisted, albeit indirectly. Thousands joined the Chemical Warfare Service, obtained some training, and were soon discharged, thereby supplying some of the skilled and semi-skilled laboratory manpower.[3] There were said to be about 22 000 chemists, including teachers, federal and state employees, pharmacologists, analysts, and not least scientific workers in industry in the early 1920s.[4] The increase in the number of professionally trained research staff in American industry was nothing short of phenomenal, and, if the statistics are to be believed, soared from 9300 in 1919 to at least 20 000 and possibly a maximum of 28 000 ten years later; chemical manufacturers employed about 3500 in 1927.[5] All these numbers appear high and conflict with other surveys, but become plausible for chemicals if one assumes that the total includes qualified though not necessarily university-trained personnel. In 1925 the National Industrial Conference Board collated the replies from 19 chemical companies, which, between them, employed about 35 000 people. Among them were 3875 with some elementary professional training working as foremen, clerks, or sales staff. A further 1636 (almost 5 per cent of the total) had been to university, and two-thirds of them had graduated as chemists or chemical engineers. Over half of the scientists, almost 600 out of 1090, were employed in laboratories or in the works, but it is interesting to note that even then a large minority were in the sales and general management departments.[6]

Several attempts were made to count the number of research workers and

[1] *Chem. Trade J.* **57** (1915), no. 1485, 435; **58** (1916), no. 1495, 45; **61** (1917) no. 1589, 367–8, summarizing the Annual General Meeting of British Dyes; *First report of the directors of British Dyes Ltd.* (1916).

[2] *Chem. Trade J.* **59** (1916), no. 1520, 27; (1916), no. 1534, 331.

[3] USTC, *Census 1918* (TIS 11), (1919), p. 69. About 2000 were discharged from the Service within a few weeks of the Armistice.

[4] The estimate was made by the secretary of the American Chemical Society, which had about 15 000 members in 1921–2. Bureau of Vocational Information. *Women in chemistry* (1922), p. 3.

[5] Cooper, Franklin S. Location and extent of industrial research activity in the U.S., in National Resources Planning Board. *Research, a national resource, Part 2, Industrial research* (1941), pp. 173–4, 180.

[6] National Industrial Conference Board. *Technical education in the chemical industry* (1925), p. 10.

laboratories. In 1921 there were about 550 laboratories, in 1926/7 almost 1000 in the United States.[1] Du Pont headed the list: its research staff numbered 1200, of whom 400 were graduates or otherwise qualified, in 1920. A year later the slump halved their numbers, but by 1926/7 they were back to 1031, of whom 308 were qualified. Union Carbide & Carbon and its chemical subsidiary employed about 70 scientists in 1921 and about 120 six years later. At American Cyanamid the personnel increased from 6 to 25 and at Hercules Powder from 25 to 45 between those years. Dow, then one of the small firms, first completed the questionnaire in 1926/7 and reported that it had 'over 100' chemists and engineers. These manpower figures compare well with those of the General Electric Co., which enlarged its research staff from 87 to 220 between 1920 and 1926.[2]

The American statistics give the impression of a veritable burgeoning of research activity in the 1920s. An especially remarkable element behind the expansion in the context of this story, and one which was unique to the United States, was the simultaneous development of the petroleum and the chemical industries. They had not yet been combined to any significant extent as is the case now, but the research work that was getting under way then was to be of the utmost importance in shaping the longer term expansion of the two industries.

The German Enquete also addressed itself to the question of research staff. The committee relied on the 1925 census of occupation and reported that the proportion of professionally qualified staff to all employees was 4 per cent in the chemical industry. Next came the electrical and the mechanical engineering industries with 3·6 per cent and 3·2 per cent respectively. In absolute numbers, 13 500 qualified people were working in the chemical industry, 19 000 in electrical, and 42 500 in mechanical engineering.[3] For chemists, the annual censuses of the Verein deutscher Chemiker are a specially valuable source. At the beginning of 1913 there were, in round numbers, 3100 in the chemical industry. Nine years later there were 4000 and, ignoring a small setback in 1926 (which may have had something to do with the reshuffles at IG Farben), they increased slowly throughout the 1920s to around 5000 in January 1930. About 15 per cent were classified as managerial and directors and would therefore not be directly engaged in research work.[4]

The status of the German research chemist was unaffected by the war and the larger employers were careful not get rid of their staff during the chaotic

[1] Allied Chemical, however, characteristically did not complete the questionnaire, although the firm operated several laboratories.

[2] Flinn, A. D. and Cobb, R. Research laboratories in the U.S. *Bull. Nat. Research Council* **1** (1920), pt. 2, 2; **3** (1921), pt. 1, 16; West, C. J. and Risher, E. L. Industrial research laboratories in the U.S. *Bull. Nat. Research Council* (July 1927), no. 60. Research staff includes qualified people in any discipline.

[3] *Enquete*, p. 26. Professionally qualified staff included technicians and people with legal and commercial qualifications.

[4] Deutsche Hochschulverwaltungen (Eds.) *Bedarf und Nachwuchs an Chemikern und Physikern* (1932), pp. 16–17.

aftermath.[1] The ABCM mission noted this and other signs that suggested to the British team that research was integrated with the other activities, and that the management was not only chemically minded but also familiar with the problems and needs of their research staff. The members became positively lyrical after visiting Bayer where '... the whole laboratory ... and the general surroundings are so congenial that they must conduce to enthusiasm for work'.[2] The observation, it may be noted, was made a few months after the end of the war when conditions were hardly congenial in Germany. The ABCM group also described some interesting aspects of research at Leverkusen: the separate laboratories for analytical and for pure chemistry, the practice (which was beginning to be followed elsewhere) of contracting out some work to universities, and the special development sections, which brought together chemists, engineers, and accountants.[3]

The elaborate organization at Leverkusen had existed well before the war and for years had provided the institutional framework for extensive research. (The other large companies had similar arrangements, and at Badische a special research centre had been built for synthetic ammonia.) But successful research requires more than buildings, manpower, and money. It needs purpose, steadfastness in direction, competent technicians and chemists with a questioning as well as a receptive mind, and, above all, luck. In all these respects, except possibly sheer numbers, Germany's rivals remained deficient. The Balfour Committee after looking at Britain's economic failings noted 'that the requirements of industry ... include not only the intensive training of the expert, but also the widespread diffusion among future industrial leaders of the capacity to appreciate the value of science to industry.'[4] In this context the experience of I.C.I.'s predecessors is particularly relevant. Soon after its formation the British Dyestuffs Corporation was employing about 100 chemists, though not all were on research work. In the financial year 1919/20 nearly £250 000 was spent on improving the laboratories at Huddersfield and Blackley, which were then being directed by Professors Robinson and Green respectively.[5] From this promising beginning, or, perhaps more accurately, continuation of the British Dyes and Levinstein policies, there was a swift decline, which impaired the quality of the work, destroyed the continuity of the research programme, and lowered the morale of the staff. The Sankey judgement and the sudden recession were the beginning, and later the negotiations with the Germans formed a long period of uncertainty.[6] What-

[1] Duisberg, C. *Meine Lebenserinnerungen* (1933), p. 107. At Bayer some scientists were temporarily assigned to commercial training courses to enable them to handle foreign enquiries more efficiently.

[2] Association of British Chemical Manufacturers. *Report of the British Chemical Mission on the chemical factories in the occupied area of Germany* (1919), p. 35.

[3] ibid., p.14.

[4] Committee on Industry and Trade, Final Report, HMSO (1929), p. 218.

[5] Sir Harry Birchenough at the Annual General Meeting of the British Dyestuffs Corporation on 21 May 1920. *Chem. Trade J.* 66 (1920), no. 1723, 704 ff.

[6] On the Sankey judgement see p. 243; the negotiations with the IG are described on pp. 273–4.

ever the precise cause, once the rot had started it was difficult to stop: the Technical Committee and the Advisory Council, which had survived the war, were disbanded and the link with academic chemistry cut. With the departure of Levinstein there was no one on the board familiar with dyestuffs chemistry. The research staff fell from 80 to 30 between 1920 and 1923 and was reported to be down to 19 at the beginning of the following year.[1] Nobels also suffered from economy measures. Their research department had been directed and reorganized from 1909 onwards by F. L. (later Sir Frederick) Nathan and W. Rintoul. Even after the outbreak of war there were only 18 qualified staff, half of them on analytical work. But there was a considerable increase in the next few years, and by 1918 there were 21 research and 13 analytical chemists. McGowan's rationalization measures and short-sightedness led to redundancies, from which the research department took several years to recover. In the latter half of the 1920s the work on explosives was supplemented by research on cellulosic compounds for paint and leather-cloth formulations. The wider scope resulted in a considerably larger department, which counted 66 research chemists, 8 analytical chemists, 11 physicists, and nearly 200 technicians, laboratory and clerical staff towards the end of the period.[2]

One has to turn to Brunner, Mond & Co. in order to find the equivalent of the fundamental and applied work that was so typical of the large German companies. Sufficient profits and a sympathetic board of directors certainly helped. But in addition the company was fortunate in having F. A. Freeth as head of research, for besides his chemical ability he was unusually far-sighted and persevering, and loyally supported his staff. An important step was the formation of a small group to study gas reactions at the Winnington laboratories and the decision to collaborate with Professor Michels of the University of Amsterdam. Work on high pressures began about 1929 and led in 1933 to the discovery of polyethylene at Winnington.[3] The long gestation period paid off in what was to be one of the most important developments in the field of plastics. The company was also heavily committed to applied research in connexion with the construction of a synthetic ammonia plant at Billingham. The work was intensified from about 1926 onwards, when investigations of coal hydrogenation processes and urea manufacture were sanctioned, and in 1927 the technical staff was greatly increased.[4]

The efficient administration of research became an important matter once chemical manufacturers went beyond routine testing and analysis of products. Here again the German firms, because they had first encountered the problem,

[1] Memorandum by Sir Frank Heath (then Secretary of the Department of Scientific and Industrial Research) to Lord Parmoor, Lord President of the Council, dated 30 January 1930. PRO, CAB/24/164. Lord Ashfield, who was then Government director of the Corporation, told the Cabinet Committee on British Dyes at its meeting on 18 March 1924 (PRO, CAB/24/165, p. 9) that there were 60 chemists. His figure probably includes technical and technical service staff with a chemical qualification, in addition to research chemists.

[2] Miles, F. D. *A history of research in the Nobel Division of I.C.I.* (1955), pp. 76, 94, 97.

[3] Allen, J. A. *Studies in innovation in the steel and chemical industries* (1967), pp. 15 ff.

[4] Parke, V. E. *Billingham—the first ten years* (1957), p. 109.

evolved some ground rules and refined them as they went along. Du Ponts were probably the first firm outside Germany to look systematically at the management of research. At first the company had only one laboratory, operated by the Chemical Department at Wilmington. As part of Du Pont's reorganization in 1921, each manufacturing department or Division was given its own research laboratory whose head, it is worth noting, carried the same rank as the Divisional production or sales director. The laboratory at Wilmington was retained, its functions were broadened, and in 1927 all studies connected with fundamental aspects of chemistry that had no immediate commercial potential were assigned to it.[1]

IG Farben and I.C.I. retained the research laboratories of their predecessors. Each of IG Farben's Betriebsgemeinschaften or regional factory groups had its own research organization, and at the cost of some duplication and extra committee work, the system maintained the old traditions and *esprit de corps* of the great research departments at Ludwigshafen, Hoechst, and Leverkusen. There was initially less or none of that tradition among the I.C.I. laboratories excepting possibly Winnington. Freeth and Rintoul were translated from Winnington and Ardeer respectively to London and promoted to joint research managers. Thus the interests of research and development were in good hands at the centre, but the directors of the British concern were at first less sympathetic to scientific work than the men at IG Farben.

The policy of the German company towards their research staff in no way differed from that of Badische or Bayer. The IG Farben chemist could look forward to a worthwhile career and a secure job: he was given the opportunity (if he was so minded) to develop the results of his work at least to the pilot-plant stage, and if he showed practical ability was transferred to the works that operated his process. This was the starting point for promotion to administrative appointments. Men who showed no aptitude for management followed a different career and were able to gain prestige and a reputation by work of a quasi-academic nature.[2] The movement of staff from laboratory to works and in general the high standing of research work were a characteristic feature of the German and the Swiss chemical industry and at least until the late 1920s were rarely encountered elsewhere.

One often reads nowadays of applied research costing millions, but in the 1920s, despite the sea-change caused by the war, most companies spent very little. There is not much material on the research expenditure of large firms, perhaps because they did not want to publicize how little they spent.[3] The

[1] Bolton, E. K. Du Pont research, Speech to the American Section of the Society of Chemical Industry on 5 January 1945.

[2] Kahan, J. L'Essor de l'IG Farben et le recrutement de ses cadres superieurs. *Chim. Ind.* **61** (1949), no. 5, 503.

[3] Bartlett, H. R. The development of industrial research in the United States, in National Resources Planning Board, op. cit., pp. 19 ff. Bartlett summarized the development of about fifty industrial research laboratories, including those of Du Pont, American Cyanamid, Dow, and Monsanto, but not one company gave Bartlett figures of their research expenditure. The change

British Dyestuffs Corporation spent £70 000 in 1919/20, a boom year. In 1926 the managing director of the Billingham works said that £200 000 would be devoted to the investigation of new synthetic processes, but he did not say over how many years the money would be spread.[1] Du Pont spent \$2 million (£0·45 million) in 1924, when its net sales were \$91 million (£21 million).[2] American Cyanamid is reported to have spent \$1·5 million on research and development from its foundation in 1907 to June 1927, which may be compared with total sales during the 20 years of around \$80 million. But after the mid-1920s the company's research effort was intensified and amounted to \$0·3 million (about £61 000) in 1926/7, when net sales were \$10·7 million (£2·2 million); the next year \$0·4 million were spent.[3]

There are more detailed, and therefore more meaningful, statistics for IG Farben. Between 1927 and 1930 there was little change in the number

TABLE 12.1

Research expenditure and sales of IG Farben, 1927–30[4]
(£ million)

		1927	1928	1929	1930
Division I: nitrogen,	Research	5·3	4·6	4·1	2·8
oil from coal, new	Sales	25·0	27·0	24·0	17·0
synthetic products	Research %	21	17	17	16
Division II: dyes,	Research	2·0	2·0	2·4	1·9
pharmaceuticals,	Sales	34·3	38·0	39·6	35·0
general chemicals	Research %	6	5	6	5
Division III: fibres,	Research	0·4	0·4	0·4	0·3
photoproducts, film	Sales	4·0	6·0	7·5	6·5
	Research %	10	7	5	5
Total	Research	7·7	7·0	6·9	5·0
	Sales	63·3	71·0	71·1	58·5
	Research %	12	10	10	9
Total research staff		1000	1050	1100	1100

Note: Figures have been rounded off.

of the total research staff, but expenditure declined from RM 154 million to RM 100 million (about £7·7 million to £5 million respectively). The cut was particularly severe in Division I, where the expenditure on pilot plants and

that has come over industrial research since that report was issued about 30 years ago is illustrated by the wide publicity given to research and to the financial resources devoted to it.

[1] Sir Harry Birchenough, art. cit.; Parke, op. cit., p. 103.

[2] Committee on Industry and Trade. *Factors in industrial and commercial efficiency*, HMSO, (1927), p. 319. The figure for research may have been provided direct by Du Pont to the Committee.

[3] Haynes, W. On the chemical frontier—the Cyanamid story, unpublished typescript (1957/8), no page number.

[4] Link, A. E. Control of IG Farben, Special Report of the Military Governor, U.S. Zone, 1 October 1945; ter Meer, F. *Die IG Farben* (1953).

semi-technical operations was virtually halved. Nevertheless almost £17 million was spent over the four years on wide-ranging investigations, which were directed to a large extent to further developments of high-pressure synthesis. The less costly research effort of Division II was maintained, and that of Division III reduced.

The United States Tariff Commission published figures annually of the research expenditure on dyestuffs and pharmaceuticals. The American figures were smaller than those of IG Farben and fluctuated considerably in line with trading conditions. The peak year was 1918 when \$4·6 million was spent, the amount being equivalent to 5·5 per cent of the sales value. Many lean years followed and in 1926 only \$2 million was spent, corresponding to 3·4 per cent of the sales value. Then, as business improved, research expenditure mounted and in 1929 reached \$3·2 million (about £650 000) equivalent to 4·3 per cent of the sales value.[1]

The American data compare unfavourably with IG Farben's expenditure on Division II, which included dyestuffs and pharmaceuticals. But the United States figures understate the position by excluding the expenditure on consultants, who collectively made a substantial contribution to the American research effort. A. D. Little of Boston remained the most important. In 1926/7 his firm employed 19 research chemists, 10 analytical chemists, 7 engineers, and several cost accountants, which represents a substantial establishment by any standard. During the 1920s A. D. Little Inc. gradually abandoned routine analytical investigations and broadened the scope of the work considerably. They advised General Motors on the establishment of a central laboratory, produced pure iso-octane for the Ethyl Corporation as a standard for the comparison of anti-knock compounds, and carried out a systematic study on the preparation of aliphatic chemicals from petroleum, thereby ushering in a most important industrial development.[2] Consultants, as distinct from analytical laboratories, were less important in Europe, and the manufacturers relied on their own research staff or bought installations on a 'turnkey' basis from firms specializing in the supply of process plant. Thus Krupps, Lurgi, and Bamag in Germany had departments for the design and installation of entire chemical works. Similarly the Nitrogen Engineering Corporation, which became part of American Cyanamid in 1929, built ammonia synthesis plants to its own design for Mathieson Alkali in the United States, Norsk Hydro in Norway, and other clients in Belgium, France, Japan, and the Soviet Union.

The Research Institutes and Associations were, so far as I know, an Anglo-American innovation. The Mellon Institute at Pittsburgh, endowed by the

[1] Annual Reports of the U.S. Tariff Commission. The Commission did not then call for particulars of research expenditure on other products. Thus the figures do not reflect the total costs of the larger chemical companies.

[2] Bulletin of the National Research Council, art. cit.; History of Arthur D. Little, typescript (1947), no page number. In 1956 A. D. Little bought the Miner Laboratories, which had begun as a consulting partnership half a century earlier.

Mellon brothers in 1913, quickly became successful and influential. From the start it was particularly active in chemical research and managed to combine the atmosphere and environment of academic work with the practical needs of the sponsoring firms. The client set the problem and paid for its solution by the creation of a specific 'fellowship' with a minimum duration or tenure of one year. The Institute in cooperation with the University of Pittsburgh found a qualified research worker, provided the equipment and laboratory personnel, and saw to it that the study was completed on time and without the disclosure of confidential information. The results of the research belonged to the client. The holder of the fellowship was able in this quasi-academic environment to obtain the advice of other fellowship holders engaged on other work. Such cross-fertilization and the skilful blending of industrial research, or more precisely work that had a definite commercial purpose, with fundamental study had an immediate appeal: initially there were facilities for 23 fellowships, but by 1920 the number had doubled, and in the later 1920s the Institute employed 79 chemists, 10 engineers, 3 biologists, and 10 specialists in ceramics. A few years later the income from the fellowships was $1 million a year. The clients were mainly small or medium-sized firms, but Carbide & Carbon Chemicals, one of the larger and certainly more enterprising firms, maintained four fellowships for most of the 1920s. Similar institutes were founded later, notably the Battelle Institute of Columbus (Ohio) in 1929. Unlike the Mellon Institute, its practice was to assign investigations to departments specializing in that particular branch of science and there was accordingly less turnover of the research staff.[1] In any comparison between the American and the German chemical research efforts, the work of these institutes must be included as part of the industrial effort.

The most important feature of the institutes was their practical, business-like approach to research: they did not guarantee the client that the work for which he was paying would yield commercial results, but he knew that it was in competent hands. By contrast research organizations that were less profit oriented were not so successful. The Smithsonian Institute's Research Corporation, nourished by the royalties on F. G. Cottrell's dust-precipitation process, lacked the funds to develop, let alone finance, the large-scale utilization of the fruits of research. Its disbursements were derisory, but in the 1930s, in association with the Chemical Foundation, it supported E. O. Lawrence at the University of California and R. J. van de Graaff at the Massachusetts Institute of Technology. Lawrence later assigned his cyclotron patent to the Corporation. The Chemical Foundation, to which the German chemical patents had been made over after the Great War, failed, contrary to expectations, to make a lot of money out of them. The reason was simple enough: under its statutes it could only grant non-exclusive licences, but in

[1] Weidlein, E. R. and Hamor, W. A. *Glances at industrial research* (1936). In 1967 the Mellon Institute combined with the Carnegie Institute of Technology (an educational institution) to form the Carnegie-Mellon University.

organic chemistry such licences made no great appeal to manufacturers wanting to develop exclusive rights to a new product or process.[1]

During the war the Federal Government had lavishly supported research and the commercial 'spin-off' was sometimes considerable. But peace and Harding's 'normalcy' were accompanied by parsimony. We have seen one instance in the decline and end of the Fixed Nitrogen Laboratory.[2] But there were exceptions to the mood of the moment. Herbert Hoover, who became Secretary of Commerce in 1921, and so responsible for the Bureau of Standards, determined to employ the extensive federal research facilities for practical ends. It was not intended to help individual firms, but to assist industry generally; for example by improving appliances, making optical glass, or developing more efficient batteries. Showing how to make things better appeals to Americans, and, despite criticism from some manufacturers, this kind of Government involvement in applied research, unlike the nitrogen work, lasted throughout Hoover's tenure of office. In addition links with industry were strengthened by persuading trade associations to appoint individuals as 'research associates' to study particular problems of their industry at the laboratories of the Bureau of Standards.[3]

The post-war involvement of Government in industrial research, exemplified in the United States by the Bureau's activities and in Britain by the Research Associations sponsored by the Department of Scientific and Industrial Research, left the chemical industry unaffected. In Britain cooperative associations to which Government and industry were to contribute equally were launched with high hopes at the end of the war. But, as on the other side of the Atlantic, money became short and the project to involve chemical manufacturers in cooperative chemical research foundered on the rock of non-exclusive licences. It was not for want of trying. When Labour and Conservative cabinets were discussing the withdrawal from the British Dyestuffs Corporation, it was suggested that the Treasury should use the loan and share repayment to encourage chemical research. Later A. J. Balfour, then chairman of the Committee on Civil Research, recommended to the Cabinet that the proceeds be used

> . . . in developing through the Department of Scientific and Industrial Research facilities for pursuing chemical investigations not necessarily to the making of dyes, but of value to the Fighting Services and to the chemical industry generally.[4]

It seemed a bit vague, and indeed nothing came of the proposal. In September 1926 the first, and to date the only, cooperative research association connected with a branch of the chemical industry was formed: it covered the manufacture of paint, colour, and varnish. The Anglo-American experience would suggest that perhaps there is no half-way stage in chemistry between

[1] Cameron, F. T. *Cottrell, Samaritan of science* (1952).
[2] cf. Chapter 8, pp. 225–6.
[3] Cochrane, R. C. *Measures for progress* (1966), pp. 224–5, 263 ff.
[4] PRO, CAB/24/165: Report of the British Dyes Committee, 26 July 1924, p. 5; CAB/24/175. Report by the Chairman of the Committee on Civil Research, 16 October 1925, p. 2.

research for profit by industry and academic research. The motivations and objectives are different, and for industry the stakes are so high that ultimate control over the results must be secure. Nevertheless the connexion with university chemistry cannot be ignored, and needs to be briefly reviewed in the context of academic science in general.

Chemistry students and university chemistry

The attitude of chemical manufacturers to research was changing after the war: support became positive, though not always sustained, and some firms made available considerable sums for the purpose. Nevertheless, except for a few years immediately after the war, the number of chemistry students remained stationary or even declined. Academic statistics were less systematic then and the breakdown by subjects is unsatisfactory, but the following figures, whatever their shortcomings, suggest that science did not retain its appeal for teenagers and that applied science and especially chemistry were positively unfashionable.

TABLE 12.2

Chemistry students in Britain, Germany, the U.S.A., and the Swiss Federal Polytechnic in the 1920s[1]

(000s)

Year	*Great Britain*		*Germany*	*U.S.A.*	*Swiss Federal Polytechnic*
	Pure science	*Applied science*			
1920/1	6·3	7·0	6·7	5·7	0·33
1922/3	7·7	5·6	7·6	n.a.	0·21
1924/5	7·3	4·2	6·5	4·1	0·18
1925/6	7·4	4·0	6·4	4·2	0·16
1928/9	7·4	4·1	4·3	4·9	0·20

Notes: Applied science students in Britain include engineering, applied chemistry, metallurgy, mining, and architecture. Pure science includes disciplines other than chemistry. In 1920/1 Oxbridge and Exeter were excluded. The German figures refer to matriculated students of chemistry in the winter semesters of the years shown. The United States data refer to chemical engineering students only, and the Swiss to chemistry students at the Zurich Polytechnic.

The contrast between the growing interest shown by industry, and the apparent lack of it among the future recruits calls for some explanation. It is clear that once the immediate post-war wave of deferred students had passed there was a falling off that was particularly pronounced among applied

[1] Sources: Great Britain—reports of the University Grants Committee; Germany—Deutsche Hochschulverwaltungen (Eds.) *Bedarf und Nachwuchs an Chemikern und Physikern* (1932), p. 44; U.S.A.—Department of the Interior (Office of Education) *Biennial survey of education, 1918–1920* (2, pp. 303–4), *1924–6* (p. 823), *1926–8* (2, pp. 697–8, 718–19); Switzerland—Guggenbühl, G. G. et al. *Eidgenössische Technische Hochschule 1855–1955* (1955), pp. 256–7.

science students in Britain and chemical engineering students in the United States. In the former country the Thomson Committee, under the chairmanship of the then Cavendish Professor of Experimental Physics at Cambridge, investigated the problem at the end of the war. It was not a very good report; the universities escaped lightly, but Oxford and Cambridge were criticized because they restricted the choice of subjects open to students who did not take honours. There were some sensible recommendations about postgraduate work and the Ph.D. The Committee examined the state of applied science and its neglect by industry. The trouble was not caused by lack of process chemists and of men to do routine analysis, but higher up in the administration. The Committee thought it was essential

> . . . to get competent works managers with enough knowledge of science, enough appreciation of the point of view of the engineer and enough power of thinking quantitatively on questions of energy.[1]

The problem was not the state of university science, but the failure of employers to specify their requirements. The former was patently untrue, the latter a gross oversimplification. In fairness to the Committee, however, it must be said that when it asked for practical suggestions from industry it got nothing useful. It may be that while businessmen were becoming prepared to pay for research, they were unwilling or unable to set clear objectives, and to direct and organize the work of their research staff. Under such conditions the staff would not be properly employed, the men would become frustrated, and good recruits would not be attracted to the work.

Towards the end of the 1920s there were few signs of change except in some of the I.C.I. Divisions. The Balfour Committee, like many others, had a look at research and summed up the situation as follows:

> . . . in order that an industry may obtain practical value from scientific research, it is absolutely necessary for at least an adequate proportion of those who occupy responsible positions in the industry to have a scientific habit of mind . . . their minds should be receptive of the results of such research and alert to its practical bearings.[2]

A recent enquiry into the careers of chemists reinforces the impression that British industry made little appeal to graduates, and that the 1920s were worse in this respect than the 1890s and 1900s. The sample is admittedly small—about 190 before and 75 after the war—and biased towards specially selected chemists, the 1851 Exhibition Scholars. About 30 per cent went into industry in the period 1891–1914, and 25 per cent in 1919–29. The universities increased their lead and in the latter period were chosen by more than two-thirds of the graduates. (The remainder opted for the Government Service as

[1] Report of the Committee . . . on the Position of Natural Science in the Educational System of Great Britain; HMSO (1918), Cd 9011, pp. 53, 59, 64.

[2] Final Report of the Committee on Industry and Trade, HMSO (1928), Cmd 3282, pp. 217–18.

scientists or administrators).[1] The proportions reflect career choices and the opinions of young chemists about future employment prospects. The attraction of university posts and the unpopularity of industry goes some way towards explaining a perplexing contrast: the evidence of greater industrial research on the one hand and of a stationary student population on the other. For the minority who went into industry the facilities and the equipment were better, but the job satisfaction lower. For the majority, the security of university life, regardless of the lack of money and the absence of growth, had obviously more appeal. Neither industry nor universities were actively recruiting, and as the demand was not noticeably rising, so the supply of science students failed to increase.

Even in the United States, where businessmen were particularly interested in technological development, there was some uncertainty about the role of science graduates. The directors of large American chemical companies were asked in the mid-1920s about the employment of such graduates in their firms. They had few suggestions to offer, apart from the hope of better relations between university and industry and—rather more to the point—of the usefulness of courses in business administration for science students.[2] The conclusion is inescapable that, despite isolated exceptions on both sides, the gap between academic and applied chemistry was not bridged in Britain or the United States in the 1920s.

Chemical engineering, which was rapidly gaining in importance, did not develop with equal speed its academic standing. In the United States the number of students reading for the subject was lower at the end of the decade than at the beginning. In Britain statistics are only available for postgraduates: in the academic year 1925/6, 380 were reading pure chemistry and 85 applied chemistry, which included chemical engineering; three years later the numbers were 375 and 66. Meantime, and in contrast, physics was booming, and the number of postgraduates rose from 122 to 185 in the three years ending 1928/9.[3] The figures tell much about the relative status of the two disciplines in this country.

In Germany the place of the research chemist and of the technologist was assured, but there were some worrying trends. In particular the demand for chemists was not rising as fast as the supply. It was found that between 1925 and 1930 an average of about 650 students a year completed their Ph.D. or diploma examination in chemistry.[4] On the other hand the chemical industry was creating 100 new jobs annually, and other industries together with teach-

[1] MacLeod, R. M. and Andrews, E. K. Scientific Careers of 1851 Exhibition Scholars. *Nature, Lond.* **218** (1968), 1015. I wish to thank Miss Andrews for her help in providing me with the break-down on which the text is based.

[2] National Industrial Conference Board, op. cit., p. 7.

[3] Reports of the University Grants Committee; Hansard **717** (1965), 618, written answers column 364.

[4] The diploma refers to the examination at the *Technische Hochschulen* that was taken after 3 to 3½ years' attendance. The number of graduates includes those who took the special teachers examination (*Staatsexamen*).

ing provided a further 150–180 new posts annually. Allowing for deaths and retirements the total annual demand for graduate chemists ranged from 420 to 480 in the late 1920s. The surplus was alarming, and indeed by 1931 the cumulative effects of excessive supply combined with the diminished job opportunities in the Depression, caused about 3000 graduate chemists out of a total of 16 800 to be earning their living in other jobs or to be out of work altogether.[1]

German chemists on the dole, or at best not working at their trade, points to a novel and shocking waste of skills, and explains the sharp fall in the student population shown in Table 12.2. There was a further complication. Before the war many German chemists worked abroad, notably in Russia, Austria–Hungary, and North America, often, though by no means exclusively, in German-owned factories. But after the war the employers in those countries gave the jobs to their own nationals. The diminished opportunities were compensated to some extent by the policy of German and also Swiss firms in concentrating all research and development work on home ground: this was good for their own nationals, but had unsatisfactory results elsewhere because the Swiss and German-owned subsidiaries in Britain, France, or the United States tended to rely entirely on the work of their parent and failed to develop research facilities of their own.

While the chemical industry changed significantly during the first thirty years of the century largely as a result of the discoveries following the development of physical chemistry, there was rather less change in the university departments where this and other branches of chemistry were taught. In short the gap between academic and industrial chemistry tended to widen, with detrimental effects on both sides. Others must be left to investigate the implications, and I will restrict myself to a summary of the position and point to the steps taken to remedy it.

Weimar Germany retained the *Lehr-* and *Lernfreiheit* of the Imperial Reich, as well as the emphasis on pure chemistry. But the adoption of the intermediate examination (*Verbandsexamen*), with its uniform requirements, necessarily obliged candidates to adhere to a prescribed curriculum. Once this oral test of general chemical knowledge had been passed the student was free to prepare this thesis when and where he wanted. Usually a complete university course lasted five to six years.[2] The distinction between the universities and *Technische Hochschulen* was maintained into the 1920s. The latter were of course orientated towards technology, but most were also providing some courses in arts subjects and even economics. They were, however, subsidiary subjects, and it is doubtful whether such additions to the curriculum were very important, let alone of benefit, to students of chemical technology.

[1] Deutsche Hochschulverwaltungen, op. cit., pp. 29–36.

[2] Paneth, F. The teaching of chemistry in German Universities. *J. chem. Educ.* 5 (1928), no. 6, 709; Deutsche Hochschulverwaltungen, op. cit., pp. 9–10. The *Verbandsexamen* had no official standing, though it was essential to anyone working for the doctorate.

Neither Britain nor France followed Germany in providing adequate teaching facilities at advanced level in applied chemistry. In Britain many committees had pointed to the need for early action in this area, but as late as 1921 the University Grants Committee observed:

There is nothing in the nature of technology which makes it necessarily unsuited to the methods and spirit of university work . . . we believe it to be urgently necessary, therefore, to define more closely the aim of university courses in engineering and technology, and to differentiate such courses from work properly assignable to technical colleges.[1]

And there the matter rested for a good many years, although one might have thought that the war had demonstrated the need for people with professional qualifications. The Institute of Chemistry played a useful role in this context through its Associateship and Fellowship for the majority who were outside the range of the universities, and by 1927 about 5200 people had obtained these titles through qualifying examinations.[2] The Institute of Chemical Engineers was founded in 1922, but in its early years was of relatively little significance in setting standards of training and qualification in its branch of chemistry.

The brilliant work of Rutherford and his school of physics at the Cavendish in the 1920s determined the scientific outlook at Cambridge and so set the tone for the rest of the country. At Oxford chemical research was handicapped for decades by the College laboratory system, which, in turn, survived on account of the totally inadequate university facilities. The Royal Commission recommended the closing of College laboratories and considerable improvement in the University-owned research laboratories; during the inter-war years these long overdue steps were taken.[3] Throughout the period Oxbridge (or rather Cambridge with Oxford trailing behind), and London, including Imperial College, were the leading British institutions for advanced scientific work, and between them accounted for three-quarters of all post-graduate students. It is worth noting that the German universities, which had attracted so many British students in the 1900s, were now completely out of the picture.[4]

The situation in France was certainly no better. After the war there were only two institutions that taught applied chemistry at advanced level: the

[1] Report of the University Grants Committee, HMSO (1921), Cmd 1163, p. 12.

[2] Chapman, A. C. *The growth of the profession of chemistry during the past half-century* (1927), p. 9.

[3] Royal Commission on Oxford and Cambridge Universities, HMSO, (1922), Cmd 1588 p. 238.

[4] MacLeod and Andrews, art. cit., p. 1014. I am indebted to Miss Andrews for providing me with a detailed analysis of the 1851 Exhibition Scholars who were doing postgraduate work on chemistry between 1919 and 1929. Of the 70 scholars, 24 went to London, 14 to Cambridge and 12 to Oxford. The rest were scattered over other British universities. By comparison, in the period 1891 to 1914 there had been 194, and London led with 31, followed by Manchester (16) and Cambridge (8). After the war not one went to a foreign university, but in the 13 years ending 1914, 100 went abroad, among them 41 to Leipzig, 20 to Göttingen, 19 to Berlin, and 11 to Paris. There is an element of double-counting because a few scholars went to more than one university.

Ecole Superieure de Chimie at Mulhouse and the Ecole de Physique et de Chimie Industrielle run by the city of Paris. The courses on chemical technology at the universities of Nancy, Lille, and Lyon appear to have been abandoned or else severely truncated. Chemical research at universities was, generally speaking, restricted by lack of funds and so failed to attract scientists. The French tradition of lectures and demonstrations accompanied by neglect of undergraduate research survived criticism and continued into the 1920s. The practice of selecting teaching staff by competitive examination handicapped those whose gifts were inventive rather than repetitive. There was indeed the Caisse des Recherches Scientifiques, but its grants administered by Civil Servants of the Ministry of Education were wholly insufficient. The weakness of French chemistry between the wars has been ascribed to the neglect of research and the miserable facilities for teaching applied chemistry.[1] They were not the sole causes, but they were important and could have been avoided.

Teaching and research cost money, but everywhere revenues fell behind expenses. In Britain, Oxford, though surely not the worst hit, was articulate in its distress. In March 1919 the dons appealed to the Board of Education for £17 200 a year plus £163 000 on capital account. They pointed to an alarming financial position, and the crisis in Oxford's scientific departments was one of the prime causes of the appointment of a Royal Commission. Cambridge also felt the pinch, and in 1921 the Vice-Chancellor told the Royal Commission that an extra £14 300 a year was needed for science.[2] In retrospect it is surprising that wealthy colleges could be so parochial in their attitude to scientific work as not to help the university administration in so crucial a matter. Where the universities failed, the State moved in. Not only Oxbridge, but also the Redbrick universities were helped. In 1925/6, the first year for which complete returns are available, the income of all universities was about £4·8 million, out of which the Government contributed £1·6 million. In 1928/9 the amounts were £5·2 and £1·8 million respectively. Details of the expenditure by departments are not available, but the science faculties certainly benefited from the Government grant.

On the Continent the Government contribution was relatively larger, but everywhere it was considered inadequate. Especially in Germany, where the war and the following inflation destroyed the value of fixed grants, emergency measures had to be taken to protect science departments. In 1920 three societies were formed to help chemical education, chemical research, and chemical literature.[3] They were backed by endowments made by the chemical

[1] Moureu, C. Les conditions générales de la recherche scientifique en France, in Gérard, J. (Ed.) *Dix ans d'efforts scientifiques et industriels* (1926), vol. 1, p. 43 ff.; Matagrin, A. *L'Industrie des produits chimiques et ses travailleurs* (1925), pp. 257–8; Wehrenalp, E. B. von, *Der Niedergang, der französischen Naturwissenschaften* (1940), p. 35.

[2] Royal Commission, op. cit., pp. 114, 118.

[3] Their full German titles were respectively Justus Liebig Gesellschaft zur Förderung des chemischen Unterrichts, Emil Fischer Gesellschaft zur Förderung der chemischen Forschung, and Adolf Baeyer Gesellschaft zur Förderung der chemischen Literatur.

industry that were exempted from the gift tax. The governors of the societies awarded scholarships, supported the chemistry institutes of the Kaiser-Wilhelm-Gesellschaft, financed the abstracting service of the German chemical society and helped to pay for the publication of the third edition of Beilstein's *Handbuch der organischen Chemie.*[1] At about the same time the Notgemeinschaft deutscher Wissenschaft (renamed Deutsche Forschungsgemeinschaft in 1930) and the Helmholtz Gesellschaft für physikalisch–technische Forschung were launched. The former was the main channel through which the Government's annual grant of RM 5 million from 1925 to 1928 and thereafter RM 8 million (£0·4 million) was channelled to universities and research institutes. The Helmholtz Gesellschaft obtained its funds primarily from trade associations. There was some overlap between the two bodies, but the 700 scientists who were assisted by the Notgemeinschaft during the 1920's, often for several years, were the chief beneficiaries, and their departmental heads welcomed the grants for equipment distributed by the Helmholtz society.[2]

The various German societies were of considerable help to academic chemistry in that they restored the position of research, which had been weakened by war and inflation. In particular they contributed to the development of the Kaiser-Wilhelm-Gesellschaft from eight research institutes in 1914 to thirty in 1928, which represented every aspect of *Wissenschaft*. Their activities were directed by about 70 senior scientists, including 26 chemists and physicists. This permanent staff, equivalent in rank to university professors, supervised the work of several hundred research assistants on leave from industry or German universities; there was also a sprinkling of foreign graduates. For instance the Institute of Physical Chemistry under Fritz Haber had three divisional heads in the second half of the 1920s and about 60 research workers who were completing their doctorates or were engaged on post-doctorate work. The more than 600 papers published by members of the institute between 1919 and 1933 testify not only to the originality but also to the diligence of the research workers.[3] The Kaiser Wilhelm institutes at first devoted themselves exclusively to pure science, but after the outbreak of war and in the 1920s several of the newly founded institutes were created expressly to tackle research for specific industries, for example coal, and iron and steel. The change in purpose did not, however, affect the standards of work expected of the members of these institutes.

Attempts were made in Russia and the United States to narrow the gap between university chemistry and the chemical industry. The Soviet Govern-

[1] Jacobson, P. Die Adolf Baeyer Gesellschaft zur Förderung der chemischen Literatur. *Z. angew. Chem.* **33** (1920), pt. 1, 177 ff.

[2] Schmidt-Ott, F. *Erlebtes und Erstrebtes* (1952).

[3] Max-Planck-Gesellschaft (Eds.), *50 Jahre Kaiser-Wilhelm-Gesellschaft und Max-Planck-Gesellschaft 1911–1961* (1961), pp. 180, 184–5; Schmidt-Ott, D. Geschichte des Fritz Haber Instituts. *Jahrbuch der Max-Planck-Gesellschaft* (1961), pt. 2, p. 374; information from Dr. J. Jaenicke.

ment proclaimed its technological interest and many institutes suddenly appeared to link industry with science. Many, perhaps even the majority, were merely the research laboratories of Czarist days under a different flag, but in this way continuity was at least maintained and the institutional framework for research safeguarded.[1] Work was seriously impaired in the first few years after the Revolution by lack of fuel and laboratory supplies, but as the situation improved the institutes were able to resume research. There was much less opportunity for research at the universities, because for some years after the Revolution higher degrees were not awarded. The Aspirantura, which required three years of postgraduate study, was introduced in 1925, but in 1929 only 69 students had been accepted for it. The doctorate, which needed a further two years of study, was reintroduced in 1934.[2]

Leningrad had two institutes for chemistry: Applied Chemistry and High Pressures; in the latter Ipatieff and twelve assistants were working in the 1920s. Moscow had the Karpov Institute of Physical Chemistry, established by L. Y. Karpov, an old Bolshevik, in 1918 as the Central Chemical Laboratory. There was also the Laboratory of General Chemistry, which went back to the eighteenth century, had been reorganized in 1918/19, and later became part of the Soviet Academy's research facilities.[3] The institutes depended on different Government departments, between which, as is often the case, there was some rivalry. The Scientific–Technical Administration (NTU) was primarily responsible for the upkeep of institutes devoted to pure research, and the Commissariat for Heavy Industry dealt with those studying problems of applied chemistry. Towards the end of the 1920s the conflict between the departments increased and carried political overtones that also involved the Academy of Sciences. That famous institution remained relatively small and retained the characteristics and even the personnel of Imperial days. In 1927, however, the Academy was reformed and given a new charter.[4]

In the United States the vigorous post-war growth of the universities is illustrated by the figures in Table 12·3 on the next page.

The graduate and collegiate departments were roughly twice as large in 1920 as in 1910, and as the table shows, grew rapidly in the 1920s. As a consequence the demand for university teachers rose, and, in contrast to western Europe, higher education was a 'growth industry'. But the engineering departments developed unevenly. All were much larger than in 1910, when only about 870 students were recorded as following courses on chemical engineering. Clearly electrical engineering was the most popular subject, mechanical engineering held its own, but chemical engineering actually

[1] Hirsch, A. *Industrialised Russia* (1934), pp. 99 ff. states there were over 50 institutes with more than 10 000 scientists. V. N. Ipatieff in Modern Science in Russia, *J. chem. Educ.* **20** (1943), no. 4, 163–4 mentions 99 institutes and 33 000 scientists. The differences between them are probably caused by different definitions.

[2] Vucinich, A. *The Soviet Academy of Sciences* (1956), p. 71.

[3] Turkevich, J. *Chemistry in the Soviet Union* (1965), pp. 59, 150.

[4] In 1927 there were only 70 academicians, 259 corresponding members (including 193 foreigners) and a staff of 1018, of whom 683 were scientists. Vucinich, op. cit., pp. 7 ff.

TABLE 12.3

U.S.A.: selected student statistics, 1920, 1924, 1926, 1928[1]
(000s)

	1920	1924	1926	1928
Students in collegiate departments	341	486	595	695
Students in graduate departments	15·6	28·8	32·5	44·2
Teaching staff	42·9	56·3	62·2	67·2
Students in engineering departments:				
Mechanical	11·8	10·6	9·7	10·4
Electrical	9·5	14·0	15·7	15·8
Chemical	5·7	4·1	4·2	4·9
Doctorates awarded in above departments	0·5	1·1	1·3	1·4

declined, and despite a recovery there were nearly 1000 fewer students in 1928 than in 1920. The standards at graduate level varied, and were often below those of Europe. Nevertheless the number of doctorates awarded trebled in the 1920s and may be taken as a good indicator of the rising importance attached to technological work at advanced level.

The content and method of teaching were criticized by the American Chemical Society on the grounds that the instruction was not thorough enough and that it failed to arouse the students' interest in research. The employers were also dissatisfied and complained that their graduate recruits lacked initiative, accuracy, and imagination, that their English was poor and that they did not read enough;[2] many would echo the complaints today. But two important factors in American higher education were about to combine and cause a major change, not in the characteristics of the students, but in science teaching generally. One was the State or City college with its emphasis on practical or even vocational training in order to help the industries of the region or the particular locality. The other was the German tradition of scholarship and pure research, which had been absorbed by hundreds of American students before 1914 and was being widely spread by them after the war. The union of the two, of practical with theoretical, which seemed to elude Europeans, had a double effect: curricula became broader and the relations between pure and applied science closer. A notable example was the transformation of the Massachusetts Institute of Technology from an excellent engineering school to an institution with outstanding postgraduate departments in other disciplines.

The qualitative improvement became noticeable in the 1930s and later. It stemmed from the universities' ability to attract and retain able teachers by means of a salary policy that was more enlightened than in Europe. The staff was now entirely drawn from American universities, in contrast to the position up to 1914 when a significant minority had been to a foreign (usually German)

[1] Source: as for Table 12.2.
[2] Bureau of Vocational Information, op. cit., pp. 203–4.

university before taking up a teaching appointment in the United States.[1] The learned and the professional societies, owing to their size and influence, were able to press for improvements. The American Chemical Society was by far the largest and towards the end of the 1920s had roughly 16 000 members. It disposed of substantial funds, its income being $370 000 in 1924 and $380 000 six years later, and used them to subsidize the publication of journals and to pay for an abstracting service. The Society kept step with the increasing specialization of chemistry by creating additional 'Divisions': in 1908 there were five: Agriculture and Food; Fertilizer Chemistry; Industrial Chemists and Chemical Engineers; Organic Chemistry; Physical and Inorganic Chemistry. Three new Divisions were added in the next ten years: Medicinal Chemistry; Biological Chemistry; Water, Sewage, and Sanitation. Another nine followed between 1919 and 1927: Rubber; Sugar; Cellulose; Petroleum; Gas and Fuel; Colloid Chemistry; Paint, Varnish, and Plastics; History of Chemistry; and, particularly relevant in this context, Chemical Education.[2] Such divisions, far from fragmenting the Society, strengthened it by creating groups for particular tasks.

There were also specialist societies with a considerable membership. For instance the American Electrochemical Society had about 1800 members and the American Institute of Chemical Engineers about 600. The latter was principally concerned with chemical technology and the professional training and qualifications of chemical engineers. The Institute was among the original sponsors of the Engineers' Council for Professional Development, established in 1932 to prepare uniform standards for university engineering courses.[3]

Particulars of university expenditure on chemical research are not available for the period. It is probable that the disbursements rose, for the 1920s witnessed a striking increase in university income: from $240 million (£65 million) in 1920 to $547 million (£112 million) eight years later.[4] The Federal Government contributed less than 10 per cent of the total, the States and the cities, together, about one-fifth. Fees, income from investments, and benefactions accounted for the rest. Support from industry was generous immediately after the war, but withered away when profits fell in the early 1920s.[5] The return of prosperity a few years later, however, was not accompanied by any significant increase in industrial contributions. Nor was help from other quarters particularly large: some money was provided by the National Research Council, and on a somewhat larger scale, by the Carnegie Institution

[1] Visher, S. S. *Scientists starred 1903–1943* (1947), pp. 176–7, 284–5. The size of Visher's sample is small, but the minor role of foreign universities among the younger chemists, i.e. those born after 1890, emerges clearly.

[2] Browne, C. A. and Weeks, M. E. *A history of the American Chemical Society* (1952), pp. 265–6.

[3] Anglo-American Council on Productivity. *Universities and industry* (1951), p. 7.

[4] *Biennial survey of education*, for the years in question.

[5] In 1918 Du Pont created 21 university fellowships in chemistry, each worth $750 a year. By 1922 they had been pruned to 13. American Cyanamid, Grasselli Chemical, and National Aniline also endowed fellowships. Hull, C. Funds available in the U.S. for the encouragement of scientific research. *Bull. Nat. Research Council*, (1920) No. 9. The second survey of funds for research, issued in 1928, omits all mention of company fellowships.

and the Rockefeller Foundation. Herbert Hoover was one of those who felt that this was inadequate and so, possibly inspired by the example of the German Notgemeinschaft, launched the National Research Fund in 1926: it was a flop, and four years later a mere \$0·4 million (about £80 000) had been given by industry and distributed by the Fund.[1] The example shows that American industry had still a long way to go before supporting university research to any measurable extent.

Towards diversification

In the past thirty years the range of chemicals has broadened enormously, and people have become increasingly familiar with the diverse products of the industry. But such diversity is a new characteristic. It had not gone very far by 1930, except in Germany and the United States, and in the main the product mix and the customers were not so very different in that year from the 1900s. Demand was higher, but the supply still consisted predominantly of alkalis, acids, dyestuffs, and fertilizers with a few new and promising accretions, especially among the pharmaceutical preparations. Nitrogen fixation had, indeed, worked a major change in sales to agriculture, and the extension of high-pressure technology to the synthesis of methanol and of motor fuel was to have similar effects in other markets.[2] In general, however, the commercial developments in the period covered by the book did not stem from fundamental work, but from the need for improving the quality of the existing range of products: examples are ester solvents for cellulose paints and special dyestuffs for cellulose fibres.

With few exceptions, the manufacture of organic chemicals remained firmly based on coal-tar derivatives. Although other raw materials were available, economics and familiarity with well-established processes delayed the introduction of other routes. An interesting instance of that occurred just before the Great War. H. Staudinger and C. Engler, while investigating the synthesis of rubber, drew attention to the by-products of petroleum cracking as suitable raw materials. Badische, which kept in touch with Engler, experimented at one time with *n*- and iso-pentane as a starting point for the preparation of isoprene, and it was even forecast that the petroleum industry would be able to supply the raw materials for the synthesis of rubber.[3] But the interest in synthetic rubber diminished and more than a quarter century was to pass before the prediction came true. Nevertheless petroleum-derived chemicals were beginning to appear during the first thirty years of the century, though hardly yet on the scale of an industry. In Europe benzene and toluene were extracted from Borneo crude oil during the war, and in America isopropanol

[1] Dupree, A. *Science in the Federal Government* (1957), pp. 341–3.

[2] Some preliminary work on the catalytic reaction of carbon monoxide and hydrogen was carried out in 1912. It was resumed in 1922, and two years later synthetic methanol was first made on a commercial scale at Leuna. By 1930 several factories were making it elsewhere in Europe and in the United States.

[3] Holt, A. Neuere Arbeiten auf dem Kautschukgebiet. *Z. angew. Chem.* **27** (1914), pt. 1, 155.

was made from propylene by reaction with sulphuric acid followed by hydrolysis (Carleton Ellis process). The alcohol was first produced in this way in 1918 and two years later the Standard Oil Co. began to synthesize substantial volumes at its Bayway (N.J.) refinery. The activities of Carbide & Carbon Chemicals were of even greater significance. The company opened a works at South Charleston (W.Va.) in 1924, which developed quickly into a diversified petroleum chemicals enterprise. Ethane was cracked to yield ethylene, from which ethanol, ethylene glycol, and acetic acid were obtained; propane was cracked to propylene whence isopropanol and *n*-butanol were derived. A few years later the Royal Dutch–Shell group established the Shell Chemical Co., whose first major intervention in chemical production was the manufacture of synthetic ammonia. At Long Beach (Cal.) the company employed the low pressure Mont Cenis process and obtained hydrogen by cracking natural gas.[1]

Thus within less than ten years three American companies were using the by-products of oil refining or natural gas as raw materials for the synthesis of chemicals. Petroleum chemicals, which now account for the bulk of chemical output, date from the 1920s: they were not new compounds since they had previously been prepared, and continued to be made for many more years, from coal tar, by fermentation routes, or wood distillation. The distinction was one of cost: chemicals from petroleum were obtained from cheap materials or even by-products that had previously been burnt and so were costed at fuel value only. Another important aspect was that these developments occurred in America, where liquid or gaseous feedstock was plentiful and the market opportunities exceptionally favourable. The motor car played a big role at that crucial point: it exercised a fast-growing demand for a variety of chemical products. Motor car finishes used to be based on vegetable oils and lacquers, and the painting routine was long and laborious. But from the early 1920s onwards nitro-cellulose paints displaced the coach builder's traditional materials. These new finishes depended on butanol and ester solvents (ethyl and butyl acetates) for quicker drying and better finish. Assembly-line methods could now be extended to the paint shop and in this way one of the last obstacles to mass production disappeared. The popular car was also an economy car, hence the demand for better motor fuels and reliable protection against frost. The answer to the first was anti-knock fluid, tetraethyl lead, to improve the performance of the fuel. It was made by reacting lead–sodium alloy with ethyl chloride; the fluid also contained ethylene dibromide and ethylene dichloride. The answer to the second was glycerine, methanol, and, more effective than either, ethylene-glycol, which had initially been used to prevent accidental dynamite explosions in winter, and was first sold to motorists as an anti-freeze by Union Carbide in 1927.

By comparison with the traditional products of the industry, the new

[1] Carbide & Carbon Chemical Co. *The Carbidea*, November 1927, no page; *Fortune*, November 1941 (reprint), pp. 15, 37–8; Haynes, W. *American chemical industry* (1948), vol. 4, p. 89.

compounds were being made on a fairly small scale throughout the 1920s. This can be illustrated by looking at the volume of chemicals used for motor fuels and cars. In the first year of tetraethyl lead manufacture in 1924, only 560 tons were sold. Production fell off in the next two years because of lead poisoning. Special safeguards were introduced and manufacture gradually resumed; sales amounted to 800 tons in 1927 and 3600 tons two years later.[1] The production of ethylene-glycol was still insignificant in 1924, but the following year reached 1000 tons and by 1927 had grown to 5300 tons. The output of ethyl acetate reflected the rising demand from makers of cellulose finishes: in 1922, 7200 tons were made, in 1924 the figure was 12 000 tons, in 1926 19 050 tons, and in 1928 just over 26 000 tons.[2]

The development of plastics, now the pace setters of the chemical industry, was uneven. Celluloid was still the principal product. The production in the leading countries was estimated at 50 000 tons in 1925, and so had more than doubled since 1913. Immediately after the war the manufacturing capacity was greatly enlarged, but the hopes of a boom were not fulfilled, and though there was plant for making 120 000 tons a year in the United States, Germany, and France, the greater part was under-utilized.[3] By contrast the business in phenolic resins was growing very rapidly indeed in the United States, where the Tariff Commission recorded production of 2600 tons in 1922, 6000 tons five years later, and 14 700 tons in 1929. At the latter date the sales value was $10·4 million (say £2 million), and the phenolics were clearly becoming major industrial products.[4] Statistics in Europe are not available, but it is recorded that 1000 tons of phenolics were made in France in 1928;[5] the German output was certainly larger.

In at least one instance, methanol, the new process of manufacture had effects out of all proportion to the tonnage involved. Badische, having developed the commercial production of synthetic methanol, began exporting the alcohol to the United States in 1924, and within a few months had established a market for their product. The American manufacturers of wood methanol found that their position was being challenged and the following year demanded higher tariffs. An enquiry was held and the President raised the duty from 12 to 18 ¢ a U.S. gal. The 50 per cent increase gave the wood distillers a respite and incidentally encouraged American chemical firms to make methanol by the synthetic route. But while the Germans could be shut out of the United States, they could not be prevented from displacing

[1] GTX 774, Exhibit 'E'. Yearly sales volume and prices of tetraethyl lead and GTX 834, tetraethyl lead profits. The sales of the Ethyl Corporation grew from $2 million in 1924 to $5·5 million in 1927, and then jumped to $19·3 million (almost £4 million) in 1929.

[2] USTC *Censuses* for the years mentioned.

[3] Schueller, E. Les matières plastiques, in Gérard, J. (Ed.), op. cit., vol. 1, pp. 1403–5. The manufacture of celluloid started in Japan and Italy shortly after the war, and by 1926 one Japanese firm was already making 4000 tons a year. Kaufman, M. *The first century of plastics*, n.d. (1964?), pp. 47–8.

[4] USTC *Censuses* for the years mentioned.

[5] Reagan, D. J. *The French chemical industry and trade in 1928* (Tariff Information Bulletin No. 652), (1929), p. 36.

American methanol in other markets; United States exports were halved between 1923 and 1929 and, to take out but one example, the business with Britain was lost to Badische, which held it until I.C.I. began producing synthetic methanol.[1]

The growing volume of new products and the diversity of their uses were tangible expressions of the industry's widening range. These new products were beginning to displace materials prepared by other processes. In a few instances synthetics were having major economic effects, as in the case of ammonia and Chilean nitrates, or methanol and wood distillation. But in 1930 the main weight of the industry still consisted of those sectors that had represented it at Paris thirty years previously: the chlor-alkali group of products, dyestuffs, and fertilizers. The depression abruptly spotlighted the weaknesses inherent in such a limited range of products. To put it in another way, diversification had not yet gone far enough and was not yet sufficiently important to compensate for the lower profits of the alkali and dyestuffs trade. As for fertilizers, despite tariffs and cartels, the makers were unable to help the farmers, and when their incomes dropped, the sales of fertilizers fell. The slump showed that chemical manufacturers were not inured to business cycles. I.C.I., one of the largest, reflected the problems in the annual report for 1930: the trade of the Alkali and General Chemical Divisions fell by 14 per cent, fertilizer sales in the United Kingdom dropped by a quarter (though exports, propped up by the cartel, showed 'a considerable advance'), dyestuffs sales were down by 13 per cent and those of limestone by 12 per cent. On the other hand the volume of the leather-cloth and cellulose paint business was maintained. The company's shares mirrored the decline in profitability and between the peak of 1929 and July 1930 the market capitalization of I.C.I. fell by 55 per cent.[2] Other firms fared no better. Even IG Farben, which was probably the most diversified of the European firms, was bowled over by the storm: comparing 1930 with 1929, the value of dyestuffs sales fell by 13 per cent, exports of nitrogenous fertilizers by almost a half because trade had to be conceded to the other members of the cartel, and the number of employees by 18 per cent.[3] In 1931 the situation became worse.

In the normal course of events research and development would have contributed to the steady broadening of the industry's product range. But events were abnormal, and the slump powerfully accelerated diversification, because manufacturers adopted it as the answer to the problems caused by the falling business in the staple products. In this respect the end of the 1920s also marks the end of one stage in the industry's growth and the beginning of another in which it extended horizontally across the entire economy instead of only a few of its sectors, and so caused a transformation of its product range and ensured a long period of prosperity.

[1] Haynes, op. cit., vol. 4, pp. 171 ff.; Waller, P. *Probleme der deutschen chemischen Industrie* (1928), pp. 214 f.

[2] The capitalisation of mergers, *The Economist* **111** (1930), no. 4535, 182.

[3] ter Meer, F. *Die IG Farben* (1953), pp. 70, 73, 80.

13

WORKERS, WORKING CONDITIONS, AND LABOUR RELATIONS

THE chemical industry was not a large employer by comparison with other basic industries. During the first thirty years of the century its manpower, though growing, remained well behind other technologically new manufactures such as electrical goods or motor vehicles. Definitions here, as elsewhere, cause difficulties, but for our purposes the rough figures for the United States and Britain indicate the size. In the former there were 15 000 chemical workers in 1899 and 56 000 in 1925.[1] The classification excludes people in paint and varnish, matches, explosives, and fertilizers, which are usually considered part of the chemical trades, and if included would more than double the number. The British industry on roughly the same definition employed about 125 000 people in 1907 and 1924.[2]

The increasing size of the works has already been noted, and the large factories with their ancillary services formed the nuclei of the chemical towns. During the later part of the nineteenth century Widnes, Northwich, St. Gobain, Dombasle, Ludwigshafen, and Hoechst emerged as single industry or even single company towns. During the twentieth century others were added, among them Billingham in England, Midland (Mich., U.S.A.), Leuna and Leverkusen in Germany, Rjukan in Norway, and St. Auban in France.

Concentration of production and of manpower went hand in hand. Both created many problems: the technical and financial aspects received much attention, but labour relations were overlooked, and were not recognized as a management function until well into the 1920s. There were a few notable exceptions. Bayer established a managers' committee in 1904 to handle policy questions on hiring, firing, and complaints, but all matters concerning wages were referred to the board meetings. The Bayer works at Elberfeld had quite a large labour department, for it employed a dozen wage clerks in 1906, but apart from making up pay packets and issuing regulations it had no executive functions.[3] Du Pont also concerned itself formally with labour relations at an early date, and in 1911 formed the 'Labor Efficiency Department' to deal with

[1] Brissenden, P. F. *Earnings of factory workers* (Dept. of Commerce Census Monographs, No. 10), (1929), pp. 377 ff.

[2] *Final Report on the Third Census of Production* (1924), HMSO (1931), p. 7. The actual numbers were 124 000 in 1907 and 127 000 in 1924, and covered the following trades: chemicals, dyestuffs, drugs, fertilizers, glue, sheep dips, paint, colour, varnish, coke and by-products of coke ovens. matches, explosives.

[3] Bonhoeffer, O., and Grabendörfer, R. *Geschichte und Entwicklung der Farbenfabriken vorm, Fr. Bayer & Co.* (1918), pp. 539–40. (Referred to hereafter as *Geschichte ... Bayer*.)

work study, develop training schemes, improve safety standards, and generally maintain 'work satisfaction'. The department's staff lectured managers on job security, promotion, and wage-rate determination, and in 1912 published *Scientific Management in Powder Making.*[1]

In the absence of a formal system of labour relations, the traditional management attitudes and methods generally prevailed. They were, in short, autocratic, harked back to Smilesian precepts of self-help, and occasionally were sympathetically paternalistic. There were many instances of cordial and even familiar relations between management and men. A German observer in pre-war America noted that graduate engineers and chemists worked alongside the men, and that there was no insistence on status. He also commented on the absence of employment contracts, pension funds, and medical help.[2] In Europe, particularly on the Continent, there were many instances of benevolence (occasionally bordering on generosity), but they were invariably accompanied by strict adherence to grades and the segregation of workers from foremen and other staff.

There was, for most of the period, no consistent labour policy. Training schemes for young workers were considered worthy of support, the more so as there was a growing demand for specialist craftsmen. But by no means all employers were equally committed to them. The German firms, Solvays, and Brunner, Mond had extensive educational facilities, but the French manufacturers were not really interested, and their critics agreed that training in chemical factories was badly organized and inadequate.

There is no clear evidence that training, welfare, and amenities on their own made for and ensured a contented work force. Badische and Bayer, who in these respects were certainly among the most progressive employers, had serious labour problems.[3] On the other hand Brunner, Mond, who were just as advanced, had no troubles in their Cheshire factories. But later, at Billingham, the environment was different: the men for the new synthetic ammonia works were recruited from the steel, shipbuilding, and mining industries of the north-east, and relations were tough. Nor were they softened by I.C.I.'s decision in the late 1920s to move men and their families, almost 2000 people, from Glasgow and Wallsend to the Billingham South (now Cassel) works. The events at Billingham took place in the early years of I.C.I. and highlighted the difficulties accompanying the rationalization moves of the newly formed concern. Other large firms in the course of reorganizing their activities probably encountered similar problems. Under the circumstances a positive and modern attitude to labour problems was called for; Sir Alfred Mond was aware of it and rose to the occasion: 'We shall devote, as soon as we can find time, a good deal of thought to labour relations', and in a more philosophical vein referring to the workers asked '. . . in what way can we make them share

[1] Dale, E. Du Pont: pioneer in systematic management. *Adm. Sci. Q.* **2** (1957), no. 1, 36–7.
[2] Rassow, B. Was haben wir Chemiker in Amerika gelernt. *Z. angew. Chem.* **26** (1913), pt. 1, 706.
[3] See below pp. 393–4.

in our prosperity? In what way can we make their lives more happy or content?'[1] A few years later H. J. Mitchell, the president of I.C.I., spoke of encouraging a sense of partnership between employees and management and hoped that despite the size of the company the identity of the individual would be retained.[2] His company appointed a director to take charge of all personnel questions and so gave practical expression to these sentiments. Despite such changes one cannot help but feel that labour policy was pursued on two planes: one at board level presented to the public a lofty and high-minded image of the good employer, the other at works level was the unvarnished and tough relationship of the local manager with his men. The directors might regret that size prevented the individual from identifying with the company, but the men on the plant would still resent the impersonal aspects of the concern. So, on the one hand labour relations were systematized and elaborated, and on the other the old, informal links between employees and managers withered away.

Chemical work was men's work. It was heavy and often round the clock. Legislation prohibited night work for women and juveniles, and so excluded them from shift jobs. In the United States they represented only 5–6 per cent of the employees in the pre-war chemical industry, compared with about 22 per cent in all industrial employment.[3] There were 10 000 women in the French chemical industry (including explosives and matches) in July 1914, and 25 000 three years later. Despite the increase, they accounted for an insignificant proportion of all women workers: 3 per cent before the war and 4 per cent at its height.[4] The British Censuses of Production show that juveniles (i.e. those under 18) represented about a tenth of the industry's employees. The number of women in the chemical trades listed at the beginning of the chapter rose from 17 000 to 25 000 and their relative importance from 14 to 20 per cent between 1907 and 1924. At the former date they were especially prominent in the match and explosives trades, at the latter they also formed a large minority of the sector comprising chemicals, dyes, and drugs, specifically of the expanding pharmaceutical preparation business.[5] Women did not become important in the offices and laboratories of chemical manufacturers until the war. At Badische the first shorthand typist was an Englishwoman in the patent section, which tells us something about the correspondence. By 1917, however, 11 per cent of the firm's office staff were women, and their numbers grew further so that in the early 1920s they accounted for 14 per cent of the office staff.[6]

[1] Meeting of I.C.I. 18 March 1927; *The Times*, 19 March 1927.

[2] Methods of inculcating modern management principles and practices in large scale undertakings (1935), typescript of paper to the Int. Management Congress, London, 1935.

[3] Brissenden, op. cit., p. 398.

[4] Fontaine, A. *L'Industrie française pendant la guerre* (1925), p. 73.

[5] *Final Report on the Third Census of Production*, op. cit., pp. 7, 8.

[6] Voigtländer-Tetzner, W. Der Mensch in der BASF, typescript, n.d. (1938–9?), vol. 3, p. 553.

From the limited information that is still available it is possible to piece together a composite picture of the labour force and working conditions generally that is probably broadly representative of the industry. In Germany before the war the turnover among the workers was surprisingly high. According to a survey carried out in 1892, men below thirty changed jobs twice a year.[1] About twenty years later the turnover was still very high at Leverkusen; for example, during 1907 5540 men were hired and 5182 left, and in 1913 the numbers were 5980 and 5292 respectively. At the latter date there were almost 8000 employees. The continuous building activity and the rapid growth of the huge factory probably contributed to the general restlessness. The turnover was also high among the craftsmen for whom special training facilities were provided: in 1912, 51 per cent of the newly hired men left within a month, and a mere 10 per cent lasted for more than 6 months.[2]

It would be unwise to generalize from the experience of the Leverkusen works, but the attitude of the craftsmen must have worried the management. Their importance was increasing and was not checked by the war. The German occupational censuses recorded that the ratio of unskilled to skilled workers was 100 : 17 in 1892 and 100 : 38 in 1925; in that year more than 32 000 mechanics, fitters, and engineers and over 10 000 foundrymen and plumbers were employed in the workshops of the chemical companies. There were also thousands of builders, carpenters, and drivers, as well as hundreds of coopers.[3] By contrast, old crafts such as those of the dye maker disappeared, as routine testing in the works laboratories became standard practice and was done by semi-skilled technical assistants. The labour statistics of Badische reproduced below illustrate the decline of the unskilled and untrained yard men, and the onward march of the mechanics.

TABLE 13.1

Job classification of Badische workers 1895–1913[4]

	1895 %	1900 %	1905 %	1909 %	1913 %
Workers on plant (incl. shift-men)	55	57	59	56	52
Mechanics and other craftsmen	30	35	33	38	43
Yard labour	15	8	8	6	5
Total	100	100	100	100	100

The growth in the number of supervisory personnel and of commercial and scientific staff was just as important in its way as the increase shown among

[1] Wenzel, O. *Die Arbeiter in der chemischen Industrie und ihre Löhne* (1904), p. 865. Seasonal work was customary in the fertilizer trades and some other branches, and men might be taken on in January only to be discharged in April and go back to casual work in towns or on farms.

[2] Beckerath, P. G. von. Personalwesen, in Jacobi (Ed.) *Beiträge zur hundertjährigen Firmengeschichte* (1963/4), p. 409; Pinnow, H. *Werksgeschichte* (1938), p. 141; Duisberg, C. *Die Arbeiterschaft der chemischen Grossindustrie* (1921), p. 13. (Curt was the son of Carl Duisberg.)

[3] Wenzel, op. cit., p. 869; *Enquete*, p. 49.

[4] I am indebted to the company for this information.

the craftsmen. The foremen and their assistants, the charge-hands, came to occupy a key role. They were usually recruited from among the older men, who made up in service and experience what they lacked in skill, and who were absolutely essential to the operation of the shift system. The foremen in the craft trades were often qualified (in Germany this seems to have been the rule in the larger works) and held a certificate or diploma of professional competence. In several companies the foremen were given staff status, and so contributed to the growth of that group. The statistics for IG Farben and its predecessor companies speak for themselves in this context.

TABLE 13.2

IG Farben and predecessor companies: number of workers and staff[1]

	1 Jan 1914		1 Jan 1927		1 Jan 1930	
	(000s)	(%)	(000s)	(%)	(000s)	(%)
Workers	37·3	81	67·8	77	74·6	77
Staff	8·8	19	20·3	23	22·9	23
Total	46·1	100	88·1	100	97·5	100

The situation was no different in Switzerland: the occupational census for 1929 showed that in the chemical industry as a whole there were 450 workers for every 100 staff. In Basle, where the dyestuffs and pharmaceutical firms were located, the significance of the staff element was even greater, the ratio being 390 : 100.[2] For Britain the Censuses of Production show that staff accounted for 12 per cent of all employees in 1907 and 18 per cent in 1924.

The absolute and relative size of staff tell only part of the story. It would be even more useful if changes in the qualifications of the staff were documented. But it appears that no comprehensive surveys of the managerial, professional, and technical manpower were made at the time. The available evidence indicates that clerical and commercial staff outnumbered the rest. In Germany many had a diploma or a certificate from a commercial college; elsewhere the expertise was gained on the job.[3]

Working conditions in the chemical factories were unpleasant at the beginning of the century. At the Cassel Cyanide Works, as H. Fern recalled fifty years later, the fitters' shop in 1907 still had an earthen floor and was lit by a single gas jet. 'Refinements such as ambulance rooms, canteens or wash basins were unknown while the "usual offices" were at best perfunctory.'[4]

[1] *Enquete*, p. 114 (memorandum by C. Bosch).

[2] Baumgartner, R. *Die wirtschaftliche Bedeutung der chemischen Industrie in Basel* (1947) pp. 43 and 147.

[3] So far as I know only Voigtländer-Tetzner has dealt with these questions in the period under review, and then only for Badische. More information would be extremely useful and could shed light on the social factors behind the industry's growth.

[4] Fern, H. Fifty years ago. *I.C.I. Magazine* (October 1957), pp. 336–7.

In these respects they ordered things better abroad, and a critical observer of the chemical manufacturers of Basle found the washing facilities in all the factories, except Geigy's, satisfactory, and those of CIBA positively good.[1] Even in the late 1920s some of the older chemical plants had a 'cheerless atmosphere', and seemed 'to be an anachronism from the point of view of efficiency and human welfare, and strongly reminiscent of the bad conditions of the early industrial revolution not as regards wages and hours, but in respect of environmental conditions of work'.[2]

Such improvement in working conditions and the abatement of occupational diseases as took place was the result primarily of social legislation. In Germany it was connected with the Bismarckian welfare measures and the formation of occupational associations (Berufsgenossenschaften), which became responsible for implementing and later policing the measures.[3] Britain had special factory legislation for dangerous trades and provision was made for occupational diseases in the Workmen's Compensation Bill of 1906. In both countries the regulations and orders set minimum standards that the conscientious employers had already adopted. The first requirement was elementary safety precautions, next came good ventilation, then proper washing facilities and the provision of suitable clothing. At the turn of the century A. P. Laurie drew a harrowing account of the dangers in the saltcake and bleaching powder trades.[4] But when the third edition of Sir Thomas Oliver's *Diseases of Occupation* appeared they had been modernized and become innocuous though not inoffensive, and he mentioned the electrolytic manufacture of alkali as a threat on account of dermatitis.[5]

Towards the end of the nineteenth century, the effects of legislation and of technological change began to be reflected in an improvement of the workers' health and a fall in the accident rate. In 1905 the Berufsgenossenschaft reported that it was 8·4 per thousand in heavy chemicals and dyes and 3·6 per thousand in explosives, where the most stringent regulations applied. Accidents, when they occurred, ended more often fatally in explosives (5·1 per thousand) than in the other sectors, where general chemicals and dyestuffs had the lowest death rates (1·2 and 0·5 per thousand).[6] The Germans extended the scope of accident notification in the 1920s, but if allowance is made for this the accident rate per 1000 insured workers fell from 7·3 in 1912 to 5·4 ten years later, and then fluctuated between 4·7 and 5·6 for the remainder of the 1920s.[7] In Britain there were 45 fatal accidents in the industry in 1908 and

[1] Wassilieff, N. *Die Arbeitsverhätnisse in den Basler Chemischen Fabriken* (1905), pp. 12–13.

[2] *The new survey of London life and labour*, vol. 5—*London industries* (1933), pt. 2, p. 205.

[3] The association for the chemical industry also covered workers in salt, soap, candle, rubber, roofing felt, petroleum, lubricating oil, and mineral water factories. In the late 1890s the association employed seven officials to study and report on safety and accident prevention. *Paris Exhibition* (1), xxvi.

[4] In Oliver, Sir Thomas (Ed.) *Dangerous trades* (1902), Chapter 40.

[5] Oliver, Sir Thomas. *Diseases of occupation* (3rd edn. 1916), p. 123.

[6] Grossmann, H. *Die Bedeutung der chemischen Technik für das deutsche Wirtschaftsleben* (1907), p. 54.

[7] *Enquete*, pp. 500–1 (Evidence of Dr. Ungewitter).

31 in the following year, corresponding to roughly 3 and 2 per thousand. Statistically speaking the chemical worker was in a less dangerous industry than the miner, the docker, and the railwayman.[1]

There was, however, no cause for complacency. New chemical compounds brought novel dangers. Carbon disulphide, which was becoming important owing to the growth of rayon manufacture, caused intoxication symptoms, and the careless manufacture of carbon tetrachloride entailed an attack of jaundice. Lead poisoning continued to take its toll: nearly thirty deaths a year were reported in the United Kingdom before the war from the manufacture and use of white-lead paint. After the war more effective regulations cut down the death rate, and abroad the International Labour Office was able to report a fall in the number of fatal cases.[2] The manufacture of tetraethyl lead in 1925–6 caused a temporary and local increase in the incidence of lead poisoning. But Standard Oil and Du Pont devised satisfactory precautions, and production was resumed.[3] There was also cancer of the bladder among dyestuffs workers. It was particularly insidious as the tumour took about eleven years to develop.[4] At first public health authorities and doctors thought that many workers were at risk, and it was only much later that the cause was specifically traced to β-naphthylamine, an important intermediate of azo-dyes.

Accidents and dangers were one important aspect of working conditions. Another, less easy to quantify, but just as significant and far reaching in its implications, was the working rhythm, and here too the changes were noteworthy. In the older plants, especially the Leblanc factories and the fertilizer works, very arduous work alternated with periods of recuperation. But the new processes demanded less physical strength of the men and no manipulative skills with wheel-barrow or long-handled rake. Shift working came to mean shorter hours and less physical exertion, but on the other hand called for regular attention to routine, careful observation, and the punctilious turning of valves and reading of instruments. The shovel was on the way out, sacks were filled by machines, and conveyor belts, introduced before the war, became general in the 1920s. These important alterations in the pattern of work, which had important effects on the productivity of the industry, began in Germany and the United States, and were adopted by chemical manufacturers elsewhere during and after the war. Everywhere they imposed a different kind of discipline and required a high degree of orderliness. Least progress in these respects was made in Russia, and working conditions in that

[1] Home Office; Cases of Industrial Poisoning . . . Accidents and Dangerous Occurrences . . . 1909, HMSO (1910), Cd 5048. Owing to different definitions the British figures are not strictly comparable with the German.

[2] Home Office, op. cit.; International Labour Office, *White lead* (1927) (ILO, Series F, No. 11); *New survey of London . . .*, op. cit., p. 219.

[3] Haynes, W. *American chemical industry* (1948), pp. 402–3.

[4] International Labour Office, *Cancer of the bladder among workers in aniline factories* (1921), (ILO, Series F, No. 1), pp. 5–6. The investigation was based on a detailed survey of the sickness records and the causes of deaths of workers in the Basle dyestuffs factories from 1871 to 1910.

country were indeed so different that they require a brief digression. The men were hired by the day or for a period of several months starting at Easter or Michaelmas. Contract workers left their families behind and were housed by the firm in barracks where the discipline was quasi-military; they drew rations and cooked for themselves. Some pay was usually withheld until the end of the contract, if only to cut down the celebrations that invariably followed the regular pay day.[1] The revolution of 1905/6 caused some changes: thus at the Bayer factory outside Moscow it was noticed that the men were bringing their families, and as their dependants could not be accommodated in the barracks they sought lodgings in the city, so contributing to the appalling housing problems. After the revolution there were fewer contract workers from the country. But the town labourers who came by the day on foot or by train were bad time-keepers.[2] A quarter century later an American chemical consultant who was friendly to the Soviet Government also complained of bad time-keeping, indiscipline, and a high labour turnover. Under the Czars there had been a bailiff or administrator in charge of labour in the foreign-owned works. This functionary disappeared in 1917/18, but the Russian plant chemists and engineers who succeeded him were inexperienced, and the discipline remained poor.[3]

Before the war the working week was usually very long. A few benevolent or far-sighted employers occasionally offered concessions, especially when three-shifts replaced two-shifts. But another and just as important factor in the shorter working week was the growing number of skilled craftsmen. The electricians, pipe welders, fitters, instrument mechanics, and boilermen were specialists, able to negotiate from strength because the manufacturers needed them. The terms were good and included fewer hours. The tradesmen's hours also affected those of the day men and the unskilled assistants, and so these two groups were gradually brought into the new order of things.

In Britain Brunner, Mond was the first large company in the chemical industry to shorten the hours of the tradesmen to 48 a week and of labourers to $49\frac{1}{4}$.[4] This was in 1895. Castner–Kellner's new works at Wallsend hired shiftmen for the sodium plant to work a 56-hour week in the 1900s when the neighbouring shipbuilders were still toiling for over 60.[5] The two firms appear to have been exceptional employers, and even immediately after the Great War shiftmen were often on 12 hours' duty, and day workers usually had a week of 53–54 hours. But by the mid-twenties the working week had become one of 7 shifts of 8 hours each (one at overtime rates) for the former, and of

[1] Hauffe, M. Die chemische Grossindustrie in Russland, geschildert auf Grund eigener Erfahrungen und Beobachtungen. *Z. angew. Chem.* **13** (1900), no. 1, 27 f.

[2] Blank and Löw in *Geschichte . . . Bayer*, op. cit., pp. 268 f.

[3] Hirsch, A. *Industrialized Russia* (1934), pp. 107 f.

[4] In winter one hour less was worked a week, and in summer one hour more. The adjustment was made by varying the length of the lunch break. Anon. *The first 50 years of Brunner, Mond & Co.* (1923).

[5] Anon. *Fifty years of progress* (*Castner–Kellner*), n.d. (1947?), pp. 45–6.

47 hours for the latter.[1] In that decade the 5½ day week became usual in chemical works.[2]

Abroad the pattern was similar. At the beginning of the century the working day for most employees at Bayer was 12 hours, including breaks aggregating 2 hours; in 1905 it was reduced to 10½ hours, including 1½ hours for breaks. Shift men, however, continued to do a 12-hour turn, though in some sections of the plant this was shortened to 8 hours.[3] In 1919 the hours of the day men were cut to 9½ a day, including half an hour for lunch, and on Saturday work ended at noon. Badische retained the 12-hour day until 1911 (including the unpaid 1½-hour lunch break). In that year the times were changed to 7.30 a.m. to 6 p.m. and the 10½-hour day continued until 1918, when the 8-hour day was introduced. At the adjoining Oppau ammonia works, however, Bosch introduced three 8-hour shifts in 1913, and as a further innovation included a 20-minute pause in each shift, for which payment was made.[4] The Swiss dyestuffs factories worked slightly shorter hours in the 1900s: from about 54 to 58 a week, excluding lunch and other breaks, which ranged from 1 to 2 hours a day.[5] In 1907 Geigy set a precedent by shortening work by 1¼ hours daily from Monday to Friday and by 1 hour on Saturday.[6] The 10-hour day law was enacted in France in 1900 and came into force four years later. Chemical firms generally worked 60 hours a week, even in 1913, though some pharmaceutical companies were working 54–55 hours a week and closed early on Saturday. As a rule the French employers were opposed to shorter hours and the 5½-day week.[7] A handful of manufacturers, however, were converted to the 8-hour day even before the end of the nineteenth century. Among them was the Deutsche Gold- und Silberscheideanstalt (DEGUSSA) of Frankfurt in the 1880s, and, more importantly, the Solvay brothers, who adopted the three-shift system of 8 hours for some of their soda works in the 1890s, and extended it to the rest in 1907.[8]

For the United States the figures in Table 13.3 afford a comparison between the hours in the chemical industry, motor vehicles, electrical machinery, and manufacturing industry generally. They also illustrate the change between pre-war and post-war practice. There is little information on hours in the principal companies. At the Du Pont factories in Delaware the working week was 60 hours in the early 1900s, though the average in the State was then 54 hours. The company's machinists secured the 54-hour week in 1901.[9] Monsanto in St. Louis worked a 12-hour day in the 1900s, while Dow intro-

[1] Committee on Industry and Trade (Balfour Committee), *Survey of industrial relations*, HMSO (1926), pp. 130, 147.

[2] *New survey of London*, op. cit., vol. 5, pt. 2, pp. 201, 212, 221.

[3] *Geschichte . . . Bayer*, op. cit., pp. 417, 543.

[4] Voigtländer-Tetzner, op. cit., vol. 1, pp. 103 ff.

[5] Wassilieff, op. cit., pp. 12–13.

[6] Geigy papers in the Wirtschaftsarchiv, Basle.

[7] Matagrin, A. *L'Industrie des produits chimiques et ses travailleurs* (1925), pp. 377 ff.

[8] DEGUSSA 1873–1923. *Chem. Ind., Düsseld*, **46** (1923), no. 7, 103; Bolle, J. *Solvay 1863–1963*, n.d. (1963?), pp 139, 170–1.

[9] Hancock, H. The industrial workman along the Brandywine, typescript, August 1958, pp. 58–60.

TABLE 13.3

U.S. average weekly hours in industry, selected years 1899–1929[1]

Year	*Manufacturing industry*	*Automobile industry*	*Electrical machinery*	*Chemical manufacture*
1899	59·6	58·3	57·4	64·4
1904	57·9	56·4	55·6	62·3
1909	56·8	55·5	54·7	61·3
1914	55·1	52·9	53·5	60·6
1919	50·8	49·6	49·1	52·6
1921	50·3	49·1	48·2	50·3
1924	49·2	45·5	44·7	52·7
1926	50·0	45·2	45·0	50·6
1928	50·3	43·4	45·7	50·5
1929	49·5	43·0	46·2	50·2

duced the 8-hour day in 1910 and was apparently the first company to do so in the mid-West.[2]

The customary working hours, often longer than in other industries, caused dissatisfaction. Some European countries provided for the 8-hour day (or 48-hour week) by law: Germany in 1918, France, Switzerland, Norway, and Belgium between 1919 and 1921. (The shift-men's time was amended correspondingly, usually to 56 hours a week.) In Britain, however, the hours of male workers were not fixed by statute, except in mining, but by collective agreements. At Brunner, Mond, the largest employers in the industry until the formation of I.C.I. the 47-hour week was introduced for day workers in 1919, and the shiftmen got a 56-hour week with every third Wednesday off.[3] As for the United States, the Federal Government laid down the maximum hours for its employees and those in transport and communications; the other workers were governed by States' legislation.[4]

Germany seems to have been the only country where the 8-hour day was challenged. The move followed the stabilization of the Mark, and at the beginning of 1924 the arbitration tribunal authorized an increase from 8 to 9 hours, which Bayer and Badische implemented in February and March respectively; the unpaid lunch break lasted for 40–60 minutes, so that the working day lasted about 10 hours. At Westeregeln they went even further and reintroduced 12-hour shifts.[5]

[1] To 1921: Brissenden, op. cit., p. 407; 1924–9: Beney, M. A. *Wages in the U.S.* (1930), pp. 51 ff. The figures refer to the fourth quarter of the year.

[2] I am indebted to Mr. B. Thomas and Dr. W. E. Bennett for the information concerning Monsanto and Dow respectively.

[3] Irvine, A. S. *A history of the Alkali Division* (1958), p. 7.

[4] International Labour Conference, 23rd Session, Reduction of hours of work in printing and kindred trades (1937). Appendix, and Reduction of hours of work in the chemical industry (1937), pp. 12–21.

[5] Beckerath, op. cit., p. 420; Voigtländer-Tetzner, op. cit., vol. 1, pp. 103–9; Anon. *50 Jahre Consolidirte Alkaliwerke Westeregeln* (1931), p. 184. At Badische the lengthening of the working day caused a short strike, which the men lost.

Between 1900 and 1930 there was some change in the attitude of employers towards the question of holidays. Before 1900 holidays were generally speaking a privilege, unpaid, and at the discretion of the directors; a generation later they were an entitlement and paid. Brunner, Mond and the Solvays were, once again, the pioneers. The former introduced a week's paid leave in 1884 and gave double pay in 1902. Castner–Kellner and Chance & Hunt followed a similar policy in Britain, and abroad most of the Solvay companies granted a week with double pay. By 1914 the large German dyestuffs companies had followed suit, but related holidays to length of service. Badische, for example, introduced paid holidays in 1908, though only at the rate of three days for 10 years' service; in 1913 this was lowered to 5 years.[1] Hoechst adopted paid holidays in 1906, and Bayer gave all weekly paid men—that is in the main the craftsmen and the plant men—four days' paid leave after 3 years' service; a week's leave was obtained after 10 years' service.[2] In 1919 the German unions secured a minimum paid holiday entitlement rising from four days to a maximum of twelve days after 10 years' service.[3] The following year the Swiss chemical manufacturers agreed to a minimum of two days after 2 years, rising to ten days after 7 years.[4] By contrast there were no formal arrangements in France and it was left to the discretion of the employer. Paid leave was rare in the British chemical industry and exceptional in the United States.[5]

Conditions and hours of work, though important, were not the chief source of friction between management and men. Wages were the crucial factor. Another element in labour relations that it will be useful to bear in mind was the emergence, at any rate in the larger companies, of a staff bureaucracy. The gulf between owners and men widened, and as new functionaries appeared in the hierarchy, the conflict sharpened. Occasionally the spread of unions and the activities of their officials further polarized the friction over pay.

There is a good deal of fragmentary information on wages and labour relations. Unfortunately much of it is imprecise, and comparisons are especially dangerous because of the widespread practice of giving fringe benefits that were entirely at the management's discretion. They usually took the form of bonuses or long-service awards and were occasionally quite substantial. A few examples will give an idea of the directors' benevolence: Bayer added 25 M (say £1. 5*s*.) a year to a man's pay after a year's satisfactory service, and raised the bonus by 25 M every 2–3 years depending on job performance and profits.[6] The Swiss firms awarded annual bonuses ranging from 50 to 100 fr. (£2–4) to men with over 5 years' service. The money was not paid out, but

[1] Voigtländer-Tetzner, op. cit, vol. 1, pp. 176–80.

[2] Duisberg, op. cit., pp. 21–2.

[3] Voigtländer-Tetzner, op. cit., vol. 1, p. 182.

[4] Jaquet, N. *Die Entwicklung und volkswirtschaftliche Bedeutung der schweizerischen Teerfarbenindustrie* (1922), pp. 124–5.

[5] Du Pont, for example, introduced a fortnight's holiday with pay in 1934.

[6] Farbenfabriken Bayer. *Entwicklung und heutiger Stand der Firma und ihrer Wolfahrtseinrichtungen* (May 1914 edn.), p. 29.

credited to a savings account and given to the employee a year after he had left the job or, on death, to the dependants.[1] Brunner, Mond introduced a similar bonus scheme, equivalent to $2\frac{1}{2}$ per cent of pay, in 1907, but six years later it was changed to a long-service supplement of 1 shilling a week after two years' service, 2 shillings after four, and a maximum of 3 shillings after six.[2] There were also a few instances of profit sharing in the chemical industry. On the Continent S. A. des Matières Colorantes & Produits Chimiques de St. Denis practised it at the turn of the century,[3] and about a generation later IG Farben combined profit sharing with long-service awards and related both to the dividend.[4] Dow Chemical started a profit-sharing scheme in 1913 and retained it until the 1930s. The General Chemical Co. considered the idea, but after the formation of Allied Chemical it was dropped.[5] Bonuses and other supplements were designed to reward the loyalty of the older men and those with a good record. That minority certainly benefited, but for the majority, the men with a year or two of service, what mattered was the hourly or weekly rate for the job and the take-home pay.

For Britain not enough material has survived to establish a trend. The Board of Trade used to publish details of changes in rates agreed collectively, that is in major industries that had a powerful union organization. But 'changes in the less organized trades escape notice', and this accounts for the omission of chemicals.[6] As far as one can tell, the wages of chemical workers tended towards the lower end of the scale. They were below the average of men's wages, which Bowley put at £1 10*s*. 9*d*. a week in 1906, and were probably close to the £1 6*s*. 4*d*. a week for workers in gas and water, though there was rather less job security in the chemical trades.[7]

For the United States there is rather more information, and the figures in Table 13.4 over afford a comparison with other rapidly growing industries. The automobile workers enjoyed a higher income, in part due to their skill as coachbuilders and mechanics, but their wages increased barely faster than those of the chemical workers. Wages in electrical engineering were quite appreciably lower than in the other two industries and increased far less, probably because the average was depressed by the lower pay of the women workers.

At Dow's works in Midland, wages for factory hands jumped from \$1 a day about 1900 to \$2.50 (say 10 shillings) in 1912; his craftsmen and engineers got about a third more. The increase owed something to the nearby automobile factories at Detroit and Flint and attempts to stop the men from

[1] Wassilieff, op. cit., pp. 32–3. There was also a small cash payment at Christmas.

[2] Irvine, op. cit., p. 8; *The first 50 years of Brunner, Mond*, op. cit. The long-service award scheme was amended in 1920 and in 1937.

[3] *Paris exhibition* (2), 25–6.

[4] ter Meer, F. *Die IG Farben* (1953), p. 79.

[5] *The General Chemical Co. after 20 years* (1919), p. 73. I am indebted to Dr. W. E. Bennett for the information on Dow.

[6] Board of Trade, 17th Annual Report on Changes in Rates of Wages and Hours of Labour in the U.K., HMSO (1910), Cd 5324, pp. 3–4.

[7] Bowley, A. L. *Wages and income in the U.K. since 1860* (1937), p. 30. The 'wage censuses' of 1886 and 1906 omitted chemical workers.

TABLE 13.4

Annual average wages in selected U.S. industries, 1899–1914[1]

	1899	1904	1909	1914	*Increase* 1899–1914 (%)
		($)			
Chemical manufacture	483	544	594	683	41
Automobile industry	589	603	647	844	43
Electrical machinery	490	527	566	625	28

moving there.[2] Elsewhere rates were somewhat lower. Hancock mentions that the average weekly pay in Delaware was \$9–10 in 1905—a little above Du Pont's 'yard hands' and equal to the pay of the more skilled 'powder men'. Coachbuilders, printers, and mechanics in Delaware were paid above the average.[3] A German chemist, familiar with the American industry, thought that \$1.50–\$2.00 for a ten-hour day fairly represented the range of a day worker's pay before the war.[4] At the Syracuse plant of the Solvay Process Co. there was little change in the pay of the day men during the 1900s, but the lime-kiln shift men had their pay raised from 17 cents to 20 cents an hour between 1901 and 1911, and their mates on the soda shift improved theirs from 20 to 25 cents.[5] In general shift operators and skilled men did rather better than other workers in the pre-war chemical industry.

The wages of chemical workers in France and Belgium remained unchanged before the war. The former were paid 3.70 fr. a day in the early 1890s and 3.95 fr. (say 3 shillings) in 1911. At the latter date women received two-thirds of the men's average pay. By comparison the average for industrial workers in the Paris area was 4.50 fr. a day in 1900, but only 3 fr. a day in the rest of the country.[6] In Belgium the usual wage of chemical workers was 3 fr. a day, but craftsmen could make up to 5 fr.[7] The low pay, absolutely and also relative to other industries, reflected in part the low productivity of the acid and fertilizer trades, which made up the bulk of the industry. The lack of organization among the workers was probably a contributory factor.

In the chemical works of Basle a short strike in 1897 and an energetic union organizer brought about some changes that took the minimum wage to 18–23 fr. a week, and after three years' service to 23–27 fr. (say 18–22 shillings) a week. Hoffmann–La Roche, the pharmaceutical manufacturers, were at the top end of the scale and also paid extra for overtime, whereas CIBA and Geigy made no allowance for overtime. By 1913 the average weekly

[1] Brissenden, op. cit., pp. 377 ff.
[2] Information from Dr. W. E. Bennett.
[3] Op. cit., p. 52. Some of the Du Pont men had free or at any rate cheap houses.
[4] Frerichs, F. W. 32 Jahre in the Vereinigten Staaten als Chemiker und Fabrikant. *Z. angew. Chem.* **26** (1913), no. 2, pt. 1, 21.
[5] Information from Mr. R. F. Hansen of Allied Chemical.
[6] Matagrin, op. cit., p. 363. One franc was then worth about 9½ pence.
[7] Ministère de l'Industrie et du Travail. *Fabrication des produits chimiques* (1905), pp. 12–13.

pay of the dye workers ranged from 26 to 31 fr. a week, and craftsmen got 4–5 fr. a week more.[1]

The wages of the German chemical workers rose appreciably, so reflecting the industry's prosperity and the need for attracting and retaining the men. The returns published by the Berufsgenossenschaft for the chemical and allied trades showed a rise from about 900 M (£45) a year in 1895 to 1000 M five years later and to an average of 1266 M (£63) in 1913.[2] There were significant regional differences between the various branches of the industry. For example, in 1892, when the average daily wage for chemical workers in the Reich was 2·92 M, those employed in the Frankfurt area averaged 3·02 M and those in Rhineland–Westphalia 3·11 M. The men in the dyestuffs industry formed the élite, and their national average of 3·11 M a day was some way above that of the fertilizer workers, whose daily average was only 2·84 M.[3] The larger firms paid more and raised wages more rapidly. The average pay at Badische rose from 1020 M (£51) a year in 1895 to 1200 M in 1904 and to 1683 (£84) in 1913.[4] There were also considerable variations in the wages of the general workers according to the length of their service, and between them and the craftsmen. In October 1913 the latter earned on average 34·81 M a week at Leverkusen and their mates 30·71 M, while the adult plant worker made 30·64 M and women exactly half that pay.[5] The wages of the German chemical workers were good, but they were not the highest paid employees. According to an estimate by IG Farben, the workers in the iron and steel industry averaged almost 35 M (£1. 15*s*.) a week on the eve of the war, the mechanical engineers 31 M, and the men in chemical trades, on a somewhat different definition from the Berufsgenossenschaft, 29 M (almost £1. 10*s*.).[6]

The war and post-war inflation raised the pay of workers in the British chemical trades. The increase, in monetary terms, was appreciable: on the eve of the war day men got about 23 shillings for a week of 53–54 hours, but in June 1925 they were paid 49 shillings for a 47-hour week. Process men on a seven-shift week had their wages increased from 35 to 72 shillings, and after the war the weekend shift was at time and a half.[7] Two points should be noted. Firstly, the increases of about 100 per cent were amongst the highest awarded in Britain, and matched those secured by workers in the boot and shoe trade and in road transport, and were exceeded only by the printers and the railwaymen.[8] This is the more remarkable if one bears in mind that union activity was not particularly effective among the men, and that some of the

[1] Jaquet, op. cit., pp. 122–3; Wassilieff, op. cit., pp. 18 ff.; Baumgartner, op. cit., p. 78. The wages exclude the bonuses to which reference has been made above.

[2] Witt, O. N. *Die chemische Industrie des deutschen Reiches* (1906), p. 18; *Enquete*, op. cit., p. 56.

[3] Wenzel, op. cit., pp. 866–7.

[4] Voigtländer-Tetzner, op. cit., vol. 1, pp. 121a, 128 f. The particulars for Badische have been obtained by dividing total wages by the number of workers. It is not clear whether bonuses are included.

[5] Duisberg, op. cit., p. 19.

[6] The information was compiled by the company's staff and was found among the company's papers in Frankfurt after the war. The source of the data was not given.

[7] Committee on Industry and Trade. *Industrial Relations*, HMSO (1926), p. 83. The figures for 1925 are those recommended by the Chemical Employers' Federation.

[8] ibid., p. 90.

industry's sectors were not prosperous after the war. Secondly, price increases eroded a substantial portion of the gains. The cost-of-living index rose by about 70 per cent between 1913 and 1925[1] so that the improvement in real wages was relatively small. But for the day men there was also a significant reduction in hours, which, if taken into account, reflected an appreciable improvement. A few years later the *New survey of London life and labour* estimated that the real earnings of the men in the general chemical trades were about 10 per cent higher than in 1900.[2] There seems to have been no further increase in wage rates in the second half of the 1920s. Indeed neither the British Dyestuffs Corporation nor the United Alkali Co. were in a position to pay more, but, as business improved, unemployment fell, and longer hours were worked, earnings rose.

Developments in the United States broadly resembled those in Britain.[3] An extensive enquiry by the National Industrial Conference Board showed that chemical workers increased their average weekly earnings in monetary terms by 100–110 per cent between July 1914 and 1925/6.[4] Higher prices, as in Britain, reduced the benefit: the Bureau of Labor Statistics consumer price index rose by 77 per cent between 1913 and 1925, and then fell back slightly, so that in the late 1920s the earnings in Table 13.5 would be about a fifth

TABLE 13.5

Average weekly earnings in selected U.S. industries ($)[5]

		1914 *July*	1925	1927	1929
Chemical industry	Skilled and semi-skilled	15·15	31·78	30·11	31·04
	Unskilled	12·48	26·53	28·81	29·02
Automobile manufacture	Skilled and semi-skilled	16·30	32·76	33·05	33·96
	Unskilled	11·17	25·42	27·64	28·73
Electrical engineering	Skilled and semi-skilled	14·78	31·55	31·84	33·89
	Unskilled	11·47	21·96	22·53	23·02
All industry	Skilled and semi-skilled	14·16	30·89	31·14	32·58
	Unskilled	10·71	23·74	24·40	25·49

higher in real terms than before the war.[6] After 1925 there was little change in the earnings of semi-skilled and skilled men in the chemical industry, but the unskilled workers raised their earnings by about a tenth. Although

[1] Mitchell, B. R. and Deane, P. *Abstract of British historical statistics* (1962), pp. 344–5. The index numbers (1914 = 100) are 102 and 175 respectively.

[2] Op. cit., p. 202. No allowance is made for shorter hours.

[3] The short post-war boom reached its peak as far as chemical workers' earnings were concerned in summer 1920. A year later earnings were down by a fifth. This was the low point for the 1920s. Browne, H. F. *Wages, hours and employment in American manufacturing industry 1914–1924* (1924), p. 80.

[4] Beney, op. cit., pp. 46–7. The sample covered 1500 establishments and 866 000 workers in every industry, predominantly in the Eastern States.

[5] Beney, op. cit., pp. 46–7.

[6] The Consumer Price Index of the BLS 1947–9 = 100 was 42·3 in 1913, 75·0 in 1925, and 73 n 1928 as well as 1929. U.S. Dept. of Commerce. *Historical statistics of the U.S.* (1961), p. 126.

chemical workers as a group did quite well by comparison with 1914, their position deteriorated slightly by comparison with skilled workers in electrical engineering and with earnings in manufacturing industry generally.

The cash earnings of the German chemical workers increased just as steeply. From the pre-war level of 1266 M (£63) a year, they rose to 1825 RM in 1925 and 2539 RM (about £127) a year in 1927.[1] The striking improvement in the second half of the 1920s owed less to the increase in time rates (which were small after 1927) than to longer hours and overtime. Regional differences, once significant, mattered little after the war because the pattern of production in central Germany hardly differed from that of the older centres in the Rhineland.[2] The skilled men in the industry did well after the re-establishment of the currency in 1923/4, and in real terms earned 19 per cent more in 1928 than in 1913/14. The unskilled men whose hourly rates were about 15 per cent lower, improved their real earnings compared with pre-war by nearly the same percentage. In June 1928 the unskilled men in the dyestuffs industry earned 49·50 RM (say £2. 10*s*.) a week, 10 RM more than electrical engineers working in Berlin.[3] It would appear that chemical workers improved their pay relative to other industrial workers, and in a large measure this was due to IG Farben's commercial policy, which ensured a high level of activity and so of employment and income.

Trade-union activities in the British and American chemical industries can be quickly described. After the demise of the Chemical and Copper Workers' Union in the 1890s, there was no special organization for the British chemical workers before the war.[4] After the war the Transport and General Workers' Union (TGWU), the Municipal Workers' Union, and to a lesser extent the Union of Shop, Distributive, and Allied Workers (USDAW) were active among the men. The first named organized the workers in the factories making dyestuffs in the London area just before the General Strike.[5] The big unions were challenged by the National Union of Drug and Chemical Workers, formed in 1920 by the amalgamation of two older unions that had started signing up people working for wholesale and retail chemists in London just before and during the war. Towards the end of 1922 there were about 2500 men on its books. The left-wing politics of the Drug and Chemical Workers caused a breach with the other unions and it left the T.U.C. From that outside position it fought the T.U.C. and rival unions over poaching, but lost members and restricted its work to London. In 1936 the name was changed to Chemical Workers' Union.[6]

[1] *Enquete*, p. 56. The earnings are the average given by the Berufsgenossenschaft for the chemical and allied trades.

[2] *Enquete*, p. 484. (Evidence of Haupt of the Fabrikarbeiterverband Deutschlands.)

[3] ibid., pp. 59–60; Statistisches Reichsamt. *Deutsche Wirtschaftskunde* (1930), p. 280.

[4] See my *Chemical industry during the nineteenth century* (1958), pp. 246–7, on this.

[5] *New survey of London*, op. cit., p. 202.

[6] It now has about 16 000 members. Lerner, S. W. *Breakaway unions and the small Trade Union* (1961), pp. 14 ff.

Joint Industrial Councils with powers to fix minimum, legally enforceable wages were set up towards the end of the war, members being drawn from among the employers and the big unions. Heavy Chemicals, and Drugs and Fine Chemicals were the first two sectors of the industry to have such Councils.[1] With the formation of I.C.I. the situation was completely transformed. The new company employed a substantial portion of the industry's labour force, but tended to make its own settlements with the unions on terms that were better than those of the other employers, let alone the Joint Industrial Councils. I.C.I. did not prevent unions from canvassing members, but sought to head them off by instituting Works Councils: the first delegate conference was held in 1929 and was attended by 800 men.[2] The Works Councils were more than mere sounding boards, and at plant level played a useful role in harmonizing labour relations. At a much higher level this was also the aim of the well-publicized, but ineffectual and short-lived Mond–Turner meetings.[3]

So far as one can tell the men in the American chemical industry did not join unions.[4] Some companies had works unions. The General Chemical Co. took matters rather further and established an Employee Association, which ran the benefit club, handled safety and vocational training, and also had a say in the conditions of work. The company paid the union's expenses and the employees' representatives met once a year as a sort of Works Council.[5] There are reports of strikes in chemical factories, but there was no proper union machinery to lead and support the strikers. Soon after the war the Monsanto Chemical Co. dealt with a strike by employing strike breakers who managed to keep production going.[6] The American chemical industry was systematically organized in the 1930s by the Congress of Industrial Organisation and the United Mine Workers of America.

Labour relations in the industry on the Continent were often acrimonious and the unions active and belligerent. The conflicts took place in a sector of the economy that usually paid well and was known for the scope of its welfare capitalism. In Switzerland such troublesome occurrences were rare, though even there a union—the Fachverein chemischer Arbeiter—was established by the chemical workers of Basle in 1896. It came under the direction of Dr. Wassilieff, an unlikely Swiss, who not only got his men out on strike, but also obtained some improvement in their wages. Then interest flagged and the union was dissolved in 1910. A little later the Basle branch of the Swiss textile workers negotiated on behalf of the chemical workers. This arrangement proved satisfactory to all parties and between the wars there was no strike

[1] I am indebted to Mr. H. Weate of USDAW and Lord Cooper of the General and Municipa Workers' Union for this information.

[2] Statement made at Annual General Meeting of I.C.I. on 18 April 1929.

[3] See p. 345.

[4] According to Brissenden, who does not give the source of his information, 1½ per cent of the employees were union members in 1910, and 0·2 per cent 10 years later.

[5] General Chemical Co., op. cit., p. 72.

[6] I am indebted to Mr. Benjamin Thomas of that company for the information.

in the industry.[1] Several unions were active in the French chemical industry at the turn of the century. They were small and of local significance, but it has been stated that a fifth of the men were members.[2] There was certainly a good deal of union activity, because the Bayer history mentions three strikes at the Flers works in 1891, 1895, and 1904; the third lasted six days, the troops were called in and it collapsed.[3] Strikes in the chemical industry were also reported in Austria (the worst year being 1909) and Russia, where once again Bayer was struck. In both countries better conditions and wages were demanded, but the Austrian workers were less successful than the Russian, who got an increase in 1907.[4]

Germany witnessed the most violent struggles. Trade-union membership grew phenomenally from about a quarter million in 1895 to about a million in 1902 and three million seven years later. Most belonged to the Verband der Fabrikarbeiter, the Socialist union. (The Christian, generally Catholic, unions were strong among the miners and the Hirsch–Duncker unions were active among the clerks and metal workers.) During the 1900s the white-collar unions recruited vigorously and by 1913 had enrolled about 800 000 members.[5] The first objective of the leaders was to obtain recognition, the second to negotiate collective agreements. The obstacles were considerable: the chemical companies were large, prosperous, concerted their labour policy, and, not least, supplemented social legislation by extensive health and housing schemes. They also paid well. But it was precisely the amenities that caused much resentment, because the men, having no say in these matters, suspected that such capitalist benevolence was a substitute for higher wages. And on the question of wages there was more friction. The hours were long, for trade was good, but the men did not share in the prosperity, because though overtime was compulsory it was generally paid at ordinary rates.

Matters came to a head at Ludwigshafen in about 1894.[6] A union branch had been formed and sought to obtain representation for the men on questions concerning the long hours and wages. The company was sympathetic on the former and offered to introduce an extra shift with a corresponding loss of pay. The union protested, but was insufficiently strong to take action and Badische settled directly with the men.[7] The next major incident took place at Leverkusen about 1904 when the metal workers from nearby Solingen began

[1] Bürgin, A. *Geschichte des Geigy Unternehmens von 1758 bis 1939* (1958), p. 172; Neumann, H. and Weckerle, E. (Eds.) *Papierfabrikation, chemische Industrie* (1947), p. 114; Jaquet, op. cit., p. 125.

[2] Matagrin, op. cit., pp. 442–3.

[3] Burgdorf in *Geschichte . . . Bayer*, op. cit., p. 262.

[4] Glaser, J. *Die chemische Industrie Oesterreichs und ihre Entwicklung* (1918), p. 207; Blank and Löw in *Geschichte . . . Bayer*, op. cit., p. 269.

[5] Clapham, J. H. *The economic development of France and Germany* (4th edn. 1936), p. 329; Voigtländer-Tetzner, op. cit., vol. 3, pp. 621–2.

[6] The first recorded strike in the industry seems to have occurred at the Elberfeld works of Bayer in 1889. It was for higher wages and failed. The Leverkusen plant did not exist at the time. Bonhoeffer in *Geschichte . . . Bayer*, op. cit., p. 541.

[7] Voigtländer-Tetzner, op. cit., vol. 2, pp. 403–5.

to enrol the fitters at the Bayer works.[1] The men then approached the management to obtain recognition and discuss their wage claim. They were rebuffed and struck work. Bayer announced that only 400 men were out, but the unions claimed that 2000 out of 3100 craftsmen were on strike. The company hired blacklegs and set up a works police. After some six weeks the strike petered out, and as the company refused to negotiate, the state arbitration tribunal (Gewerbegericht) could not intervene. Initially no concessions were made, but in November 1904 overtime rates 40 per cent above the basic rates of pay were introduced.[2]

Despite the unions' failure there were several other short and limited strikes. In 1905/6 it was the turn of Badische where in 1905 only 5 per cent of the men were union members. Some of them were employed in the company's gas works. They demanded higher wages and were sacked. The following January the men on the acid plant also demanded higher wages. When their leaders were sacked, the rest struck. Strike breakers were called in and with their help the staff were able to keep acid production going. Badische refused to take back the strikers and this attitude caused real trouble, because the men in the indigo section of the works now struck in sympathy. By the end of January 1906 about 800 men were out, at a time when stocks were low and demand brisk. The management now decided to negotiate, recognized the unions, re-engaged the strikers, awarded a 10 per cent increase, and agreed on overtime and holiday rates. The settlement marked the high point of the unions' achievement at Badische before the war. In 1907 they failed to secure a collective bargain and the company retained the initiative by raising wages in November of that year and again in March 1910. There was more trouble in 1911, this time over working hours, and in July the company was hit by a strike of about 2000 unskilled men and general labourers in the dispatch department and the warehouses. This strike lasted for a fortnight and then petered out of for want of support among the skilled hands. No concessions were made by the firm.[3]

These manifestations of deteriorating labour relations were due not only to the vigour of the unions seeking membership and recognition in a hitherto non-unionized industry, but also to the incompetence of the management. In this respect the smaller companies could and sometimes did adapt more quickly to changing conditions: DEGUSSA of Frankfurt created a special committee to consider wages in 1896 and in 1908 started a works council, elected by the men, one of whose functions was to advise on conditions of employment.[4] These two bodies were useful in that grievances could be aired and solutions discussed formally and regularly. Bayer and Badische tried a different approach. The former established a works council to handle safety,

[1] They belonged to the Metallarbeiterverband and a Hirsch-Duncker 'Gewerkverein'.

[2] Grabendörfer in *Geschichte . . . Bayer*, op. cit., pp. 554–6. On this occasion some of the men at the Elberfeld works struck briefly in sympathy.

[3] Voigtländer-Tetzner, op. cit., pp. 423 ff.

[4] DEGUSSA 1873–1923, art. cit., p. 103.

welfare and complaints. The directors also appointed a Sozialsekretär whose advisory duties were as extensive as those of the Citizens' Advice Bureaux in contemporary Britain. This functionary was also *ex officio* chairman of the numerous societies where men of the same trade got together to improve their skills and, as the company put it, 'develop a comradely spirit'. There were altogether 18 of these societies—one for the railwaymen, another for the sulphuric acid workers, a third for the men of the intermediates plant, a fourth for the laboratory staff and so on, and it was not long before trade unionists called the members 'yellow' or 'company' unionists. Bayer achieved little by forming these groups and may have harmed itself by sponsoring their activities and openly favouring the members, so that the 'comradely spirit' was only for the comrades.[1] As if this were not enough to wean the men from the union and tie them to the firm, the company introduced staff status some years before the war: those qualified and selected became entitled to a week's notice, were paid by the week, and secured longer holidays as well as larger sick pay.[2]

Badische reacted even more aggressively against the unions by launching a fully-fledged company union in 1911. The members of the Arbeiterverein enjoyed such useful privileges as buying food cheaply from the company and using the club house with its extensive facilities. The Verein prospered and its membership grew from 800 to between 4000 and 5000 in 1914 when it accounted for roughly half the employees. The various trade unions spoke for 30 per cent of the workers and the remainder were not connected with any union.[3]

At Hoechst they refused to recognize unions altogether. G. von Brüning, the company's chief executive from 1901 to 1913, was an entrepreneur of the old school. In 1911, when Badische was in some trouble, he caused a notice to be put up that stated that his company would not be in the least influenced by the demands of the unions, but would continue to take account of the 'justified desires of our workers if brought directly to our attention'.[4] This splendidly regal message was duly noted, for as far as I know Hoechst remained free from strikes.

At the end of the war such old-fashioned attitudes were swept away. Works councils were established by law in Germany in 1916 and the company unions were dissolved in November 1918. The following year Badische had to abandon piece-work rates, yielded to a strike of its clerks, and even had to submit to the censure of the arbitration tribunal over a lock-out at Oppau.[5]

[1] Duisberg, op. cit., pp. 39 ff.

[2] Bayer, *Entwicklung . . . Wohlfahrtseinrichtungen*, op. cit., pp. 37, 46.

[3] Voigtländer-Tetzner, op. cit., vol. 2, pp. 435–6, 448.

[4] Brüning's note said *inter alia*: '. . . wir erinnern daran dass wir gerechten Wünschen unserer Arbeiter, wenn diese uns direkt vorgebracht werden stets Rechnung getragen haben und Rechnung tragen werden.' Fischer, E. Die Gründer der Farbwerke Hoechst AG. *Nassauische Lebensbilder* 6 (1961), 260.

[5] Voigtländer-Tetzner, vol. 2, pp. 453–6; vol. 3, pp. 621 ff. The Allgemeine freie Angestelltenbund (AfA), a white-collar union, had almost 3500 members at Badische at that time.

From the settlement of legitimate grievances and the abolition of quasi-feudal rules and regulations, labour relations deteriorated to a state of intermittent anarchy that reflected the intensity of the wider struggle of the left-wing parties to gain control of the German economy and of its political apparatus. There were frequent strikes at Leuna and Ludwigshafen between 1919 and 1922, and at the former site the troops had to be called in. Even Hoechst had its share of troubles: threatened by a strike the management declared a lock-out, only to find that the town council of Hoechst sided with the men.[1] Strikes were also reported from Leverkusen in August 1920 and again in February 1921; on the second occasion the men claimed that the management was abandoning the 8-hour day safeguarded by statute since 1918. There was also a lock-out of 4 days in mid-August 1923.[2]

When the new Mark was introduced at the end of 1923 and business suddenly contracted, employers judged that it was time to enforce discipline, the more so as public opinion was being antagonized by the strong-arm tactics of some unions. The show-down came in March 1924 when the Communists called out the men at Ludwigshafen. The strike was over the longer working week since the abandonment of the 8-hour day and the piece-work rates, which had been reintroduced during the inflation. The strike lasted 9 weeks, was violent, and finally collapsed for lack of funds.[3] Although the unions had been defeated, IG Farben, unlike some of its predecessors, decided to work with them. The new concern accepted collective bargaining, but successfully insisted on tightening up production schedules, which significantly raised the work load.[4] The men lost ground to the extent that they had to accept the tough piece-work schedules and the payment of production bonuses at the employer's direction. At any rate there was little fight left in them: in 1926 no strikes were recorded in the German chemical industry, the following year there were eight involving the loss of 27 000 working days, but in 1928 only 1800 days were lost. In a quiet pre-war year, 1905, there were nine strikes, involving 19 factories and altogether 1002 workers, and although the days lost were not recorded we can safely assume that the figure was higher than in the late 1920s.[5]

In the first thirty years of the century the layer of commercial and technical staff between management and men became thicker. Despite their increasing importance we know very little about the white-collar workers. Their jobs were more secure, holidays longer, and amenities more agreeable than those of the workers. The commercial staff got the same pay as clerks and salesmen in other industries, and from what little information has survived it would

[1] Wickel, H. *IG Deutschland* (1932), pp. 95–6. Wickel, a left-wing journalist omitted to mention the outcome.

[2] Beckerath, op. cit., p. 406.

[3] Voigtländer-Tetzner, op. cit., vol. 2, pp. 466 ff.

[4] *Enquete*, pp. 484, 494, 493 (evidence of union leaders Haupt, Roth, and Sparre).

[5] ibid., p. 54; Grossmann, op. cit., p. 61.

appear that the pay of the counting-house clerk was probably not very different from that of the experienced plant worker.[1]

On chemists there is rather more information. In pre-war Germany they did well: they were taken on at 3000 M (£150) a year, and after a year's probation were given long-term contracts and salary increments that enabled the majority to look forward to 5000 M (£250) a year.[2] Five thousand Marks was about three times the average worker's income at Badische and four times the pre-war average of the Berufsgenossenschaft members. In the United States the differentials were about the same: Frerichs reported[3] that qualified chemists started at $720–900 a year (about £150–180) and reached $2000 (just over £400) after 8–10 years. That was the peak for research chemists, but chemists in charge of manufacturing operations could and did obtain more. In Britain the salary of assistant chemists ranged from £125 to £175, and older men could make up to £300 a year.

From the limited evidence available it appears that in the United States chemists were able to command much higher salaries after the war. In the early 1920s, the minimum salary of research chemists was $1500, rising to about $5000 a year.[4] The U.S. Tariff Commission made sample surveys of salaries between 1924 and 1928, which showed that the average was gradually rising. Towards the end of the period a third of the chemists in the sample were earning over $75 a week (say £800 a year). No corresponding British surveys exist, but it is likely that £250 a year would represent a broad average in the 1920s. French chemists were also pressing for higher pay and towards that end formed the Syndicat Professional des Ingenieurs Chimistes Français in 1919. They were unable to obtain recognition from the employers, but were able to negotiate an appreciable increase in salary. A census taken by the Syndicat showed that 54 per cent of the members got salaries above 12 000 fr. a year in 1920, and a year later the proportion was 69 per cent.[5]

The absence of comprehensive salary data in the 1920s is particularly unfortunate because it prevents a comparison of the earnings of professionally qualified manpower. Such a comparison would shed useful light on the career choice and rewards, and so on the staff structure of chemical companies.

Chemical manufacturers, though not generous employers, could afford supplementary benefits. Such welfare capitalism was sometimes far-reaching and contributed to the impression of a well-meaning and indeed socially

[1] Before the war Chance & Hunt paid their junior clerks 15 shillings a week and gave them free lunch. Pratt, P. C. The not-so-good old days. *I.C.I. Magazine*, September 1958, p. 310. According to Matagrin (op. cit., p. 310), clerks earned from 150 to 300 fr. (£6 to £12) a month; chemical workers earned about 120 fr. a month.

[2] ABCM. *Report of the British mission on the chemical factories in . . . Germany* (1919), p. 33.

[3] Art. cit., p. 22.

[4] Bureau of Vocational Information. *Women in chemistry* (New York 1922), p. 233. At $3.80 to the £ (the average for 1921) the sterling equivalents were 400 and 1300 respectively.

[5] International Labour Office. *Engineers and chemists: status and employment in industry* (1924). ILO, Series L, No. 1, pp. 68–9. Matagrin, op. cit., pp. 349–50. At the ruling rate of exchange 12 000 fr. was about £230.

progressive management, at a time when deteriorating labour relations somewhat tarnished this image. The directors' benevolence was often perfectly genuine. But sometimes it was handled clumsily and dispensed patronizingly, so having the opposite effect from that intended and leaving behind cynicism and distrust. Brunner, Mond, for example, made it a condition of employment from 1890 onwards that all juveniles up to the age of 18 attend evening classes. The intent was praiseworthy, but why force teenagers to spend part of their leisure in this prescribed manner? The rule was abandoned in 1905 and the youths given paid leave two afternoons a week.[1] The du Ponts were equally insensitive when they established the 'Brandywine Club' in 1892. This teetotal club, which required an admission fee, was supposed to encourage off-duty conviviality among the powder-men, many of whom were Irish. It collapsed for lack of support.[2] And what are we to think of Hooker Electrochemical's health services? They included a hospital and a visiting nurse 'who not only cares for sick employees, their wives and families, but investigates absenteeism and keeps the foreman informed'.[3] It is likely that the men would resent welfare combined with sneaking. Perhaps it was stopped, as other practices ended when no longer consonant with contemporary opinion.

Subsidized housing was certainly the most conspicuous manifestation of welfare capitalism. As a rule it benefited only a minority of the employees, but that minority assuredly lived better and more cheaply than the majority who rented from a private landlord.[4] Many of the terrace, semi-detached or duplex houses have survived and testify to advanced designs and high standards. A family house on the Helmshof estate owned by Badische consisted of two rooms, a kitchen, attic, cellar and a garden of 30′ × 36′ (approx. 120 sq. metres) and was let for 1.80 M a week. A slightly larger house cost 2.30 M a week, which, according to the company, was one-third of the rent charged by private landlords. At Hoechst the rents were about the same, and the larger houses—consisting of 3 rooms, double attic, kitchen, and cellar—were let for 4.20 M a week, which was equivalent to about 15 per cent of a skilled man's wages.[5]

Some companies were forced to turn builders because factories such as Leverkusen and Leuna were at some distance from the nearest towns. Within a few years hamlets became sizeable company towns. A striking example was Rjukan in Norway, which Norsk Hydro made the centre of its operations. It consisted of a few farmsteads in 1907, but within 16 years the company had built 1260 flats and houses there and also provided sites for schools and other

[1] *The first 50 years of Brunner, Mond & Co.*, op. cit. [2] Hancock, op. cit., pp. 180–1.

[3] Anon. *Hooker Electrochemical Co. after 25 years* (1929), p. 10.

[4] At Leverkusen 22 per cent of the workers lived in company houses in 1913 and 30 per cent seven years later. The proportion was a little lower at Hoechst and in the Cheshire factories of Brunner, Mond. At the Rjukan works well over half the labour force lived in company-owned properties owing to the lack of other accommodation.

[5] Grossmann, op. cit., p. 58. The housing estates of the German dyestuffs companies may be compared with Port Sunlight where rents (including rates) represented on average a quarter to a fifth of the weekly wage. Wilson, C. *The history of Unilever* (1954), vol. 1, p. 147.

needs.[1] Even in built-up areas it remained common practice for Continental manufacturers to build houses or more usually blocks of flats. Thus Geigy and CIBA owned property in Kleinbasel, and Kuhlmann rebuilt the damaged company houses at Loos, which had been in the battle zone. In 1925 the company owned, outright or together with non-profit making housing associations, about 900 flats and houses, on which the rents covered the outgoings, but yielded no return on capital.[2]

In Britain Brunner, Mond owned over 700 houses on the Winnington Estate in Cheshire by 1922,[3] and a few years later I.C.I. started a large building programme at Billingham. In Cheshire the company's houses were close to the alkali works and did not have the style that lifted some of the German housing well above average mediocrity. But the Winnington settlement was superior to contemporary working-class housing and met a real need, because council housing schemes were inadequate.

At the opening of the century American companies did not, with few exceptions, provide houses for their employees. Du Pont had built some for its powder men above the Brandywine mills, and the General Chemical Co., owing to the remote location of some of its factories, provided accommodation at Capelton and Sulphide in Canada and in the United States at Pulaski (Va.) and Bay Point (Cal.). Between 1916 and 1918 the company spent about $2 million on the Overlook Colony at Marcus Hook between Wilmington and Philadelphia and subsidized the rents.[4] Du Pont also provided emergency housing for the war-time munition works, and Dow Chemical put up many houses at Midland, which were valued at $250 000 in 1921.[5] So far as I know these housing schemes ended with the war and were not resumed later.

The company as landlord and the employee as tenant was inevitably a relationship engendering occasional friction. In Germany the courts refused to allow the companies to evict tenants who ceased to be employed without proper notice. Before the war, four weeks' notice was the rule at Leverkusen; after the war the tenant's rights were protected by statute. Similar arrangements existed elsewhere except in the United States.

Besides the 'off-site' benefits there were considerable 'on-site' amenities, which included baths, laundries, and subsidized canteen meals. The German works provided rooms with tables and benches where the men, as was customary, could eat the food brought by their families. In France, however, communal feeding was unpopular and the individualism of the men caused them to resist change and retain the long lunch break for eating at home or in a café.[6] American employers did nothing along these lines until the war, when they too began to provide baths and canteens. In the late 1920s, a survey by the National Industrial Conference Board found that in the Eastern States

[1] *Norsk Hydro history*, pp. 471–3.
[2] Anon. *Les Etablissements Kuhlmann 1825–1925* (1925), p. 68.
[3] *The first 50 years of Brunner, Mond & Co.*, op. cit.
[4] *The General Chemical Co.*, op. cit., pp. 75–7.
[5] Accounts of the Dow Chemical Co.
[6] Matagrin, op. cit., pp. 410, 421–2.

and as far west as Ohio and Michigan all firms with over 5000 employees had canteens. Of the 210 questionnaires completed by chemical manufacturers, 26 per cent had canteens. This proportion was only exceeded by manufacturers of rubber and of foodstuffs.[1]

A valuable fringe benefit was the provision of free medical services and help during sickness. Accident insurance became compulsory in most countries during the first quarter of the century, but there was no protection against sickness and the consequential loss of pay drove men to public assistance. The practice among Continental firms varied greatly. In Basle CIBA paid half wages for six weeks and Geigy for twelve; the latter also maintained a convalescent home. Hoffmann–La Roche was unique in paying full wages during the first fortnight of sickness.[2] Badische had a convalescent home and also a hospital for TB. The company did not pay wages during sickness, but ran a contributory sick-benefit club that provided assistance. St. Gobain was reported to be paying sick benefit and Matières Colorantes & Produits Chimiques de St. Denis paid full wages for up to three months. If true this was exceptionally generous even by the standards of the 1960s.[3]

In Anglo-Saxon countries illness forced the worker to eat into his savings and rely on the sick clubs. There was usually one in every factory, and in Widnes membership was compulsory in about half the works. Their payments ranged from 7 to 10 shillings a week for up to six months and thereafter at half that rate; even at its most generous the support represented less than half-pay.[4] Brunner, Mond's compulsory sick scheme dated back to 1873. Men paid 4*d*. a week, youths 2*d*. In return they received free medical attention and after an improvement in the scheme in 1897, 10*s*. weekly. Five years later the company took over the entire cost, but when the National Insurance Act of 1911 came into operation all company sick benefit, except in special hardship cases, ceased.[5] In the United States employers left sickness assistance to mutual societies or the churches. Hancock reports that the Roman Catholic church took care of the sick at the Brandywine powder mills, but the Du Pont family, who were the local gentry, often made *ex gratia* payments.[6] The General Chemical Co. gave its employees an annual medical check-up, and in 1910 the board considered the introduction of a company-run sickness and injury benefit scheme, but the project was not carried out.[7] The Solvay Process Co. and Semet–Solvay launched the Solvay Mutual Benefit Society in 1888, which owed something to the advanced social ideas of the Belgian founders. Each employee contributed 20–30 cents a month, which was matched by the companies. The Society paid benefit for up to six months,

[1] N.I.C.B. *Industrial lunch rooms* (1928), pp. 53–4.
[2] Wassilieff, op. cit., pp. 28 ff. This was at the turn of the century.
[3] *Paris Exhibition* (2), 25–6. The information was supplied by the companies named to Haller who compiled the report.
[4] Hardie, D. W. F. *A history of the chemical industry in Widnes* (1950), pp. 121–2.
[5] *The first 50 years of Brunner, Mond*, op. cit.
[6] Op. cit., pp. 176–7.
[7] *The General Chemical Co.*, op. cit., pp. 67 ff.

contributed to the cost of operations, and also granted death benefit.[1] With the exception of the Solvay Mutual Benefit Society, sick clubs were usually small mutual societies and their resources were strained to the limit when there was an unexpected call for help from many members. After the war some of the large American companies intervened directly by taking out group insurance on behalf of the men. (They may also have shouldered the premiums, but it is more likely that the men were called upon to pay their contributions.) This businesslike approach to a social need at a time when there was no state or federal help was certainly of some value. Du Pont adopted group insurance in 1919 and within four years 11 000 employees were in the scheme.[2] Union Carbide, in addition to free medical services, introduced a life and accident group insurance scheme in 1927.[3]

Many companies operated pension schemes. As they were usually non-contributory, few were generous and most covered only a small minority of the employees. They were meant to supplement private savings or the weekly payments under the German and later the British state old-age pensions for workers.[4] Thus Badische had only 149 pensioners at the end of 1921, although it was then employing many thousands of workers.[5] The Mond Pension Fund of Brunner, Mond & Co. was restricted to 30 beneficiaries until 1924.[6] The Oldbury Works Pension Fund, established in 1895 and financed by Albright & Wilson and the families of the founders, covered the majority of the workers, but its resources were small, so that the pension was initially only £1. 5*s*. 0*d*. a month for a married man.[7] The dyestuffs makers in Basle were exceptional in this respect, because their pensions, though at the discretion of the management, could reach up to half of the terminal wage. For example Geigy paid 600–1000 fr. (say £24–40) a year at 60 after 25 years' service, and CIBA's pensions were up to 40 fr. a week in the early 1930s.[8] Even where the companies had contributory pension schemes, the benefit was small. The pension fund of Etablissement Malétra paid 240 fr. a year (say £10) to men over 60.[9] The larger companies introduced more generous payments after the war, though even they were forced to make occasional subventions in order to balance the fund actuarily. Solvay Process in the United States and Norsk Hydro in Norway were instances of companies pursuing such a policy, but

[1] I am indebted to Mr. R. F. Hansen of Allied Chemical for this information.

[2] Annual Report for 1923.

[3] *The Carbidea*, November 1927, no page reference. It is likely, though neither the Du Pont report nor the Union Carbide journal mention it, that the premiums for group pensions were substantially below those quoted to individuals.

[4] Salaried employees were excluded from the state pensions in both countries. The German pensions began at 70 and ranged from 2 to 4 M a week (2 to 4 shillings) depending on the contributions, which were paid equally by employer and employee. The British old age pension also began at 70 and the initial rate was 5 shillings a week.

[5] The minimum qualifying period was 5 years. *Die BASF* (company book), undated (*c*. 1923), p. 130.

[6] Irvine, op. cit., p. 10.

[7] Threlfall, R. E. *100 years of phosphorus making* (1951), pp. 375–6.

[8] Wassilieff, op. cit., pp. 28–9; *CIBA 1884–1934* (brochure), p. 77.

[9] *Paris Exhibition* (1), 47.

neither they nor others exceeded half pay, and usually the pension ranged from a quarter to a third of the final wage.[1]

Staff pensions, at least in German-speaking countries, were positively generous. Some were non-contributory, none were transferable; they were in short a reward for long service and a deterrent to movement between companies. The BASF staff pension operated from 1888 to 1925, the employee contributing 4 per cent of his salary, which was matched by the company. Maximum entitlement was reached after 40 years' service, when the employee would receive his full salary or 4500 M (£225) a year, whichever was the lower; in 1912 the limit was raised to 5000 M.[2] CIBA's fund was started in 1891, and like that of BASF paid the full salary after 40 years' service. Widows got 60 per cent of their husband's pension and CIBA's foremen became members of the staff pension fund in 1918.[3] Norsk Hydro, however, had a non-contributory staff pension scheme; it was launched in the financial year 1912/13 and the maximum entitlement was 60 per cent of the leaving salary.[4] Outside Europe, staff pension schemes were infrequent, though it appears that Du Pont and some of Solvay's licensees had them. Staff in the United States chemical industry tended to be mobile, and their length of service depended more on their current pay and career prospects than on the expectation of a comfortable retirement.

Although the documentation on pensions and other benefits is inadequate t points in the same direction, and one can legitimately generalize from what has been said above. In Europe all large companies and many of the medium-sized ones had taken steps by the 1920s to provide their employees with some fringe benefits. 'On-site' benefits contributed marginally but still usefully to the improvement in working conditions of all employees. 'Off-site' benefits, where they existed in the form of sick pay, discretionary pensions, and cheap housing, were valuable and certainly helped a minority among the chemical workers and so distinguished them from those in other trades who did not have these advantages. In a few cases details are available of the cost of the company welfare services. At Badische, for example, in the period 1905–13 the number of workers and foremen increased by 44 per cent and the wage payments doubled to 17·6 million M (say £0·9 million). The company's payments to the state social security schemes rose from 267 000 M to 609 000 M (£30 000), but the cost of the various welfare services jumped from 0·9 million M to 2·9 million M (£45 000 to £145 000) between 1905 and 1913.[5] On the eve of the war the value of these benefits represented roughly £14 per worker and was equivalent to 16–17 per cent of their average annual pay. For Kuhlmann, incomplete data suggest that the company's expenditure on social services averaged about £10 a head in 1925.[6]

[1] Information from Mr. Hansen of Allied Chemical and *Norsk Hydro history*, pp. 475–6.
[2] Voigtländer-Tetzner, op. cit., vol. 3, pp. 589–90.
[3] CIBA brochure, op. cit., p. 77.
[4] *Norsk Hydro history*, pp. 475–6.
[5] Information from the company.
[6] Kuhlmann, op. cit., p. 68.

Practical philanthropy could be expensive. But it engendered some good-will for most of the time, and so helped to sustain the impression that chemical manufacturers were also good employers. Carefully nurtured it became part of the tradition of enlightened labour relations in the industry. The tradition survives to the present day and plays some role in the negotiations with the unions.

BIBLIOGRAPHICAL NOTE

THE literature on the chemical industry is large and the list of references in the bibliography contains more than 500 entries. As it is difficult to distinguish between the major and the minor sources, I have thought it useful to mention the principal works and indicate their qualities or failings. Hence this note. It should not, however, be considered a critical, let alone comprehensive, bibliography, but an annotated reading list.

For the sake of convenience I have divided the note into three sections: first, official publications; next books, pamphlets, and articles; and finally company histories and brochures. Names or abbreviations (for which see p. ix) in CAPITALS will be found under that entry in the bibliography.

Official publications

In the nineteenth century few official reports were specifically concerned with chemical manufacture, but the education commissions of Victorian England contained many references to the weaknesses of chemistry teaching. A generation later the picture was entirely different. Government involvement in industrial production during the Great War was accompanied by numerous enquiries and documents. The investigations were often very detailed and frequently provide important material on the chemical industry. As a 'key' industry it continued to attract a good deal of official attention throughout the 1920s.

At the outset two publications, dealing with the industry generally, require mention. The Paris Exhibition of 1900 included a large chemical section, and HALLER's report to the Jury for Class 87 is an invaluable review of the situation at the turn of the century. A quarter-century later the Preparatory Committee for the International Economic Conference called for studies on certain major industries, among them chemicals. The LEAGUE OF NATIONS published the memoranda submitted by five countries in 1927, and that contributed by Dr. C. Ungewitter for Germany contains many statistics on chemical production and trade.

For the United Kingdom the annual reports of the ALKALI INSPECTOR and the Censuses of Production for 1907 and 1924 carried out by the BOARD OF TRADE are the chief statistical sources. The BOARD OF TRADE was the sponsoring department for the chemical industry and some of its reports are extremely useful, notably those of the DEPARTMENTAL COMMITTEE ON THE POST-WAR POSITION OF THE SULPHURIC ACID AND FERTILISER TRADES, of the SUPERPHOSPHATE COMMITTEE, and of the DYESTUFFS INDUSTRY DEVELOPMENT COMMITTEE. The MINISTRY OF MUNITIONS published the detailed study of the NITROGEN PRODUCTS COMMITTEE, which reflects the careful attention then being paid to the different nitrogen fixation processes. Additional information, usually inadequate but invariably outspoken, will be found in the reports of the STANDING COMMITTEE ON TRUSTS. The high intellectual calibre of its members makes all the reports eminently readable; those on soap, explosives, dyestuffs, and salt are particularly relevant in this context. The COMMITTEE ON INDUSTRY AND TRADE under Mr. (later Sir) Arthur Balfour carried out a major inquiry in the late 1920s. The reports should be consulted for their treatment of various aspects of the British industrial scene; they contain many references to chemical manufacture and also to industrial research.

Anyone wishing to follow the development of the modern American chemical industry will have to look at the growth of organic chemicals. The *Censuses of dyes and other synthetic organic chemicals* published by the USTC (U.S. TARIFF COMMISSION) are the most comprehensive source of information on the subject. The first was taken in 1917. The scope of the censuses was enlarged as more firms entered the business, and was further extended to include particulars of employment and expenditure on research. The situation of the American makers and users of dyestuffs in the early part of the Great War is described by NORTON in a DEPARTMENT OF COMMERCE report. After that war, the protection of the U.S. dyestuffs industry became a live political issue. The HOUSE OF REPRESENTATIVES COMMITTEE ON WAYS AND MEANS HEARINGS in 1919 provided ample opportunity for all parties to air their views on the proposed legislation, which, with modifications, governs the tariff on organic chemicals to the present day.

For the inorganics sector the studies of the FEDERAL TRADE COMMISSION supplement the USTC, although they do not go into the same detail. Some of the TNEC reports refer to chemical manufacturers by name, but a much better insight into the affairs of the industry can be gained from the charges brought by the U.S. Department of Justice against Allied Chemical, I.C.I., and Du Pont. The complaint against Du Pont was filed on 30 September 1949, and the judge gave his verdict over five years later. The documents in the case—trial transcripts, oral depositions, and trial exhibits—fill 29 volumes and more than 20 000 pages, and represent the most valuable documentation on the growth of the company.

The state of the chemical industry in Germany and France in the 1920s was reviewed by DELAHANTY and REAGAN respectively in reports of the U.S. DEPARTMENT OF COMMERCE. The German ENQUETE devoted an entire volume of over 500 pages to the chemical industry. The report, while not controversial, is most informative, except on cartels and international agreements. The evidence submitted by the principal manufacturers (which takes up about four-fifths of the book) is interesting and shows clearly the dominant role of IG Farben.

As far as education, research, labour problems, and occupational diseases are concerned, few publications deal specifically with chemical aspects of these problems. On British higher education generally the reports of the UNIVERSITY GRANTS COMMITTEE are useful, and so in a narrower context is the DEPARTMENTAL COMMITTEE ON THE ROYAL COLLEGE OF SCIENCE (of which Haldane was chairman) and that of the ROYAL COMMISSION ON OXFORD AND CAMBRIDGE. The INTERNATIONAL LABOUR OFFICE was the first official body to look at the professional status of chemists: *Engineers and Chemists, Status and Employment* was published in 1924. It is not a comprehensive study, and unfortunately was not followed up. The DEUTSCHE HOCHSCHULVERWALTUNGEN first investigated systematically the supply of and the demand for chemistry graduates, and so was the forerunner of the now common scientific manpower surveys. Its report painted an alarming picture of unemployment or underemployment in the profession. Official surveys of industrial research facilities were made in the United States by the NATIONAL RESEARCH COUNCIL only in the 1920s, and they reveal how little was done at the time even by chemical manufacturers. The International Labour Office also concerned itself with occupational diseases, and its report on *Cancer of the Bladder*, which was then a serious health hazard in the manufacture of certain kinds of dyestuffs intermediates, was the outcome of careful studies over a period of many years.

Books, pamphlets, articles

Industrial history, ideally, should knit together technological and commercial developments and should aim to assess the industry's changing role in the economy. The attainment of these objectives depends critically on adequate and reliable statistics, which in turn presuppose the existence of a generally accepted definition of the industry. So far as I know international agreement on what constitutes chemical manufacture has only been reached within the last ten to fifteen years. Certainly such statistics as were collected in the 1900s and later were not on a basis that allows many comparisons between countries. Within each country the supply and quality of the data vary greatly. Nevertheless it is usually possible (provided one does not seek excessive accuracy) to illustrate the industry's growth by relying on a few 'indicators'—the production of sulphuric acid and of superphosphate, employment, and foreign trade in chemicals—for which the series go back to the turn of the century. These indicators ought to be revised from time to time in line with technological change, but this has not been possible, because the official statistics have never kept in step with the alterations in the industry's product mix.

Enough warning has been given of the uncertain and unsatisfactory statistical foundation. But this obstacle has not deterred those who decided to use the tools of contemporary economic analysis for their examination of the industry's development. They have not, in my opinion, proved their case, because the hypotheses and models, albeit theoretically sound, cannot be properly tested for lack of data. There are, however, several original and stimulating contributions, on some of which I have drawn. Among the pre-war authors are HEMPEL in the United States and HOPPMANN and WALLER in Germany. They describe, but do not quantify. Since the war, BAUMGARTNER in Switzerland has broken fresh ground: his monograph demonstrates with a wealth of statistics the position of the chemical industry in Basle between the wars, and of its kind is a pioneering work. More recently MARKOVITCH, HOFFMANN *et al.*, and NUTTER *et al.* have assembled historical production series for France, Germany, and Russia respectively. Although industrial chemicals play a relatively subordinate role in the compilations, the time series (subject to the reservations mentioned above) enable us to follow the course of the industry in the first thirty years of the century and beyond. HOHENBERG has extended the work and has looked at the factors of production, technical change, and price movements for chemicals in western Europe to 1914. His approach and his conclusions, despite many shortcomings, are interesting and should lead to further investigations.

STOCKING and WATKINS are the indispensable guides to world cartels in the 1920s and 1930s. Their chapters on nitrogen and on other aspects of the chemical industry, though written about 25 years ago, remain authoritative.

Histories of the chemical industry have been published in several countries. HAYNES's six volumes (of which 1 to 4 are particularly relevant to this period) are based on detailed research supplemented by interviews with the leading personalities. Regrettably there is nothing on the same scale elsewhere. MIALL and HARDIE and DAVIDSON PRATT have written on the British chemical industry in 1931 and 1966 respectively, but their books are neither comprehensive nor scholarly.

RICHARDSON's account of chemicals in ALDCROFT's collection of essays stops in 1914, but his article on dyestuffs takes the story to 1939 and is a good introduction to the subject.

Germany dominated the European chemical scene, but the diversity of its industry and the details of its growth appear to have daunted historians. The story has therefore to be pieced together from various sources. There is a short but useful account of dyestuffs in BEER; UNGEWITTER and RASSOW (*Die Chemische Industrie*), though both were well informed, wrote badly and their stories are incomplete. That criticism cannot be levelled at SCHULTZE, but his book ends in the 1900s and so covers only part of my period. A history of the German chemical industry in the first half of the century is badly needed. Meanwhile there is a separate literature on IG Farben and its predecessors. Except for TER MEER's apologia, the books are emotive and often pure fiction. The alphabetical list of authors reads BOYLAN, DUBOIS, SASULY, SCHENZINGER, and WICKEL, but calm appraisal of IG Farben's role in the world chemical industry will not be found in their works. It should be attempted before long, the more so as many of the documents survived the war and are now in American and British archives.

The French industry is better served: BAUD and MATAGRIN are clear and informative. The two monumental volumes (they run to 3000 pages and weigh 16 lb) edited by GERARD are well padded, but also contain much that is useful and not available elsewhere. An American study on post-war France by OGBURN and JAFFE has many references to the chemical industry.

Descriptive histories, often containing statistics, exist for several other countries. GLASER has done the job for Imperial Austria; the Republic hardly had a chemical industry. For Belgium the reader is referred to LUCION (pre-war) and LEPERSONNE (post-war). In Switzerland the chemical enterprises of Basle represent the bulk of that country's chemical industry. JAQUET has dealt thoroughly and most competently with their development up to about 1920; the company histories are mentioned in the next section. NEUMANN and WECKERLE have written a short account which takes in heavy chemicals as well as dyestuffs. PARRAVANO edited the contributions on the various sectors of the Italian chemical industry. The book appeared in 1938, and as it is almost the only source on post-war Italy its boastfulness and political overtones have to be tolerated. WARRINGTON and NICHOLLS describe the growth of the Canadian industry, but fail to illustrate the great changes in the scale of manufacture, presumably owing to the absence of official and other statistics. By contrast, this problem never worried those authors who chose Russia as their subject. It is not the welter of figures, but their credibility which causes some difficulties. I have drawn on DYES, FEDOTIJEV, and KATS among others. LUKIANOV's history ends with the nineteenth century, but occasionally spills over into the 1900s.

I have usually relied on contemporaneous sources for the technological aspects of the story. In some cases (notably the syntheses of ammonia and methanol) operating conditions have changed, and a description of current practice will be found in recent editions of the standard technical encyclopaedias. The evolution of industrial chemistry from the earliest days to the present is described by SHERWOOD TAYLOR. GAIT provides a short and clear guide through the complexities of petroleum chemicals, the origins of which go back to the 1920s.

The fixation of atmospheric nitrogen is well documented. HABER (*Fünf Vorträge*) and Bosch have described it in their respective Nobel prize lectures; Waeser, Grossmann, and Pollitt deal comprehensively with the techno-commercial aspects. For cyanamide specifically Franck is the chief authority. Lambert's statistics on fertilizers are detailed and valuable.

Higgins and Morton are prime sources on dyestuffs. A comprehensive history of plastics is still lacking; meanwhile Kaufman's little book represents an extremely useful introduction. Freeman's systematic and analytical treatment of plastics is excellent, but is relevant to post-1945 rather than to pre-1930. Two other authors call for mention, although their subjects are really outside the period covered by this book: Howard has written the history of synthetic rubber, Allen that of polythene and of 'Terylene' polyester fibre. The latter is of special interest owing to his description of the links between academic and applied chemistry and of the transition from the laboratory stage to commercial development.

The connexion between research, development, and large-scale production has frequently been discussed in recent years, Many discoveries and their commercial exploitation have been studied, but it is still too early to give a definitive verdict on the importance of the different factors that determine the speed and extent of the industrial utilization of a discovery. After the Great War, Government sponsored research became significant in some countries, but its effect on industry has not yet been assessed. In Britain the official accounts of the Research Associations and of the Department of Scientific and Industrial Research by Heath and Hetherington, and more recently by Melville, while reliable, are not critical.

In order to round off the story of the industry I have frequently drawn on the biographies and reminiscences of those connected with it. Germans figure prominently in the list because their industry played a particularly conspicuous role. Among chemists Emil Fischer and Willstatter are informative on the connexion between university and industry. Weizmann, Ipatieff, and Cottrell, whose biography was written by Cameron, illustrate the public role of chemists and their increasingly close involvement with Government. The lives of the British soda makers in the nineteenth century were extolled by Fenwick Allen (*Some founders of the chemical industry*, 2nd ed., 1907), and Haynes set out to do the same for their American rivals in *Chemical pioneers*. But in general the manufacturers were reticent or maybe just inarticulate. Bolitho's life of Alfred Mond (first Lord Melchett) is incomplete and inadequately describes that unhappy man and herald of industrial reorganization; Armstrong's short study of the Monds is altogether more revealing. Duisberg wrote his reminiscences when he was the Grand Old Man of the German chemical industry and had mellowed; his biography by Flechtner, written nearly thirty years later when the emotions had cooled, is useful. Holdermann's life of Bosch is sympathetic. The close links between Government and science in Imperial and Weimar Germany are observed through the eyes of a senior civil servant in Schmidt-Ott's autobiography.

An account specifically devoted to the changing relations between academic and industrial chemistry in the leading countries has still to be written. Observers noted that the two were moving apart, or at least were indifferent to each other's needs, and their proposals to remedy this state of affairs are relevant in the context of my book. Thus Frerich's report is an important source on the training of American chemical engineers. F. Fischer in Germany, Haller and after him Croiset in

France lamented the neglected state of chemical education. The situation improved in Germany, but in France, GRANDMOUGIN was still complaining during the Great War. There are several monographs on the principles and objectives of higher scientific education systems: FLEXNER's comparison of the United States, Britain, and Germany, though dated now, was extremely relevant at the time. GUGGENBÜHL's detailed history of that centre of scientific excellence, the Swiss Federal Polytechnic in Zurich, is a model of its kind.

The appropriateness of scientific education to the needs of the times has attracted the attention of social scientists. CARDWELL and COTGROVE have considered the subject in historical terms, while MACLEOD and ANDREWS have concerned themselves with a particular group of scientists. More work along these lines is bound to be of interest and value.

Labour relations in the chemical industry are insufficiently documented, and it is probable that a history of chemical workers can only be built up from information in the archives of the large firms. The numbers involved were not large enough, nor were the men sufficiently discontented to attract the attention of historians. However, OLIVER has described the dangers of the chemical trades, and LERNER and WASSILIEFF the attempts at unionizing the workers in Britain and Switzerland respectively. A minority benefited from extensive welfare services which are amply illustrated in many company histories.

Company histories and brochures

The eighty or so titles listed in this section are not a complete record, but a representative selection. I have classified them under the names of the company or group (shown in capitals) because this makes identification easy and avoids the problem posed by the many anonymous authors.

Generally speaking company histories do little credit to the management that commissioned them. The books or brochures are often badly put together and the contents blatantly slanted, so that it is difficult to distinguish them from public relations handouts. Occasionally splendid illustrations redeem, in part at least, the textual shortcomings—one such example is Bolle's two presentation volumes for SOLVAY's centenary. MONTECATINI's seventy-fifth anniversary was marked, *inter alia*, by the publication of an interesting illustrated history of Italy since the 1880s, with hardly a reference to the company, the country's largest chemical enterprise! All too few histories approach in quality Coleman's scholarly and numerate work on COURTAULDS.

BADISCHE, BAYER, and HOECHST celebrated their centenaries in the 1960s, and their *Festschriften* afford an interesting contrast. That of BADISCHE is a disappointment. Jacobi edited the BAYER history, which is often informative, but less revealing than the fifty-year history, a huge tome, of which (so far as I know) only a single copy is still available for consultation at Leverkusen. Bäumler has done a useful job for HOECHST. The chapter on the company's financial history was contributed by Muthesius and repays study. It has been reported that a quarter of a million copies were printed, which would suggest that there is publicity value even in business history. An English edition has recently become available.

In the United Kingdom, Joseph Crosfield & Sons, a subsidiary of UNILEVER, commissioned A. E. Musson to write their history. Part 3 of his book contains an extremely clear account of the relations between soap- and alkali-makers in the

1900s and 1910s. A history of I.C.I. and of its forerunners has been needed for many years, and the first volume of Reader's book takes the story to about 1927. It contains a great deal of useful information and many statistics, and fills a gap in our knowledge of the development of the British industry.[1]

Of the numerous American company publications, I will single out Dutton's history of Du Pont, as being characteristic of the entire group. The author describes his book as 'an effort toward a better understanding of one American corporation', and so it is up to a point, but he tends to gloss over those aspects which the official investigations emphasize.

The three Swiss companies have adopted different policies towards historiography: CIBA added to their annual report for 1958 an interesting review of the period 1884–1959; Sandoz does not reveal much except on the contentious issue of the break-up of the Swiss IG; Bürgin's history of Geigy is a model of its kind and a major contribution to industrial history. Gignoux has written on Pechiney and Choffel on St. Gobain: both companies are large and have long been in business (the latter for about 300 years); they deserved fuller treatment.

[1] At the time of writing this note the book had not yet been published, but Mr. Reader kindly allowed me to read the typescript of the final draft, on which I have based my comments.

BIBLIOGRAPHY

I: OFFICIAL PUBLICATIONS

(a) United Kingdom[1]

(i) *Annual or periodical*

Reports of the Alkali Inspector.
Reports of the University Grants Committee.
Reports of the Reparations Commission.

(ii) *Reports of Departments, etc.*

Treasury

Scientific societies (Government grants), 1906, P.P. 1906, **65**.

Report of University Colleges (Great Britain) Grant in Aid, 1907, P.P. 1907, **64**.

Import Duties Advisory Committee, Report dated 10 July 1933, submitted to the Chancellor of the Exchequer by the Import Duties Advisory Committee on the position that will arise on the expiry of the Dyestuff (Import Regulation) Act 1920, 1933 (Cmd 4411), P.P. 1932/33, **11**.

Removal of certain excise restrictions on distillation and of allowances on industrial alcohol and exported spirits, 1945 (Cmd 6622), P.P. 1944/45, **10**.

Board of Trade

Census of Production 1907, preliminary tables, 1910 (Cd 5005 for Part II, Cd 5162 for Part III), P.P. 1910, **109**.

17th Annual Report on changes in rates of wages and hours of labour in the U.K., 1910 (Cd 5334), P.P. 1910, **84**.

Report of the Departmental Committee appointed by the Board of Trade to consider the position of the textile trades after the war (Birchenough Committee), 1918 (Cd 9070), P.P. 1918, **13**.

Departmental Committee on the post-war position of the sulphuric acid and fertilizer trades (war-time version which omits some passages and statistics), 1918 (Cd 8994), P.P. 1918, **13**.

Scheme for the allocation and administration of the funds provided by Parliament for the development of the dye industry by means of financial assistance to companies and firms in aid of developments, extensions and research, 1918 (Cd 9194), P.P. 1918, **22**.

Report of a Departmental Committee on the post-war position of the sulphuric acid and fertilizer trades (Complete version of Cd 8994 above), 1919 (Cmd 23), P.P. 1919, **29**.

Memorandum by the Board of Trade on the articles of which the importation into the U.K. is prohibited except under licence, 1919 (Cmd 454), P.P. 1919, **45**.

[1] Unless otherwise stated all U.K. documents are published by Her Majesty's Stationery Office. Parliamentary Papers (P.P.) are followed by year and volume number. Each volume has a table of contents.

Safeguarding of Industries Act, Report of the Superphosphate Committee, 1925 (Cmd 2475), P.P. 1924/25, **15**.

Safeguarding of Industries Act, Report of a committee appointed by the Board of Trade (Chadwick Committee), 1926 (Cmd 2631), P.P. 1926, **15**.

Report of the Dyestuffs Industry Development Committee on the present position and development of the dyestuffs manufacturing industry in Great Britain, 1930 (Cmd 3658), P.P. 1930/31, **11**.

Statistical tables relating to British and foreign trade and industry, 1931 (Cmd 3849), P.P. 1930/31, **31**.

Final report of the Third Census of Production [1924], vol. 4, The chemical and allied trades, 1931.

Third report of the Dyestuffs Industry Development Committee on the present position and development of the dyestuffs manufacturing industry in Great Britain, 1932 (Cmd 4191), P.P. 1931/32, **7**.

Home Office/Home Department

Preliminary tables of cases of industrial poisoning, fatal and non-fatal accidents and dangerous occurrences in factories, workshops, etc. during the year 1909, 1910 (Cd 5048), P.P. 1910, **83**.

Departmental Committee on Celluloid, Report and minutes of evidence, 1913 (Cd 7158 and 7159), P.P. 1914, **15**.

Ministry of Reconstruction

Committee on the Chemical Trade, 1917 (Cd 8882), P.P. 1917/18, **18**.

Report of the Machinery of Government Committee (Haldane Committee), 1918 (Cd 9230), P.P. 1918, **12**.

Report of Committee on Trusts, 1919 (Cd 9236), P.P. 1918, **13**.

Ministry of Munitions

Report of the British Mission appointed to visit enemy chemical factories in the occupied zone engaged in the production of munitions of war, 1919, confidential. (The Introduction was published in 1921 (Cmd 1137), P.P. 1921, **20**.)

Review of munitions output 1914–1918, n.d. (1919 ?), released in 1960.

Nitrogen Products Committee (Goold–Adams Committee), 1919 (Cmd 482), P.P. 1919, **26**.

History of the Ministry of Munitions, vol. 7: The control of materials, 1921.

Board of Education

Report for the year 1913–14 from those Universities and University Colleges in Great Britain which are in receipt of grants from the Board of Education, 1915 (Cd 8137), P.P. 1914/16, **19**.

Diplomatic and Consular Reports

Rose, F. Report on the rise, progress and present condition of the carbide and acetylene industries in Germany, 1900 (Cd 353–5), P.P. 1900, **91**.

Rose, F. Report on chemical instruction in Germany and the growth and present condition of the German chemical industries, 1901 (Cd 430–16), P.P. 1901, **80**.

Rose, F. Supplementary report on chemical instruction in Germany and the growth and present condition of the German chemical industries, 1902, (Cd 787–16), P.P. 1902, **103**.

(iii) *Other*

Departmental Committee on the Royal College of Science (Haldane Committee), 1906 (Cd 2872), P.P. 1906, **31**.

Imperial College of Science, First Annual Report of the Governing Body of the I.C.S., 1909 (Cd 4602), P.P. 1909, **19**.

Committee appointed by the Prime Minister to enquire into the position of natural science in the educational system of Great Britain (Thomson Committee), Report, 1918 (Cd 9011), P.P. 1918, **9**.

Committee on commercial and industrial policy after the war (Lord Balfour of Burleigh Committee), Final Report, 1918 (Cd 9035), P.P. 1918, **13**.

Standing Committee on Trusts, findings and decisions of a sub-committee appointed by the Standing Committee on Trusts to enquire into the existence and effect of a ring in the salt trade, 1920 (Cmd 832), P.P. 1920, **23**.

Report on the soap industry prepared by a sub-committee appointed by the Standing Committee on Trusts, 1921 (Cmd 1126), P.P. 1921 (Session 1), **16**.

Standing Committee on Trusts, report on the explosives industry, 1921 (Cmd 1347), P.P. 1921 (session 1), **16**.

Report on dyes and dyestuffs prepared by a sub-committee appointed by the Standing Committee on Trusts, 1921 (Cmd 1370), P.P. 1921 (Session 1), **16**.

Royal Commission on Oxford and Cambridge, Report, 1922 (Cmd 1588), P.P. 1922, **10**.

Papers relating to the liquidation of the Government interests in the British Dyestuffs Corporation Ltd., 1925 (Cmd 2545), P.P. 1924/25, **22**.

Committee on Industry and Trade (Balfour Committee), Survey of industrial relations, 1926.

Committee on Industry and Trade (Balfour Committee), Factors in industrial and commercial efficiency, 1927.

Committee on Industry and Trade (Balfour Committee), Survey of textile industries, 1928.

Committee on Industry and Trade (Balfour Committee), Further factors in industrial and commercial efficiency, 1928.

Committee on Industry and Trade (Balfour Committee), Final Report, 1929 (Cmd 3282), P.P. 1928/29, **7**.

Monopolies and Restrictive Practices Commission, Report on the supply of certain industrial and medical gases, 1956 (HC 13), P.P. 1956/57, **17**.

Monopolies Commission, Report on the supply of chemical fertilizers, 1959, P.P. 1958/59. **17**.

Combined Intelligence Objectives Sub-Committee, IG Farben (Item No. 30, File 23–15), 1945.

HARDING, I. *Patent Office Centenary* (1953).

NORRIS, W. G. *Chemical Service in Defence of the Realm*, Ministry of Supply (1957).

(b) United States[1]

(i) *Reports of Departments, etc.*

USTC (U.S. Tariff Commission)

Census of Dyes and other synthetic organic chemicals (The first Census, for 1917, was published in 1918. The Censuses have appeared annually since as part of the Tariff Information Series.)

Commissioner of Education/Dept of the Interior

Office of Education

Reports (annual or biennial).

Biennial Survey of Education.

Department of Commerce

13th Census [1909], vol. 10, Manufactures (1913).

NORTON, T. H. *Dyestuffs for American textile and other industries* (Special Agents Series No. 96), (1915).

DELAHANTY, T. W. *The German dyestuffs industry* (Miscell. Series No. 126), (1924).

REAGAN, D. J. *The French chemical industry and trade in 1928* (Tariff Information Bulletin No. 652), (1929).

The American chemical industry (Trade Promotion Series No. 78), (1929).

BRISSENDEN, P. F. *Earnings of factory workers 1899–1927* (Bureau of the Census Monographs No. 10), (1929).

COCHRANE, R. C. *Measures for progress* (National Bureau of Standards), (1966).

Federal Trade Commission

Report on international phosphate cartels (1946).

The sulphur industry and international cartels (1947).

International cartels in the alkali industry (1950).

The fertilizer industry (1950).

Department of State

Report of the Mission on Japanese Combines, Part 1: Analytical and technical data (Far Eastern Series, publication No. 2628), (1946).

Bureau of the Census

Historical statistics of the U.S., colonial times to 1957 (1960).

(ii) *Other*

National Research Council

Research laboratories in the U.S. *Bull. natn. Res. Coun., Wash.* **1** (1920) 2 and **3** (1921) 1.

Funds available in the U.S. for the encouragement of scientific research. *Bull. natn. Res. Coun., Wash.* **1** (1920), No. 9.

Industrial research laboratories in the U.S. *Bull. natn. Res. Coun., Wash.* (July 1927), No. 60.

National Science Foundation

Scientific personnel resources (1955).

[1] Unless otherwise stated all U.S. documents are published by the Government Printing Office, Washington.

House of Representatives Committee on Ways and Means, Hearings on H.R. 2706 and 6495, *Dyestuffs* (1920).

TNEC (Temporary National Economic Committee)

WILCOX, C. *Competition and monopoly in American industry* (1941).

PEARCE, C. A. *et al. Trade Association survey* (1941).

National Resources Planning Board

Research—A national resource, pt. II Industrial research (1941).

U.S. Military Government of Germany

LINK, A. E. *Economic Study of IG* (Document TP/200/P/6), (August 1945).

U.S. *v.* Allied Chemical & Dye Corporation

Civil Action 14–320 in the District Court of the U.S. for the Southern District of New York, Complaint filed 29 May 1941, New York 1941.

U.S. *v.* I.C.I. *et al.*

Civil Action 24–13 in the District Court for the Southern District of New York, Complaint filed 6 January 1944, New York 1944.

U.S. *v.* Du Pont, General Motors *et al.*

Civil Action 49 C–1071 in the District Court for the Northern District of Illinois, Complaint filed 30 September 1949. The proceedings comprise: Trial Transcripts (11 vols.), Oral Arguments (2 vols.), Deposition of I. du Pont (1 vol.), Deposition of A. P. Sloan jr. (1 vol.), Defendants' Trial Exhibits (abbr. DP) (6 vols.), Government Trial Exhibits (abbr. GTX) (7 vols.), Post Trial Brief for the U.S. (2 vols.), Judge's Opinion, 3 December 1954. Chicago, 1949–1954.

(c) Other countries

Belgium

Ministère de l'Industrie et du Travail

Monographies industrielles, fabrication des produits chimiques, J. Lebègue & Cie, Brussels (1905).

France

Ministère de Travail

Resultats statistiques du recensement général de la population 1906, Imprimerie Nationale Paris, 3 vols. (1908–10).

HALLER, A. *Les industries chimiques et pharmaceutiques* (Rapport du Jury International, Class 87, Chemicals, at the Paris Exhibition of 1900) Gauthier Villars, Paris, 2 vols. (1903).

FLEURENT, E. C. A. *Les grandes industries chimiques à l'Exposition Universelle de 1900*, Conservatoire des Arts et Métiers, Paris (1900).

VERLOT, C. *Rapport fait au nom de la Commission du Commerce et de l'Industrie sur le projet et les propositions de loi relatifs à l'enseignement technique, industriel et commercial*, Imprimerie de la Chambre des Députés, Paris (1912).

FLEURENT, E. C. A. *Les industries chimiques en France et en Allemagne*, Conservatoire des Arts et Métiers, Paris, 3rd series, vol. 3 (1915/16).

Ministère du Commerce

Rapport général sur l'industrie Française, sa situation, son avenir, Imprimerie Nationale, Paris, (1919).

Germany

Reichsamt des Inneren

Denkschrift über das Kartellwesen, Siemenroth, Berlin, (1906).

Kontradiktorische Verhandlungen über Kartelle, Siemenroth, Berlin, 6 vols. (1903–6).

Kriegsministerium

Bewirtschaftung von kriegswichtigen Stoffen, Kriegsrohstoff Abteilung, Berlin 13 vols. (1914–18).

Kriegsernährungsamt

GROSSMANN, H. *et al. Düngemittel im Kriege*, Kriegsernährungsamt, Berlin (1917).

Reichsarchiv

Der Weltkrieg 1914–18: Kriegsrüstung und Kriegswirtschaft, E. S. Mittler, Berlin, 2 vols. (1930).

Statistisches Reichsamt

Deutsche Wirtschaftskunde, R. Hobbing, Berlin (1930).

Deutsche Hochschulverwaltungen

Bedarf und Nachwuchs an Chemikern und Physikern, Struppe & Winckler, Berlin (1932).

Ausschuss zur Untersuchung der Erzeugungs- und Absatzbedingungen der deutschen Wirtschaft

Die deutsche Chemische Industrie, ENQUETE (3. Unterausschuss), E. S. Mittler, Berlin (1930).

Die deutsche Seifen- und Parfumerieindustrie (1. Unterausschuss, 5. Arbeitsgruppe), E. S. Mittler, Berlin (1931).

Der deutsche Aussenhandel (1. Unterausschuss), E.S. Mittler, Berlin, 2 vols. (1930).

(d) International organizations

League of Nations[1]

Economic and Financial Section

The chemical industry (Economic and Financial 1927, II, 4), (1927).

MAGREGOR, D. H. *International cartels* (Economic and Financial 1927, II, 16), (1927).

WIEDENFELD, K. *Cartels and combines* (Economic and Financial 1926, II, 70), (1927).

DECUGIS, H. *et al. Review of the legal aspects of industrial agreements* (Economic and Financial 1930 II, 11), (1930).

International Labour Office (I.L.O.)[1]

Cancer of the bladder among aniline workers (Studies and Reports, Series F, No. 1), (1921).

White lead (Studies and Reports, Series F, No. 11), (1927).

[1] Published in Geneva by the organization.

Engineers and chemists: status and employment (Studies and Reports, Series L, No. 1), (1924).

International Labour Conference (23rd Session)

Report 4. Reduction of hours of work in printing and kindred trades (1937).

Report 5. Reduction of hours of work in the chemical industry (1937).

Institut International d'Agriculture/Food and Agriculture Organization

Yearbook of agricultural statistics, Rome, annual.

Production yearbook, Rome, annual.

II: BOOKS, PAMPHLETS, ARTICLES

AKERS, SIR WALLACE. The Research Laboratories of I.C.I. Ltd. *Proc. Journées Internationales de Chimie Industrielle*, Université du Travail, Charleroi (1948).

ALIEN PROPERTY CUSTODIAN (A. M. PALMER) Summary of Report *J. ind. Engng. Chem.* (April 1919), 352–65.

ALDCROFT, D. H. (ED.) *The development of British industry and foreign competition 1875–1914*, Allen & Unwin, London (1968).

ALLEN, J. A. *Studies in innovation in the steel and chemical industries*, University Press, Manchester (1967).

ANGLO-AMERICAN COUNCIL ON PRODUCTIVITY. *Universities and industry*, Council on Productivity, London (1951).

ANON. *The Federal anti-trust laws 1890–1915*, Commerce Clearing House, New York (1952).

ANON. *50 Jahre Kaiser-Wilhelm-Gesellschaft und Max-Planck-Gesellschaft zur Förderung der Wissenschaften*, Max-Planck-Gesellschaft, Göttingen (1961).

ANON. *The jubilee of the Chemical Society 1891*, Chemical Society, London (1896).

ANTWERP CHAMBER OF COMMERCE. *Annuaire 1927* (Special Issue), Chambre de Commerce, Antwerp (1928).

ARGLES, M. *South Kensington to Robbins*, Longmans, London (1964).

ARMSTRONG, H. E. The Monds and chemical industry, a study in heredity *Nature, Lond.* **127** (14 February 1931), 238 ff.

ARMYTAGE, W. H. G. *Civic Universities*, Benn, London (1953).

ASHBY, SIR ERIC. Technological humanism. *J. Inst. Metals* **85** (1956/7), 464 f.

ASHBY, SIR ERIC. *Technology and the academics*, Macmillan, London (1958).

ASSOCIATION OF BRITISH CHEMICAL MANUFACTURERS. Report of the British Chemical Mission on the chemical factories in the occupied area of Germany, A.B.C.M., London (1919).

ASSOCIATION NATIONALE D'EXPANSION ECONOMIQUE. *Index to French production* (in English), publ. by Association, Paris (1920 ed.).

AUST, R. WIZINGER. Schweizerische Forschung über künstliche Farbstoffe in den vergangenen fünf Jahrzehnten. *Neue Zürcher Zeitung* (21 July 1955), Fernausgabe.

BALDWIN, R. T. History of the chlorine industry. *J. Chem. Educ.* **4** (1927), 313 ff.

BARTLETT, H. R. *The development of industrial research in the United States*, see under Official Publications, U.S., National Resources Planning Board.

BASLER CHEMISCHE INDUSTRIE KOLLEKTIV-GRUPPE. *Schweizerische Landesausstellung Bern 1914*, priv. publ. Basle (1914).

BAUD, P. *Les industries chimiques régionales en France*, O. Doin, Paris (1922).

BAUD, P. *L'Industrie chimique en France*, Masson & Cie, Paris (1932).

BAUMANN, E. G. F. du Bois. obituary in *Neue Zürcher Zeitung* (15 November 1953), Fernausgabe.

BAUMGARTNER, R. *Die wirtschaftliche Bedeutung der chemischen Industrie in Basel*, A. Francke, Bern (1947).

BEDFORD, S. *A legacy*, Weidenfeld & Nicolson, London (1956).

BEDHOME, R. L'Evolution de la Carbochimie dans le Bassin du Nord et du Pas-de-Calais. *Chim. Ind.* **87** (1962), 4, 435–9.

BEER, E. J. *The beginning of rayon*, priv. publ. Paignton (1962).

BEER, J. J. *The emergence of the German dye industry*, University of Illinois Press, Urbana (1959).

BELLOT, H. HALE *The history of University College*, University of London Press, London (1929).

BENEDIKT, H. *Die Wirtschaftliche Entwicklung in der Franz Joseph-Zeit*, Herold Verlag, Vienna (1958).

BENEY, M. A. *Wages in the United States 1914–1929*, National Industrial Conference Board, New York (1930).

BERDAHL, R. O. *British Universities and the State*, University of California Press, Berkeley (1959).

BERENSON, C. (ED.) *The chemical industry: Viewpoint and perspectives*, J. Wiley & Sons, New York (1963).

—— *Administration of the chemical enterprise*, J. Wiley & Sons, New York (1963).

BERLE, A. A. and MEANS, G. C. *The modern corporation and private property*, Macmillan, New York (1932).

BERNTHSEN, A. Über Luftsalpetersäure. *Z. angew. Chem.* **22** (1909) 24, 1167–78.

—— Die synthetische Gewinnung des Ammoniaks. *Z. angew. Chem.* **26** (1913) pt. 1, 1 ff.

BOLITHO, H. *Alfred Mond*, Martin Secker, London (1933).

BOLTON, E. K. *Du Pont research* (reprint of the Perkin Lecture to the American Section of the Society of Chemical Industry, 1945).

BOSCH, C. *Sozialisierung und chemische Industrie*, Verlag Chemie, Berlin (1921).

—— *Über die Entwicklung der chemischen Hochdrucktechnik bei dem Aufbau der neuen Ammoniakindustrie* (Nobel Prize lecture, 21 May 1932), Imp. Royale, Stockholm (1933).

BOSSHARD, E. Aus den Anfängen der schweizerischen chemischen Industrie, *Zofinger Tagblatt*, Zofingen (1932).

BOWLEY, A. L. *Wages and income in the U.K. since 1860*, University Press, Cambridge (1937).

BOYLAN, J. *Sequel to the Apocalypse*, Booktab Inc., New York (1942).

BRAND, C. J. Some fertilizer history connected with World War I. *Agricultural History*, **19** (1945), 2, 104 ff.

BRICARD, S. L'Industrie de l'azote dans l'economie Française. *Chim. Ind.* **68** (1952), 6, 937–945.

BRITISH DYESTUFFS CORPORATION. Review of the dyestuffs industry, typescript (1927).

BROUN, J. Die chemische Industrie Russlands. *Chem. Ind., Düsseld.* **23** (1900), 24, 541 ff.

BROWNE, C. A. and WEEKS, M. E. *A History of the American Chemical Society*, American Chemical Society, Washington (1952).

BROWNE, C. A. *Liebig and the Law of the Minimum*, see under MOULTON F. R. (ED.), *Liebig and after Liebig. . . .*

BROWNE, H. F. *Wages, hours and employment in American manufacturing industries 1914–1924*, National Industrial Conference Board, New York (1924).

BRUBACHER, J. S. and RUDY, W. *Higher education in transition*, Harper & Brothers, New York (1958).

BULLOCK A. *The life and times of Ernest Bevin*, vol. i, Heinemann, London, (1960).

BUREAU OF VOCATIONAL INFORMATION. *Women in Chemistry*, publ. by Bureau, New York (1922).

BURNS, C. DELISLE. *A History of Birkbeck College*, University of London Press, London (1924).

BUSEMANN, M. (ED.) *Die wirtschaftliche Bedeutung der Ukraine*, R. von Decker, Berlin (1918).

CALVERT, A. F. *Salt in Cheshire*, Spon, London (1915).

CAMERON, F. T. *Cottrell, samaritan of science*, Doubleday & Co, Garden City N.Y. (1952).

CAMPBELL, E. D. *History of the Chemical Laboratory of the University of Michigan*, Michigan University Press, Ann Arbor (1916).

CAMPBELL, M. and HATTON, H. *H. H. Dow*, Appleton-Century-Crofts, New York (1951).

CARDWELL, D. S. L. *The organisation of science in England*, Heinemann, London (1957).

CARR, W. H. A. *The Du Ponts of Delaware*, F. Muller, London (1965).

CASTELEIN, E. *Les requisitions 'en masse' à Anvers*, A. de Tavernier, Antwerp (1918).

CHAMBERS, SIR PAUL. My twenty years with ICI. *I.C.I. Magazine*. **46** (1968), 341, 36–42.

CHAPMAN, A. CHASTON. *The growth of the profession of chemistry during the past half-century*, Institute of Chemistry, London (1927).

CHARLTON, H. B. *Portrait of a university*, University Press, Manchester (1951).

CHRISTIANSEN, C. C. *Chemische und Farbenindustrie*, see under WEBER, A. (ED.), *Über den Standort der Industrien.*

CIBA AG (ED.) *Herkunft und Gestaltung der Industriellen Chemie in Basel*, Urs-Graf Verlag, Olten-Lausanne (1959).

CLAPHAM, J. H. *The economic development of France and Germany 1815–1914*, University Press, Cambridge (4th ed. 1936).

CLARKSON, G. B. *Industrial America in the World War: the strategy behind the lines*, Houghton Mifflin, Boston (1923).

CLEMENT R. and RIVIÈRE, C. *Les matières plastiques*, see under GÉRARD, J. (ED.), *Dix ans d'efforts scientifiques et industriels.*

COATES, J. E. *The Haber Memorial Lecture* (in vol. 4 of Memorial Lectures delivered before the Chemical Society), Chemical Society, London (1951).

COCHRANE, R. C. *Measures for Progress*, see under Official Publications, U.S., Department of Commerce.

COMBE, P. *Niveau de vie et progrès technique en France 1860–1939*, Presses Universitaires de France, Paris (1956).

COOMBS, WHITNEY. *The wages of unskilled labor in manufacturing industries in the U.S. 1890–1924*, Columbia University Press, New York (1926).

COOPER, FRANKLIN S. *Location and extent of industrial research activity in the U.S.*, see under Official Publications, U.S., National Resources Planning Board.

COTGROVE, S. T. *Technical education and social change*, Allen & Unwin, London (1958).

COTTRELL, A. *Nitric acid and nitrates*, Gurney & Jackson, London (1923).

COTTRELL, F. G. The Research Corporation. *J. ind. Engng. Chem.* **4** (1912), 864 f.

CRANE, J. E. Development of the synthetic ammonia industry in the U.S. *J. ind. Engng. Chem.* **22** (1930), 7, 795–8.

CROISET, A. (ED.) *L'Éducation de la Démocratie*, F. Alcan, Paris (1907).

DALE, E. Du Pont; pioneer in systematic management. *Admin. Science Q.* **2** (1957), 1, 25–59.

—— *The great organizers*, McGraw Hill, New York (1960).

DELABY, R. (ED.) *Centenaire de la Société Chimique de France*, Masson & Cie, Paris (1957).

DENTERGHEM, C. DE KERCHOVEN DE. *L'Industrie Belge pendant l'occupation Allemande*, Presses Universitaires de France, Paris (1927).

DERUGINSKY, V. T. *L'Instruction publique*, see under KOVALESKY, M. W. DE (ED.), *La Russie à la Fin du XIXème Siècle*.

DEWAR, SIR JAMES. *Presidential Address to the British Association for the Advancement of Science meeting at Belfast in 1902*, British Association, London (1903).

DIETRICH, W. *et al.* Im Dienst der Farbe. *Die BASF 6* (N.S.) (1956), 6, 245 ff.

DIGGLE, G. E. *A history of Widnes*, Corporation of Widnes (1961).

DUBOIS, J. E., JNR. *Generals in grey suits*, Bodley Head, London (1953).

DUCKWORTH, A. Rise of the pharmaceutical industry. *Chemist Drugg.* reprint of the Centenary issue, November 1957.

DUISBERG, CARL. Die Abnahme der allgemeinen Bildung bei den Chemiestudierenden, *Z. angew. Chem.* **13** (1900), 6, 131–3.

—— Die Wissenschaft und Technik in der chemischen Industrie mit besonderer Rücksichtigung der Teerfarbenindustrie. *Z. angew. Chem.* **25** (1912), **1**, 6 f.

—— *Fortschritte und Probleme der Chemischen Industrie*, O. Spamer, Leipzig (1913).

—— *Meine Lebenserinnerungen*, P. Reclam, Leipzig (1933).

DUISBERG, CURT. *Die Arbeiterschaft der chemischen Grossindustrie*, C. Heymann, Berlin (1921).

DUPREE, A. HUNTER. *Science in the Federal Government*, Harvard University Press, Cambridge (1957).

DYES, W. A. Die Production, Ein- und Ausfuhr chemischer Producte in Russland. *Chem. Ind., Düsseld.* **22** (1899), 17/18, 389 f.

—— Produktion und Rentabilität der Russischen Chemischen Industrie. *Chem. Ind., Düsseld.* **28** (1905), a series of four articles appearing in Nos. 11, 12, 13, and 14.

EBERT, W. *Die chemische Industrie Deutschlands*, Verlag Chemie, Berlin–Leipzig (1926).

EDWARDS, B. *Chemicals—servant or master?* National Labour Press, London (1947).

EDWARDS, SIR RONALD. *Co-operative industrial research*, Pitman, London (1950).

ERBAN, F. *Kartelle und Konventionen in der chemischen Industrie*, Appretur Zeitung Verlag, Gera (1910).

EYRE, J. V. *Henry Edward Armstrong*, Butterworth, London (1958).

FAIRLIE, J. A. *British war administration*, Oxford University Press, New York (1919).

FARGHER, R. G. The Perkin family and the textile industry, *Manchester Guardian*, 7 May 1956.

FAUQUE, M. *L'Évolution économique de la grande industrie chimique en France*. (dissertation for the Doctorat d'État), Les Editions Universitaires, Strasbourg (1932).

FEDOTIJEV, P. P. The Russian chemical industry, *Chem. Trade J.* **20** (1897), 521, 332–3, and 523, 364–5.

—— Die chemische Grossindustrie Russlands. *Z. angew. Chem.* **15** (1902), 34, 865–7.

FENOGLIO, G. *Die volkswirtschaftlichen Folgen des Weltkrieges in Italien*, Puttkammer & Mühlbrecht, Berlin (2nd ed. 1915).

FERN, H. Fifty years ago. *I.C.I. Magazine*, **35** (October 1957), 336–7.

FISCH, R. *Les industries chimiques de la région Lyonnaise* (dissertation for the Doctorat en Droit), priv. publ. Mâcon (1923).

FISCHER, EMIL. *Aus Meinem Leben*, J. Springer, Berlin (1922).

FISCHER, EMIL and BECKMANN, E. *Das Kaiser-Wilhelm-Institut für Chemie, Berlin-Dahlem*, Vieweg, Braunschweig (1913).

FISCHER, ERNST. Die Gründer der Farbwerke Hoechst. *Nassauische Lebensbilder*, **6** (1951), 252–61.

FISCHER, F. *Das Studium der Technischen Chemie an den Universitäten und Technischen Hochschulen Deutschlands und das Chemiker Examen*, Vieweg, Braunschweig (1897).

—— *Chemische Technologie an den Universitäten und Technischen Hochschulen Deutschlands*, Vieweg, Braunschweig (1898).

FLECHTNER, H.-J. *Carl Duisberg*, Econ, Düsseldorf (1960).

FLEXNER, A. *Universities, American, English and German*, Oxford University Press, New York (1930).

FLINN, A. D. and COBB, R. *Research Laboratories in the U.S.* see under Official Publications, U.S., National Research Council.

FONTAINE, A. *L'Industrie Française pendant la Guerre*, Presses Universitaires de France, Paris (1925).

FORSTER, M. O. *Emil Fischer Memorial Lecture* (in vol. 3 of Memorial Lectures delivered before the Chemical Society), Chemical Society, London (1933).

FRANCK, H. H. *et al. Der Kalkstickstoff in Wissenschaft, Technik und Wirtschaft*, F. Enke, Stuttgart (1931).

FRANK, A. The utilization of atmospheric nitrogen (paper read before the Faraday Society), *Chem. Trade J.* **42** (1908), 1100, 581–2 and 1101, 605–6, and **43**, 1102, 5 and 1103, 30.

FREEMAN, C. The plastics industry: a comparative study of research and innovation. *Natn. Inst. econ. social Res. Rev.* (1963), 26, 22–62.

FREETH, F. A. Explosives for the First World War, *New Scientist* **23** (1964), 402, 274 ff.

FRENCH, J. C. *A history of the University founded by Johns Hopkins*, Johns Hopkins Press, Baltimore (1946).

FRERICHS, F. W. Report of the Committee on Chemical Engineering Education. *Trans. Am. Inst. Chem. Engrs.* **3** (1910), 122–65.

—— 32 Jahre in den Vereinigten Staaten als Chemiker und Fabrikant. *Z. angew. Chem.* **26** (1913), 2, 17–24.

FRITSCH, J. *The manufacture of chemical manures*, Scott, Greenwood & Son, London (1911).

GAIT, A. J. *Heavy organic chemicals*, Pergamon Press, Oxford (1967).

GALLO, A. Les industries chimiques de l'Aquitaine et des Pyrénées. *Chim. Ind.* **48** (1961), 523, 42–50.

GEORGE, F. *La rénovation de l'industrie chimique Française*, Albin-Michel, Paris n.d. (1919 ?).

GÉRARD, J. (ED.) *Dix ans d'efforts scientifiques et industriels*, Chimie & Industrie, Paris, 2 vols. (1926).

GINORI-CONTI, P. L'Industrie chimique en Italie—son développement. *Chim. Ind.* special number, February 1929.

GLASER, C. H. von Brunck. *Ber. dt. chem. Ges.* **46** (1913), pt. 1, 353–89.

GLASER, J. *Die chemische Industrie Oesterreichs und ihre Entwicklung*, Bornträger, Berlin (1918).

GRABOWER, R. *Die finanzielle Entwicklung der Aktiengesellschaften in der deutschen chemischen Industrie*, Duncker & Humblot, Leipzig (1910).

GRANDMOUGIN, E. *L'Enseignement de la chimie industrielle en France*, Dunod & Pinat, Paris (1917).

—— *L'Essor des industries chimiques en France*, Dunod & Pinat, Paris (1917).

GRAY, A. N. *Phosphates and superphosphates*, International Superphosphate Manufacturers Association, London (2nd ed. 1944).

GREILING, W. *75 Jahre Chemieverband*, Chemieverband, Frankfurt (1952).

GROSS, H. *Material zur Aufteilung der IG Farbenindustrie AG*, Institut für Weltwirtschaft, Kiel (1950).

GROSSMANN, H. and WEICKSEL, P. *Die Stickstoffindustrie der Welt*, Allgemeiner Industrie Verlag, Berlin (1930).

GROSSMANN, H. *et al. Düngemittel im Kriege*, see under Official Publications, Germany, Kriegsernährungsamt.

—— *Die Bedeutung der chemischen Technik für das deutsche Wirtschaftsleben*, W. Knapp, Halle (1907).

—— A. Frank zu seinem 80. Geburtstag, *Chem. Ind., Düsseld.* **37** (1914), 2, 32 ff.

GUGGENBÜHL, G. *et al. Geschichte der eidgenössischen Technischen Hochschule* 1855–1955, Neue Zürcher Zeitung, Zurich (1955).

GUILLET, L. *État actuel de l'industrie des produits chimiques inorganiques en France*, L'Association des Chimistes de Sucrerie et de Distillerie, Paris, 2 vols. (1902)

(Reprint of the papers presented to the 4th International Congress of Applied Chemistry, Paris 1900).
GUNTHER, A. E. *Early science in Oxford*, vol. 15, R. T. Gunther, priv. publ., Oxford (1967).
GÜRBER, H. Die elsässische Kaliindustrie und das Kaliweltkartell, Ph.D. thesis, Zurich University (1942).

HABER, F. Die Elektrochemie auf der Pariser Weltausstellung, *Z. angew. Chem.* **15** (1901), 8 and 9.
—— *Fünf Vorträge*, Springer, Berlin (1924).
—— *Aus Leben und Beruf*, Springer, Berlin (1927).
HABER, L. F. *The chemical industry during the nineteenth century*, Clarendon Press, Oxford (1958).
HALDANE, R. B. *An autobiography*, Hodder & Stoughton, London (1929).
HALE, A. J. *The applications of electrolysis in chemical industry*, Longmans, London (1918).
HALLER, A. L'Enseignement pratique en général et en particulier celui de la chimie dans nos universités. *Revue gen. Matièr. color.* **2** (July 1898), 257 ff.
HANCOCK, H. The industrial workman along the Brandywine (Hagley Foundation) Du Pont Archives, typescript (1958).
HANSEN, K. Die Kennedy-Runde vor ihrer Bewährungsprobe. *Chem. Ind., Düsseld.* **16** (1964), 12, 837.
HARDIE, D. W. F. *A history of the chemical industry in Widnes*, I.C.I. General Chemicals Division, Widnes (1950).
—— *Electrolytic manufacture of chemicals from salt*, Oxford University Press, London (1959).
—— *Acetylene manufacture and use*, Oxford University Press, London (1965).
HARDIE, D. W. F. and PRATT, J. DAVIDSON. *A history of the modern British chemical industry*, Pergamon Press, Oxford (1966).
HART, P. E. Business concentration in the U.K. *Jl R. statist. Soc.* **123** (1960), pt. 1, 56 f.
HASENCLEVER, M. Zur Lage der chemischen Grossindustrie. *Chem. Ind., Düsseld.* **28** (1905), 3, 52 ff.
HAUFFE, H. Die chemische Grossindustrie in Russland geschildert auf Grund eigener Erfahrungen und Beobachtungen. *Z. angew. Chem.* **13** (1900), 1, 26–30.
HAYNES, WILLIAMS. *Chemical pioneers: the founders of the American chemical industry*, D. van Nostrand, New York (1939).
—— *American chemical industry*, D. van Nostrand, New York, 6 vols. (1945–54).
—— *Brimstone, the stone that burns*, D. van Nostrand, Princeton–New York (1959).
HAYNES, WILLIAMS and GORDY, E. L. (EDS.) Chemical Industry's Contribution to the Nation 1635–1935 (Supplement to *Chemical Industries* for May 1935), Chemical Markets Inc., New York (1935).
HEARNSHAW, F. J. C. *Centenary history of Kings College London*, G. Harrap, London (1929).
HEATH, SIR H. FRANK and HETHERINGTON, A. L. *Industrial research and development in the U.K.*, Faber, London (1946).
HEITZ, W. *Erinnerungen aus den Jahren 1890–1930*, priv. publ. Basle, n.d. (1937 ?).

HELBIG, E. Pharmazie und chemische Grossindustrie, Ph.D. thesis, Tübingen (1922).

HELFERICH, B. Aus der Geschichte der Chemie der Universität Leipzig. *Chemikerzeitung—chem. Appar.* **84** (1960), 6, 169–72.

HEMPEL, E. H. *The economics of chemical industries*, John Wiley, New York (1939).

HERNANDEZ, R. *El Salitre*, Asociacion de Productores de Salitre de Chile, Santiago (1930).

HESSE, A. and GROSSMANN, H. (EDS.) *England's Handelskrieg und die chemische Industrie*, Enke, Stuttgart (1915, rev. ed. 1917).

HIGGINS, S. H. *Dyeing in Germany and America with notes on colour production*, University Press, Manchester, (1st ed. 1907, 2nd ed. 1916).

HILGER, G. and MEYER, A. G. *The incompatible allies, a memoir history of German–Soviet relations*, Macmillan, New York (1953).

HIMES, N. E. (ED.) *Economics, sociology and the modern world*, Harvard University Press, Cambridge (1935).

HIRT, F. L. Growth characteristics of the economy illustrated by the chemical industry. *Survey of Current Business*, September 1954.

HIRSCH, A. *Industrialized Russia*, Chemical Catalog Co., New York (1934).

HJELT, EDV. F. K. Beilstein. *Ber. dt. chem. Ges.* **40** (1907), 4, 5051–4.

HOFFMANN, W. G., GRUMBACH, F., and HESSE, H. *Das Wachstum der deutschen Wirtschaft seit der Mitte des 19. Jahrhunderts*, Springer, Berlin (1965).

HOFSTADTER, R. and HARDY, C. D. *The development and scope of higher education in the U.S.*, Columbia University Press, New York (1952).

HOHENBERG, P. M. *Chemicals in western Europe 1850–1914: an economic study of technical change*, Rand McNally–North Holland Publishing Co., Chicago (1967).

HOLDERMANN, C. *Carl Bosch*, Econ, Düsseldorf (1953).

HOLT, A. Neuere Arbeiten auf dem Kautschukgebiet. *Z. angew. Chem.* **27** (1914), 1, 153 ff.

HOPFF, H. Zur Entwicklung der Kunststoffchemie. *Chem. Ind., Düsseld.* **4** (1952), 10, 719–25.

HOPPMANN, H. *Die rationelle Gestaltung der chemisch-technischen Produktion* (Ph.D. thesis, Halle), Verlag-Chemie, Berlin (1934).

HORNE, A. *The price of glory*, Macmillan, London (1962).

HORSTER. *Die volkswirtschaftliche Bedeutung der chemischen Industrie am Oberrhein*, Helbing & Lichtenhahn, Basle (1922).

HOWARD, F. A. *Buna Rubber*, D. van Nostrand, New York (1947).

HUFBAUER, G. C. *Synthetic materials and the theory of international trade*, G. Duckworth, London (1966).

HULL, C. *Funds available in the U.S. for the encouragement of scientific research*, see under Official Publications, U.S., National Research Council.

I.C.I. DYESTUFFS DIVISION. *An outline of the chemistry and technology of the dyestuffs industry*, I.C.I. Dyestuffs Division, Manchester (1954).

I.C.I. *Sulphuric acid manufacture and use*, I.C.I., London (1955).

ILGNER, M. Die Rohstoffversorgung der deutschen Schwefelsäure-Industrie, Ph.D. thesis, Frankfurt University, (1923).

IOFFE, Y. A., KOZLOV, A. I., and ALMAN, G. L. *Ekonomika i Planirouaniya Khimi-*

cheskoi Promyshlennosti, Molotov Institute of Economics and Engineering, publication No. 11, Leningrad (1940).

IPATIEFF, V. N. Modern science in Russia. *J. chem. Educ.* **20** (1943).

—— *The life of a chemist*, University Press, Stanford (1946).

JACOBSON, P. Die Adolf-Baeyer-Gesellschaft zur Förderung der chemischen Literatur. *Z. angew. Chem.* **33** (1920), 1, 177–80.

JAMES, A. F. B. *Nitrate facts and figures*, F. C. Mathieson & Co., London, annual 1904–14.

JAMES, M. *Alfred I. du Pont*, Bobbs Merrill, New York (1941).

JAQUET, N. *Die Entwicklung und volkswirtschaftliche Bedeutung der schweizerischen Teerfarbenindustrie*, Helbing & Lichtenhahn, Basle (1922).

JEWKES, J. *et al. The sources of invention*, Macmillan, London (1958).

KAHAN, J. L'Essor de l'I.G. Farben et le recrutement de ses cadres supérieurs. *Chim. Ind.* **61** (1949), 5, 502–4.

KAP-HERR, J. DE. *Les matières colorantes*, see under GÉRARD, J. (ED.), *Dix ans d'efforts scientifiques et industriels.*

KASELITZ, O. Die Vorbildung der Chemiker. *Z. angew. Chem.* **21** (1908), 1, 5–7.

KATS, A. M. *Voprosy Ekonomiki Osnovnoi Khimicheskoi Promyshlennosti SSR*, Khimizdat, Moscow (1939).

KAUFMAN, M. *The first century of plastics*, The Plastics Institute, London, n.d. (1964 ?).

KEEBLE, SIR FREDERICK. *Fertilizers and food production*, Oxford University Press, London (1932).

KESSLER, COUNT HARRY. *Walther Rathenau*, Klemm, Berlin (1928), Howe, London (1929).

KLIPPERT. Die Entwicklung der Technik in der Düngerindustrie von Anfang bis auf die heutige Zeit. *Z. angew. Chem.* **18** (1904), 9, 322–5.

KOESTLER, A. *The act of creation*, Hutchinson, London (1964).

KOVALEVSKY, M. W. DE (ED.) *La Russie à la fin du XIXème siècle*, P. Dupont, Paris (1900).

KRISCHAN, A. IG Entflechtung im Spiegel der Literatur. *Chem. Ind., Düsseld.* **4** (1952), 10, 899–904.

KRISCHE, P. *Das Kali*, F. Enke, Stuttgart (1923).

KUBIERSCHKY, K. Die künstlichen Düngemittel und ihre Bedeutung. *Z. angew. Chem.* **26** (1913), pt. 1, 721–9.

LACHMANN-MOSSE, R. Die Stickstoffindustrie und ihre internationale Kartellierung, Ph.D. thesis, Zurich university (1940).

LAIBLE, W. Die Entwicklung der BASF, Ph.D. thesis, Erlangen (1924).

LAMBERT, E. and M. *Annuaire statistique des engrais et produits chimiques*, Librairie Agricole de la Maison Rustique, Paris (1912).

LAMBERT, M. *Les engrais chimiques et produits chimiques utiles à l'agriculture*, Association Nationale d'Expansion Economique, Paris (1917).

LANDES, D. S. Technological change and development in western Europe 1750–1914, in Habbakuk, H. J. & Postan, M. M. (eds.) *The Cambridge economic history of Europe*, University Press, Cambridge, vol. 6, pt. 1 (1966).
—— *The unbound Prometheus*, University Press, Cambridge (1969).
LEMOINE, R. J. *La concentration des entreprises en Belgique*, Goemare, Brussels (1929).
LEPERSONNE, H. Quelques Aspects de l'Industrie Chimique en Belgique. *Chim. Ind.*, Special Issue, March 1931.
LERNER, S. W. *Breakaway unions and the small trade union*, Allen & Unwin, London (1961).
LESSING, L. P. Du Pont. *Fortune*, October, 1950.
LE SUEUR, E. L. Inception and development of electrolytic alkali manufacture. *Trans. electrochem. Soc.* **63** (1933), 187–93.
LEVINSTEIN, H. From within the Dyestuffs Industry. *J. Soc. Chem. Ind., Lond.* **50** (1931), transactions, 253–6.
LE WITA, H. *Autour de la guerre chimique*, J. Tallandier, Paris (1928).
LIEFMANN, R. *Beteiligungs- und Finanzierungsgesellschaften*, G. Fischer, Jena (1909).
LÖFFL, V. K. *Die Chemische Industrie Frankreichs*, F. Enke, Stuttgart (1917).
LONDON and CAMBRIDGE ECONOMIC SERVICE. *The British economy, key statistics 1900–1966*, The Times Publishing Co., London, n.d. (1966 ?).
LUCION, R. *La grosse industrie chimique en Belgique*, L'Independence Belge, Brussels (1908).
LUKIANOV, P. M. *Istoria Khimicheskikh Promyslov i Khimicheskoi Promyshlennosti Rossii do kontsa XIX Veka*, Akademia Nauk (Institute of Historical Studies), Moscow, 4 vols. (1948–52).
LUNGE, G. *Die chemische Industrie und die chemisch-technischen Hochschulen in Nord Amerika*, Michel & Büchler, Berne (1894).
—— *Zur Geschichte der Entstehung und Entwicklung der chemischen Industrien in der Schweiz*, Orell Füssli, Zurich (1901).
—— Über den gegenwärtigen Stand der Schwefelsäurefabrikation. *Z. angew. Chem.* **16** (1903), 29, 694–5.
—— Zur Geschichte der Entwicklung des Kontaktverfahren für Darstellung von Schwefelsäure in den Vereinigten Staaten. *Z. angew. Chem.* **23** (1910), 16, 720–1.

MACLEOD, R. M. and ANDREWS, E. KAY. Scientific careers of 1851 Exhibition Scholars. *Nature, Lond.* **218** (15 June 1968), 1011–16.
MAHONEY, T. *The merchants of life*, Harper & Brothers, New York (1959).
MARCUS, A. *Die grossen Chemiekonzerne*, S. Hirzel, Leipzig (1929).
MARKOVITCH, T. J. L'Industrie française de 1879 à 1964—sources et méthodes, Institut de Science Economique Appliquée, Paris (1965) (*Cahiers de l'ISEA*, Série AF 4, No. 163).
—— L'Industrie française de 1789 à 1964—conclusions générales, Institut de Science Economique Appliquée, Paris (1966) (*Cahiers de l'ISEA*, Série AF 7, No. 179).
MARSHALL, A. *Explosives*, J. & A. Churchill, London, 2 vols. (2nd ed. 1917).
MARTEN, A. Farbstoffindustrie und Farbstoffzölle in den Vereinigten Staaten von Amerika, Ph.D. thesis, Berlin university (1928).

MATAGRIN, A. *L'Industrie des produits chimiques et ses travailleurs*, Doin, Paris (1925).

MATIGNON, C. L'Azote, see under GÉRARD, J. (ED.), *Dix ans d'efforts scientifiques et industriels*.

MAYER, M. *Young man in a hurry*, University of Chicago Alumni Association, Chicago (1957).

MELVILLE, SIR HARRY, *The Department of Scientific and Industrial Research*, Allen & Unwin, London (1962).

MENDELEEFF, D. I. Industrie du naphthe et des produits chimiques, see under KOVALEVSKY, M. W. DE, *La Russie à la fin du XIXème siècle*.

MERCK, J. H. *Entwicklung und Stand der pharmazeutischen Grossindustrie*, (Ph.D. thesis), Stilke, Berlin (1923).

MEYER, E. VON. *A history of chemistry*, Macmillan, London (3rd Eng. ed. 1906).

MIALL, S. *A history of the British chemical industry*, E. Benn, London (1931).

MITCHELL, B. R. and DEANE, P. *Abstract of British historical statistics*, University Press, Cambridge (1962).

MITCHELL, H. J. *Methods of inculcating modern management principles and practices in large scale undertakings*, International Management Congress, London (1935).

—— *Relationship between headquarters' and departments and the cooperating groups of I.C.I. Ltd.*, VIIth International Management Congress, Washington (1938).

MITTASCH, A. *Geschichte der Ammoniaksynthese*, Verlag Chemie, Weinheim (1951).

MOND, SIR ALFRED. *Industry and politics*, Macmillan, London (1927).

MORANDI, L. Aspects et perspectives de l'industrie chimique italienne. *Chim. Ind.*, **63** (1950), 1, 79 ff.

MORTON, J. History of the development of fast dyeing and dyes, priv. publ. Cramond Bridge, Scotland (1929).

MOULTON, F. R. (ED.) *Liebig and after Liebig, a century of progress in agricultural chemistry*, American Association for the Advancement of Science, Washington (1942).

MOULTON, LORD J. F. *Science and war*, University Press, Cambridge (1919).

MOUREU, C. *La chimie et la guerre science et avenir*, Masson & Cie, Paris (1920).

—— Les conditions générales de la recherche scientifique en France, see under GÉRARD, J. (ED.), *Dix ans d'efforts scientifiques et industriels*.

MUELLER, W. F. The origins of the basic inventions underlying Du Pont's major product and process innovations 1920 to 1950, see NATIONAL BUREAU OF ECONOMIC RESEARCH,

MUIR, J. *John Anderson*, J. Smith, Glasgow (1950).

NAGEL, A. VON. *Indanthren, Komplexfarbstoffe, Tenside*, BASF, Ludwigshafen (1968).

—— *Stickstoff*, BASF, Ludwigshafen (1969).

NATHAN, SIR FREDERIC. The manufacture of acetone. *J. Soc. chem. Ind., Lond.* **38** (1919), 14, transactions 271–3.

NATIONAL BUREAU OF ECONOMIC RESEARCH. *The rate and direction of inventive activity: economic and social factors*, University Press, Princeton (1962).

NATIONAL INDUSTRIAL CONFERENCE BOARD. *Technical education in the chemical industry*, N.I.C.B., New York (1925).

NATIONAL INDUSTRIAL CONFERENCE BOARD. *Industrial lunch rooms*, N.I.C.B., New York (1928).

NEKRASSOV, N. N. *Ekonomika Khimicheskoi Promyshlennosti*, Moscow (1957).

NELSON, J. *The Mine Workers District 50*, Exposition Press, New York (1955).

NEUMANN, H. and WECKERLE, E. (EDS.) *Die Schweizerische Industrie und ihre Arbeiter, No. 8: Papierfabrikation, Chemische Industrie*, Schweizerischer Gewerkschaftsbund, Olten (1947).

NEW SURVEY OF LONDON LIFE AND LABOUR. *London industries*, vol. 5, pt. 2, P. S. King, London (1933).

NORGES INDUSTRI (PUBL.). *Norwegian hydro-electric industry*, Norges Industri, Oslo (1932).

NORTON, T. H. The chemical industries of Norway. *Chem. Trade J.* **52** (1913), 1348, 311 and 1349, 339–40.

—— The chemical industries of Belgium. *Chem. Trade J.* **52** (1913), 1339, 51–2.

—— *Dyestuffs for American textile and other industries*, see under Official Publications, U.S., Department of Commerce.

NUTTER, G. W., BORENSTEIN, I., and KAUFMANN, A. *Growth of industrial production in the Soviet Union*, University Press, Princeton (1962).

OBERDORFFER, K. (ED.) *Ludwigshafener Chemiker*, Econ, Düsseldorf (1958).

OGBURN, W. F. and JAFFE, W. *The economic development of post-war France*, Columbia University Press, New York (1929).

OLIVER, SIR THOMAS (ED.) *Dangerous trades*, John Murray, London (1902).

OLIVER, SIR THOMAS. *Diseases of occupation*, Methuen, London (3rd ed. 1916).

OST, H. Die chemische Technologie an den Technischen Hochschulen, *Z. angew. Chem.* **13** (1900), 27, 659–63.

—— *Lehrbuch der Chemischen Technologie*, Jaenicke, Leipzig (14th ed. 1925).

PACKARD, W. G. T. *The history of the fertilizer industry in Britain*, Fertilizer Society, London, n.d., proceedings No. 19.

PANETH, F. The teaching of chemistry in German universities. *J. chem. Educ.* **5** (1928), 6, 706–9.

PANTYUCHOFF, N. Ten years of Russian chemical industry 1898–1908, *Chem. Trade J.* **66** (1910), 1182, 63–5.

PARIS EXHIBITION. See under Official Publications, France, A. HALLER.

PARRAVANO, N. (ED.) *La Chimica in Italia*, Xth International Chemical Congress, Rome (1938).

PARTINGTON, J. R. and PARKER, L. H. *The nitrogen industry*, Constable, London (1922).

PAULSEN, F. *The German universities and university study*, Longmans, London (1906).

PAXMAN, H. *Die deutsche Kaliwirtschaft*, Hiehold & Co., Berlin (1929).

PILCHER, R. B. *History of the Institute of Chemistry 1877–1914*, Institute of Chemistry London (1914).

PLANCK, M. *25 Jahre Kaiser-Wilhelm-Gesellschaft*, Springer, Berlin, 3 vols. (1936).

POCHITONOW, W. *Der gegenwärtige Stand der russischen chemischen Industrie*, F. Enke, Stuttgart (1915).

POLLITT, G. P. The synthetic ammonia industry (in *Trans. second Wld Pwr Conf.*, Berlin, 1930, vol. 2, pp. 145–64), VDI Verlag, Berlin (1931).
POUND, A. Salt of the earth: the story of J. B. Ford. *Atlantic Monthly*, Boston (1940).
PRATT, P. C. The not-so-good old days. *I.C.I. Magazine* (September 1958) 310–12.
PRECHT, H. Die Entwicklung der Kaliindustrie. *Z. angew. Chem.* **19** (1906), 1, 1–4.
PRING, J. N. *Some electro-chemical centres*, University Press, Manchester (1908).

RAAB, F. *et al. Die Technische Hochschule Fridericiana Karlsruhe, Festschrift zur 125 Jahrfeier*, Technische Hochschule, Karlsruhe (1950).
RABATÉ, E. Les engrais, see under GÉRARD, J. (ED.), *Dix ans d'efforts scientifiques et industriels.*
RASSOW, B. (ED.). *Geschichte des Vereins Deutscher Chemiker*, O. Spamer, Leipzig (1912).
—— Was haben wir Chemiker in Amerika gelernt? *Z. angew. Chem.* **26** (1913), 1, 705–8.
—— Die Chemische Industrie see under WIEDENFELD, K. (ED.), *Die Deutsche Wirtschaft und ihn Führer.*
RATHENAU, W. *Deutschlands Rohstoffversorgung*, S. Fischer, Berlin (1917).
RICHARDSON, H. W. The development of the British dyestuffs industry before 1939. *Scott. J. polit. Economy* **9** (1962), 2, 110–29.
—— Chemicals, see under ALDCROFT, D. H. (ED.), *The development of British industry and foreign competition.*
RICHTER, F. (ED.) *75 Jahre Beilstein's Handbuch der Organischen Chemie*, Springer, Berlin–Göttingen (1957).
RODERICK, G. W. *The emergence of a scientific society in England 1800–1955*, Macmillan, London (1967).
ROSTOW, W. W. *The stages of economic growth*, University Press, Cambridge (1960).
ROWE, L. S. *The early effects of the European War upon the finance, commerce and industry of Chile*, Oxford University Press, New York (1918).
RUSKE, W. *100 Jahre Deutsche Chemische Gesellschaft*, Verlag Chemie, Weinheim (1967).

SASULY, R. *IG Farben*, Boni & Gaer, New York (1947).
SAFTIEN, K. 50 Jahre Indanthren Farbstoffe, *Die BASF* **1** (1951), 3, 63 ff.
SCHALL, H. Die chemische Industrie Deutschlands, Ph.D. thesis, Hochschule für Wirtschafts- und Sozialwissenschaften, Nürnberg (1959).
SCHENZINGER, K. A. *Bei IG Farben*, W. Andermann, Munich–Vienna (1953).
SCHICHT, H. The development and present situation of the chemical industry in Czechoslovakia. *J. Soc. chem. Ind., Lond.* **50** (1931), transactions 276–7.
SCHMIDT-OTT, D. Geschichte des Fritz Haber Instituts *Jahrbuch der Max-Planck-Gesellschaft* (1961), II, 374 ff.
SCHMIDT-OTT, F. *Erlebtes und Erstrebtes*, F. Steiner, Wiesbaden (1952).
SCHMITT, R. Die pharmazeutische Industrie und ihre Stellung in der Weltwirtschaft, Ph.D. thesis, Frankfurt (1932).

SCHREIB, H. *Die Fabrikation der Soda nach dem Ammoniakverfahren*, Springer, Berlin (1905).

SCHUELLER, E. Les matières plastiques, see under GÉRARD, J. (ED.), *Dix ans d'effort scientifiques et industriels.*

SCHULTZE, H. *Die Entwicklung der chemischen Industrie Deutschlands*, Tausch & Grosse, Halle (1908).

SCHUMPETER, E. B. (ED.) *The industrialization of Japan and Manchukuo 1930–1940*, Macmillan, New York (1940).

SCHWARZENBACH, G. W. D. Treadwell. *Neue Zürcher Zeitung* (7 August 1955), Fernausgabe.

SÉVIN, R. L'usine électrochimique de Bellegarde. *J. Four élect.* **63** (1959), 4, 163 f.

SIMONIN, G. (ED.). *L'Industrie Chimique Française: Lyon et le Sud-Est*, Association Amicale des Anciens Elèves de l'Ecole Supérieure de Chimie Industrielle, Lyon (1959).

SQUARZINA, F. *Produzione e commercio dello zolfo in Sicilia nel secolo XIX*, ILTE, Turin (1963).

STEEL, L. Forty years on. *I.C.I. Magazine* **41** (1963), 310, 8–10.

STOCKING, G. W. and WATKINS, M. W. *Cartels in action*, Twentieth Century Fund, New York (1946).

—— —— *Cartels or competition?* Twentieth Century Fund, New York (1948).

STONE, I. F. *The aniline color, dyestuff and chemical conditions from 1 August 1914 to 1 April 1917*, priv. publ. New York (1917).

STRUMILIN, S. Khimicheskaya Promyshlennost SSR. *Problem Ekonomiki* (1935), 2, 118 ff.

SULFRIAN, A. and PELTZER, J. *Betriebs- und gesamtwirtschaftliche Probleme der chemischen Produktion*, Enke, Stuttgart (1938).

TAYLOR, F. SHERWOOD. *A history of industrial chemistry*, Heinemann, London (1957).

TER MEER, F. *Die IG Farben*, Econ, Düsseldorf (1953).

TIMM, B. Die Anwendung hoher Drucke in der chemischen Technik, *Die BASF* **17** (1967), 1, 23–5.

TOCQUEVILLE, A. DE. *Democracy in America*, Longmans, London, 2 vols. (1875).

TRUMP, E. N. Looking back at 50 years in ammonia soda alkali. *Chem. metall. Engng* **40** (1933), 3, 126–8.

TURKEVICH, J. *Chemistry in the Soviet Union*, University Press, Princeton (1965).

TURNER, D. M. *History of science teaching in England*, Chapman & Hall, London (1927).

UNGEWITTER, C. (ED.) *Ausgewählte Kapital aus der chemisch-industriellen Wirtschaftspolitik 1877–1927*, Verein zur Wahrung der Interessen der chemischen Industrie Deutschlands, Berlin (1927).

—— *Chemie in Deutschland*, Junker & Dünnhaupt, Berlin (1938).

URQUHART, A. R. Cellulose derivative rayons, past, present and future (in *The development of some man-made fibres*), The Textile Institute, Manchester (1952).

VERSHOFEN, W. *Wirtschaftsgeschichte der chemisch-pharmazeutischen Industrie*, Editio Cantor, Aulendorf, Germany, 3 vols. (1958).

VISHER, S. S. *Scientists starred 1903–1943*, Johns Hopkins Press, Baltimore (1947).

VUCINICH, A. *The Soviet Academy of Sciences*, University Press, Stanford (1946).

WAESER, B. *Die Luftstickstoffindustrie*, O. Spamer, Leipzig (1st ed. 1922, 2nd ed. 1932).

WALDMEYER, E. *Die schweizerische Salz- und Soda Industrie* (Ph.D. thesis, Berne), Neuenschwander, Weinfelden (1928).

WALLER, P. *Probleme der deutschen chemischen Industrie* (Ph.D. thesis, Hamburg, 1927), H. Meyer, Halberstadt (1928).

WARRINGTON, C. J. S. and NICHOLLS, R. V. V. *A history of chemistry in Canada*, Pitman & Sons, Toronto (1949).

WASSILIEFF, N. *Die Arbeitsverhältnisse in den Basler chemischen Fabriken*, Genossenschaftsdruckerei, Basle (1905).

WEBER, A. (ED.) *Über den Standort der Industrien*, J. C. B. Mohr, Tübingen (1914).

WEIDLEIN, E. R. and HAMOR, W. A. *Glances at industrial research*, Reinhold Publishing, New York (1936).

WEIZMANN, CHAIM. *Trial and error*, Hamish Hamilton, London (1949).

WEHRENALP, E. B. von. *Der Niedergang der französischen Naturwissenschaften*, Junker & Dünnhaupt, Berlin (1940).

WENZEL, O. Die Arbeiter in der chemischen Industrie und ihre Löhne (in vol. 4 of the *Proc. 5th Int. Cong. appl. Chem.*, Berlin, 1903), Deutscher Verlag, Berlin (1904).

WEST, C. J. and RISHER, E. L. *Industrial research laboratories in the U.S.*, see under Official Publications, U.S., National Research Council.

WHEELER-BENNETT, J. W. *John Anderson*, Macmillan, London (1962).

WHITTAKER, C. M. Some early stages in the renaissance of the British dyemaking industry, tales from Turnbridge, Huddersfield 1899–1920, *J. Soc. Dyers Colour.* **72** (1956), 12, 557–63.

WICKEL, H. *IG Deutschland*, Der Bücherkreis, Berlin (1932).

WIEDENFELD, K. (ED.) *Die Deutsche Wirtschaft und Ihre Führer*, Im Flamberg, Gotha (1925).

WILLIAMS, T. I. *The chemical industry*, Penguin Books, London (1953).

WILLSTÄTTER, R. *Aus meinem Leben*, Verlag Chemie, Weinheim (1949).

WINKLER, J. K. *The Du Pont dynasty*, Reynal & Hitchcock, New York (1935).

WITT, O. N. Über die Ausbildung der Chemiker für die Technik, *Z. angew. Chem.* **14** (1901), 26, 633–40.

—— *Die Chemische Industrie auf der Internationalen Weltausstellung zu Paris 1900*, R. Gaertner, Berlin (1902).

—— *Die chemische Industrie des deutschen Reiches*, R. Gaertner, Berlin (1906).

WOHLGEMUTH, L. M. Studium der Ausländer an deutschen Hochschulen. *Z. angew. Chem.* **20** (1907), 35, 1506–8.

YAMASAKI, K. and OGAWA, G. *The effect of the World War upon the commerce and industry of Japan*, Yale University Press, New Haven (1929).

YERKES, R. M. *The new world of science*, The Century Co., New York (1920).

ZIMMERMANN, P. A. *Patentwesen in der Chemie*, BASF, Ludwigshafen (1965).

III: COMPANY HISTORIES AND BROCHURES[1]

AFRICAN EXPLOSIVES and CHEMICAL INDUSTRIES

Cartwright, A. P. *The Dynamite Co.*, Macdonald, London (1964).

ALLIED CHEMICAL CO.

Cosby, J. History of Allied Chemical, typescript (1958).

ALBRIGHT and WILSON LTD.

Threlfall, R. E. *The story of 100 years of phosphorus making* (1951).

AMERICAN CYANAMID CO.

McMurray, J. H. The history of Calco, typescript (1943).

Haynes, Williams. On the chemical frontier—the Cyanamid Story, typescript (1957/8).

BASF

Anon. *Die Badische Anilin- & Soda-Fabrik*, n.d. (c. 1922).

Schuster, C. *Badische Anilin- & Soda-Fabrik*, n.d.

Voigtländer-Tetzner, W. Der Mensche in der BASF, typescript, n.d. (1938–9 ?).

Voigtländer-Tetzner, W. and Pinnow, H. Geschichte der BASF: Kaufmännische Entwicklung, typescript, n.d. (1938–9 ?).

Bayer, H. Das Werden eines Grossunternehmens. *Die BASF* **14** (1964), 4. 172–83.

Steinert, O. and Roggersdorf, W. *Im Reiche der Chemie—100 Jahre BASF*, Econ, Düsseldorf (1965).

BAYER

Entwicklung und heutiger Stand der Firma und ihrer Wohlfahrtseinrichtungen (1914 ed.)

Anon. (Ed.) *Geschichte und Entwicklung der Farbenfabriken vorm. Friedrich Bayer & Co. Elberfeld in den ersten 50 Jahren* (1918).[2]

Blank, A. and Löw, W. Geschichte der Fabrik . . . in Moskau.

Bonhoeffer, O. Die Arbeiterverhältnisse im Werk Elberfeld.

Burgdorf, K. Geschichte der Fabrik Flers.

Grabendörfer, R. Die Arbeiterverhältnisse in Leverkusen.

Doermer, O. Die Konventionen und ihr Einfluss auf Farbenfabriken.

Pinnow, H. *Werksgeschichte*, Bruckmann KG, Munich (1938).

Stenger, E. *100 Jahre Photographie und die Agfa*, Knorr & Hirth, Munich (1939).

Kramer, H. O. R. *et al. Revolution im Unsichtbaren*, Econ, Düsseldorf (1963).

Jacobi, F. (Ed.) *Beiträge zur hundertjährigen Firmengeschichte* (1964).

Beckerath, P. G. von. Personalwesen.

Brand, P. Photographische Erzeugnisse.

Flechtner, H.-J. Vom Dreibund bis zur 'kleinen IG'.

—— Von der 'kleinen IG' zur IG Farbenindustrie.

Schultheis, H. Synthetische Hochpolymere.

Hauser, H. *Geschichte des Werkes Uerdingen* (1956).

BRUNNER, MOND and CO.

Anon. *The first fifty years of Brunner, Mond & Co.* (1923).

[1] Most of the titles in this list were published privately by the companies concerned. The name of the publisher and the place of publication are shown only where the book has been published commercially.

[2] The book, though printed, was not published. I have been unable to discover how many copies were bound and distributed. One copy is kept at Leverkusen.

CHEMICAL ALLIANCE INC.

Anon. *Historical review of the object, organization and activities of the Chemical Alliance Inc. during the World War* (1920).

CHEMISCHE FABRIKEN DR. KURT ALBERT GMBH

Anon. *Zur Geschichte der Firma Dr. Kurt Albert* (1935).

CHEMISCHE FABRIK KALK (VORM. VORSTER & GRUNEBERG)

Anon. *Festschrift zur Feier des 50 Jährigen Bestehens* (1908).

Greiling, W. *100 Jahre Chemische Fabrik Kalk* (1958).

CHEMISCHE FABRIK SCHWEIZERHALL

Anon. *Denkschrift zum fünfzigjährigen Bestehen der AG Chemische Fabrik Schweizerhall in Basel 1890–1940* (1940).

CIBA

Anon. *Histoire et développement de la Société pour l'Industrie Chimique à Bâle 1864–1924*, n.d. (1926 ?).

Anon. *Gesellschaft für chemische Industrie 1884–1934*, n.d. (1934 ?).

Anon. *From the Annals of the Society of Chemical Industry in Basle* (1939).

Anon. *Cinquantenaire de l'Usine de Monthey*, n.d. (1954 ?).

Bericht zum 75. Geschäftsjahr 1884–1959 (1959).

CONSOLIDIRTE ALKALIWERKE WESTEREGELN

Anon. *50 Jahre 1881–1931* (1931).

COURTAULDS LTD.

Ward-Jackson, C. H. *A history of Courtaulds*, Curwen Press, London (1941).

Coleman, D. C. *Courtaulds*, Clarendon Press, Oxford, 2 vols. (1969).

DAVISON CHEMICAL CO.

Anon. *A century of progress* (1926).

DOW CHEMICAL CO.

Gerstacker, C. A. The Dow story. *Armed Forces chem. J.* (July/August 1956).

DU PONT

Smokeless Powder Department. *A history of the Du Pont Company's relations with the U.S. Government 1802–1927* (1928).

Anon. *Du Pont, the autobiography of an American enterprise* (1925).

Dutton, W. S. *Du Pont*, Scribners', New York (1st ed. 1942, 3rd ed. 1951).

GENERAL CHEMICAL CO.

Anon. *The General Chemical Co. after 20 years* (1919).

GEIGY AG

Bürgin, A. *Geschichte des Geigy-Unternehmens von 1758 bis 1939*, Birkhäuser, Basle (1958).

HAËN, CHEMISCHE FABRIK E. DE

Anon. *Hundert Jahre Chemische Fabrik E. de Haën* (1961).

HERCULES POWDER CO.

Anon. *A history* (1947).

HOECHST and subsidiary companies

Anon. *Farbwerke vorm. Meister Lucius & Brüning*, n.d. (1913 ?).

Raschen, H. and Hoffmann, P. *75 Jahre Chemische Fabrik Griesheim Elektron* (1938).[1]

Pinnow, H. *Die Farbwerke Hoechst* (1938).[1]

Greisenheyner, M. *Schaffende Gedanken—Schaffende Hände* (1952).

[1] Published by IG Farben.

Niebelschütz, W. von. *Knapsack* (1957).
Pistor, G. *100 Jahre Griesheim*, publ. by author (1958).
Bäumler, E. *Ein Jahrhundert Chemie*, Econ, Düsseldorf (1963).

HOOKER ELECTROCHEMICAL CO.
Anon. *Hooker Electrochemical after 25 years* (1925).
Thomas, R. E. *Salt and water, power and people* (1955).

I.C.I.
Anon. *A short account of the activities of the Company*, n.d. (1935 ?).
Anon. *This is your concern* (1947).
Anon. *Fifty years of progress: the story of the Castner-Kellner Alkali Co.*, n.d. (1947 ?).
Trotter, R. (Ed.) *The history of Nobel's Explosives Co. Ltd.* (1938).
Miles, F. D. *A history of research in the Nobel Division of I.C.I.* (1955).
Parke, V. E. *Billingham—the first ten years* (1957).
Irvine, A. S. *A history of the Alkali Division*, n.d. (1959 ?).
Hill, M. F. *Magadi*, Kynoch Press, Birmingham (1964).
Reader, W. J. *History of I.C.I.*, vol. 1, Oxford University Press, London (1970).

IG FARBEN
Greiling, W. 25 Jahre Leuna 1918–1943, typescript, n.d. (1944 ?).

KALI-CHEMIE AG
Spiegel, L. *100 Jahre Kunheim* (1925).
Rüsberg, F. *Fünfzig Jahre Kali-Chemie AG* (1949).

KUHLMANN
Anon. *Les Etablissements Kuhlmann 1825–1925* (1925).

A. D. LITTLE
Little, A. D. *The handwriting on the wall*, Little, Brown & Co., Boston (1928).

LINDE'S EISMASCHINEN
Rosenberg, E. von (Ed.) *75 Jahre Linde* (1954).

MATHIESON ALKALI WORKS
Anon. *Fifty years of chemical progress* (1942).

METALLGESELLSCHAFT AG
Däbritz, W. *Fünfzig Jahre Metallgesellschaft* (1931).

MONTECATINI EDISON
Anon. *Settanta cinque anni di vita Italiana* (1963).

NORSK HYDRO KVAELSTOF A/S—SOC. NORVÉGIENNE DE L'AZOTE
Anon. *La Société Norvégienne de l'Azote, 1905–1955* (1957).

OESTERREICHISCHER VEREIN
Anon. *Der Oesterreichische Verein für Chemische und Metallurgische Production 1865–1906* (1906).

PECHINEY
Gignoux, C. J. *Histoire d'une entreprise française*, Hachette, Paris (1955).

PENNSALT
Leavitt, R. K. *Prologue to tomorrow* (1950).

F. RASCHIG GMBH
Ihle, E. A. *75 Jahre Raschig* (1966).

SANDOZ AG
Anon. *Sandoz 1886–1961: 75 years of research and enterprise* (1961).

SCHERING AG

Holländer, H. *Geschichte der Schering AG* (1955).

SAINT GOBAIN

Choffel, J. *Saint Gobain*, Plon, Paris (1960).

SOLVAY & CIE

Anon. *Cérémonie de Commémoration du 50ème Anniversaire de la Fondation* (1913).

Hoff, L. Le Centenaire des Soudières Réunies. *Chim. Ind.* **73** (1955), 5, 998–1001.

Bolle, J. *Solvay, L'Invention, L'homme, L'entreprise industrielle* 2 vols. (1963).[1]

TENNANT GROUP OF COMPANIES

Anon. *Enterprise* (1945).

UNILEVER GROUP

Wilson, C. *The history of Unilever*, Cassell & Co., London, 2 vols. (1954).

Musson, A. E. *Enterprise in soap and chemicals*, University Press, Manchester (1965).

UNITED ALKALI CO.

Anon. *The struggle for supremacy* (1907).

Anon. *Centenary of the alkali industry* (1923).

[1] A one-volume paperback edition was published by Weissenbruch, Brussels.

INDEX